# Contact Angle, Wettability and Adhesion
# Volume 3

# CONTACT ANGLE, WETTABILITY AND ADHESION

## VOLUME 3

*Editor:*
## K.L. Mittal

## CRC Press
Taylor & Francis Group
Boca Raton London New York

CRC Press is an imprint of the
Taylor & Francis Group, an **informa** business

First published 2003 by VSP Publishing

Published 2018 by CRC Press
Taylor & Francis Group
6000 Broken Sound Parkway NW, Suite 300
Boca Raton, FL 33487-2742

First issued in paperback 2019

No claim to original U.S. Government works

ISBN-13: 978-0-367-44663-5 (pbk)
ISBN-13: 978-90-6764-391-7 (hbk)

**Visit the Taylor & Francis Web site at**
**http://www.taylorandfrancis.com**

**and the CRC Press Web site at**
**http://www.crcpress.com**

# Contents

*Contact Angle, Wettability and Adhesion*, Vol. 3, pp. ix–x
Ed. K.L. Mittal
© VSP 2003

# Preface

This volume chronicles the proceedings of the Third International Symposium on Contact Angle, Wettability and Adhesion held under the aegis of MST Conferences in Providence, Rhode Island, May 20–23, 2002. The premier symposium with the same appellation was held in 1992 in honor of Prof. Robert J. Good as a part of the American Chemical Society meeting in San Francisco, and the second event was held under the auspices of MST Conferences in Newark, New Jersey, June 21–23, 2000. The proceedings of these two earlier symposia have been properly documented as hard-bound books [1, 2].

Since the second symposium held in the year 2000 was a huge success, so we decided to hold the third event to provide a forum to update and consolidate the research activity on this topic of immense technological import. Even a cursory look at the literature will evince that there is brisk research activity apropos of contact angle and wettability from both fundamental and applied points of view. The world of wettability is very wide as it plays an extremely important role in a legion of technological areas.

The technical program for this symposium was comprised of 50 papers which included both overviews and original research contributions. Many ramifications of contact angle and wettability were addressed in this symposium, and certain controversial issues were also highlighted. On occasions, there were quite intense, but illuminating, discussions which provided a forum for attendees to express quite divergent and discordant views.

Now coming to this volume, it contains a total of 25 papers covering myriad aspects of contact angle and wettability. It must be recorded that all manuscripts were rigorously peer-reviewed – each manuscript was sent to three reviewers – and all were revised (some twice or even thrice) and properly edited before inclusion in this volume. Concomitantly, this book should not be considered in the same league as many other proceedings volumes which are usually neither peer-reviewed nor adequately edited. In other words, this book represents an archival publication of the highest standard.

This book (called Volume 3) is divided into three parts: Part 1: General Papers; Part 2: Contact Angle Measurements/Determination and Solid Surface Free Energy; and Part 3: Wetting and Spreading: Fundamental and Applied Aspects. The topics covered include fundamental aspects of contact line region; effect of adsorbed vapor on liquid–solid adhesion; molecular origin of contact angles; various factors influencing contact angle measurements; different kinds of contact angles; various ways to measure contact angles; contact angle hysteresis; determination of solid surface free energies via contact angles; contact angle measurements on various materials (smooth, rough, porous, heterogeneous); factors influencing/dictating wetting and spreading phenomena; ultrahydrophobic polymer surfaces; switchable wettability; reactive wetting; wetting by nanocrystallites; dewetting; wetting of self-assembled monolayers; reversible wetting of structured surfaces; wetting in granular and porous media; relationship between wetting and adhesion; relevance/importance of wetting and surface energetics in technological applications, including food industry.

I sincerely hope this volume and its predecessors [1, 2] containing bountiful information (about 2000 pages) will be of great interest and value to everyone interested in the contemporary R&D activity in the fascinating world of contact angles and wettability. Yours truly further hopes that the information garnered in these volumes will serve as a fountainhead for new research ideas and applications.

## Acknowledgements

It is always a pleasure to write this particular segment of a Preface. First, I am thankful to Dr. Robert H. Lacombe for taking care of the organizational aspects of this symposium. Special thanks are extended to professional colleagues who served as reviewers and provided many valuable comments which are a prerequisite for a high standard publication. The interest, enthusiasm and contribution of the authors must be acknowledged as these were the essential elements in making this book a reality. Finally my appreciation goes to the staff of VSP (publisher) for the excellent job done in producing this book.

K.L. Mittal
P.O. Box 1280
Hopewell Jct., NY 12533

1. K.L. Mittal (Ed.), *Contact Angle, Wettability and Adhesion*. VSP, Utrecht (1993).
2. K.L. Mittal (Ed.), *Contact Angle, Wettability and Adhesion*, Vol. 2. VSP, Utrecht (2002).

# Part 1
# General Papers

*Contact Angle, Wettability and Adhesion*, Vol. 3, pp. 3–23
Ed. K.L. Mittal
© VSP 2003

# Adsorption, evaporation, condensation, and fluid flow in the contact line region

PETER C. WAYNER, Jr.[*]

*The Isermann Department of Chemical Engineering, Rensselaer Polytechnic Institute, Troy, NY 12180-3590*

**Abstract**—Since the chemical potential is a function of both pressure and temperature, a dynamic Kelvin-Clapeyron (K-C) model for the contact line region, which includes these effects on the local vapor pressure, gives an enhanced understanding of both equilibrium and non-equilibrium processes. A general equation for the equilibrium film thickness profile is obtained by combining an isothermal change in the interfacial pressure jump (Kelvin effect) with an isobaric change in the interfacial temperature jump (Clapeyron effect). For the non-equilibrium case, interfacial kinetic theory connects the variation of the local vapor pressure to interfacial mass transfer and, therefore, gives the coupling of phase change, liquid film profile and contact line motion. The strong coupling between molecular dynamics, temperature, and pressure suggests that complex nano-scale flux fields including phase change are hidden within macroscopic experimental measurements. An overview of past experimental and theoretical research on the K-C model is presented.

The relative importance of the interfacial temperature jump to the interfacial pressure jump is evaluated and found to be significant. In essence, due to the strong coupling between temperature and pressure, it is very difficult to avoid phase change in the contact line region. The infinite stress at the contact line for viscous flow can be easily relieved by evaporation-adsorption in many systems. Depending on the system, the effect of evaporation followed by multi-layer adsorption is larger than that of surface diffusion. Theoretical and experimental examples with large and extremely small excess surface temperature are discussed.

*Keywords*: Spreading; contact angle; condensation; capillarity; disjoining pressure; contact line; excess free energy.

## 1. INTRODUCTION

Intermolecular interactions in the three-phase contact line region, where a liquid-vapor interface intersects a solid substrate, have been extensively studied because of their importance to many equilibrium and non-equilibrium phenomena such as contact angle, adsorption, spreading, evaporation, condensation, boiling, wetting and stability. Since the chemical potential is a function of both pressure and temperature, the Kelvin-Clapeyron (K-C) continuum model for the contact line re-

*Phone: 518-276-6199, Fax: 518-276-4030, E-mail: wayner@rpi.edu

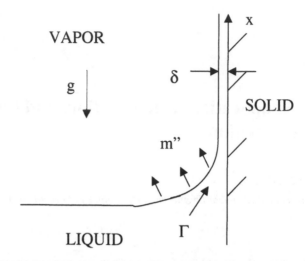

**Figure 1.** Vertical extended meniscus with fluid flow and phase change.

gion, which includes these effects on the local vapor pressure, gives an enhanced understanding of both equilibrium and non-equilibrium processes. Conceptually, the contact line region can be modeled as a meniscus in which the shape gives the interfacial pressure jump [1-17], with the equilibrium vapor pressure being a function of the film thickness and the radius of curvature [e.g., 1-4, 15, 17-20]. Experimentally, with completely wetting systems, the adsorbed film thickness, curvature, and the related disjoining pressure isotherm are of principal importance. With partially wetting systems, the observable apparent contact angle and curvature are of principal importance. Since the molecular exchange process at interfaces is very dynamic, kinetic theory connects the variation of the local vapor pressure to interfacial mass transfer [e.g., 4, 15, 19, 20]. These observations led to the development of the K-C model for the dynamic contact line region [4, 7, 15, 21–24], which is discussed herein. Although only simple examples are emphasized herein, the presented equations can also be solved numerically for additional detail. At the outset, the use of a simple, one-dimensional models is noted. For example, at the monolayer scale, a continuum model is used to describe the effects of the average film thickness and profile. However, we believe that the resulting theoretical insights, which are consistent with many observations, justify this approximate average approach used by many modelers.

A schematic drawing of the generic vertical extended meniscus for a completely wetting fluid with possible liquid flow and phase change at the liquid-vapor interface is presented in Figure 1. The following terms for an equilibrium system can be defined: *extended meniscus* – the combined intrinsic meniscus, transition region and the thin film extending above it; *intrinsic meniscus* – that portion of the extended meniscus which is described by capillarity while excluding the effects of disjoining pressure; *thin film* – that portion of the extended meniscus above the transition region defined by the disjoining pressure concept; *transition region* – the region where both capillarity and disjoining pressure are important [4]. With appropriate conceptual and model modifications, these concepts will also be applied to partially wetting and non-equilibrium systems.

For the small slope region, a local one-dimensional mass balance for the system in Figure 1 leads to

$$\rho_{lm} \frac{\partial \delta}{\partial t} = -\frac{\partial \Gamma}{\partial x} - m'' \tag{1}$$

The average value of the surface flow rate per unit width, $\Gamma$, and the evaporation/condensation mass flux, $m''$, are a function of the chemical potential field and interfacial kinetic exchange, which are discussed next.

## 2. EQUILIBRIUM SYSTEMS

The chemical potential and the change in the chemical potential energy per unit volume of a single component liquid in a gravitational field are given by Eqs. (2-4) [e.g., 4 and 15]. Modifications can, of course, be added for mixtures. Since experience tells us that most systems are mixtures, even if only extremely dilute due to contamination, the following models obviously need to be extended. Examples of initial results using mixtures are given in [25-27].

$$\mu_g = \mu(T_l, P_l) + \rho_{lm} \, g \, x \tag{2}$$

$$\Delta\mu_g = -[\sigma_{lv} K + \Pi - \rho_{lm} \, g \, x] + \frac{\rho_{lm} \Delta h_m}{\bar{T}} [T_{lvx} - T_{vx}] \tag{3}$$

$$\Delta\mu_g = \rho_{lM} \, R \, T_{lv} \ln \frac{P_{vlvx}}{P_{vx}} \tag{4}$$

The pressure in the bulk vapor at x is $P_{vx}$, whereas $P_{vlvx}$ is the vapor pressure at the liquid-vapor interface at x. The interfacial vapor pressure is affected by capillarity (the product of the bulk surface tension, $\sigma_{lv}$, and interfacial curvature, K: $\sigma_{lv} K$), disjoining pressure ($\Pi$), the hydrostatic head in a gravitational field g ($\rho_{lm}$ gx), and temperature, T. The product of the density, $\rho_{lm}$, and the latent heat of vaporization, $\Delta h_m$, is the volumetric heat of vaporization at the average phase change temperature, $\bar{T}$. Using Eq. (3), three equilibrium results can be readily obtained and verified experimentally. For isothermal equilibrium, the extended Young-Laplace equation is obtained which includes the definition of the disjoining pressure, $\Pi$, when K = 0 [28, 29]:

$$\sigma_{lv} K + \Pi = \rho_{lm} \, g \, x = P_{vx} - P_{lx} \tag{5}$$

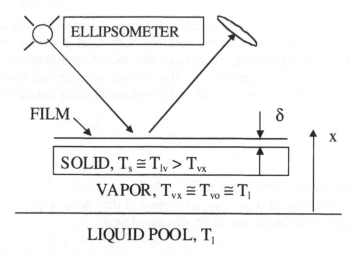

**Figure 2.** Optical measurement of adsorbed film thickness on a solid with an excess surface temperature suspended above a pool of liquid.

In Figure 1, $\Pi > 0$ and $P_l < P_v$. Experimentally, the shape of the extended meniscus gives the interfacial free energy per unit volume or the pressure jump at the interface. For an isothermal horizontal flat completely wetting (Hamaker constant, $A < 0$) film of a simple fluid, the equilibrium adsorption isotherm for a film with a thickness $\delta < 15$ nm with $K = 0$ is

$$\rho_{lM} RT \ln \frac{P_{vlv}}{P_{sat}} = -\Pi = \frac{A}{6\pi\delta^3} \tag{6}$$

The Hamaker constant and the effective pressure jump at the interface can be obtained by measuring the adsorbed film thickness as a function of the pressure in the vapor, $P_{vlv}$, relative to saturation, $P_{sat}$ [e.g., 29, 30]. These equations demonstrate that there is substantial potential for flow in the contact line region. For example, with the pentane/quartz system at 298 K, $A = -1.28 \times 10^{-20}$ J, and at $\delta = 0.3$ nm, $\Pi = 2.5 \times 10^7$ N/m$^2$. The gradient of this pressure jump (chemical potential) can give surface flow and/or phase change in the contact line region. The related reduction in vapor pressure at 298 K is $P_{vlv}/P_{sat} = 0.31$, which is available for flow in the vapor from a thick region where the vapor pressure is saturated to the thin film region, where the vapor pressure at the liquid-vapor interface, $P_{vlv}$, is substantially less due to changes in the temperature and pressure in the film. Substantial potential for flow in both the liquid and vapor is available.

Marmur [14] has discussed an alternate form of the pressure jump at the liquid-vapor interface, Eq. (5), in which the interfacial tensions in the contact line region are a function of the thickness and the inclination of the interface and thereby account for the molecular interactions. In our work, the bulk value of the liquid-vapor interfacial tension is used along with the disjoining pressure to account for the molecular interactions. These various pressure jump models need to be reconciled in future work.

Since our particular interest is in non-isothermal systems, the effects of both temperature and pressure changes are important. The non-isothermal *equilibrium* use of Eq. (3) leads to Eq. (7) and the suggested experimental set-up given in Fig. 2 for the measurement of the thickness of an adsorbed thin film as a function of a temperature jump at the liquid-vapor interface with the heated substrate at $T_s \cong T_{lv}$ [e.g., 31].

$$\sigma_{lv} K + \Pi - \rho_{lm} g x = \frac{\rho_{lm} \Delta h_m}{\bar{T}} [T_{lvx} - T_{vx}] \tag{7}$$

We find that the thickness and, therefore, thickness profile are strongly coupled with the temperature and excess temperature. For an apolar flat ($K = 0$) wetting film ($A < 0$) with $\delta < 15$ nm and $x \cong 0$, Eq. (7) leads to

$$\delta_o^3 = \frac{-A\bar{T}}{6\pi \rho_{lm} \Delta h_m [T_{lvx} - T_{vx}]} \tag{8}$$

The cube of the thickness of the adsorbed film is inversely proportional to the liquid-vapor interfacial temperature jump, a substrate superheat. Using Eq. (8), two extremes are: (1) for an approximately continuous monolayer of pentane on quartz resulting from a large superheat, $\Delta T = 40$ K, $\delta_o = 0.28$ nm; (2) for a moderately small temperature difference example, $\Delta T = 10^{-2}$ K, $\delta_o = 4.4$ nm. An interesting result discussed below is that the excess surface temperature at the thickness of a monolayer correlates with the excess surface temperature at the critical heat flux in nucleate boiling heat transfer. This indicates that thin film adsorption is important in the evaporating microlayer formed under a bubble in boiling, which is very turbulent in the bulk region. This is possible because the disjoining pressure for a monolayer is significant. For extreme case (1), $\Pi = 3.07 \times 10^7 \, \text{N/m}^2$. For the thicker example, the small value of $\Delta T$ demonstrates that it is extremely difficult to have isothermal experimental conditions, if not impossible. We are concerned with transport phenomenon in this broad range of thickness. Equating the change in chemical potential with respect to pressure to the change in chemical potential with respect to temperature gives Eq. (9) for the equilibrium interfacial profile of a superheated thin film on a rough surface. Equations (8 and 9) can be used as the boundary conditions at the contact line of an evaporating meniscus on a superheated substrate. The large pressure difference between the adsorbed thin film at the contact line and the thicker portion of the meniscus can give substantial flow.

$$K - \frac{A}{6\pi\sigma_{lv}\delta_o^3} = \frac{\rho_{lm}\Delta h_m}{\sigma_{lv}\bar{T}} [T_{lv} - T_v] \tag{9}$$

For a partially wetting system like a sessile drop, the usual measurements to obtain the interfacial shape and pressure jump are the observable macroscopic apparent contact angle, $\theta$, and macroscopic curvature. Using ellipsometry, there have been some measurements of the thickness of the extremely thin adsorbed film ahead of the contact line [e.g., 29, 32, 33]. For partially wetting systems at isothermal equilibrium, Sharma presented the following useful set of equations for the excess free energy, $\Delta G$, apparent contact angle, $\theta$, disjoining pressure, $\Pi$, and stability [17, 34, 35].

$$\sigma_{lv}(\cos\theta - 1) = \Delta G(\delta_e) \tag{10}$$

$$\Delta G(\delta_e) = S^{LW}\left[\frac{d_o^2}{\delta_e^2}\right] + S^P \exp\left[\frac{d_o - \delta_e}{l}\right] \tag{11}$$

$$\Pi = -\frac{\partial \Delta G}{\partial \delta} \tag{12}$$

$S^{LW}$ and $S^P$ are the apolar and polar components of the spreading coefficient, $l$ is a correlation length for a polar fluid, $d_o$ is the equilibrium cut-off distance of 0.158 nm and $\delta_e$ is the equilibrium thickness of the thin film a large distance from $\theta$ where it becomes flat. The thickness span over which a flat film is unstable is obtained from the second derivative of the free energy, Eq. (13). The functional form of Eq. (13) is given in Fig. 3. Since the film is unstable when $\Delta G_{\delta\delta} < 0$, there is a region between $\delta_1$ and $\delta_2$ where the film is unstable. The lower value represents the thickness of an extremely thin adsorbed film between thicker regions like the lower limit of the thin film thickness between drops in dropwise condensation.

$$\Delta G_{\delta\delta} = 6 S^{LW}\left[\frac{d_o^2}{\delta^4}\right] + \frac{S^P}{l^2}\exp\left(\frac{d_o - \delta}{l}\right) \tag{13}$$

Using Eq. (14), Eq. (15) with $\theta(\delta)$ for the shape of the thin film in the immediate vicinity of the contact line has been obtained [e.g., 8, 10, 13, 36, 37].

$$\sigma_{lv}K + \Pi = const. \tag{14}$$

$$\cos\theta = 1 - \frac{A}{12\pi\sigma_{lv}\delta_o^2}\left[1 - \frac{\delta_o^2}{\delta^2}\right] \tag{15}$$

The interfacial slope is a function of the film thickness in the contact line region [e.g., 6-14, 36, 37]. Based on this model, the shape of the transition region has been found to be that of a concave (towards the vapor for $\theta > 0$) meniscus with the initial interfacial slope of $\delta' = 0$ and a final slope consistent with the observable contact angle, $\theta$ at $\delta \to \infty$. The concave shape at the nanoscale, $K > 0$, is

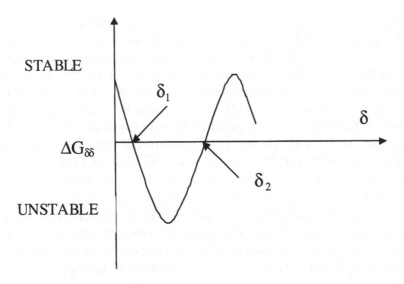

**Figure 3.** The functional form of Eq. (13).

consistent with the Kelvin concept that the reduction in vapor pressure due to cur-vature is needed to offset the increase in vapor pressure due to the disjoining pres-sure in a partially wetting system, i.e., $\Pi < 0$, $A > 0$. This also indicates that the leading edge of a completely wetting film should have the pancake shape of a convex meniscus so that the increase in vapor pressure due to shape offsets the vapor pressure decrease due to the disjoining pressure [12]. I hasten to note that large values of the "curvatures" sometimes resulting from the modeling could fit into "crevices" or over "peaks" associated with surface roughness [e.g., 22].

## 3. NON-EQUILIBRIUM SYSTEMS

For $\Delta\mu \neq 0$, Eqs. (3 and 4) can be used with interfacial kinetic theory to obtain Eq. (16) (Kelvin-Clapeyron model) for interfacial mass flux written as a net heat flux associated with interfacial condensation or evaporation, $q''_{lv}$ [e.g., 4, 7, 15, 21, 24, 35]. The classical interfacial heat transfer coefficient $h_{lv}$, is replaced by heat trans-fer coefficients associated with the Kelvin, $h_{lv}^{kl}$, and Clapeyron, $h_{lv}^{cl}$, effects.

$$q''_{lv} = h_{lv}\,[T_{lvx} - T_{vx}] = h_{lv}^{cl}\,[T_{lvx} - T_{vx}] - h_{lv}^{kl}\,[\Pi + \sigma_{lv}K - \rho_{lm}gx] \qquad (16)$$

These coefficients are defined by Eq. (17) in which M is the molecular weight, $V_{lM}$ is the liquid molar volume and C is a condensation coefficient.

$$h_{lv}^{cl} = \left(\frac{C^2 M}{2\pi RT_{lv}}\right)^{0.5} \frac{M P_v \Delta h_m^2}{R T_v T_{lv}}; \quad h_{lv}^{kl} = \left(\frac{C^2 M}{2\pi RT_{lv}}\right)^{0.5} \frac{V_{lM} P_v \Delta h_m}{R T_{lv}} \qquad (17)$$

The net mass flux is a function of the interfacial temperature jump and the interfacial pressure jump due to the shape. The classical heat transfer coefficient, $h_{lv}$, which is viewed as a function of only the temperature jump is replaced in Eq. (16) by coefficients for the temperature and pressure jumps. Due to the effect of temperature on the vapor pressure, the coefficients are a strong function of the temperature level and increase rapidly with temperature. The values of the coefficients demonstrate that it is much easier to measure (or simply, observe) the interfacial pressure jump via the profile than the interfacial temperature jump. For example, Eq. (18) is obtained for ethanol at T = 47°C, $P_v$ = 2.68 x $10^4$ Pa. For a partially wetting ethanol sessile drop with a radius of curvature of r = 60 μm, ΔP = 680 N/m². For an equivalent temperature jump effect in Eq. (18), ΔT = 3.26 x $10^{-4}$°C. In addition, small temporal and/or spatial temperature variations and/or fluctuations are always present. Therefore, we conclude that the effect of extremely small temperature gradients, which are impossible to measure, is significant in many non-isothermal and apparently "isothermal" systems.

$$q_{lv}'' = 3.66x10^6 \cdot (T_{lv} - T_v) + 1.753 \cdot (P_l - P_v) \tag{18}$$

Fluid flow in a curved thin film is also controlled by the pressure gradient. Using the lubrication approximation as an example, while neglecting liquid-vapor interfacial shear stress, the one-dimensional steady state mass flow rate per unit width, Γ, can be approximated by Eq. (19) [e.g., 7, 15]

$$\Gamma = -\frac{\delta^3}{3v} \frac{d}{dx} [P_l + \rho_{lm} g x] \tag{19}$$

in which, $v$ is the kinematic viscosity. Using Eqs. (5, 6 and 19) with K = 0 and g ≅ 0 gives the very approximate, but insightful, Eq. (20) for the steady state flow rate. As δ → 0, the stress becomes extremely large and another mechanism for flow at the contact line is needed to avoid the singularity due to the "no-slip" condition. Although the curvature gradient was neglected in Eq. (20), we note that large curvature gradients are also possible and probably present in many situations [e.g., 4, 21].

$$\Gamma = \frac{A}{6\pi v \delta} \frac{d\delta}{dx} \tag{20}$$

Using appropriate values, Eq. (19) can be compared with Eq. (21) for surface diffusion, $\Gamma_s$, in which $D_s$ is the surface diffusivity, $n$ is the molecular density and $k$ is Boltzmann's constant [38, 39].

$$\Gamma_s = -\frac{\rho_l D_s \delta}{n k T} \frac{dP_l}{dx} \tag{21}$$

In [39], we found that surface diffusion became relatively important only at thicknesses less than a monolayer for the decane-silicon system and, therefore, analyzed the effect of the relatively more important evaporation/vapor-diffusion/adsorption process in spreading near the contact line. The results discussed below demonstrate that the infinite stress at the contact line can be removed by this adsorption process [e.g., 22, 36, 39].

Equations (1, 16 and 19) can be combined to describe the details of the connection between interfacial heat flux due to phase change, temperature jump and pressure jump. The resulting Eq. (22) gives the steady state dimensionless shape, $\eta$, of a meniscus with phase change, which can be measured optically [e.g., 21]. The experimental results were found to be consistent with the modeling [e.g., 21, 29, 33]. These and similar equations have been solved numerically to describe many transport processes [e.g., 4, 7, 35, 21-48].

$$\frac{d}{d\xi}(\frac{\eta^3}{3}\frac{d\phi}{d\xi}) = \frac{1}{1+\kappa\eta}(1+\phi) \qquad (22)$$

$$\phi = -\frac{1}{\eta^3} - \varepsilon\frac{d^2\eta}{d\xi^2}, \qquad \eta = \frac{\delta}{\delta_o}, \qquad \xi = \sqrt{\frac{6\pi x^2 v h_{lv}^{cl}\Delta T_o}{-A\Delta h_m}}$$

$$\kappa = \frac{h_{lv}^{cl}\delta_o}{k_l}, \qquad \varepsilon = -\frac{6\pi\sigma v\delta_o h_{lv}^{kl}}{A\Delta h_m}$$

For a very simple approximate but insightful example of the use of Eq. (22), we look at the variation of the interfacial heat transfer coefficient for phase change in the immediate vicinity of the contact line. Neglecting the resistance due to conduction in the liquid, $k_l \rightarrow \infty$, and curvature effects, $\eta'' = 0$, Eq. (23) for the variation of the dimensionless heat transfer coefficient with thickness can be obtained from Eq. (22) [7].

$$\frac{h_{lv}}{h_{lv}^{cl}} = 1 - \eta^{-3} \qquad (23)$$

Therefore, we can quickly demonstrate that the evaporation rate vanishes when $\delta = \delta_o$, $\eta = 1$ (the adsorbed equilibrium thickness, $\delta = \delta_o$, is given by Eq. 8). Further, the evaporation rate reaches 96% of the value calculated using only the Clapeyron effect on vapor pressure, $h^{cl}_{lv}$, when $\delta = 3\delta_o$ and curvature effects are negligible. As a numerical example, for the completely wetting pentane/glass system for the very large excess surface temperature at Critical Heat Flux in boiling, mass transfer equilibrium occurs at $\delta = \delta_o = 0.28$ nm for $\Delta T = 44$ K. Whereas, a very large evaporative heat transfer coefficient occurs a very short distance away at $\delta = \delta_o = 0.84$ nm where adsorption effects essentially vanish. The important point is that the singularity to heat transfer in the contact line is thereby avoided

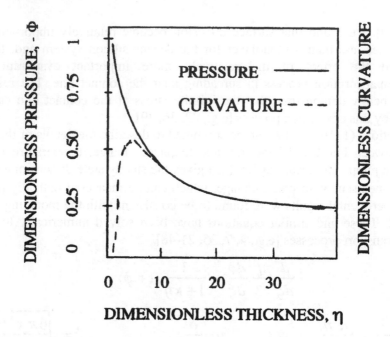

**Figure 4.** Schematics of dimensionless pressure and curvature distributions.

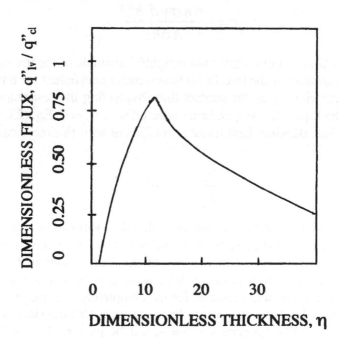

**Figure 5.** Schematic of dimensionless interfacial heat flux distribution.

by adding the Kelvin effect. It is, of course, also possible to have lower heat fluxes due to the effects of capillarity on the vapor pressure and thermal conductive resistances in the liquid and solid [see, e.g., 4, 21, 24, 45].

In Fig. 4, schematics of the dimensionless pressure jump, $\phi$, and curvature, $\eta''$, for the contact line region of a stationary evaporating meniscus are presented. The schematics show that, initially, fluid flow results from a disjoining pressure gradient. In this region the curvature builds up to a maximum value at a relatively small thickness. Then, the curvature and disjoining pressure gradients cause the flow for a short distance. In the thicker region, only the curvature gradient causes the flow towards the contact line. Details of this flow field are presented in [21, 36]. A schematic of the dimensionless interfacial heat flux is presented in Figure 5. At the interline, $\eta = 1$, the dimensionless pressure jump is given by $\phi = -1$. Therefore, the flux is equal to zero. Initially, the flux increases with an increase in film thickness due to a decrease in the Kelvin effect. Then, the flux reaches a maximum and starts to decrease due to an increase in the resistance to conduction across the thicker liquid film. The numerical study by Stephan and Busse [45] included the additional resistance to conduction in the substrate.

A more complicated problem is the rewetting of a heated surface. Although the results of extensive research on contact line motion have been presented, the relative importance of viscous stresses, slip, surface diffusion, molecular kinetics/dynamics, evaporation/condensation, excess free energy, contact angle and film shape has not been resolved, experimentally or theoretically, at the contact line because of the molecular scale of the transport processes. Usually, spreading has been described using two types of isothermal non-volatile liquid spreading models based on molecular kinetics/dynamics and/or liquid phase viscous flow [12, 49-73]. However, since vapor phase transport is also possible, a third type of isothermal model for contact line motion based on shape-induced condensation (Kelvin effect) was initially proposed by Wayner [22, 36, 74] and recently readdressed by Shanahan [75, 76]. This Kelvin-Clapeyron model of spreading was suggested by the pioneering experimental results on the effect of vapor pressure on spreading by Hardy and Doubleday [77], Bangham and Saweris [78], and Bascom *et al.* [79]. For the isothermal case, a seminal paper by Deryaguin *et al.* [2] demonstrated the use of the disjoining pressure [1] gradient to evaluate the effect of surface film transfer on enhanced evaporation of liquid from capillaries. The stability characteristics of thin film domains have also been addressed using the Kelvin-Clapeyron type model [e.g., 35, 43]. The K-C model effectively relieves the no-slip condition of fluid mechanics at the contact line by incorporating a mechanism for condensation.

Schematic views of the equilibrium and non-equilibrium liquid-vapor interfacial profiles in the contact line region are given in Fig. 6. In this case, to obtain a simpler system to analyze, we take $\Gamma = 0$ in equation (1) and obtain a shape preserving system based on only condensation in the immediate vicinity of the contact line where we presume viscous effects are extremely large for viscous flow.

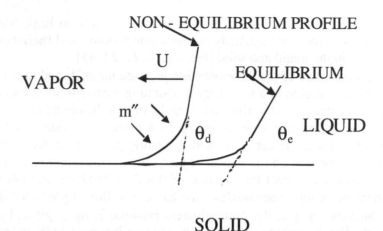

**Figure 6.** Conceptual views of the equilibrium and non-equilibrium contact line region. Forced change in profile for non-equilibrium system represented by $\theta_e \to \theta_d$ with the increase in curvature causing condensation. The roughness of the solid surface, which could be the size of the curved region, is not shown.

In [36], Eq. (24) was obtained for the velocity, U, as a function of the dynamic contact angle at the thickness $\delta$, $\theta_{d\delta}$. In essence, a forced change in the apparent contact angle, $\theta_e \to \theta_d$, due to viscous flow in the thicker region (adjacent to the region of interest) leads to condensation in the immediate vicinity of the contact line because of a forced change in the curvature and, therefore, contact line motion. As a result, the change in apparent contact angle, which can be measured at a location away from the contact line, is related to the phase-change process near the contact line, which cannot be directly measured because of the small size of the region. Recently, Shanahan has addressed the effect of dissipation along with condensation during spreading [76].

$$U = \left(\frac{\partial x}{\partial t}\right)_{\delta} = \frac{-\left(\dfrac{\partial \delta}{\partial t}\right)_x}{\left(\dfrac{\partial \delta}{\partial x}\right)_t} = \frac{-\dfrac{m''}{\rho_l}\cos\theta_{d\delta}}{\tan\theta_{d\delta}} \tag{24}$$

Using Eqs. (16 and 24), Eq. (25) for the dimensionless shape was obtained.

$$\Psi - (b\,\eta^3)^{-1} = \Delta\tau - \Phi \tag{25}$$

$$\Psi = \left[1 + \left(\frac{d\eta}{d\zeta}\right)^2\right]^{-1.5} \frac{d^2\eta}{d\zeta^2}$$

$$\Delta \tau = \frac{\rho_{lM}\, \delta_o\, \Delta h_M}{\sigma_{lv}\, T_v}\, (T_{lv} - T_v)$$

$$\Phi = a_1 \ln \frac{P_{lv}}{P_v} = a_1 \ln (1 + \gamma)$$

$$a_1 = \frac{\rho_{lM}\, R T_{lv}\, \delta_o}{\sigma_{lv}}\ , \ b = \frac{6\pi \sigma_{lv}\, \delta_o^2}{A}\ , \ x = \delta_o\, \zeta$$

$$\gamma = \frac{a_2\, U}{C_1 \cos \theta_{d\eta}\, \cot \theta_{d\eta}}$$

$$a_2 = \frac{\rho_l}{P_v} \left( \frac{2\pi R T}{M} \right)^{0.5}$$

Using $\Psi - (b\eta^3)^{-1} = c$, a constant, Eq. (25) gives the effect of the dimensionless temperature jump on the dimensionless velocity, which is presented in Fig. 7 [22]. The possibility of re-circulation in the nano-scale region at the contact line due to evaporation/condensation is discussed in [74].

Using a simplification of these equations, Reyes and Wayner [80] were able to predict experimental results measured in a very turbulent, complicated process: the effect of pressure on the excess surface temperature at departure from nucleate boiling at the Critical Heat Flux, CHF, $(T_{lv} - T_v)_{CHF}$. Using Eq. (25) with $U = 0$, Eq. (26) was obtained for the interfacial temperature jump at the CHF. All the terms on the l.h.s. of Eq. (26) can be measured. To determine the theoretical value of the dimensional constant, const., information on the intermolecular force field at the liquid-solid interface is needed. For previously reported data on the effect of pressure level on the excess temperature, the value of this unknown constant was obtained using one experimental measurement of the excess temperature for each liquid-solid system. Then, using a set of measurements at one pressure to determine the dimensional constant for each system, Eq. (26) was used to predict the value of $\Delta T_{CHF}$ at other pressures. We believe that the results presented in Fig. 8 support the validity of these concepts.

$$\frac{\rho_{lM}\, \Delta h_M\, (T_{lv} - T_v)_{CHF}}{T_v\, \sigma_{lv}^{1.5}} \cong const. \tag{26}$$

## 4. RECENT EXPERIMENTS USING THE CONSTRAINED VAPOR BUBBLE

Recently, we found that a constrained vapor bubble (CVB) in a hollow parallelepiped of quartz with a square cross section was a robust and convenient experimental system for the study of the effects of temperature and pressure on interfacial phase change [23, 81, 82]. Significant information on the equilibrium

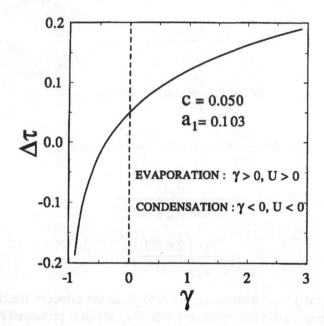

**Figure 7.** Dimensionless temperature versus dimensionless velocity obtained using Eq. (25).

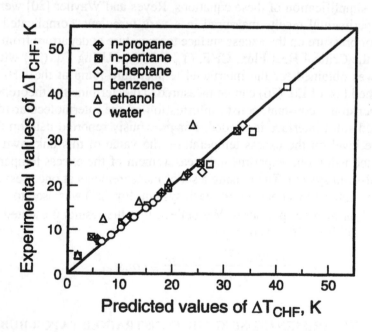

**Figure 8.** Comparison of experimental and predicted values of $\Delta T_{CHF}$ for polar and apolar fluids obtained by Reyes and Wayner [80] using Eq. (26).

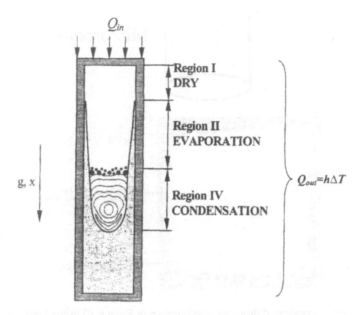

**Figure 9.** Schematic of the CVB experimental cell showing inside front suface.

characteristics of interfacial phenomena can also be obtained using a CVB [29]. A schematic drawing of the inside glass surface of the system used in [81] with a partially wetting fluid is presented in Figure 9. The cell had a square cross section with sharp internal corners and internal dimensions of 3 x 3 x 40 mm. The CVB was formed by under-filling with liquid ethanol an evacuated closed fused quartz cell, which had been heated to a high temperature during fabrication. The high temperature removal of surface hydroxyl groups gave a partially wetting system. Under equilibrium conditions, the menisci in the corners wick to the top of the cell and experimental measurements of the film profile give the interfacial properties of the system [e.g., 29]. The sum of the capillary and disjoining pressures is equal to the isothermal excess free energy per unit volume, which is a function of shape (curvature and thickness). With heating, the heat input at one end, $Q_{in}$, was lost to the external surroundings by natural convection along the entire length of the CVB. Due to the surface heat flux, there were temperature gradients in the cell with local regions of condensation (IV), evaporation from the corners (II), and dryness (I). Both fluid flow and phase change were due to the interfacial temperature jump and the shape dependent stress field (pressure) in the liquid resulting from the imposed external temperature field. The processes of nucleation, growth, and surface removal of a condensing sessile droplet on the flat surface in Region III (in the immediate vicinity of the junction of Regions II and IV) were discussed using interfacial concepts in [81-83]. This heat transport system is a large-scale version of a wickless heat pipe designed for microgravity, where gravity does not interfere with small capillary pressure systems.

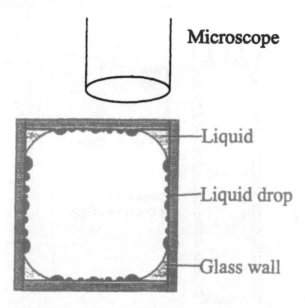

**Figure 10.** Schematic drawing of dropwise condensation in Region III at the junction of Regions II and IV.

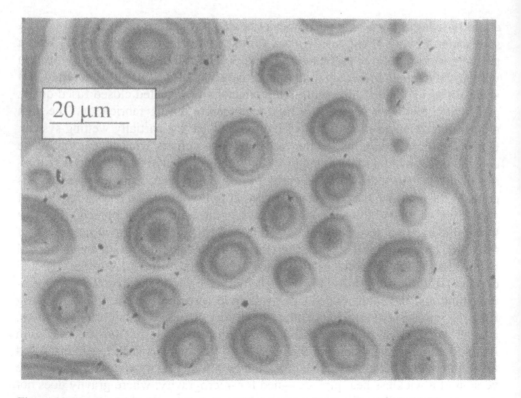

**Figure 11.** Picture of dropwise condensation with corner meniscus on the right hand side.

Although the transport processes vary in the axial direction in the CVB, different slices at various axial positions can be selected for study. Due to the large cross-sectional area of the vapor space, viscous effects in the vapor are negligible and pressure gradients in the vapor can be neglected. The change in chemical potential at the interface is modeled as a change in vapor pressure at the interface. For a complete description of the chemical potential difference causing interfacial phase change for a pure fluid, the Kelvin part of the Kelvin-Clapeyron model of the excess free energy gives both the capillary and disjoining pressure effects on the interfacial vapor pressure and the Clapeyron part gives the temperature effect on the interfacial vapor pressure.

During condensation and at equilibrium, the film thickness profiles were optically measured and recorded through the microscope shown in Fig. 10. In essence, the naturally occurring interference fringe patterns were recorded using a video camera attached to a microscope [29, 81, 82]. A picture of the naturally occurring interference fringes, which gives the thickness profiles of the corner meniscus on the right hand side and the growing droplets, is presented in Fig. 11 from [81, 82]. When the drops touched they merged. Before they touched, there was no apparent motion besides the symmetrical movement of the contact line due to condensation. The values of the local curvature (therefore, pressure field), contact angle, heat flux, vapor pressure, and approximate temperature level were obtained from the fringe pattern and Eq. (16). The measured substrate temperature in Region III was consistent with the measured thermodynamic vapor pressure. Using the average heat flux from the measured rate of volumetric condensation and the interfacial curvature, the extremely small interfacial temperature jump (which is too small to measure) can be calculated using the Kelvin-Clapeyron model of phase change heat transfer, Eq. (16). The values of the contact angle and the interfacial pressure jump give additional information about the excess free energy, the spreading coefficient, and the adsorbed film thickness using Eqs. (10-15) [82, 83].

Finally, we note that the experimental measurements and the clarity of the experimental measurements are both a function of the indexes of refraction of both the liquid and solid. More importantly, since the intermolecular force field, the thermophysical properties, and the interferometric technique all depend on the dielectric properties (electromagnetic field) of both the liquid and solid, we find that a basic unified model of the contact line region is possible.

## 5. CONCLUSIONS

Since the chemical potential is a function of both temperature and pressure, the Kelvin-Clapeyron model for the contact line region, which includes these effects on the local vapor pressure, gives an enhanced understanding of both equilibrium and non-equilibrium processes.

Interfacial kinetic theory connects the variation of the local vapor pressure to interfacial mass transfer and, therefore, demonstrates the coupling of phase change and liquid film profile with contact line motion.

The strong coupling between molecular dynamics, temperature, and pressure suggests that complex nano-scale flux fields including phase change are hidden within macroscopic experimental measurements.

Since the intermolecular force field, the thermophysical properties, and the interferometric technique all depend on the dielectric properties (electromagnetic field) of both the liquid and solid, we find that a basic unified model of the contact line region is possible.

## Acknowledgement

This material is based on work supported by the National Aeronautics and Space Administration under grant # NAG3-2383. Any opinions, findings, and conclusions or recommendations expressed in this publication are those of the author and do not necessarily reflect the view of NASA.

## NOMENCLATURE

A = Hamaker constant, area
a = constant in Eq. (25)
b = constant in Eq. (25)
C = condensation coefficient
c = constant
$D_s$ = surface diffusivity
d = distance, derivative
G = excess surface free energy
g = gravitational acceleration
h = enthalpy/mass, heat transfer coefficient
k = thermal conductivity, Boltzmann's constant
K = curvature
$l$ = correlation length
M = molecular weight
$m''$ = interfacial mass flux
n = molecular density
P = pressure
Q = heat flow rate
$q''_{lv}$ = heat flux
R = gas constant
S = spreading coefficient
T = temperature
t = time

U    = velocity
V    = volume
x    = parallel to flow direction
$\Gamma$    = mass flow rate per unit width of film, adsorbate per unit area
$\gamma$    = dimensionless velocity in Eq. (25)
$\Delta$    = difference
$\delta$    = liquid film thickness
$\varepsilon$    = dimensionless group in Eq. (22)
$\eta$    = dimensionless film thickness in Eq. (22)
$\theta$    = apparent contact angle
$\kappa$    = dimensionless group in Eq. (22)
$\mu$    = chemical potential
$\nu$    = kinematic viscosity
$\xi$    = dimensionless position in Eq. (22)
$\Pi$    = disjoining pressure
$\rho$    = fluid density
$\sigma$    = surface free energy per unit area of liquid
$\tau$    = dimensionless temperature in Eq. (25)
$\Phi$    = dimensionless chemical potential in Eq. (25)
$\phi$    = dimensionless pressure difference in Eq. (22)
$\Psi$    = dimensionless curvature in Eq. (25)
$\zeta$    = dimensionless location in Eq. (25)

*Subscripts and superscripts*

CHF    = critical heat flux
cl    = Clapeyron effect
d    = dynamic
e    = equilibrium
g    = includes gravity
kl    = Kelvin effect
LW    = Lifshitz-van der Waals
M    = molar
*l*    = liquid
m    = unit mass
o    = contact line, equilibrium, cut-off distance
P    = polar
s    = solid substrate
sat    = saturation
t    = time
v    = vapor
x    = evaluated at x
d    = derivative, thickness

'      = derivative
δ      = thickness
∞      = infinity
-      = average

## REFERENCES

1. B. V. Derjaguin, *Colloid J. USSR (English Translation)*, **17**, 191-197 (1955).
2. B. V. Derjaguin, S. V. Nerpin and N. V. Churayev, *Bulletin RILEM No.* **29**, 93-98 (1965).
3. J. C. P. Broekhoff and J. H. de Boer, *J Catalysis*, **10**, 391-400 (1968).
4. M. Potash, Jr. and P. C. Wayner, Jr., *Int. J. Heat Mass Transfer*, **15**, 1851-1863 (1972).
5. C. A. Miller and E. Ruckenstein, *J. Colloid Interface Sci.*, **48**, 368-373 (1974).
6. B. V. Deryagin, V. M. Starov and N. V. Churaev, *Kolloid. Zh.*, **38**, 875-879 (1976).
7. P. C. Wayner, Jr., Y. K. Kao and L.V. LaCroix, *Int. J. Heat Mass Transfer*, **19**, 487-491 (1976).
8. L. R. White, *J. Chem. Soc. Faraday Trans. I*, **73**, 390-398 (1977).
9. F. Renk, P. C. Wayner, Jr. and G. M. Homsy, *J. Colloid Interface Sci.*, **67**, 408-414 (1978).
10. P. C. Wayner, Jr., *J. Colloid Interface Sci.*, **77**, 495-499 (1980).
11. V. M. Starov and N. V. Churaev, *Colloid J. USSR (English Translation)*, **42**, 585-591 (1981).
12. P. G. de Gennes, *Rev. Mod. Phys.*, **57**, 827-863 (1985).
13. G. J. Hirasaki, in *Interfacial Phenomena in Petroleum Recovery*, N. R. Morrow (Ed.), pp 77-99, Marcel Dekker, New York (1990).
14. A. Marmur, *J. Adhesion Sci. Technol.*, **6**, 689-701 (1992).
15. P. C. Wayner, Jr., *Colloids Surfaces*, **52**, 71-84 (1991).
16. F. Brochard-Wyart, J. M. di Meglio, D. Quere and P. G. de Gennes, *Langmuir*, **7**, 335-338 (1991).
17. A. Sharma, *Langmuir*, **9**, 3580-3586 (1993).
18. J. H. Keenan, *Thermodynamics*, John Wiley & Sons, New York (1941).
19. V. P. Carey, *Liquid-Vapor Phase-Change Phenomena: An Introduction to the Thermophysics of Vaporization and Condensation Processes in Heat Transfer Equipment*, Hemisphere Publishing Corp., Washington, D. C. (1992).
20. A. Sharma, *Langmuir*, **14**, 4915-4928 (1998).
21. S. DasGupta, I. Y. Kim and P. C. Wayner, Jr., *Trans. ASME J Heat Transfer*, **116**, 1007-1015 (1994).
22. P. C. Wayner, Jr., *Colloids Surfaces A*, **89**, 89-95 (1994).
23. P. C. Wayner, Jr., *AIChE J*, **45**, 2055-2068 (1999).
24. S. Moosman and G. M. Homsy, *J. Colloid Interface Sci.*, **73**, 212-223 (1980).
25. R. Cook, C. Y. Tung and P. C. Wayner, Jr., *Trans. ASME J. Heat Transfer*, **103**, 325-330 (1981).
26. C. J. Parks and P. C. Wayner, Jr., *AIChE J*, **33**, 1-10 (1987).
27. S. M. Troian, E. Herbolzheimer and S. A. Safran, *Phys. Rev. Lett.*, **65** , 333-336 (1990).
28. B. V. Deryagin and N. V. Churaev, *Kolloid. Zh.*, **38**, 438-448 (1976).
29. S. DasGupta, J. L. Plawsky and P. C. Wayner, Jr., *AIChE J*, **41**, 2140-2149 (1995).
30. M. L. Gee, T. W. Healy and L. R. White, *J. Colloid Interface Sci.*, **131**, 18-23 (1989).
31. P. C. Wayner, Jr., in *Boiling Heat Transfer*, R.T. Lahey, Jr. (Ed.) pp. 569-614, Elsevier (1992).
32. D. Ausserre, A. M. Picard and L. Leger, *Phys. Rev. Lett.*, **57**, 2671-2674 (1986).
33. I. Y. Kim and P. C. Wayner, Jr., *J. Thermophysics Heat Transfer*, **10**, 320-325 (1996).
34. A. Sharma, *Langmuir*, **9**, 861-869 (1993).
35. A. Sharma, *Langmuir*, **14**, 4915-4928 (1998).
36. P. C. Wayner, Jr., *Langmuir*, **9**, 294-299 (1993).
37. A. Sharma, *Langmuir*, **9**, 3580-3586 (1993).
38. E. Ruckenstein and C. S. Dunn, *J. Colloid Interface Sci.*, **59**, 135-138 (1977).

39. P. C. Wayner, Jr. and J. Schonberg, *J. Colloid Interface Sci.*, **152**, 507-519 (1992).
40. Y. Kamotani, *Proceedings of 3rd International Heat Pipe Conference*, Palo Alto, CA, American Institute of Aeronautics and Astronautics, pp. 128-130 (1978).
41. F. W. Holm and S. P. Goplen, *Trans. ASME J. Heat Transfer*, **101**, 543-547 (1979).
42. A. V. Mirzamoghadam and I. Catton, *Trans. ASME J. Heat Transfer*, **110**, 208-213 (1988).
43. S. G. Bankoff, *Trans. ASME J. Heat Transfer*, **112**, 538-546 (1990).
44. K. P. Hallinan, H. C. Chebaro, S. J. Kim and W. S. Chang, *J. Thermophys. Heat Transfer*, **8**, 709-716 (1994).
45. P. Stephan and C. A. Busse, *Int. J. Heat Mass Transfer*, **35**, 383-391 (1992).
46. L. W. Swanson and G. P. Peterson, *Trans. ASME J. Heat Transfer*, **115**, 195-201 (1995).
47. D. M. Anderson and S. H. Davis, *Phys. Fluids*, **7**, 248-265 (1995).
48. S. J. S. Morris, *J. Fluid Mech.*, **432**, 1-30 (2001).
49. T. D. Blake and J. M. Haynes, *J. Colloid Interface Sci.*, **30**, 421-423 (1969).
50. E. B. Dussan V and S. H. Davis, *J. Fluid Mech.*, **65**, 71-95 (1974).
51. B. W. Cherry and C. M. Holmes, *J. Colloid Interface Sci.* **29**, 174-176 (1969).
52. J. Lopez, C. A. Miller and E. Ruckenstein, *J. Colloid Interface Sci.* **56**, 460-468 (1976).
53. C. Huh and L. E. Scriven, *J. Colloid Interface Sci.* **35**, 85-101 (1971).
54. L. M. Hocking, *J. Fluid Mech.*, **79**, 209-229 (1977).
55. P. Neogi and C. A. Miller, *J. Colloid Interface Sci.*, **86**, 525-538 (1982) .
56. R. L. Hoffman, *J Colloid Interface Sci.*, **94**, 470-486 (1983).
57. C. A. Miller and E. Ruckenstein, *J. Colloid Interface Sci.* **48**, 368-373 (1974).
58. E. B. Dussan V., *Annu. Rev. Fluid Mech.*, **11**, 371 (1979).
59. V. M. Starov, *Colloid J. USSR (Eng. Transl.)* **45**, 1009-1016 (1983).
60. A. M. Cazabat, *Contemp. Phys.*, **28**, 347-364 (1987).
61. P. A. Thompson and M. O. Robbins, *Phys. Rev. Lett.*, **63**, 766-769 (1989).
62. J. Koplik, J. R. Banavar and J. F. Willemsen, *Phys. Fluids A.*, **1**, 781-794 (1989).
63. P. Ehrhard and S. H. Davis, *J. Fluid Mech.*, **229**, 365-388 (1991).
64. E. V. Gribanova, *Adv Colloid Interface Sci.*, **39**, 235-255 (1992).
65. R. A. Hayes and J. Ralston, *Langmuir*, **10**, 340-344 (1994).
66. J. D. Chen and N. Wada, *J. Colloid Interface Sci.*, **148**, 207-222 (1992).
67. J. Koplik and J. R. Banavar, *Annu. Rev. Fluid Mech.*, **27**, 257-292 (1995).
68. L. Leger and J. F. Joanny, *Rep. Prog. Phys.* **55**, 431-486 (1992).
69. K. Stoev, E. Rame and S. Garoff, *Physics Fluids*, **11**, 3209-3215 (1999).
70. P. G. Petrov and J. G. Petrov, *Langmuir* **11**, 326-3268 (1995).
71. M. Schneemilch, R. A. Hayes, J. G. Petrov and J. Ralston, *Langmuir* **14**, 7047-7051 (1998).
72. T. D. Blake, A. Clarke, J. DeConinck and M. J. deRuijter, *Langmuir*, **13**, 2164-2166 (1997).
73. L. M. Pismen and B. Y. Rubinstein, *Langmuir*, **17**, 5625-5270 (2001).
74. P. C. Wayner, Jr., *Trans. ASME J Heat Transfer*, **116**, 938-945 (1994).
75. M. E. R. Shanahan, *Langmuir*, **17**, 3997-4002 (2001).
76. M. E. R. Shanahan, *Langmuir*, **17**, 8229-8235 (2001).
77. W. B. Hardy and I. Doubleday, *Proc. Royal Soc. Ser. A*, **200**, 550-574 (1922).
78. D. H. Bangham and Z. Saweris, *Trans. Faraday Soc.*, **34**, 554-570 (1938).
79. W. D. Bascom, R. L. Cottington and C. R. Singleterry, in *Contact Angle, Wettability and Adhesion*, Adv. Chem. Ser. No. 43, pp 355-379, American Chemical Society, Washington, D. C. (1964).
80. R. Reyes and P. C. Wayner, Jr., *Trans. ASME J Heat Transfer*, **117**, 779-782 (1995).
81. Y. X. Wang, J. L. Plawsky and P. C. Wayner, Jr., *Microscale Thermophysical Eng.*, **5**, 55-69 (2001).
82. P. C. Wayner, *Jr. Colloids Surfaces A*, **206**, 157-165 (2002).
83. L. Zheng, Y.-X. Wang, J. L. Plawsky and P. C. Wayner, Jr., *Langmuir*, **18**, 5170-5177 (2002).

*Contact Angle, Wettability and Adhesion*, Vol. 3, pp. 25–37
Ed. K.L. Mittal
© VSP 2003

# Experimental study of contact line dynamics by capillary rise and fall

PO-ZEN WONG * and ERIK SCHÄFFER

*Department of Physics, University of Massachusetts, Amherst, MA 01003*

**Abstract**—We studied the dynamics of contact lines near the depinning transition by capillary rise and fall experiments. The water column height $h$ was measured as a function of time $t$ by video imaging. The data in the late stage of movement were analyzed in terms of critical behavior associated with a dynamical phase transition, $dh/dt \propto (P - P_c)^\beta$. We found that $\beta \approx 1$ in the falling experiments but $\beta > 1$ in the rising experiments. The exact value of $\beta$ varied from run to run, depending on the waiting time before the rise. These results suggest that the microscopic details of the wetting film at the molecular level have important effects on the macroscopic dynamics.

*Keywords*: Contact lines dynamics; depinning transition; dynamical phase transitions; self-organized critical phenomena.

## 1. INTRODUCTION

The depinning of contact lines is an old fluid mechanics problem that has received considerable new interest in recent years. The commonly observed stick-slip motion is believed to be a generic feature found in a broad range of seemingly unrelated physical systems. Some well known examples include the motion of magnetic domain walls in response to changing magnetic fields [1], and the dynamics of flux lines in high-$T_c$ superconductors with increasing electric current [2]. In all cases, pinning results from the inherent disorder in the system, and depinning occurs when the applied force exceeds a certain threshold. It is commonly postulated that the pinned state and the moving state are separated by a dynamical phase transition that exhibits universality in its critical behavior, i.e., independent of the details of the systems [3, 4]. The advantage of studying contact line dynamics is that one can make direct visual observations of the critical behavior. The challenge, however, is that one has to apply a mechanical driving force in such a way that it does not externally introduce stick-slip motion or other fluctuations. We have studied the depinning transition by simple capillary rise and fall of a water column in glass

---

*To whom all correspondence should be addressed. Phone: (1-413)545-3288, Fax: (1-413)545-1691, E-mail: pzwong@physics.umass.edu

tubes [5, 6]. Geometric roughness and chemical inhomogeneities on the tube wall provide the random pinning forces. The motion of the contact line is driven by its own interaction with the environment and no external device is needed to push it. So we may describe the system as self-organized. This paper is a summary report of our work.

## 2. BACKGROUND

In 1985, Koplik and Levine [7] first suggested a model to describe the dynamics of fluid interfaces in random porous media. In this model, the interface is described by a single-valued function $z = f(\mathbf{x}, t)$, where $\mathbf{x}$ is a $(d - 1)$-component transverse vector. The equation of motion is

$$\frac{\partial f}{\partial t} = P + \nabla^2 f + Y(\mathbf{x}, f), \qquad (1)$$

where $P$ is a uniform driving force, $\nabla^2 f$ is an effective surface tension that minimizes the surface area, and $Y$ represents quenched random forces acting on the interface. They noted that the same equation was used by Bruinsma and Aeppli [8] to study domain wall dynamics in the random-field Ising model (RFIM). The underlying connection is that Eq. (1) is the time-dependent Ginzburg-Landau equation

$$\frac{\partial f}{\partial t} = P - \frac{\delta F}{\delta f} \qquad (2)$$

for the RFIM interfacial free energy $F$ that was derived by Grinstein and Ma [9] as

$$F(f) = \int d^{d-1}x \left[ \frac{1}{2} |\nabla f|^2 - \int_\infty^f dz\, Y(\mathbf{x}, z) \right]. \qquad (3)$$

For small values of $Y$, Koplik and Levine find that the interface translates smoothly, but, for large values of $Y$, it can be pinned in rough configurations.

The critical behavior of the depinning transition in the above model was later studied by computer simulation [9–12] and analytical calculations [3, 4]. The key result is that the interface advances via a sequence of avalanches. The average size of an avalanche $\xi$ (measured along the interface) diverges as

$$\xi \propto (P - P_c)^{-\nu} \qquad (4)$$

as the critical force $P_c$ is approached. The root mean squares interface width $w$ obeys self-affine scaling

$$w \propto \xi^\alpha \propto (P - P_c)^{-\nu\alpha} \qquad (5)$$

with $\alpha = 2 - 1/\nu = \varepsilon/3$, where $\varepsilon = (5 - d)$. The waiting time $\tau$ for an avalanche of size $\xi$ follows dynamic scaling

$$\tau \propto \xi^z \propto (P - P_c)^{-\nu z}. \qquad (6)$$

Consequently, the interface velocity is given by

$$v \propto w/\tau \propto \xi^{\alpha-z} \propto (P - P_c)^{-v(\alpha-z)} \equiv (P - P_c)^{\beta} \tag{7}$$

where $\beta = (z - \alpha)v \approx 1 - \varepsilon/9 + O(\varepsilon^2)$. In this scaling picture of the transition, there are only two independent scaling exponents, $v$ and $z$, or equivalently $\alpha$ and $\beta$. In two dimensions, the theoretical predictions are $\alpha = 1$ and $\beta \approx 2/3$, and in three dimensions $\alpha = 2/3$ and $\beta \approx 7/9$. Experimentally, the exponent $\alpha$ can be measured by analyzing the interface roughness according to Eq. (5) (see, e.g., Refs. [13, 14] and references therein) and $\beta$ may be determined by analyzing the interface velocity $v$ as a function of the applied force $P$ according to Eq. (7) [15, 16].

The depinning of contact lines follows the same scaling picture, even though the energy associated with a distorted line is not given by Eq. (3). This is because the contact line is the intersection of three interfaces: vapor–liquid, vapor–solid and liquid–solid. The distortion of the contact line affects the shapes and areas of all three interfaces and their associated energies. Ertaş and Kardar [17] assumed that the contact line moved on a solid surface with randomly fluctuating interfacial energies. Their analysis gave $\alpha = 1/3$ and $\beta \approx 2/3$ in three dimensions. Prior to their work, Robbins and Joanny also obtained $\alpha = 1/3$ [18]. Raphaël and de Gennes [19] considered contact lines passing over surface defects one at a time. They predicted $\beta = 3/2$. Joanny and Robbins [20] considered the case in which the contact line passed over periodic stripes of chemical heterogeneities parallel to the line, and they also found $\beta = 3/2$. A common assumption in these models is that they focus on the interfacial energies and ignore the excess viscous dissipation caused by the roughness. The reason is that viscous effects diminish as the fluid flow velocity approaches zero. However, Sheng and Zhou [21] argued that the breakdown of the no-slip boundary condition at the contact line could lead to significant dissipation. By analyzing different slip models, they predicted $\beta > 2$, where the exact value depends on the details of the slip action.

Experimental studies of contact line pinning have found a range of values for the exponent $\beta$. Stokes *et al.* [22] used an AC method to study the nonlinear response of the meniscus between mineral oil and a glycerol-methane mixture in a capillary tube. Analyzing the harmonics in terms of Eq. (7) gave $\beta \approx 2.5 \pm 0.4$. More recently, Kumar *et al.* [23] used the same technique to study water–decane and water–hexadecane menisci. They found $\beta \approx 5$ instead. Ström *et al.* [24] measured the dynamic contact angle visually by dipping a solid plate into a liquid at different speeds $v$. Using the capillary number $N_{cap} \equiv v\eta/\gamma$ as a dimensionless measure of the velocity, where $\gamma$ is the surface tension and $\eta$ is the viscosity, they found that the data were consistent with $\beta \approx 3/2$ for $N_{cap} < 10^{-4}$. An earlier work of Mumley *et al.* [25] analyzed capillary rise data for various liquid-liquid menisci in glass tubes with treated surfaces. They suggested $\beta \approx 2$ with data in the range of $N_{cap} > 10^{-5}$. The differences in these results suggest that many experimental details such as $N_{cap}$ and the chemical nature of the liquids and solids can all affect the results. In our experiment, we let $N_{cap}$ reach below $10^{-9}$ to get as near to the pinning threshold as

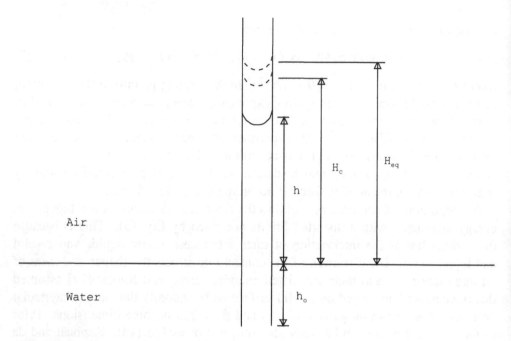

**Figure 1.** Schematics of a capillary rise experiment.

possible. We used deionized water and cleaned glass tubes to keep the chemistry less complicated.

It is interesting to note that while the theoretical predictions and experimental results give a wide range of values for $\beta$, the general form of Eq. (7) is widely adopted. Comparing experiments to theories is difficult because the nature of the pinning defects in the experiments is usually unknown and the size of the critical region is also ambiguous. In the capillary rise experiment, the velocity can vary over a wide range. In order to assess the importance of pinning effects, we need to first understand the behavior without pinning [26]. Figure 1 shows the basic geometry of such an experiment. When a tube of radius $r$ is dipped into water, in the absence of pinning, we expect the water column to rise to an equilibrium height $H_{eq} = 2\gamma \cos\theta / r\rho g$ where $\theta$ is the contact angle, $\rho$ the water density and $g$ the gravitational acceleration. Assuming that $\theta$ is independent of velocity, the upward driving pressure is $\rho g(H_{eq} - h)$ when the column height is $h$. If we further assume that the resistance to flow comes only from the viscous loss in Poiseuille flow, the equation of motion is

$$\frac{8\eta(h + h_o)}{r^2}\frac{dh}{dt} = \rho g(H_{eq} - h). \tag{8}$$

Here $h_o$ is the length of the tube submerged in water, as shown in Fig. 1. The left-hand side of this equation is the viscous pressure drop in the tube and the right-hand side is the net driving pressure due to surface tension and gravity. This equation assumes that the inertial term associated with water acceleration is negligible, which amounts to an overdamp approximation. It can be shown numerically that this assumption is valid for water near room temperature if the tube radius $r$ is less than 0.25 mm [26]. To solve this equation, we define $z = h + h_o$ and $Z_{eq} = h_o + H_{eq}$, giving

$$\frac{8\eta z}{\rho g r^2}\frac{dz}{dt} = Z_{eq} - z. \tag{9}$$

Using a normalized height $y = z/Z_{eq}$ and a normalized time $x = t/\tau_w$, where

$$\tau_w = 8\eta Z_{eq}/\rho g r^2, \tag{10}$$

we obtain a dimensionless equation $y(dy/dx) = 1 - y$. Integrating with the initial condition $y = y_1$ at $x = x_1$ gives

$$x - x_1 = -(y - y_1) - \ln\left(\frac{1-y}{1-y_1}\right) \tag{11}$$

or

$$\frac{t - t_1}{\tau_w} = -\frac{h - h_1}{H_{eq} + h_o} - \ln\left(\frac{H_{eq} - h}{H_{eq} - h_1}\right). \tag{12}$$

This is known as the Washburn equation [27]. It describes capillary rise ($h_1 < H_{eq}$) as well as capillary fall ($h_1 > H_{eq}$). The motion is characterized by the time constant $\tau_w$ given by Eq. (10). The column height $h$ approaches $H_{eq}$ exponentially at long times.

With pinning effects, we expect the air-water interface to stop at a height $H_c < H_{eq}$ in capillary rise (but $H_c > H_{eq}$ in capillary fall). So the threshold pressure for pinning is $P_c = \rho g(H_{eq} - H_c)$. Equation (7) leads to

$$\frac{dh}{dt} = \upsilon_0\left(\frac{H_c - h}{H_{eq} - H_c}\right)^{\beta}. \tag{13}$$

For $\beta \neq 1$, integrating from initial time $t_1$ and height $h_1$ yields

$$h(t) = H_c - (H_c - h_1)[1 + A(t - t_1)]^{1/(1-\beta)}, \tag{14}$$

where $A = (\beta - 1)\upsilon_0(H_c - h_1)^{\beta-1}/(H_{eq} - H_c)^{\beta}$. If $\beta < 1$, we have $A < 0$, and $h$ reaches $H_c$ after a finite time given by $\tau_0 = t - t_1 = -1/A$. If $\beta > 1$, $h$ approaches $H_c$ algebraically as $t \to \infty$. If $\beta = 1$, the integral of Eq. (13) gives,

$$h(t) = H_c - (H_c - h_1)e^{-(t-t_1)/\tau_1}, \tag{15}$$

with $\tau_1 = (H_{eq} - H_c)/\upsilon_0$. Hence the three cases $\beta < 1$, $\beta = 1$, and $\beta > 1$, give qualitatively different behaviors in the way $H_c$ is approached. Figure 2 illustrates this difference. We note in particular that for $\beta < 1$, $h$ stops at $H_c$ abruptly. For $\beta \gg 1$, the rise is almost logarithmic. Equations (12), (14) and (15) are the main expressions we use to analyze our data.

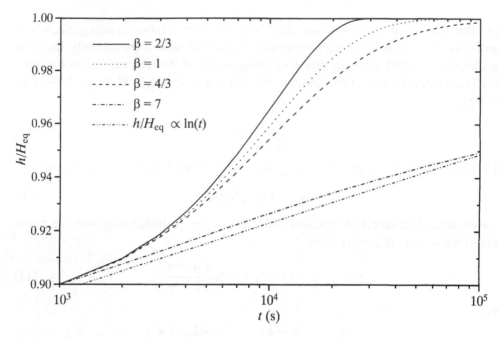

**Figure 2.** Effects of different values of $\beta$ on the approach of $h$ to $H_c$.

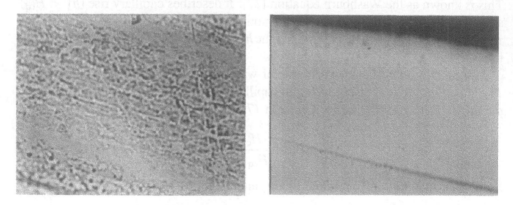

**Figure 3.** Under a microscope, the interior wall of a fractured 410 $\mu$m diameter Type R capillary tube (left) can be seen to have a much higher level of roughness compared to that of a 355 $\mu$m diameter Type S tube (right). The size of each image is $110 \times 85$ $\mu$m$^2$.

## 3. EXPERIMENTAL DETAILS

We used two types of glass tubes with diameters between 180 and 410 $\mu$m. Figure 3 shows the difference in wall roughness between two types, which we refer to as Types R and S (rough and smooth). They were cleaned immediately before the experiment using the following procedure: 400 ml of 1 M hydrochloric acid was flowed through the tube in about an hour and it was rinsed with 400 ml of deionized water, then finally boiled in deionized water for 3 h or more. Both capillary rise and fall experiments were carried out. A small syringe was used to draw the liquid to an initial height above or below $H_{eq}$. A three-way valve was used to switch the tube from the syringe to an ambient pressure with 100% humidity. A video microscopy system was used to observe the meniscus position. The difference in final heights between the rise and fall experiments is an indication of the pinning strength. Each run lasted only about $10^3$ s. Much of the movement occurred in the first few seconds, followed by very slow creeping. To observe the latter, we typically imaged with a resolution of about 5 $\mu$m/pixel. At this resolution, we found that a change in room temperature $\Delta T$ by 1 K would affect the surface tension enough to change the equilibrium height $H_{eq}$ by about 57 pixels (see Fig. 8). Consequently, the apparatus was enclosed in a thermally insulated chamber so that $\Delta T$ was less than 10 mK during a run and $H_{eq}$ was stabilized to better than half a pixel.

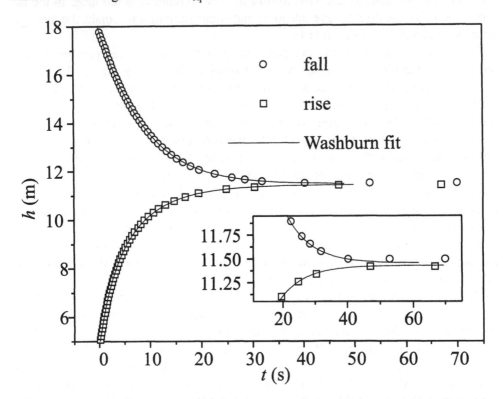

**Figure 4.** An example of capillary rise and fall behavior in 250 $\mu$m inner diameter $S$ tube observed with a coarse spatial resolution.

## 4. RESULTS AND DISCUSSION

The capillary rise experiments for glass tubes were always performed after the tube was prewetted in a capillary fall experiment. Figure 4 shows the capillary rise and fall data for a 250 $\mu$m diameter S-tube without adequate cleaning. The final heights differ by about 1 mm, evidence of pinning effects. However, the data before reaching the pinned height were well described by the Washburn equation, so the exponent $\beta$ could not be analyzed. The time constant $\tau_w$ was about 5 s. To investigate the pinning effects, we used the microscope to enhance the spatial resolution. Figure 5a shows the capillary fall data within 0.5 mm of pinning for a 310 $\mu$m diameter R-tube. We can clearly observe stick-slip behavior. Although the data are different for each consecutive run, they are not statistically independent because there are clearly some strong pinning locations in the tube. As a result, we cannot fit the data to Eq. (14) to find $\beta$. The roughness is too large in the R-tubes such that we do not have adequate statistical sampling to obtain the average behavior to compare with Eq. (14).

To find $\beta$, we used properly cleaned S-tubes with much weaker pinning effects so that the individual avalanches were not resolved by our apparatus. Figure 5b shows the rise and fall data for a 250 $\mu$m diameter S-tube. We note that the final height difference is only about 100 $\mu$m, which confirms that the pinning effect is weak. The Washburn equation was found to fit the falling data extremely well and the rising data poorly. More importantly, the time constant $\tau_w$ obtained from the falling data agreed with our expectation but that for the rising data was much larger. The latter were in the range of 30–50 s and the first term in Eq. (12) made little difference to the fit. Without the first term, Eq. (12) is reduced to Eq. (15) which corresponds to $\beta = 1$, so the observed time constant is actually $\tau_1$ and not $\tau_w$. As discussed in conjunction with Fig. 4, $\tau_w$ was only about 5 s and that applies to the first several centimeters of meniscus movement. In contrast, the fit in Fig. 5b is for the last millimeters of movement. This is the first evidence that the rising and falling data do not behave the same way.

The next step in our analysis was to test how Eq. (15) compared to Eq. (14) in describing the data. Figure 6 shows such a comparison. For the capillary fall data at late time, we find that the best fit to Eq. (14) gave $\beta = 1.09(1)$ with a standard deviation of $\delta = 0.14$ pixel. If Eq. (15) is used, it actually fits both the early and late time data, though $\delta$ is larger by a factor of 2 to 3. For the two time ranges shown in Fig. 6a, $\delta$ is 0.41 and 0.33 pixels, respectively. Hence the capillary fall data are consistent with pure Washburn exponential behavior without pinning. Such is not the case for the capillary rise data. In Fig. 6b, we find that (14) is the only equation that can successfully describe the data, but the result depends on the range of the fit. If the starting time $t_1$ in Eq. (14) was chosen to be 300 s, we obtained $\beta = 1.10(2)$ and the standard deviation was $\delta = 0.17$ pixel. Reducing $t_1$ to 200 s and 180 s gave larger values for $\beta$, with $\delta$ staying about the same. This variation of $\beta$ suggests that there is no true universal critical behavior and Eq. (14) is just an empirical form that happens to capture the behavior. If instead Eq. (15) were to describe the data, we

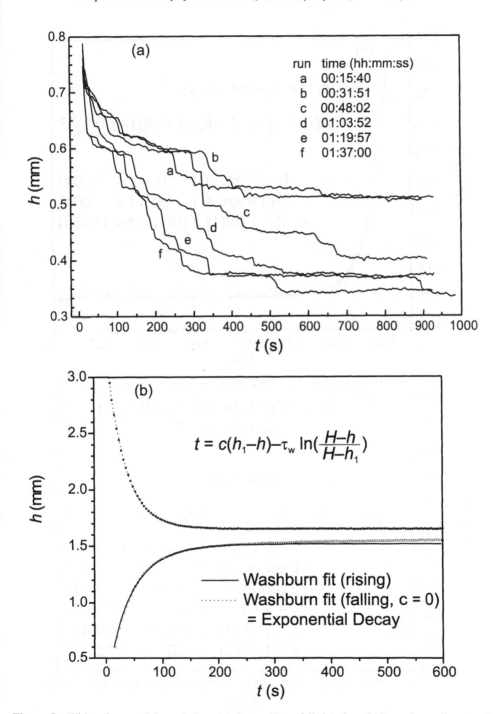

**Figure 5.** With a fine spatial resolution, (a) the capillary fall data in a 310 $\mu$m inner diameter R-tube show stick-slip behavior and cannot be used to analyze the exponent $\beta$. (b) The rise and fall data obtained with a 250 $\mu$m diameter S-tube are well described by Eq. (15) in the first 200 s. Only the rise data show noticeable deviation at later times. The fall data fit the equation with the constant $c$ equal to zero and it corresponds to a pure exponential function.

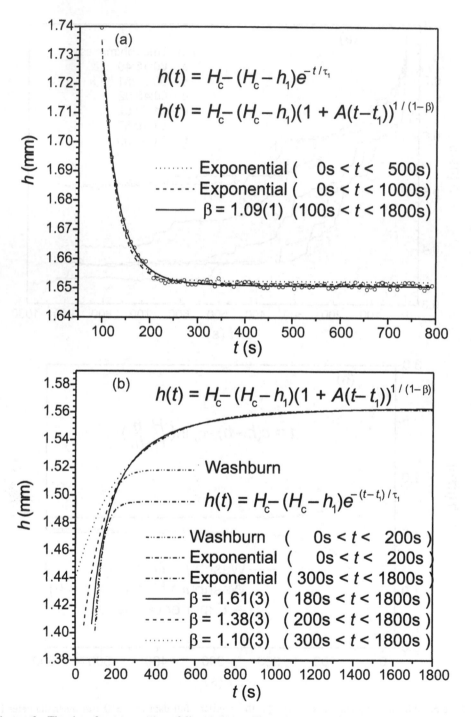

**Figure 6.** The data for (a) capillary fall and (b) capillary rise are analyzed in terms of Eqs. (14) and (15) for the 254 $\mu$m diameter S-tube. Note that the data are all within 200 $\mu$m of the pinning threshold. The numbers in parentheses represent the standard errors of the parameters obtained in the fits.

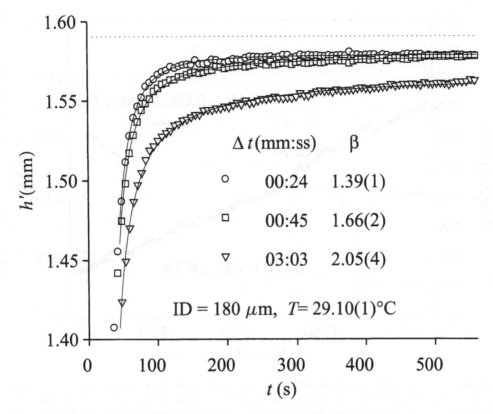

**Figure 7.** Capillary rise is fastest when it follows immediately a capillary fall experiment, and the rise is slower if the waiting time $\Delta t$ after the fall is increased. The slower rise is reflected in the increase of the exponent $\beta$.

find $\tau_1 = 37$ s in the first 200 s and $\tau_1 = 269$ s after 300 s. Clearly, Eq. (15) would not be useful even as an empirical form.

The nonuniversal nature of the exponent was confirmed as we studied different tubes and performed different runs with the same tube. We found values of $\beta$ that ranged between 1 and 4 in capillary rise, but the capillary fall data were always consistent with Washburn behavior with no pinning. This difference led us to believe that the presence of an invisible water film on the surface was playing an important role in the dynamics. Such a film could shield the weak surface heterogeneity completely in the falling experiment but its effects on the rising experiment would depend on the exact nature of the film when the meniscus is rising. Since we always carried out a rising experiment after the meniscus was pushed downward, the interior wall of the tube was prewetted with a thin water film that continued to drain or evaporate for some time. This would clearly affect the subsequent rise because a thinner film would expose more disorder and a thicker film can facilitate the flow of liquid to advance the contact line. To test this hypothesis, we performed rise experiments with a variable waiting time $\Delta t$ after the meniscus was pushed downward by a fixed distance. Figure 7 shows an example

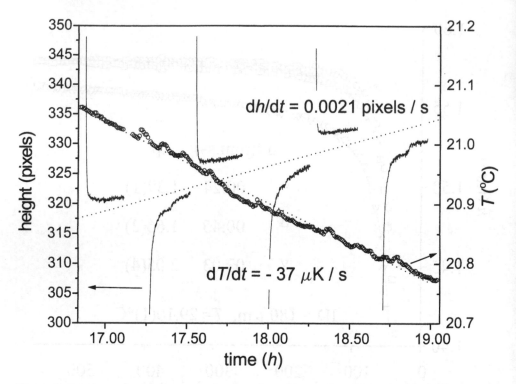

**Figure 8.** Room temperature change can have a serious effect on the result because the surface tension varies with temperature. For our system, a change of $\Delta T = 1$ K moves the meniscus position by about 57 pixels ($\approx 283$ $\mu$m) which exceeds the size of the critical region analyzed in Fig. 6. The data were taken with a 305 $\mu$m diameter S-tube.

of the results. We note that there is a noticeable slowdown in the water rise when the waiting time is increased from 24 s to about 3 min. This is consistent with our expectation that an inrease in waiting time would result in a thinner film on the surface and stronger pinning effects.

## 5. SUMMARY

The main finding in our study is that the behavior of capillary rise appears to be inherently different from capillary fall. We attribute this difference to the presence of an invisible liquid film on the surface. A capillary fall experiment would leave behind such a film and the contact is shielded from the surface disorder; a capillary rise experiment depends on the microscopic state of such a film and thereby causing the results to vary from sample to sample and even from run to run with the same sample. There appears to be no universal behavior in the latter case even though Eq. (7) appears to be a valid empirical form. Analyzing the capillary rise data with this function, we obtained generally larger values for $\beta$ compared to previous studies. There are a number of reasons that could account for the difference. In Fig. 8, we show that the temperature stability can have a very important effect because surface tension is quite sensitive to temperature change. We only used data from runs with total temperature variation below 0.01 K. Many previous experiments were less stringent, e.g., in Mumley *et al.* [25], the stated stability was $\pm 2$ K. A second factor is the size of the critical region. We show in Fig. 4 that the Washburn equation works well to less than 1 mm from the pinned height $H_c$. The highest velocity in Fig. 6 is only about 2 $\mu$m/s, which corresponds to a maximum $N_{cap}$ of $3 \times 10^{-8}$. In contrast, many studies did not find Washburn behavior at all and Eq. (7) was used to analyze data at capillary numbers $N_{cap}$ several orders of magnitude higher than ours. A third factor is that many studies used two different liquids with complicated surface interactions. As we inferred from the data in Fig. 7, the draining of an invisible wetting film can influence the macroscopic observations [28]. Hence the choice of liquid and the protocols for cleaning the tube can all affect the results.

## REFERENCES

1. J. S. Urbach, R. C. Madison and J. T. Markert, *Phys. Rev. Lett.* **75**, 276 (1995).
2. S. Bhattacharya and M. J. Higgins, *Phys. Rev. Lett.* **70**, 2617 (1993).
3. T. Nattermann, S. Stepanow, L.-H. Tang and H. Leschhorn, *J. Phys. II* **2**, 1483 (1992).
4. O. Narayan and D. S. Fisher, *Phys. Rev. B* **48**, 7030 (1993).
5. E. Schäffer and P.-z. Wong, *Phys. Rev. Lett.* **80**, 3069 (1998).
6. E. Schäffer and P.-z. Wong, *Phys. Rev. E* **61**, 5257 (2000).
7. J. Koplik and H. Levine, *Phys. Rev. B* **32**, 280 (1985).
8. R. Bruinsma and G. Aeppli, *Phys. Rev. Lett.* **52**, 1547 (1984).
9. G. Grinstein and S.-k. Ma, *Phys. Rev. B* **28**, 2588 (1983).
10. H. Ji and M. O. Robbins, *Phys. Rev. A* **44**, 2538 (1991); *Phys. Rev. B* **46**, 14519 (1992).
11. N. Martys, M. O. Robbins and M. Cieplak, *Phys. Rev. B* **44**, 12294 (1991).
12. J. P. Sethna, K. Dahmen, S. Kartha, J. A. Krumhansl, B. W. Roberts and J. D. Shore, *Phys. Rev. Lett.* **70**, 3347 (1993).
13. S. He, G. L. M. K. S. Kahanda and P.-z. Wong, *Phys. Rev. Lett.* **69**, 3731 (1992).
14. P.-z. Wong, *MRS Bull.* **19**, No. 5, 32 (1994).
15. T. Delker, D. B. Pengra and P.-z. Wong, *Phys. Rev. Lett.* **76**, 2902 (1996).
16. P.-z. Wong, T. Delker, M. Hott and D. B. Pengra, *Mater. Res. Soc. Symp. Proc.* **407**, 27–32 (1996).
17. D. Ertaş and M. Kardar, *Phys. Rev. E* **49**, R2532 (1994).
18. M. O. Robbins and J.-F. Joanny, *Europhys. Lett.* **3**, 729 (1987).
19. E. Raphaël and P. G. de Gennes, *J. Chem. Phys.* **90**, 7577 (1989).
20. J. F. Joanny and M. O. Robbins, *J. Chem. Phys.* **92**, 3206 (1990).
21. P. Sheng and M. Zhou, *Phys. Rev. A* **45**, 5694 (1992).
22. J. P. Stokes, M. J. Higgins, A. P. Kushnick, S. Bhattacharya and M. O. Robbins, *Phys. Rev. Lett.* **65**, 1885 (1990); see also J. P. Stokes, A. P. Kushnick and M. O. Robbins, *Phys. Rev. Lett.* **60**, 1386 (1988).
23. S. Kumar, D. H. Reich and M. O. Robbins, *Phys. Rev. E* **52**, R5776 (1995).
24. G. Ström, M. Fredriksson, P. Stenius and B. Rodoev, *J. Colloid Interf. Sci.* **134**, 107 (1990).
25. T. E. Mumley, C. J. Radke and M. C. Williams, *J. Colloid Interf. Sci.* **109**, 398 (1986).
26. T. Delker, Senior Honors Thesis, University of Massachusetts at Amherst (1995) (Available from P.-z. Wong).
27. E. W. Washburn, *Phys. Rev.* **17**, 273 (1921).
28. D. J. Durian, K. Abeysuriya, S. K. Watson and C. Franck, *Phys. Rev. A* **42**, 4724 (1990).

*Contact Angle, Wettability and Adhesion*, Vol. 3, pp. 39–52
Ed. K.L. Mittal
© VSP 2003

# Condensation transport in triple line motion

MARTIN E.R. SHANAHAN*

*Ecole Nationale Supérieure des Mines de Paris, Centre des Matériaux P.M. Fourt,
CNRS UMR 7633, B.P. 87, 91003 Evry Cédex, France*

**Abstract**—Although the macroscopic behaviour of liquids during spreading is largely understood, the fine scale phenomena occurring near the triple line remain somewhat obscure. Both hydrodynamic and molecular kinetics explanations have had some success, as indeed have hybrid models. This contribution introduces a third basic mechanism capable of accounting for mass transfer near the triple line, viz., condensation. The excess curvature in the proximity of the three-phase line during motion leads to condensation from the vapour phase onto the meniscus near the triple line. This local mass transfer contributes to triple line advance and thus spreading.

*Keywords*: Condensation; contact line; hydrodynamics; modified Kelvin equation; molecular kinetics; spreading; wetting.

## 1. INTRODUCTION

Triple line (TL) motion during wetting or dewetting involves, as yet, incompletely understood phenomena occurring very near the "line" itself (clearly, at a molecular scale, the concept of a "line" breaks down). A range of models has been proposed in the literature with two principal bases corresponding to (a) hydrodynamic resistance (Poiseuille shear flow) dynamically balancing the capillary force (unequilibrated Young force) [1-4] and (b) liquid molecular transfer obeying a molecular kinetics approach [5-12]. The latter approach stems from application of the Eyring theory for activated processes [13] and, in the general sense, could be applicable to any mass transfer process involving sufficiently small amounts of matter (and energy) for Maxwell-Boltzmann statistics to apply [9]. The original paper by Cherry and Holmes [5] suggested a mechanism of activated viscous flow and later work propounded rate theories to explain the TL motion in terms of molecular transfer by "jumping" [6-7].

Some attempts have been made to reconcile hydrodynamics (H) and molecular kinetics (MK) approaches, good examples of which are given in [14, 15]. Indeed,

---

*Current address: Centre Matériaux de Grande Diffusion, Ecole des Mines d'Alès 6, avenue de Clavières, 30319 Ales Cedex, France.

recent work suggests the possibility of a transition between the two [16, 17], depending on spreading conditions.

A third, distinct, potential mode of liquid transfer during spreading was suggested in a qualitative manner over 80 years ago [18]. This involves evaporation (from a sessile drop) and recondensation of the vapour of the liquid just ahead of the TL. Derjaguin's concept of disjoining pressure (see [19]) has been used by Wayner and co-workers to consider the behaviour near the TL [20-25] and so the idea of mass transfer by evaporation/condensation is not entirely new.

The present contribution is somewhat complementary to the last mentioned [26-28] but, interestingly, leads to scaling laws very similar to those emanating from the MK theory [6, 8-12]. The basic physics can be summarised quite succinctly: as predicted by the Kelvin equation, the equilibrium vapour pressure above a liquid meniscus increases with convexity, other factors being equal. An advancing TL has reduced convexity (or increased concavity) compared to the equilibrium (static) situation. This leads to reduced *equilibrium* vapour pressure and, therefore, to local condensation. The added liquid promotes the advancement of the TL.

## 2. KELVIN'S EQUATION NEAR A SOLID SURFACE

Kelvin's classic equation for vapour pressure [29] shows that the equilibrium value of this quantity, $p_{vo}$, above a convex (concave) liquid meniscus is larger (smaller) than the equivalent value, $p_o$, above a flat liquid:

$$p_{vo} = p_o \exp\left[\frac{2\gamma \overline{V}}{r_o R T}\right] \qquad (1)$$

where $\gamma$ is liquid surface tension, $\overline{V}$ liquid molar volume, $r_o$ local (spherical) radius of curvature, R the gas constant and T absolute temperature. Equation (1) applies to a spherical, convex meniscus, such as a suspended droplet, and a classic context for this equation is in the treatment of fogs or mists.

Let us now consider Kelvin's equation in a modified form. We shall assume that the meniscus presents only one radius of curvature, i.e. it is a cylindrical meniscus of radius of curvature $r_o$. Secondly, this meniscus is in close proximity to a solid surface which, for simplicity, is considered "ideal" (flat, smooth, rigid, homogeneous, isotropic, etc.). The situation is schematised in Fig. 1. We shall derive the modified Kelvin equation using the principle of "virtual condensation", by analogy to the methods of virtual work in mechanics.

Considering the portion of the meniscus delimited locally by the sector of (small) angle $\phi$ in Fig.1 at equilibrium with (unique) radius of curvature, $r_o$, and at distance $h_o$ from the solid, we impose an infinitesimal decrease of $r_o$, to $(r_o - \delta r_o)$, leading to an infinitesimal increase in film thickness, from $h_o$ to $(h_o + \delta h_o) = (h_o + \delta r_o.\cos\alpha)$. This leads to a decrease in free energy corre-

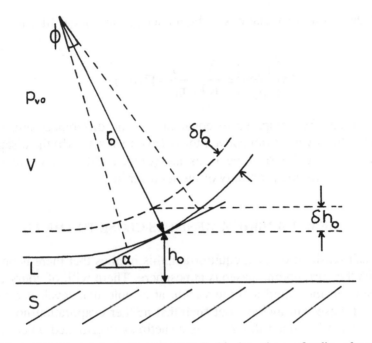

**Figure 1.** Schematic representation of a concave, cylindrical meniscus of radius of curvature $r_0$, at distance $h_0$ from solid surface and inclined at angle $\alpha$.

sponding to a reduction in the liquid/vapour interface area within the sector delimited by angle $\phi$ and is equal to $\gamma\,\phi.\delta r_0$ (assuming unit width perpendicular to Fig. 1). However, if film thickness is small (of the order of nanometers), we must also allow for the change in free energy associated with $\delta h_0$ and due to long-range interactions. For unit area of interface *parallel* to the solid surface, this is given by a decrease in free energy of $\Pi(h_0).\delta r_0.\cos\alpha$ where $\alpha$ is the local profile inclination angle and $\Pi(h_0)$ is Derjaguin's disjoining pressure at film thickness $h_0$ [30, 31]. The area considered (parallel to the solid) is $\phi r_0/\cos\alpha$ and thus the free energy variation is given by $\Pi(h_0)\phi r_0.\delta r_0$ (note that the effect of gradient of the interface cancels out).

These changes in free energy of the local system are brought about by the work supplied in condensing vapour onto the meniscus:

$$\int_{p_{vo}}^{p_o} p\,dV = nRT\int\frac{dV}{V} = nRT\,\ell n\left(\frac{p_{vo}}{p_o}\right) \qquad (2)$$

where $p$ and $V$ are variables, pressure and volume, and $n$ is the number of moles condensed ($=\phi r_0.\delta r_0/\overline{V}$). Equating expression (2) to the free energy changes discussed above due to variations in liquid surface and local film thickness, we obtain:

$$\ell n\left(\frac{p_{vo}}{p_o}\right) = \frac{\overline{V}}{RT}\left[\frac{-\gamma}{r_o} - \Pi(h_o)\right] \tag{3}$$

Equation (3) may be compared to equation (1): the differences are: (a) a sign change due to the switch from convex to concave meniscus, (b) the disappearance of the factor 2 for a cylindrical meniscus, and (c) the addition of a disjoining pressure term to allow for the proximity of the solid surface.

## 3. SHAPE OF THE EQUILIBRIUM MENISCUS CLOSE TO THE TL

When a small sessile drop is at equilibrium, this implies that there is no (net) exchange with the surrounding gaseous atmosphere. There will, of course, be interchanges between the liquid and the vapour at a molecular level, due to thermal agitation, but these will average out such that neither evaporation nor condensation occurs overall. Indeed this point is sometimes overlooked when measuring contact angles: without an atmosphere saturated in the vapour of the liquid, drop evolution may readily occur [32, 33].

Let us assume the existence of a small sessile drop on a solid in an atmosphere in the vicinity of a reservoir of the liquid (under isothermal conditions). The reservoir may be taken to have a flat liquid/vapour interface and therefore the saturated vapour pressure will locally be $p_o$. If the sessile drop is to coexist at equilibrium, $p_o$ must also correspond to its equilibrium vapour pressure, and considering equation (3), we infer that a necessary condition is:

$$\frac{\gamma}{r_o} + \Pi(h_o) = 0 \tag{4}$$

In equations (3) and (4), we have assumed a concave meniscus: if the meniscus were to be convex, the term in $\gamma / r_o$ would simply change sign. Taking x to represent distance parallel to the solid surface, it is convenient to resort to Cartesian coordinates, in which case equation (4) becomes:

$$\frac{\gamma}{r_o} + \Pi(h_o) = \frac{\gamma h_o''}{(1 + h_o'^2)^{3/2}} + \Pi(h_o) = P_A - P = 0 \tag{5}$$

where $h_o'$ and $h_o''$ have their usual meanings of first and second derivatives of $h_o$ with respect to x, and $P_A$ and $P$ are, respectively, the atmospheric pressure and hydrostatic pressure in the liquid.

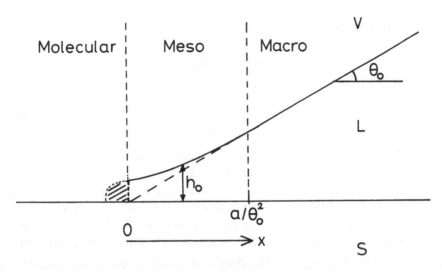

**Figure 2.** Region near the triple line (TL) for an equilibrium drop of negative spreading coefficient and positive Hamaker constant. The macro(scopic) profile is approximately linear with contact angle, $\theta_o$, and the meso profile is hyperbolic [equation (8)]. Molecular considerations at the very tip of the leading edge are not treated.

Equation (5) is a form of the augmented Young-Laplace equation (AYL) which has been presented several times in the past [34-37]. It expresses the offset of the Laplace pressure near the TL by the disjoining pressure.

In principle, relation (5) may be used to elucidate the fine structure of the liquid meniscus profile in the vicinity of the TL, provided the disjoining pressure isotherm is known. The latter is open to discussion for many polar liquids, especially water (e.g. [28, 38-40]), and so in the remainder of this article, we shall restrict our attention to apolar, van der Waals liquids.

We assume a system of negative spreading coefficient, to ensure a finite but small value of *macroscopic* equilibrium contact angle $\theta_o$. With a small meniscus slope, $h_o' \ll 1$, we can thus simplify the Cartesian expression for profile curvature in equation (5). For a van der Waals liquid, we have:

$$\Pi(h_o) = \frac{-A}{6\pi h_o^3} = \frac{-a^2\gamma}{h_o^3} \tag{6}$$

where the (positive) Hamaker constant, A, can be expressed as $A \approx 6\pi a^2\gamma$, a being a molecular diameter.

Inserting equation (6) into the (simplified) equation (5) yields:

$$\frac{d^2 h_o}{dx^2} = \frac{a^2}{h_o^3} \tag{7}$$

which has the exact integral:

$$h_0^2 = \theta_0^2 x^2 + \frac{a^2}{\theta_0^2} \tag{8}$$

where the requisite boundary conditions are constancy of gradient, $h_0'$, for macroscopic distances $(x \geq a/\theta_0^2)$ and $h_0' = \theta_0$. (The origin of x is the *mesoscopic* tip of the TL, just after the first few liquid molecules, see Fig. 2.)

Interestingly, equation (8) has been derived previously [41] but using a rather different approach. Thus, the modified Kelvin equation [equation (3)] may be used to deduce the equilibrium profile near the TL, at least on a mesoscopic scale: the molecular tip cannot be treated by continuum methods [41]. We shall employ this static solution below in a perturbation approach to the dynamic problem. Note from equation (8) that $h_0'(0) = 0$: despite a finite macroscopic contact angle ($\theta_0$), the final mesoscopic contact angle is zero, as previously shown by White [34, 42].

## 4. THE DYNAMIC PROBLEM: CONDENSATION CURRENT

We restrict ourselves to van der Waals liquids of small, but finite, contact angle, as stated above, although the basic physics can be applied to more complex systems [28]. Also to simplify (an already complex problem !), we assume steady-state flow conditions, with low spreading speeds, U.

Four basic equations are required to obtain a differential equation describing the condensation contribution to spreading, under conditions in which we assume that the TL is *already* in motion.

(a) Under steady-state conditions, net flow must be maintained constant. For a non-volatile liquid, this implies constancy of the Poiseuille current, $J_p$ [3]:

$$J_p = \frac{h^3}{3\eta}(\frac{-\partial P}{\partial x}) \tag{9}$$

where $h = h(x)$ is the *dynamic* film thickness, $\eta$ is liquid viscosity and P is pressure within the liquid.

(b) However, with the potential existence of a condensation current, $J_c$, liquid is added to the meniscus from the vapour phase, near the TL. To maintain a constant overall current, a continuity condition must be satisfied (see Fig. 3):

$$\frac{\partial J_p}{\partial x} + \frac{J_c}{\cos\alpha} \approx \frac{\partial J_p}{\partial x} + J_c = 0 \tag{10}$$

where the approximation is allowed since, by hypothesis, the profile gradient is low $(\cos\alpha \approx 1)$.

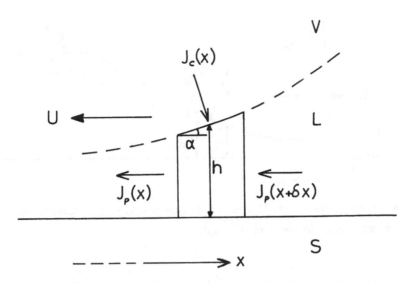

**Figure 3.** Small section of a liquid drop in the mesoscopic region near the TL showing modification of the Poiseuille current, $J_p$, during flow by the addition of a condensation current, $J_c$.

(c) We take r to be the local *dynamic* radius of curvature of the meniscus [r = r(x)] and with an advancing TL, the dynamic contact angle, $\theta$, will be greater than the equilibrium value, $\theta_0$, such that $r < r_0$, and thus the local curvature is greater under dynamic conditions (see Fig. 4). From equations (3), (5) and (6), and assuming that dynamic gradient $h' \ll 1$, we obtain an expression for $\tilde{p}_0$, the appropriate partial vapour pressure, *if* the drop profile were to be static and at equilibrium with this geometry. The excess vapour pressure, $\Delta p$, leading to condensation may thus be expressed as:

$$\Delta p = p_0 - \tilde{p}_0 = p_0 \left\{ 1 - \exp\left[ \frac{\overline{V}}{RT} \left( \frac{-\gamma}{r} + \frac{A}{6\pi h^3} \right) \right] \right\}$$

$$\approx F\gamma \left[ \frac{d^2h}{dx^2} - \frac{a^2}{h^3} \right] \approx F(P_A - P) \tag{11}$$

where $F = p_0 \overline{V} / (RT)$. If $\overline{V}$ were the liquid *vapour* molar volume, then F would simply be the molar fraction of the vapour of the liquid in the surrounding atmosphere. This was (perfunctorily) assumed earlier [26-28]. In principle, $\overline{V}$ is the liquid molar volume. However, the distinction between pure liquid and pure vapour at this mesoscopic level, near a transient inter*phase*, is not clear and so the exact form for F remains a moot point at present.

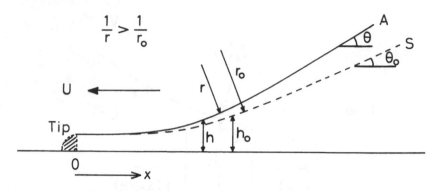

**Figure 4.** Advancing profile, A, near TL compared to static profile, S, showing higher contact angle $(\theta > \theta_0)$ and greater meniscus curvature $(r^{-1} > r_0^{-1})$.

(d) The condensation current, $J_c(x)$, a function of position x in the mesoscopic zone $(0 \le x \le a/\theta_0^2)$ is given by:

$$J_c(x) = K\Delta p \approx \frac{F\overline{V}\gamma}{(2\pi MRT)^{1/2}} \left[ \frac{d^2 h}{dx^2} - \frac{a^2}{h^3} \right] \qquad (12)$$

where $K = \overline{V}/(2\pi MRT)^{1/2}$, M being the molar mass of the liquid [43]. Again, this assumes a perfectly abrupt interface between the liquid and the vapour, but there is likely to be some indistinct inter*phase* in reality. Thus the value of K is probably underestimated above.

By combining the four basic equations, (9) to (12), we obtain a second-order differential equation which, when expressed in integro-differential form, reads as:

$$\frac{\partial P}{\partial x} + \frac{3K\eta F}{h^3} \int_x^{a/\theta_0^2} (P_A - P)\,dx \approx \frac{3\eta U_T}{h^2} \qquad (13)$$

where $U_T$ is the total speed of TL advance including initial motion and condensation effect. (Its presence here is justified by realising that the current in the frame of reference of the solid is $-U_T h$.)

A direct, analytical solution of equation (13) seems unlikely, and so a perturbation technique has been applied. By inserting the static solution for the profile, equation (8), into equation (11), we see that the integral of equation (13) becomes identically zero, facilitating an approximate solution. Realising that $U_T$ reduces to U in this limit, we can now integrate equation (13) directly to obtain:

$$(P - P_A) \approx \frac{3\eta U}{a} \tan^{-1}\left(\frac{\theta_o^2 x}{a}\right) - \text{constant} \approx \frac{3\eta U}{a}\left(\frac{\theta_o^2 x}{a} - 1\right) \qquad (14)$$

where the integration constant is determined from the disappearance of the condensation effect at macroscopic distances $(x > a / \theta_o^2)$.

By using solution (14) in conjunction with equations (11) and (12), we obtain for the condensation current:

$$J_c(x) \approx \frac{3\eta F \bar{V} U}{a(2\pi M R T)^{1/2}}\left(1 - \frac{\theta_o^2 x}{a}\right) \qquad (15)$$

or, if we prefer not to mix molar and molecular units, we can write:

$$J_c(x) \approx \frac{3\eta F a^2 U}{(2\pi m k T)^{1/2}}\left(1 - \frac{\theta_o^2 x}{a}\right) \qquad (16)$$

where m is liquid molecular mass, k is Boltzmann's constant and we have taken molecular volume $\approx a^3$.

## 5. SPREADING AND DISSIPATION

In the above model, we assume an intrinsic TL motion to which is added a condensation effect. The overall energy expenditure during spreading will be the sum of two terms:

$$T\dot{S} = T\dot{S}_M + T\dot{S}_m \qquad (17)$$

$T\dot{S}_M$ represents hydrodynamic dissipation (M = Macroscopic) due to viscous effects. In the hydrodynamic picture [3, 44], we may write :

$$T\dot{S}_M \approx 3\eta U^2 \int_0^{x_{max}} \frac{dx}{h} = \frac{3\eta U^2 I}{\theta_o} \qquad (18)$$

I represents an integral which, for the classic wedge-shaped profile near the TL leads to a logarithmic singularity [3]. However, if we adopt the static profile given by equation (8) for a van der Waals liquid, as a first approximation, we obtain:

$$I \approx \ln\left(\frac{2\theta_o^2 x_{max}}{a}\right) \qquad (19)$$

where $x_{max}$ is some macroscopic distance (contact radius of a sessile drop). We thus obtain a natural cut-off for the logarithmic flow singularity as was done previously [44].

The mesoscopic contribution (m = mesoscopic) involves energy expenditure during condensation, essentially near the TL, and can be expressed as:

$$T\dot{S}_m \approx \int_0^{a/\theta_o^2} J_c.\Delta p.dx = \frac{1}{K} \int_0^{a/\theta_o^2} J_c^2 \, dx \qquad (20)$$

which, on using equation (16), leads to:

$$T\dot{S}_m \approx \frac{3\eta^2 F^2 a^2 U^2}{(2\pi m k T)^{1/2} \theta_o^2} \qquad (21)$$

However, it is instructive to write expression (21) differently. From the hydrodynamics [3], we can write the energy balance as:

$$T\dot{S}_M \approx U\gamma(\cos\theta_o - \cos\theta) \qquad (22)$$

with $\theta$ as the dynamic angle $(\theta > \theta_o)$. Thus isolating U from equations (18) and (22) and using equation (21), we arrive at:

$$T\dot{S}_m \approx \frac{F^2 a^2 \gamma^2 (\cos\theta_o - \cos\theta)^2}{3(2\pi m k T)^{1/2} I^2} \qquad (23)$$

and finally, for the total energy expenditure we have:

$$T\dot{S} = T\dot{S}_M + T\dot{S}_m$$

$$\approx \frac{3\eta U^2 I}{\theta_o} + \frac{F^2 a^2 \gamma^2 (\cos\theta_o - \cos\theta)^2}{3(2\pi m k T)^{1/2} I^2} \qquad (24)$$

This will be discussed below.

The actual spreading speed, $U_T$, is higher than U, due to condensation, so we can write:

$$|J_p| + \int_0^{a/\theta_o^2} J_c \, dx \approx \frac{U_T a}{\theta_o} \qquad (25)$$

$U_T$ is evaluated at x = 0 and contains both hydrodynamic and condensation contributions. Evaluation of equation (25) using equation (16) leads to:

$$U_T \approx U \left\{ 1 + \frac{3 \eta F a^2}{2 \theta_0 (2 \pi m k T)^{1/2}} \right\} \tag{26}$$

where the second term on the right hand side corresponds to the condensation perturbation imposed on the intrinsic motion given by U.

## 6. DISCUSSION

Since for natural wetting, the "motor" is $\gamma (\cos \theta_0 - \cos \theta)$, by comparing equations (22) and (23) and neglecting, for the moment, the hydrodynamic contribution, we can deduce that:

$$U \approx \frac{F^2 a^2}{3 (2 \pi m k T)^{1/2} I^2} \cdot \gamma (\cos \theta_0 - \cos \theta) \tag{27}$$

This is precisely the scaling form predicted by the MK theory, in which we may write [6]:

$$U \approx \frac{2 \tilde{K} \lambda^3}{k T} \cdot \gamma (\cos \theta_0 - \cos \theta) \tag{28}$$

where $\tilde{K}$ is a molecular oscillation frequency and $\lambda$ is a "hopping" distance for molecules between adsorption sites, near the TL. It would be presumptuous to say that the two theories are equivalent, but there is a certain complementarity without doubt. In fact, the work of Blake and Haynes [6] in 1969 was done with a two-liquid system (water/benzene) and so gaseous transport is precluded as an explanation. Nevertheless, there are cases of volatile liquids where the mechanism described here is a distinct possibility for mass transfer during spreading. Nature tends to "take the easy way out" and reduce energy expenditure as much as possible during an operation. Thus, for a given system, MK processes, evaporation/condensation and hydrodynamics (and possibly others ?) are all contenders as explanations for mass transfer, but the predominant mechanism will be that which reduces energy consumption.

Equation (24) may be expressed as a scaling law:

$$T\dot{S} = T\dot{S}_M + T\dot{S}_m \sim \frac{1}{\theta_0} + (\cos \theta_0 - \cos \theta)^2 \tag{29}$$

Thus, for small contact angles, $T\dot{S}_M$, the hydrodynamic component, dominates. This will tend to be at low speeds. On the contrary, for large contact angles, $T\dot{S}_m$, the MK component, will dominate, typically corresponding to high wetting speeds. These qualitative assessments fit in well with certain experimental observations [14, 45, 46]. It would seem that silicone oils obey the hydrodynamic theory rather well: this could well be related to their low volatility limiting mass transfer by evaporation/condensation (F becomes effectively zero since $p_0 \to 0$). Also dewetting is better described by hydrodynamics [47]. Contact angles during dewetting tend to be close to the equilibrium value, thus limiting the meniscus curvature near the TL and therefore condensation. De Ruijter et al. [16] presented four stages to droplet spreading, the middle two being successively MK and H dominated. This is consistent with relation (29). Spreading is faster initially and with a higher contact angle, thus MK dominates, but subsequently, as the process slows down and the contact angle diminishes, hydrodynamics becomes increasingly important.

We have used a perturbation approach (of limited range of applicability) in the present development, but it does emphasise the interplay between dynamic contact angle and flow field as discussed by Shikhmurzaev [48]. However, there are probably circumstances in which the meniscus curvature is exacerbated without *intrinsic* motion, yet this curvature will itself suffice to obtain spreading simply by condensation [21-24].

A quantitative comparison between equations (27) and (28) is unfortunately difficult at present due to the apparent lack of experimental data for van der Waals liquids. In the case of aqueous systems [49, 50], the prefactor of equation (28), i.e. $2\tilde{K}\lambda^3/(kT)$, is typically in the range of $2 \times 10^{-3}$ to $2\,(Pa \cdot s)^{-1}$, although most values group towards the lower end. This large variability may be, at least partly, due to the importance of wetting conditions (wetting, dewetting, hysteresis, etc.) [12].

Notwithstanding, we shall estimate the prefactor of equation (27) for comparison, taking n-hexane $(C_6H_{14})$ as an example, being a volatile apolar liquid. The dimension a is of the order of $6 \times 10^{-10}\,m$ and with a drop of equilibrium contact angle, $\theta_0 \approx 10°$ (0.17 rad) and contact radius $x_{max} \approx 10^{-3}\,m$, we obtain $I \approx 11.5$ (being a logarithmic function, I is quite insensitive to exact values of the parameters). The molecular mass is ca. $1.4 \times 10^{-25}\,kg$ and with $k = 1.38 \times 10^{-23}\,JK^{-1}$. We take T = 290 K. The problem revolves around the uncertainty in the value of F from equation (11) [and also in K from equation (12)]. If $\overline{V}$ in the expression for F is taken in its strict sense to be liquid molar volume ($\approx 10^{-4}\,m^3$), with $p_0 \approx 1.5 \times 10^4\,Pa$, then we obtain a prefactor for equation (27) of the order of $10^{-5}\,(Pa \cdot s)^{-1}$, which is rather small. At the other extreme, if we assume the vapour molar volume when calculating F, we obtain a prefactor of ca. 0.4. The considerable difference between the two estimates is, of course, related to the Clausius–Clapeyron equation (see [51]). The "real" value is probably somewhere between the two extremes and if we take $10^{-3}\,(Pa \cdot s)^{-1}$ as a (logarithmic) average, we find similar values for the prefactor as those obtained experimentally for equation (28) (for aqueous systems, admittedly). Semal et al. [52] obtained values in the range $10^{-2}$ to $10^{-1}$ $(Pa \cdot s)^{-1}$ for a non-polar liquid, squalane, on glass, but due to the very low volatility of this liquid, condensation transport is thought unlikely as a major transport mechanism.

A comparison of equations (27) and (28) shows a different temperature dependence, and this suggests a method for comparing the two experimentally. However, the difference may be small, certainly in the light of the considerable ranges of values of the prefactor discussed above both from an experimental and a theoretical standpoint. Clearly, more experimental data are required.

## 7. CONCLUSION

We have considered potential effects of condensation from the vapour phase onto the liquid meniscus near the triple line (TL) as a contributory method for mass transfer during spreading.

A modified form of Kelvin's equation has been obtained for the TL region in which long-range interactions are taken into account. This equation can be used to obtain static meniscus profiles in the vicinity of the TL if the disjoining pressure isotherm is known.

The perturbation of the local meniscus due to intrinsic motion leads to greater concavity and thus condensation ensues, contributing further to TL motion in the present model. However, extant motion is not a prerequisite: simple excess curvature would suffice for the condensation related TL motion.

A consideration of the dissipation phenomena and a comparison with the molecular kinetics (MK) approach shows a certain complementarity of the two basic theories.

Doubt still exists concerning the values of the prefactor in the flow formulae (both for the present model and the MK theory) and thus more experimental data are necessary for further probing of the suggested mechanisms, particularly for apolar liquids.

*Acknowledgements*

The author thanks P.G. de Gennes and E. Ramé for useful exchanges.

**REFERENCES**

1. C. Huh and L.E. Scriven, J. Colloid Interface Sci., **35**, 85 (1971).
2. O.V. Voinov, Fluid Dynamics, **11**, 714 (1976).
3. P.G. de Gennes, Rev. Mod. Phys., **57**, 827 (1985).
4. R.G. Cox, J. Fluid Mech., **168**, 169 (1986).
5. B.W. Cherry and C.M. Holmes, J. Colloid Interface Sci., **29**, 174 (1969).
6. T.D. Blake and J.M. Haynes, J. Colloid Interface Sci., **30**, 421 (1969).
7. E. Ruckenstein and C.S. Dunn, J. Colloid Interface Sci., **59**, 135 (1977).
8. J.G. Petrov and B.P. Radoev, Colloid Polym. Sci., **259**, 753 (1981).

9. T.D. Blake, Wetting Kinetics – How Do Wetting Lines Move ? Paper presented at the AIChE International Symposium on the Mechanics of Thin Film Coating, Paper 1a., New Orleans (1988).
10. J.G. Petrov and P.G. Petrov, Colloids Surfaces, **64,** 143 (1992).
11. R.A. Hayes and J. Ralston, J. Colloid Interface Sci., **159**, 429 (1993).
12. R.A. Hayes and J. Ralston, Langmuir, **10**, 340 (1994).
13. S. Glasstone, K.J. Laidler and H. Eyring, *The Theory of Rate Processes*, McGraw Hill, New York (1941).
14. F. Brochard-Wyart and P.G. de Gennes, Adv. Colloid Interface Sci., **39**, 1 (1992).
15. P.G. Petrov and J.G. Petrov, Langmuir, **8**, 1762 (1992).
16. M.J. De Ruijter, J. De Coninck and G. Oshanin, Langmuir, **15**, 2209 (1999).
17. M.J. De Ruijter, M. Charlot, M. Voué and J. De Coninck, Langmuir, **16**, 2363 (2000).
18. W. Hardy, Philos. Mag., **38**, 49 (1919).
19. B.V. Derjaguin, Y.I. Rabinovich and N.V. Churaev, Nature, **272**, 313 (1978).
20. M. Potash, Jr. and P.C. Wayner, Jr., Inter. J. Heat Mass Transfer, **15**, 1851 (1972).
21. P.C. Wayner, Jr., Colloids Surfaces, **52**, 71 (1991).
22. J. Schonberg and P.C. Wayner, Jr., J. Colloid Interface Sci., **152**, 507 (1992).
23. P.C. Wayner, Jr., Langmuir, **9**, 294 (1993).
24. P.C. Wayner, Jr., Colloids Surfaces, A, **89**, 89 (1994).
25. P.C. Wayner, Jr., this volume.
26. M.E.R. Shanahan, C.R. Acad. Sci., Paris, **2** (IV), 157 (2001).
27. M.E.R. Shanahan, Langmuir, **17**, 3997 (2001).
28. M.E.R. Shanahan, Langmuir, **17**, 8229 (2001).
29. A.W. Adamson and A.P. Gast, *Physical Chemistry of Surfaces*, 6th ed., p. 53, John Wiley, New York (1997).
30. B.V. Derjaguin, Zh. Fiz. Khim., **14**, 137 (1940).
31. B.V. Derjaguin, Kolloidn. Zh., **17**, 191 (1955).
32. M.E.R. Shanahan and C. Bourgès, Intl. J. Adhesion Adhesives, **14**, 201 (1994).
33. C. Bourgès-Monnier and M.E.R. Shanahan, Langmuir, **11**, 2820 (1995).
34. L.R. White, J. Chem. Soc., Faraday Trans. I, **73**, 390 (1977).
35. P.C. Wayner, Jr., J. Colloid Interface Sci., **77**, 495 (1980).
36. P.C. Wayner, Jr., J. Colloid Interface Sci., **88**, 294 (1982).
37. A. Sharma, Langmuir, **9**, 3580 (1993).
38. J.N. Israelachvili, *Intermolecular and Surface Forces*, Academic Press, New York (1985).
39. I. Langmuir, J. Chem. Phys., **6**, 873 (1938).
40. R.M. Pashley, J. Colloid Interface Sci., **78**, 246 (1980).
41. F. Brochard-Wyart, J.M. di Meglio, D. Quéré and P.G. de Gennes, Langmuir, **7**, 335 (1991).
42. Y. Solomentsev and L.R. White, J. Colloid Interface Sci., **218**, 122 (1999).
43. J.B. Hudson, *Surface Science*, Wiley, New York (1998).
44. P.G. de Gennes, X. Hua and P. Levinson, J. Fluid Mech., **55**, 212 (1990).
45. A.M. Cazabat and M.A. Cohen Stuart, J. Phys. Chem., **90**, 5845 (1986).
46. M.E.R. Shanahan, M.C. Houzelle and A. Carré, Langmuir, **14**, 528 (1998).
47. C. Redon, F. Brochard-Wyart and F. Rondelez, Phys. Rev. Lett., **36**, 715 (1991).
48. Y.D. Shikhmurzaev, J. Fluid Mech., **334**, 211 (1997).
49. P.G. Petrov and J.G. Petrov, Langmuir, **11**, 3261 (1995).
50. E.V. Gribanova, Adv. Colloid Interface Sci., **39**, 235 (1992).
51. Y. Pomeau, C.R. Acad. Sci., Paris, **328**(IIb), 411 (2000).
52. S. Semal, M. Voué and J. De Coninck, Langmuir, **15**, 7848 (1999).

*Contact Angle, Wettability and Adhesion*, Vol. 3, pp. 53–65
Ed. K.L. Mittal
© VSP 2003

# Inertial dewetting: Shocks and surface waves

A. BUGUIN,* X. NOBLIN and F. BROCHARD-WYART

*Institut Curie, P.C.C - UMR 168, 11, Rue Pierre et Marie Curie, 75231 Paris cedex 05, France*

**Abstract**—We describe in the present paper the fast dewetting of a water film (thickness, e) from a hydrophobic substrate. At high velocity we observe ripples emitted ahead of the rim. We show that these ripples are the signature of a shock. The water film dewets by the nucleation and growth of a dry patch (velocity V). The film is surrounded by a rim which collects the liquid. We show that the velocity of dewetting V(e) follows the Culik law for the bursting of soap films, with a driving force including both capillarity and gravity. We have built an experimental setup, which uses the deflection of a laser beam at the air/water interface, to measure the profile of the rim. The shock is observed when the rim surfs on the immobile film at velocities $V^* > \sqrt{ge}$ (where g is the gravitational acceleration), the velocity of gravity waves in shallow water.

*Keywords*: Liquid films; dewetting; surface waves; shocks.

## 1. INTRODUCTION

For a thin film of water deposited on a hydrophobic substrate, dry regions appear and grow: this process is called dewetting. It has important practical applications, for example, it governs:
- spontaneous drying [1]
- the hydroplaning of cars [2] (dewetting between asphalt and rubber)
- the operation of 4-color offset printing [3] (dewetting between roller and paper)
- the coating of liquid films (for example, the protection of plants by fungicides in water where dewetting must be blocked).

In the last ten years, most studies have focused on dewetting of viscous liquids on solid substrates. In this case the liquid is collected into a rim around the dry region (the lack of rim formation has been reported on viscoelastic solids [4]). The dynamics which results from a balance between the viscous dissipation in the rim

---

*To whom all correspondence should be addressed. Phone: 33 (0)1 42 34 67 78,
Fax: 33 (0)1 40 51 06 36, E-mail: a.buguin@curie.fr

and the gain in surface energy is now well understood [5, 6]. To be more precise, two regimes have been studied in great details:
-   metastable regime: below a critical macroscopic thickness the liquid film is metastable and dewets by nucleation and growth of a dry patch.
-   spinodal regime: below a microscopic thickness, the film becomes unstable.

For more information about this regime, see for instance reference [7].

In this paper, we focus on the case where viscous losses become negligible, i.e., where the liquids are nearly inviscid. It occurs for very hydrophobic substrates and for relatively large film thicknesses, e, leading to large velocities [8, 9]. We will show that the dynamics is greatly modified and leads to completely different water profiles from those of viscous liquids.

## 2. DEWETTING VELOCITY

### 2.1. Static properties

Usually, the first step in the dewetting experiments consists in preparing a well-defined substrate. In our experiments, it is a large sheet (20x20 cm) of float glass silanized with octadecyltrichlorosilane [10]. This process minimizes the physical and chemical heterogeneities and we obtain a surface with a small contact angle hysteresis (~ 15°). The main parameter is the spreading parameter [11]. It is negative, expressing that the solid surface prefers to be dry:

$$S = \gamma_{SO} - (\gamma + \gamma_{SL}) = -\frac{1}{2}\rho g e_c^2 = -82 \, mJ / m^2 \qquad (1)$$

In equation (1) $\gamma_{SO}$, $\gamma_{SL}$, $\gamma$ are, respectively, the substrate/air, substrate/liquid and liquid/air interfacial energies; $e_C$=3.9 mm is the critical thickness ($\rho$ = density; g = gravitational acceleration ). For larger thicknesses, gravity dominates and the film remains flat [5]. For e<$e_C$, our films are metastable and dewet by nucleation and growth of a dry patch. We induce the nucleation (at t=0) either by dipping a teflon tip in or by blowing air with a small pipet at the centre of the film.

### 2.2. Theoretical description

We assume that the front velocity $V^*$ of the advancing rim and the velocity V for the growth of the dry patch are time independent. There are two relevant dimensionless parameters to describe the dynamics of the system
-   a Reynolds number

$$Re = \frac{V_e e}{\nu} = \frac{g^{1/2} e^{3/2}}{\nu} \cong 10^2 \qquad (2)$$

where $V_e = \sqrt{ge}$ is the velocity of capillary waves in shallow water (characteristic velocity) and $v=\eta/\rho$ ($\eta$ being the water viscosity).
– a Froude number

$$Fr = \frac{V^*}{\sqrt{ge}} \tag{3}$$

If Fr is larger than unity, the front velocity is larger than the velocity of capillary waves in shallow water and a shock appears. This shock is similar to the hydraulic jump observed at the bottom of a sink.

We call h the thickness of the rim, V its velocity and $V^*$ the front velocity (Figure 1-insert), related to V by mass conservation. We can write $h(R_S^2 - R_d^2) = eR_d^2$ ($R_d$ is the dry patch radius and $R_S$ the shock radius). It gives:

$$h(V^{*2} - V^2) = eV^{*2} \tag{4}$$

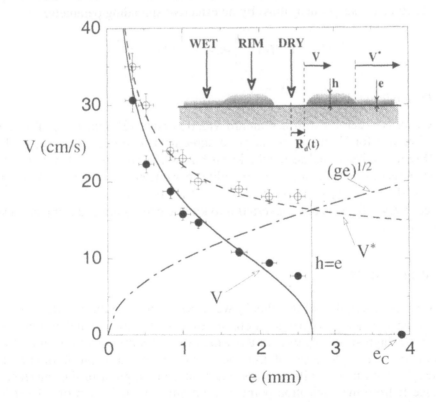

**Figure 1.** Evolution of velocity V (solid circles) for the growth of the dry patch and front velocity $V^*$ (open circles) of the advancing film versus the film thickness e. Our results are fitted to equations (6) (solid line and dotted line). $\sqrt{ge}$ is shown for comparison. The insert shows a schematic representation of the film during dewetting.

The Rankine-Hugoniot conditions for the shock are the mass conservation (equation 4), and the momentum flux conservation. A detailed description of this calculation is given in reference [12] and leads to:

$$eV^*V_1 = \frac{1}{2}g(h^2 - e^2)$$  (5)

where $V_1 = V R_d/R_S = V^2/V_S$ is the liquid velocity in the rim at radius $R_S$. From equations (4) and (5), one can calculate V and $V^*$ versus h and e:

$$\begin{cases} V^2 = \frac{1}{2}g\frac{h^2 - e^2}{e} \\ V^{*2} = \frac{1}{2}g\frac{h(e+h)}{e} \end{cases}$$  (6)

If we impose $h=e_c$ in equations (6), the velocity of dewetting V is similar to the Culik [13] law for the bursting of a soap film, where we have replaced $2\gamma$ (the driving force for suspended films) by an effective spreading parameter:

$$-\tilde{S} = \gamma_{SL} + \gamma - \gamma_{SO} - \frac{1}{2}\rho g e^2 = \frac{1}{2}\rho g(e_c^2 - e^2)$$  (7)

## 2.3. Dynamics of dewetting

Dewetting is observed with a standard video camera (25 images/s). The experimental results for V and $V^*$ are in good agreement with equations (6), for h=2.7 mm (Figure 1). Nevertheless the thickness h is smaller than $e_c$=3.9 mm. The Culik law is observed, but the apparent spreading parameter ($S_a$=-35 mJ/m$^2$) is significantly different from the static value (S=-82 mJ/m$^2$). It means that a part of the surface energy ($S_a$-S) is not converted into kinetic energy but is dissipated in some other process.

## 3. RIM PROFILES

For a better description of the shock, we need to measure the profile of the film around the expanding dry patch. There are two main problems in determining this profile: the high speed (up to 0.5 m/s),which excludes the use of a standard video camera and the high range of thickness, which excludes interferometry techniques. This led us to a different approach; at a given point of the interface at a distance R from the nucleation point, we measure the deflection of a laser beam by the rippled water surface [14]. The setup is sketched in Figure 2. The local slope angle $\theta$(R,t) (Figure 2) is measured versus time t. Because the front velocity $V^*$ is constant, we may reconstruct the profile h(t) near the shock as follows [14]:

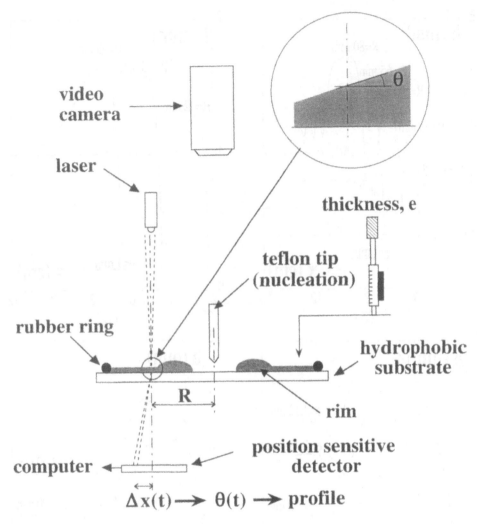

**Figure 2.** Experimental set-up to observe and characterize the profile of the bursting film on a silanized (hydrophobic) glass plate.

$$h(R, t) = e + \int_0^t \tan(\theta(R, t)) V^* dt \qquad (8)$$

In Figure 3 are shown typical profiles during the dewetting at various distances from the nucleation point and for different thicknesses, e of the liquid film. With these profiles we can verify the volume conservation: the liquid volume which has dewetted ($V_{th} = \pi R_d^2 e$) is compared to the volume measured with the profiles ($V_{mes} = 2\pi \int_{R_d}^{\infty} (h - e) r dr$). Volume conservation is verified within 4% (Figure 4).

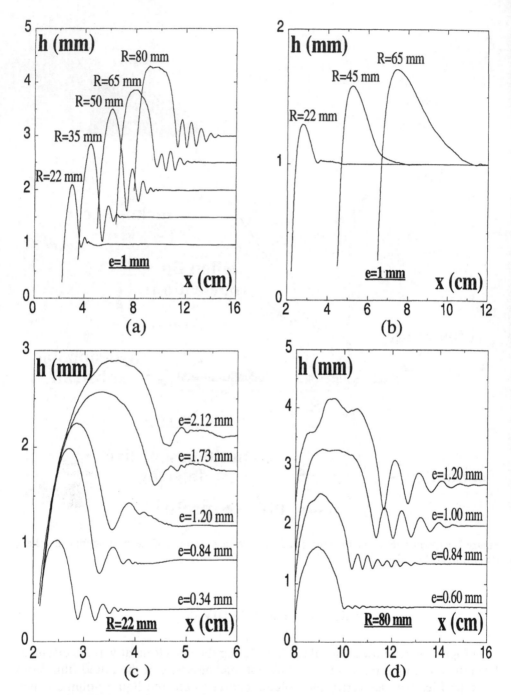

**Figure 3.** Profiles h(x) of rims and waves: a) versus time (or equivalently distance from the nucleation point). In this case the profiles are shifted by Δh=0.5 mm for clarity; b) at Froude number Fr<1. Notice the absence of ripples; c) for different thicknesses at a distance R=22 mm from the nucleation point; d) for different thicknesses at a distance R=80 mm from the nucleation point. In this case the profiles are shifted by Δh=0.5 mm for clarity.

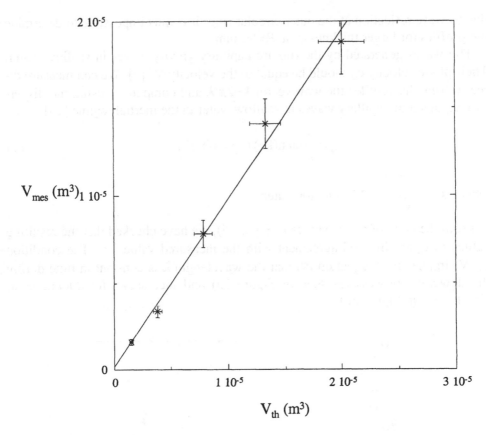

**Figure 4.** The volume of liquid calculated from the rim profiles ($V_{mes}$) compared to the volume of the dry patch ($V_{th}$). The volume conservation is verified: the liquid film is collected into the rim during dewetting.

To follow the evolution of the shock with time, we first analyse the rim profile versus the distance R from the nucleation point. The profiles at five distances (from 22 to 80 mm) are shown in Figure 3a for a film of initial thickness $e = 1.0 \pm 0.1$ mm. The rim generates capillary waves in the native film. The extension of these ripples increases with time, but the wavelength remains constant ($\lambda = 2\pi/k = 7.6$ mm). The rim is almost circular at short distances and gets larger and becomes flat: its thickness (2.3 mm at a distance R=80 mm) is close to the value h=2.7 mm derived from the fit of V and $V^*$ to equations (6).

We then investigate the influence of varying the film thickness on the shock. The profile shapes at a distance R=22 mm from the nucleation point for 5 thicknesses are shown in Figure 3c. The capillary waves are clearly apparent only for small thicknesses, e. The same measurements performed at R=80 mm from the nucleation point are shown in Figure 3d. Now the waves ahead of the rim are well formed. One notices that both the amplitude and the wavelength decrease as the thickness e decreases. For the thickest film (e=1.2 mm), the rim is not flat. This is due to wave reflection at the outer circular ring. It is thus impossible to determine the profiles for larger thicknesses at R=80 mm.

The waves generated by the rim are capillary gravity waves in shallow water. Their phase velocity $c_\varphi$ should be equal to the velocity $V^*$ [9]. We can measure directly from the profiles the wavevector $k=2\pi/\lambda$ and compute $c_\varphi$ using the dispersion equation of capillary waves in shallow water in the inertial regime [15]:

$$c_\varphi^2 = \frac{g}{k} \tanh(ke)\left(1+(k/\kappa)^2\right) \tag{9}$$

where $\kappa^{-1} = \sqrt{\dfrac{\gamma}{\rho g}} = 2.7$ mm for water.

From the plots of $c_\varphi$, $V^*$ versus k (Figure 5), we have checked that the resulting values of $c_\varphi$ are in good agreement with the measured values $V^*$. The condition $c_\varphi=V^*$ implies from equation (9) that the wavelength $\lambda$ is constant in time during the dewetting process (as seen in Figure 3a) and increases with thickness (as clearly seen in Figure 3d).

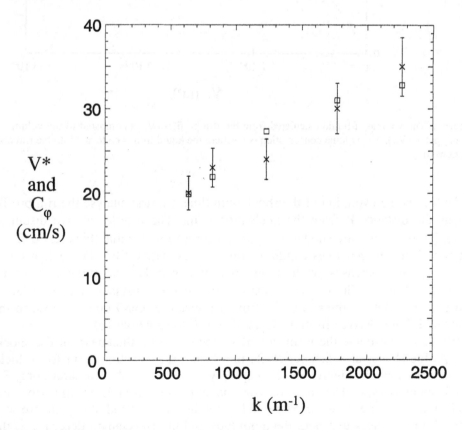

**Figure 5.** Shock velocity $V^*$ (x) and phase velocity $c_\varphi$ (□) versus wavevector k.

The influence of the Froude number is obvious if we compare Figures 3a and 3b. Figure 3b shows the rim profile obtained with a more viscous liquid for which the dewetting velocity does not satisfy the shock conditions. The liquid used is a water/glycerol (50%/50%) solution which is about six times more viscous than water. For a film thickness e=1 mm, the measured velocity for the external radius of the rim is $V^*$=6 cm/s< $\sqrt{ge}$ =10 cm/s. The Froude number is now smaller than one, whereas the Reynolds number is still large (of the order of 20). The shock and the waves have disappeared. Notice also that the shape of the rim is very asymmetric and much smoother on the liquid side.

## 4. LIQUID SUBSTRATES

We have pointed out in Section 2 the necessity to use solid surfaces free of defects (small hysteresis) in order to obtain reproducible results. One way to obtain "perfect" interfaces is to use a liquid substrate [16]. For two immiscible liquids the interface is flat on the atomic level.
But in the case of liquid substrates new physical features have to be taken into account:
     The interface between the two liquids is deformable.
     The opening of a hole in the film will induce a flow field in the substrate and
     dissipation will occur both in the film and in the substrate.
For preliminary experiments we have chosen a liquid/liquid system, in which
     the two liquids are immiscible,
     the liquid of the film does not wet the liquid substrate,
     the liquid of the substrate is denser,
     and the system is in partial wetting conditions (*i.e.* S<0).
     For the liquid substrate we used carbon tetrachloride ($\rho$=1.59 g.cm$^{-3}$) and for the liquid film water. This system is convenient because it is possible to vary the viscosity of the film (3 orders of magnitude) by adding glycerol in water without changing the other relevant parameters. With this system we obtain a spreading parameter S=-91 mJ/m$^2$ which corresponds to a critical thickness $e_C$=7.1 mm.
     Preliminary experiments show that inertial dewetting generates ripples ahead of the rim (Figure 6). In this case, the water film (thickness 500 $\mu$m) is deposited on a CCl$_4$ bath (depth about 1 cm). Moreover it is possible to observe these ripples behind the rim. In fact, the shock emits advancing and receding surface waves in the reference frame moving at the shock velocity $V^*$. These waves do not obey the same dispersion equation because the surface tension of the bare substrate (CCl$_4$) differs from the surface tension when it is covered by a water film.
     Though we have checked that the Culik law is always valid to describe the dewetting velocity (which stays smaller than the velocity predicted by the theory), quantitative experiments are still underway to measure the wavelength and the shape of the ripples. We expect a special behavior for the receding waves which should focus near the nucleation point because the shock front moving at V* is axisymmetric.

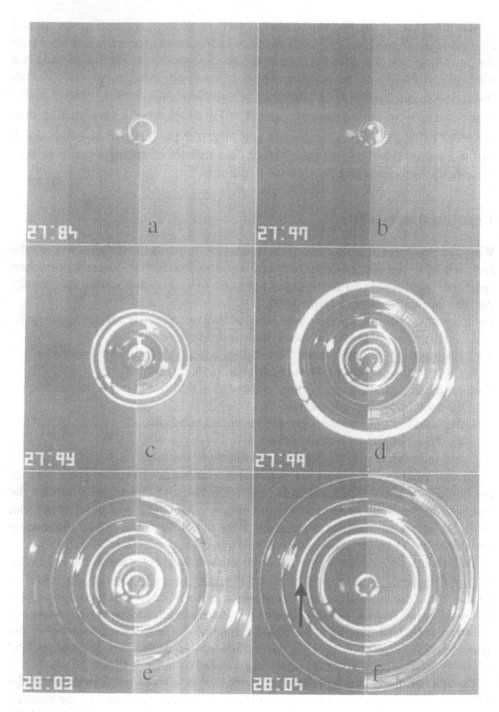

**Figure 6.** Dewetting of a liquid film (water) on a liquid substrate (CCl$_4$) observed with a video camera (between 27.84 s and 28.04 s). The width of each photograph is 5 cm. One can observe ripples ahead and behind the dewetting front (black arrow in the last photograph).

## 5. DISCUSSION

From all these measurments, we conclude that the front velocity $V^*$ and the rim thickness h do obey the Hugoniot relation (equations 6), but the magnitude of h does not follow the predictions of reference [12], which postulates that $h = e_C$ (=3.9 mm). We have investigated a number of reasons for this discrepancy.

i) *Viscous losses* in the boundary layer and at the contact line. To compute the dewetting velocity by the Culik law, we neglected the viscous forces exerted by the substrate on the rim assuming a very large Reynolds number. Nevertheless, with a boundary layer model, it is possible to evaluate this force. The dissipation in this viscous boundary layer per unit length [17] and per unit time is

$$T\dot{S} = f_v V \cong \eta \int_0^\lambda \left( \frac{V}{\delta(x)} \right)^2 \delta(x) dx \tag{10}$$

where $\delta(x) = \sqrt{vx/V}$ is the thickness of the boundary layer at distance x from the front and $\lambda = R_S - R_d$ is the width of the rim.

Integration of equation (10) leads to $f_v \cong \rho V^2 \sqrt{v\lambda/V}$. Taking a water rim 1 cm long and a dewetting velocity of 20 cm/s gives a viscous force of the order of 10 mN/m. This term becomes significant for large rims. On the other hand, we have checked that the viscous losses in the liquid wedge at the contact line, which dominate in viscous dewetting (Tanner's law), are entirely negligible.

ii) *Wave resistance* [9]. The capillary waves generated ahead of the rim carry a part of the surface energy. This energy per unit time $\dot{E}$ is [18]

$$\dot{E} = E(c_g - c_\varphi) \tag{11}$$

where $c_g$ is the group velocity (velocity of the energy), $c_\varphi$ is the phase velocity of the capillary waves of amplitude a and E is the wave energy per unit area,

$$E = \frac{1}{2} \rho c_\varphi^2 \frac{k}{th(ke)} a^2.$$

From equation (11), we can define the wave resistance $\Re$ acting on the rim moving at velocity $c_\varphi$ by the relation $\dot{E} = \Re c_\varphi$, which leads to

$$\Re = \frac{1}{2} \rho c_\varphi^2 \frac{k}{th(ke)} a^2 \left( \frac{c_g}{c_\varphi} - 1 \right) = \frac{1}{2} \gamma \kappa^2 a^2.$$ With e=1 mm, k≈1 mm⁻¹, a≈0.1 mm and

$\gamma$=72 mJ/m², we obtain $\Re = 0.5\,mJ / m^2$! This wave resistance is too weak to change the dewetting kinetics.

iii) *Nature of the driving force* $F_d$ [19]. We have evaluated $F_d$ by taking into account the interfacial tensions acting on the rim and the hydrostatic pressure on the forward side, giving $F_d = \frac{1}{2} \rho g(e_c^2 - e^2)$. This may be naive because the rim is a

deformable object which extends with time; by calculating the change of potential
energy versus time, we have reached [16] another expression for the driving
force, $F_d = \frac{1}{2}\rho g(e_c^2 - eh)$. Used in conjunction with the Hugoniot relation, equations (6), this leads to lower thicknesses h defined by the condition
$h^2 - e^2 = e_C^2 - eh$. For e=1 mm the predicted h is then h=3.5 mm as compared to
$h_{exp}$=2.7 mm. Moreover, the Culik law ($F_d = \frac{dM}{dt}V$) assumes that the rim moves
at a uniform velocity. In the limit $V^*>V$, one cannot neglect the spatial variation
of velocity when evaluating the momentum of the whole rim. The net effect is
also a reduction of h, which may be more significant [19].

## 6. CONCLUSION

Our central conclusion is that the dewetting of millimetric thickness films (on
solid or liquid substrates) is dominantly inertial (the main part of the surface energy is converted into kinetic energy), and is associated with a shock wave. It is of
interest to compare the bursting of our supported films with the inertial bursting
of (non-supported) soap films. At first sight, the supported films appear more
complex; they have a laminar boundary layer near the solid wall or a coupling
with the liquid substrate. However, the soap films are more delicate: the surfactant monolayers on their surface transport another form of shock waves, first analysed by Frankel and Mysels [20]. The dynamics depends on the pressure relations inside the monolayer.

For these suspended films the Froude number is always very large, i.e., they are
insensitive to gravity. The shock condition is always satisfied and ripples surrounding the film would be present. They have not been observed experimentally,
but they show up in the numerical simulations of the bursting of "bare" (with no
surfactants) suspended films [21]. Supported films have an adjustable Froude
number Fr. We can vary the driving force $\tilde{S}$ and the Froude number by changing
the thickness. When Fr>1, we have a shock with ripples. When Fr<1, there is no
shock. Supported films provide us with a tunable generator of shocks.

*Acknowledgments*

We are indebted to L. Vovelle and P.G. de Gennes for fruitful discussions.

## REFERENCES

1. M. Daoud and C. Williams (Eds.), *Soft Matter Physics*, Springer Verlag (1998).
2. P. Martin and F. Brochard-Wyart, Phys. Rev. Lett. **80**, 3296 (1998).
3. P.G. de Gennes, C.R. Acad. Sci. II b Paris **318**, 1033 (1994).

4. A. Carre and M. Shanahan, Langmuir **11**, 3572 (1995).
5. C. Redon, F. Brochard-Wyart and F. Rondelez, Phys. Rev. Lett. **66**, 175 (1991).
6. F. Brochard-Wyart and P.G. de Gennes, Adv. Colloid Interface Sci. **39**, 1 (1992).
7. G. Reiter, Science **282**, 888 (1998).
8. C. Andrieu, C. Sykes and F. Brochard-Wyart, J. Adhesion **58**, 15 (1996).
9. F. Brochard-Wyart, E. Raphael and L. Vovelle, C.R. Acad. Sci. II b Paris **321**, 367 (1995).
10. J. Sagiv, J. Am. Chem. Soc. **102**, 92 (1980).
11. P.G. de Gennes, Rev. Mod. Phys. **3**, 57 (1985).
12. F. Brochard-Wyart and P.G. de Gennes, C.R. Acad. Sci. II b Paris **324**, 257 (1997).
13. F.E. Culik, J. Appl. Phys. **31**, 1128 (1960).
14. A. Buguin, L. Vovelle and F. Brochard-Wyart, Phys. Rev. Lett. **83**, 1183 (1999).
15. J.P. Hulin, E. Guyon and L. Petit, *Hydrodynamique Physique*, EDP Sciences/CNRS, Paris (2001).
16. P. Martin, A. Buguin and F. Brochard-Wyart, Europhys. Lett. **28**, 421 (1994).
17. L. Landau and E. Lifshitz, *Fluid Mechanics*, MIR, Moscow (1971).
18. J. Lighthill, *Waves in Fluids*, Cambridge University Press (1978).
19. A. Buguin and F. Brochard-Wyart, C.R. Acad. Sci. II b Paris **327**, 809 (1999).
20. S. Frankel and K. Mysels, J. Phys. Chem. **73**, 3028 (1969).
21. M.P. Brenner and D. Gueyffier (MIT) – Private discussions.

*Contact Angle, Wettability and Adhesion*, Vol. 3, pp. 67–91
Ed. K.L. Mittal
© VSP 2003

# Effect of adsorbed vapor on liquid–solid adhesion

MALCOLM E. SCHRADER[*]

*Department of Inorganic and Analytical Chemistry, The Hebrew University of Jerusalem,
91904 Jerusalem, Israel*

**Abstract**—An overview is presented of a series of papers published during the last decade which show that the conventional thermodynamic approach to liquid–solid adhesion requires some fundamental changes. It is pointed out that it has been a long-neglected fact that adsorption, as generally measured in adsorption isotherms, is actually surface excess, so that it can, in principle, be negative as well as positive. As a result, the free energy of adsorption, $\Delta F$, can be positive as well as negative. Small amounts of water vapor adsorbing onto previously evacuated poly(tetrafluoroethylene) could, in principle, therefore be increasing the free energy of the low-energy polymer surface. It is further pointed out that from the strictly thermodynamic point of view, changing the free energy of a surface by adsorption of the vapor of a liquid does not necessarily change the contact angle. Resulting changes in contact angle can, however, theoretically occur from changes in the intermolecular force interaction term (proposed work of adhesion), such as those terms proposed by Good and Girifalco, Fowkes and others, where such changes would be speculative. In addition, it is pointed out that an accurate thermodynamic representation of liquid–solid adhesion should take into account the shape of the drop to be deposited (or drop that has been detached), as well as the resulting contact angle. An equation is presented for the free energy of adhesion of a spherical drop.

*Keywords*: Contact angle; work of adhesion; free energy of adsorption; surface excess; negative adsorption; net free energy of adhesion.

## 1. INTRODUCTION

### 1.1. Previous work

At the First International Symposium on Contact Angle, Wettability, and Adhesion [1], held in honor of Robert J. Good, we addressed the problem of the adsorption isotherm, that describes the formation of sessile drops as the end product of adsorption and condensation, of vapor on a solid surface. The isotherm was originally presented by Derjaguin and Zorin [2], and later quantified and modified by Adamson and co-workers [3–5]. These two groups must be credited with an outstanding contribution to sessile drop science by advancing toward the goal of integrating basic thermodynamics with the empirical and semi-empirical ap-

---

[*]Phone: (972-2) 563-7314, Fax: (972-2) 658-5319, E-mail: Schrader@vms.huji.ac.il

proaches used by experimentalists in the field. The approach adopted by Derjaguin nevertheless had a puzzling aspect, with which he and others struggled valiantly. The adsorption isotherm "crossed the $p°$ line", i.e., the vapor pressure had to exceed the saturation vapor pressure, $p°$, of the liquid in order to bring about condensation. However, since the totality of sessile drops that had condensed on the surface was regarded as a single mass of liquid, it was expected that its vapor pressure would be $p°$ (saturation vapor pressure of liquid) and that the isotherm would rapidly "return to the $p°$ line" and possibly cross it again [4]. This presented difficulties which were bypassed when the thermodynamics of formation of a single drop was developed [1, 6], as compared to treating the totality of multiple drops as a single quantity of formless condensate. It turned out that, because of the curvature at the liquid–vapor interface, the vapor pressure must exceed $p°$ for drop formation (on thermodynamic grounds, not merely because of activation energy), and should remain more than $p°$ as long as the drop exists. Thus, due to the well-known relationship between the free energy change and the logarithm of vapor pressure, it was shown that sessile drop formation is accompanied by an increase in free energy [1, 6], not a decrease as many had thought.

## 1.2. Problem addressed in this overview

The next question that arises naturally in studying the adsorption isotherm is the region starting from zero or near-zero vapor pressure and ending at $p°$. The popular approach had it that as an evacuated surface (taken to be initially devoid of molecules of vapor, or any gas), is exposed to vapor, or any gas, adsorption automatically takes place, driven by a reduction in free energy accompanying the adsorption [7]. In this view, then, if water vapor is introduced to a clean Teflon-like surface, some molecules will be adsorbed, the surface free energy will be lowered and liquid water will find it more difficult to spread on the surface. Of course, the amounts here are very small, and the effect would be very small, but nevertheless, in principle, that is what would happen. Even if the liquid were argon, whose atoms surely have no orientation effect, this approach predicts that its spreading would be impeded by its own adsorbed atoms.

Paradoxically, however, in the absence of any orientation-driven auto-phobic effect [8, 9] the presence of adsorbed molecules of a high-energy liquid on a low-energy surface should, on intuitive grounds as well as on the basis of all proposed IMF (intermolecular force) interaction terms [10–13], enhance wettability by that liquid, not impede it. As a result of adsorption of its vapor, the liquid would confront a greater proportion of high-energy sites on the solid surface, thus resulting in more spreading and lower contact angle. The present overview is based on a number of papers we have published on this subject during the last decade.

## 2. RESULTS

In this section we discuss (a) the effect of adsorbed vapor on solid surface free energy, (b) the effect of solid surface free energy on contact angle and (c) the role of IMF (intermolecular force) theory in approximating solid surface free energies.

### 2.1. Effect of adsorbed vapor on solid surface free energy

#### 2.1.1. The Young equation

During the last seven decades or so, it has been a common perception that adsorption of the vapor of a liquid onto a clean solid surface must result in a lowering of the surface free energy of the solid [7], which, in turn, must result in an increase in the contact angle of the liquid on that solid. The reason for this widely held view can be seen from the following.

The original Young equation [14] for contact angle is

$$\gamma_l \cos \theta = \gamma_s - \gamma_{sl} \tag{1}$$

where $\gamma_l$ and $\gamma_s$ are the free energies per unit area of the liquid and solid surfaces, respectively, and $\gamma_{sl}$ is the free energy per unit area of the interface formed when the solid surface is covered with the liquid. The sessile drop, of course, is assumed to be in complete mechanical equilibrium and the contact angle is assumed to be completely reversible. However, the effect of vapor pressure is not included. At this early stage of the development of contact angle theory, it was simply not taken into account. The lower case subscripts signify that a clean surface is being dealt with because vapor adsorption onto the solid is not taken into consideration or does not exist.

#### 2.1.2. The Young–Bangham equation and $\Pi_e$

In 1937, Bangham and Razouk [15] pointed out that for the case of any liquid with a significant vapor pressure, where the vapor molecules adsorb onto the solid surface (Fig. 1), $\gamma_s$ in the Young equation should be replaced with $\gamma_{SV}$, where $\gamma_{SV}$ is the free energy per unit area of the "contaminated" solid surface, i.e., the solid surface which contains an adsorbed layer of vapor molecules. They proceeded to describe the relationship between the clean and adlayer (an "adlayer", in this overview concerned with reversible adsorption, is the adsorbed layer of vapor molecules present on a solid surface in equilibrium with the vapor of the liquid at its equilibrium pressure) covered solid as

$$\gamma_L \cos \theta = \gamma_{SV} - \gamma_{SL} \tag{2}$$

$$\gamma_{SV} = \gamma_s - \Pi_e \tag{3}$$

where

$$\Pi_e = -\Delta F \tag{4}$$

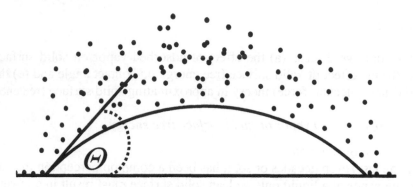

**Figure 1.** Equilibrium contact angle of liquid on solid containing adlayer of vapor. Dots are molecules of vapor; when shown close to solid surface they are adsorbed.

and $\Delta F$ is the increase in surface free energy on transforming the clean solid surface to the adlayer covered solid surface in equilibrium with vapor at saturation. Since the vapor adsorption is spontaneous, it would seem that $\Delta F$ must be negative, $\Pi_e$ positive and $\gamma_{SV}$ must be smaller than $\gamma_S$, so that vapor adsorption has decreased the free energy of the solid surface.

### 2.1.3. $\Pi_e$ as the integral of the Gibbs adsorption isotherm

However, Bangham and Razouk pointed out, furthermore, that $\Pi_e$ (which they named $F_V$) could be determined [15] by integrating the Gibbs isotherm [16] for adsorption of the vapor onto the solid surface from zero pressure to saturated vapor pressure, so that

$$\Pi_e = RT \int_0^{p^0} \Gamma \mathrm{d} \ln p \tag{5}$$

The use of the Gibbs "adsorption" equation for gas–solid systems was an innovation of Bangham and Razouk and can be understood by comparison with the well-known Gibbs treatment of a liquid solvent/solute system where the solute is added incrementally to the solvent [17], and its distribution between bulk solvent and solvent surface is treated in terms of surface excess of the solute at the solvent–air interface (Fig. 2). According to the Gibbs equation

$$\Gamma = -(c / RT) \, \mathrm{d}\gamma / \mathrm{d}c \tag{6}$$

where $\Gamma$ is the surface excess, $c$ the solute concentration, $R$ the gas constant and $T$ the absolute temperature. When the solvent (or solution) surface free energy decreases with addition of solute, the solute segregates preferentially at the surface to yield a positive surface excess (Fig. 2a). When there is no change in surface free energy of the solution with addition of solute, the surface excess is zero (Fig. 2b). When the surface free energy of the solvent (or solution) increases with addition of solute, the solute shuns the surface and concentrates preferentially in the bulk, to yield a negative surface excess (Fig. 2c).

## Solvent / Solute

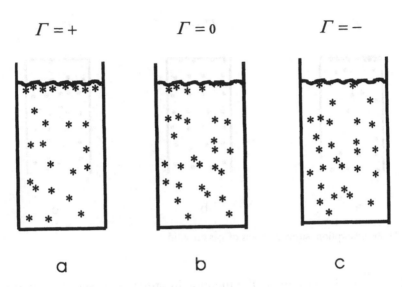

$$\Gamma = + \qquad\qquad \Gamma = 0 \qquad\qquad \Gamma = -$$

a            b            c

**Figure 2.** Gibbs adsorption: surface excess of solute in solvent.

For the case of the gas–solid adsorption isotherm, the vacuum, or space, plays the role of the solvent in the problem described above. The gas molecules in the space have the role of the solute in the solvent, and the interface between the space and the solid sample surface has the role of the interface between the solvent and the air (Fig. 3).

The choice of boundaries is, of course, always important in Gibbs surface-phase calculations. First, Bangham [18] adopted the Guggenheim–Adam [19] method of surface boundary selection. Instead of choosing an imaginary boundary where the concentration of one of the two components disappears, as Gibbs recommends, Guggenheim and Adam choose a boundary with physical reality. In this case it is the boundary of the condensed phase, i.e., the atomic surface of the solid adsorbent. The solute concentrations in moles per unit volume are replaced by pressure for the gas–solid system (concentration instead of activity, and pressure instead of fugacity, are approximations). Bangham and Razouk [15] then introduce the approximation that the gas solubility in the bulk solid is negligible, and an equation for gas/solid systems analogous to the solute/solvent case is set up as follows

$$\Gamma = - (p / RT)\, d\gamma / dp \tag{7}$$

where $p$ is the gas pressure. Solving for $d\gamma$ and integrating, we have equation (6), where $\Pi_e$ is the integral of $(-d\gamma)$. In Fig. 3a, we have a decrease of surface free energy of the solid with addition of gas molecules, i.e., increase in pressure, so the surface excess is positive. In 3b, there is no interaction of gas with the surface and

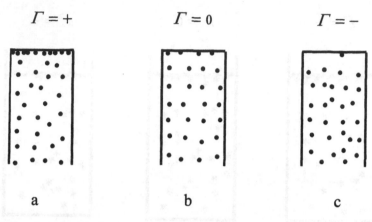

**Figure 3.** Gibbs adsorption: surface excess of gas on solid.

the surface excess is zero. In 3c, the increase in pressure causes an increase in free energy of the solid surface, so the gas molecules prefer the bulk phase and the surface excess is negative. It is important to note here that although the solid sample surface of the adsorbent in Fig. 3c is depleted in gaseous molecules (negative surface excess), those molecules of gas that are adsorbed (absolute adsorption) increase the free energy of the solid surface over that of an evacuated (zero absolute adsorption) surface.

### 2.1.4. Negative surface excess

The above discussion indicates rather clearly that a negative surface excess is to be expected for certain combinations of gas adsorbate with solid adsorbent. (It should be noted that all volumetrically and gravimetrically measured gas-solid adsorption isotherms measure surface excess (unpublished discussion) [20, 21]. From hereon the word "adsorption" will be used interchangeably with "surface excess", while molecular adsorption will be called "absolute adsorption" or "molecular adsorption"). The question is whether any such interaction from one of these combinations has actually been observed.

#### 2.1.4.1. Critical phenomena

It turns out that negative adsorption in the field of critical phenomena [22] has been known for more than two decades [23–28]. In the example given in Fig. 4, $SF_6$ is adsorbed onto graphitized carbon black [27] in the region of the critical temperature. At higher than the critical temperature the adsorption excess goes through a maximum followed by a decrease in total "adsorption", i.e., surface excess. This is, then, negative "adsorption", where increasing the concentration (pressure) lowers the total amount adsorbed. A more general treatment [28], also in the critical region, based on statistical mechanical calculation using the mean

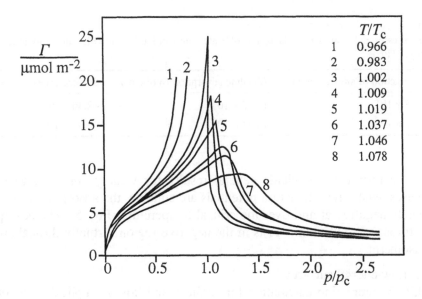

**Figure 4.** Adsorption in critical region: $SF_6$ on graphitized carbon black (experimental). Reprinted with kind permission from Ref. [24].

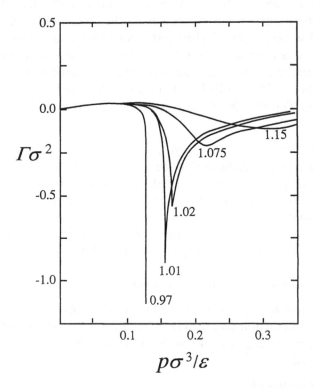

**Figure 5.** Weakly adsorbed gases in critical region (calculated). Curves are labeled with the values of $T / T_C$, where $T_C$ is the critical temperature, $\sigma$ is the molecular diameter of the gas, $p$ the pressure and $\varepsilon$ the exponential attractive potential of the surface. Reprinted with permission from Ref. [28].

**Table 1.**
Surface excesses calculated for adsorption of water onto polyethylene (PE) and poly(tetrafluoro-ethylene) (PTFE)

| Solid | Fraction of sites covered | Absolute adsorption (mol/cm$^2$) | Surface excess (mol/cm$^2$) |
|-------|---------------------------|----------------------------------|------------------------------|
| PE    | $7 \times 10^{-6}$        | $1.202 \times 10^{-14}$          | $-2.46 \times 10^{-14}$      |
| PTFE  | $6 \times 10^{-7}$        | $1.031 \times 10^{-15}$          | $-3.55 \times 10^{-14}$      |

spherical approximation, yields general results for weakly adsorbed gases. In some cases, when the interaction potentials are suitable, the adsorption is actually completely negative at and over the critical temperature (Fig. 5). The adsorption passes through a minimum and is in the negative region virtually throughout the entire isotherm.

### 2.1.4.2. Low-energy surfaces

A strikingly interesting calculation for surfaces and liquids applicable to contact angle measurement was performed on the basis of nearest neighbor interaction potentials [29]. The results of the calculations are given by the author in terms of fraction of sites covered (Table 1), where it can be seen that the adsorption of the relatively "high-energy" water molecules to the low-energy polyethylene (PE) and poly(tetrafluoroethylene) (PTFE) surfaces is quite sparse, amounting only to a few millionths of a monolayer. We have converted these values to absolute adsorption and surface excess and it turns out that the adsorption of water vapor at room temperature to PE and PTFE yields a negative surface excess according to the calculations. The details of the conversion are given in Ref. [23] and references therein.

### 2.1.4.3. Consequences of negative adsorption

The consequence of the above is that the presence of some adsorbed vapor molecules on the surface (absolute adsorption), as compared to a hypothetical situation where they have been removed, can either [23] (a) increase the free energy of the solid surface, (b) decrease it, or (c) have no effect on it. This, as mentioned above, is in contrast to the previously accepted view that absolute adsorption of vapor of any sort must decrease the free energy of a clean solid surface.

## 2.2. Effect of adsorbed vapor on contact angle

The effect of vapor adsorption on the surface free energy of the solid is dependent on the specific interaction parameters of the vapor molecules and the solid surface. In the following, the influence of the solid surface free energy on the contact angle is examined.

### 2.2.1. Previous approach

The ubiquitous perception since 1937 has been that a decrease in free energy of the solid surface occurs due to adsorption of vapor molecules and that this creates a thermodynamic requirement that there be an increase in contact angle. Once again, the reason for this misconception is obvious from examination of the Young equation as modified by Bangham

$$\gamma_L \cos \theta = \gamma_{SV} - \gamma_{SL} \tag{2}$$

where

$$\gamma_{SV} = \gamma_S - \Pi_e \tag{3}$$

If $\Pi_e$ is positive, then $\gamma_{SV}$ is less than $\gamma_S$, and $\cos \theta$ decreases linearly (likewise, if $\Pi_e$ is negative, then $\cos \theta$ will increase linearly, but this possibility was not considered until recently [23]). It seems then that the contact angle will always increase with decrease of solid surface free energy, and will always decrease if there is an increase in solid surface free energy.

### 2.2.2. Present approach

The problem with this old and well-established mirage is that it is impossible to change the characterization of the solid surface in this manner, from clean to vapor equilibrated, without simultaneously changing the chosen position of the imaginary (in general) boundary representing the solid–liquid interface. In equation (1) $\gamma_s$ is the free energy of the clean solid and $\gamma_{sl}$ is the free energy of the interface (virtual boundary) formed when the clean solid is covered by the liquid, which will give rise to a clean solid surface plus a liquid surface when it is separated again. This is not because there is anything intrinsic to the history of the interfacial region which forces this to happen (of course, the interfacial region is, by assumption, always the same). It is because, in this particular "thought experiment", we choose to have the interfacial region split in this manner. In equation (2), $\gamma_{SV}$ is the free energy of the adlayer "contaminated" solid and $\gamma_{SL}$ is the free energy of the interface formed when the adlayer contaminated solid is covered by the liquid, which will give rise to an adlayer on the solid surface plus a liquid surface when it is separated again. In this case, we choose the interface which will separate in this latter manner. Thus, the definition, or chosen position, of the interface has changed (in addition, the area of the interface has changed, and we will discuss this in more detail later on).

### 2.2.3. Dupré analysis

The significance of this can be seen when the sessile drop is subjected to a Dupré analysis [30, 31]. In Fig. 6 on the left, the Young equation describes the horizontal balance, at equilibrium, between the tendency to contract the drop to a smaller $\cos \theta$ (larger contact angle) and expand it to a larger $\cos \theta$ (smaller contact angle). For the same sessile drop the Dupré analysis, which originally did not include the possible effect of vapor pressure of the liquid, can be used to express the work of adhesion between the drop and the solid surface which is given by

$$W = \gamma_s + \gamma_l - \gamma_{sl} \tag{8}$$

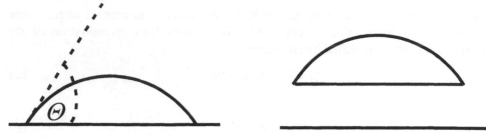

**Figure 6.** Contact angle and work of adhesion. To the left is a sessile drop with contact angle $\theta$. YDP defines the work (free energy) necessary to separate the liquid drop to an infinite distance from the solid, leaving behind a clean solid surface, as shown to the right. The process is, of course, reversible. Note that the detached sessile drop retains its original shape, i.e., YDP does not take into account the free energy change on rearrangement of the detached drop to the spherical shape of minimum area.

As has recently been pointed out [32], this gives the work, or free energy input necessary to separate, to infinity, the drop from the surface while maintaining the drop in its original shape. Combining the Young equation (1) with the Dupré equation (8), as suggested by Pockels [33], $\gamma_s - \gamma_{sl}$ cancels out and we have the Young–Dupré–Pockels (YDP) equation

$$W = \gamma_l(1 + \cos \theta) \tag{9}$$

which gives the work of adhesion of the sessile drop (of given shape) to a solid surface as a function of the liquid surface tension and contact angle alone!

Interestingly, according to N. K. Adam [34], the idea of this equation has been stated only in words, without derivation, by T. Young. According to Adam, "When the adhesion (to the solid) is less than the self-cohesion of the liquid, there is a contact angle.... When the adhesion is equal to or greater than the cohesion, the angle is zero." At any rate we will continue to refer to the equation as Young–Dupré–Pockels (YDP), in accordance with its derivation.

We remind the reader at this point that equation (9), as for the case of equations (1) and (8), from which it is derived, does not deal with the issue of vapor pressure. In present day terminology, we would say that the equation deals with a liquid which has negligible vapor pressure.

In their discussion of the effect of vapor pressure on the surface free energy of the solid in the Young equation, Bangham and Razouk [15] also discuss the Dupré and YDP equations. They point out that since the solid must be in equilibrium with the vapor pressure of the liquid at saturation, the YDP equation gives $W$, the work of adhesion, for a liquid to an adlayer-coated solid surface (Fig. 7). This, they point out (although in somewhat confusing language), is in contrast to the work of adhesion as "usually understood", which is that of a liquid to a clean solid surface. Of course, an adlayer-coated solid surface here does not necessarily imply a densely-packed monomolecular layer, but rather an adsorbed "layer" of molecules whose magnitude is determined by equilibration with the saturated va-

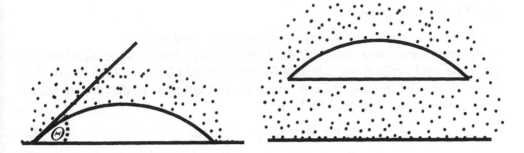

**Figure 7.** Effect of liquid vapor on work of adhesion. This figure repeats the reversible process which in Fig. 6 described a liquid with negligible vapor pressure. Here it describes a liquid with significant vapor pressure. YDP now gives the work to detach the drop from a solid surface under conditions of equilibrium with the vapor of the liquid, so that the solid surface retains an adsorbed layer of vapor after removal of the liquid drop.

por, which could involve adsorption onto as few as one in a billon or less of the available sites on the solid surface, or to a dense multilayer. Bangham and Razouk, therefore, derive the YDP equation by combining the Bangham-modified Young equation [15] with the Bangham-modified Dupré equation [15]

$$\gamma_L \cos \theta = \gamma_{SV} - \gamma_{SL} \tag{2}$$

$$W = \gamma_{SV} + \gamma_L - \gamma_{SL} \tag{10}$$

It can be seen that now $\gamma_{SV} - \gamma_{SL}$ cancels out, resulting in an expression for $W$ which is the same as that in equation (9)

$$W = \gamma_L (1 + \cos \theta) \tag{11}$$

Of course, the $W$ in equations (11) and (9) apply to different systems, with two different sets of conditions and, of course, different contact angles if the works of adhesion are different. It may be recalled that in the contemporary view [15] where sessile drop equilibrium includes quasi-vapor equilibrium, equations (1) and (8) apply to situations where the liquid has no vapor pressure, while equations (2) and (10) apply where the system is in equilibrium with the vapor at saturation (Fig. 7). Consequently, equation (9), which is derived from equations (1) and (8), and equation (11), which is derived from equations (2) and (10), refer to sessile drops composed of two different liquids, one without vapor pressure and one with vapor pressure, each one in equilibrium with the solid surface. The equation for the work of adhesion is then the same for both, and yields a work of adhesion $W$ applicable to both under their own conditions and depends on the values of $\gamma_L$ and $\theta$ in each case.

## 2.3. The Harkins equation

### 2.3.1. Works of adhesion for clean and adlayer-covered surfaces

Suppose, however, that we derive the YDP equations for the same liquid drop on the same solid surface, one equation for the work of adhesion of the sessile drop to the clean solid surface, and one for the work of adhesion of that drop to the adlayer covered solid surface [35]. Then,

$$\gamma_L \cos \theta_C = \gamma_S - \gamma_{SL} \quad \text{Young} \tag{12}$$

$$W_C = \gamma_S + \gamma_L - \gamma_{SL} \quad \text{Dupré} \tag{13}$$

$$W_C = \gamma_L ( 1 + \cos \theta_C) \quad \text{YDP} \tag{14}$$

and

$$\gamma_L \cos \theta_A = \gamma_{SV} - \gamma_{SL} \quad \text{Young} \tag{15}$$

$$W_A = \gamma_{SV} + \gamma_L - \gamma_{SL} \quad \text{Dupré} \tag{16}$$

$$W_A = \gamma_L (1 + \cos \theta_A) \quad \text{YDP} \tag{17}$$

where $\theta_C$ and $\theta_A$ are the contact angles measured on clean and adlayer covered surfaces, respectively, and $W_C$ and $W_A$ are the works of adhesion of the sessile drop to the clean and adlayer covered surfaces, respectively. Note that we now have a YDP equation for $W_C$ and a YDP equation for $W_A$. The YDP equation for $W_A$ is that derived by Bangham and Razouk [15], for adlayer-covered solid surfaces, where $\theta_A$ is measured at vapor, as well as mechanical, equilibrium. The YDP equation for $W_C$ is derived for the clean solid surface, where the contact angle $\theta_C$ is somehow or other measured for the drop on the solid surface in the absence of adsorbed vapor (we note at this point that the existence of such a measurement is problematical). The form of the two equations is the same, but the works of adhesion are different, as determined by the two different contact angles. It is important, at this point, to observe that we have not derived any relationship between the works of adhesion $W_A$ and $W_C$. Each must be determined through separate measurement of its contact angle, $\theta_A$ or $\theta_C$.

### 2.3.2. Conversion of $W_A$ and $W_C$ to each other?

In 1942, Harkins and Livingston [36, 37] published a treatment of the work of adhesion concept which would seem to allow the conversion of $W_A$ to $W_C$ and vice versa, through the addition or subtraction of $\Pi_e$. The essence of the treatment may be given as follows:

$$\gamma_L \cos \theta_C = \gamma_S - \gamma_{SL} \tag{12}$$

$$W_C = \gamma_S + \gamma_L - \gamma_{SL} \tag{13}$$

$$\gamma_L \cos \theta_A = \gamma_{SV} - \gamma_{SL} \tag{15}$$

$$W_A = \gamma_{SV} + \gamma_L - \gamma_{SL} \tag{16}$$

Equations (12) and (13) are the Young equation and the Dupré equation, respectively, for the clean solid surface, and equations (15) and (16) are the Young equation and the Dupré equation, respectively, for the adlayer-covered solid surface.

Harkins was interested in $W_C$, the work of adhesion for a clean surface. However, he also knew, what was emphasized in Bangham's treatment of the subject a few years earlier, that the contact angle could be measured at equilibrium only in the presence of the saturated vapor. He, therefore, took the Dupré equation for $W_C$ of the clean surface equation (13), and combined it with the Young equation for $\theta_A$ of the adlayer covered surface equation (15), to allegedly obtain a YDP equation for $W_C$ based on measurement of the contact angle $\theta_A$. The result was

$$W_C = \gamma_L(1 + \cos\theta_A) + \Pi_e \tag{18}$$

since $\Pi_e$ equals $\gamma_S - \gamma_{SV}$. Now, intuitively it is obvious that this is not a proper derivation [35]. Certainly the Young equation for measuring the contact angle under a given set of conditions has to be combined with the Dupré analysis for the same set of conditions, i.e., equation (12) with equation (13) or equation (15) with equation (16). Translating this into algebra, the symbols in any two equations that are solved together must have the same meaning. The obvious culprit here, then, is $\gamma_{SL}$. This symbol in equation (13) is chosen to be the free energy of the interface formed from covering the clean solid surface with the liquid of the sessile drop. In equation (15) it is defined as the free energy of the interface formed from covering the adlayer-containing solid surface with the liquid of the sessile drop. Furthermore, the surface areas of the two interfaces are not the same, and we shall discuss this later in more detail.

A derivation for equation (18) has been given many times in the literature but the essence has always been the same as that in the original Harkins–Livingston formulation. The fact that the derivation is not correct certainly implies that the result is in error. However, many surface chemists have come to regard equation (18) as intuitively obvious — not requiring a derivation. In the following, therefore, we point out that the equation is not operable, and cannot be accurate, regardless of derivation [35].

### 2.3.3. Why the equation due to Harkins cannot be correct?
As stated by Bangham and Razouk [15], "...W is the work necessary to pull apart the liquid and solid phases and produce, instead of 1 cm$^2$ of solid–liquid interface, 1 cm$^2$ each of solid–vapor and liquid–vapor interfaces..." . The works of adhesion are all defined per unit area (in the YDP equation, of course, $\gamma_L$ provides the unit). For any area, the right-hand sides of equations (19) and (20) contain the actual surface area to give the total work of adhesion. Thus,

$$W_A' = \pi r_A^2 \gamma_L (1 + \cos\theta_A) \tag{19}$$

gives the total work of adhesion, $W_A'$, for the adlayer-covered solid, where $r_A$ is the radius of the solid–liquid interface of the drop at contact angle $\theta_A$ and

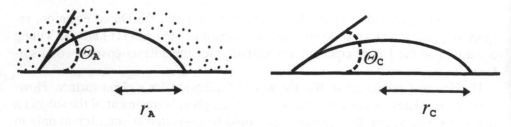

**Figure 8.** $\pi\, r^2\, \Pi_e$ correction for adlayer: variation with contact angle. This figure shows, on the left side, the case of a liquid with significant vapor pressure and the contact angle $\theta$ which it establishes at equilibrium with the solid surface, which contains a reversibly formed adsorbed layer of the vapor of the liquid. On the right side it is shown that if the adlayer is forcibly removed, such as, for example, by vacuum pumping, $\theta$ may change due to the change in nature of the solid surface which interacts with the liquid. In that case the radius of the solid–liquid interface must change as $\theta$ changes.

$$W_{\text{C}}' = \pi r_{\text{c}}^2 \gamma_{\text{L}}(1+\cos\theta_{\text{c}}) \tag{20}$$

gives the total work of adhesion, $W_{\text{C}}'$, for the clean solid, where $r_{\text{C}}$ is the radius of the solid–liquid interface of the drop at contact angle $\theta_{\text{C}}$. For a drop of constant volume where $\theta_{\text{C}}$ does not equal $\theta_{\text{A}}$, $r_{\text{C}}$ cannot equal $r_{\text{A}}$.

Now, the "intuitively obvious" assertion would have it that the addition of $\Pi_e$ can transform $W_{\text{A}}$ to $W_{\text{C}}$. However, $\Pi_e$ is the free energy of vapor adsorption per unit area, determined from an adsorption isotherm. For a solid–liquid interface of any area, $\Pi_e$ must be multiplied by $\pi r^2$.

There is then

$$\pi r_{\text{A}}^2 \gamma_{\text{L}}(1+\cos\theta_{\text{A}}) + \pi r_{\text{X}}^2 \Pi_e = \pi r_{\text{C}}^2 \gamma_{\text{L}}(1+\cos\theta_{\text{C}}) \tag{21}$$

where the X as subscript of the $r^2$ coefficient of $\Pi_e$ indicates the following problem. Is $r_{\text{A}}$ or $r_{\text{C}}$ used in the coefficient of $\Pi_e$? That is, is this free energy of desorption applied to the surface area covered by the drop with contact angle $\theta_{\text{A}}$ or to the area covered by the drop with contact angle $\theta_{\text{C}}$ (Fig. 8)? There is, of course, no answer. It turns out that the correction term $\pi\, r^2\, \Pi_e$ in the Harkins equation, which can be written as

$$W_{\text{C}}' = W_{\text{A}}' + \pi r^2\, \Pi_e \tag{22}$$

is supposed to convert the clean and adlayer covered surface contact angles back and forth to each other, is itself a function of contact angle for a drop of given volume. Clearly the Harkins approach does not work [35].

## 2.4. Intermolecular force approach and adsorbed vapor

While the Young and YDP equations provided a convenient framework for visualizing the thermodynamic relationship between the contact angle and the various interfacial free energies involved in sessile drop equilibrium, they did not provide a way to predict contact angles of an untried liquid–solid combination based on the separate surface characteristics of the components. This was due to the fact that neither solid surface free energy nor solid–liquid interfacial free energy could be measured experimentally by the techniques used for surface tension and contact angle measurements.

### 2.4.1. Critical surface tension

The first successful systematic attempt at predicting wettability by characterization of the solid surface was the approach of Zisman and co-workers [8]. Its main achievement was the characterization of solids by the empirical quantity "critical surface tension" which enabled the prediction of which liquids would or would not "wet" (i.e., zero contact angle) a solid completely. In practice it was limited, however, to dispersion (London forces) type interactions.

### 2.4.2. Intermolecular force approach

In 1960 Good and Girifalco [10] proposed a semi-empirical approximation, based on an "intermolecular force (IMF)" approach, utilizing a geometric mean interaction, for the thermodynamic quantity $W$ (work of adhesion) in the YDP equation, thereby allowing the calculation of solid surface "free energies" (actually approximate energies) from a plot of $\cos \theta$ vs. reciprocal of surface tension of liquid. Their proposed interaction term (i.e., work of adhesion), however, had an arbitrary constant $\varphi$ to compensate for nondispersion forces. Thus, the usefulness of the approximation was limited. In 1964 Fowkes [11] proposed to resolve surface free energies into dispersion force and other (e.g., hydrogen bonding, metallic) components. He then treated pure dispersion interactions by the geometric mean, and calculated solid surface dispersion free energies by using dispersion free energies along with total free energies in a Good–Girifalco reciprocal-surface–tension plot [11]. The resulting IMF system, known as GGF (Good–Girifalco–Fowkes), has been quite successful in treating interactions (and thereby predicting contact angles) of nonpolar vs. nonpolar or nonpolar vs. polar materials. That is, for contact angles, a liquid and solid where at least one of the two undergoes dispersion interaction only. The essence of the derivation can be described as follows, using the GGF interaction term, i.e., their approximation for $W_C$

$$W_C = 2\,(\,\gamma_S^{\,d}\,\gamma_L^{\,d}\,)^{1/2} \tag{23}$$

Their method, called intermolecular force theory, is based on the interaction of assumed adjacent layers of the two phases, here solid and liquid. Thus the assumption is clean solid vs. liquid, which gives the work of adhesion for a clean solid surface, $W_C$. From the Young–Dupré equation, this is equal to $(1 + \cos \theta)$. Now, the only way the equilibrium contact angle of a liquid with significant vapor pressure can be measured is at saturation vapor pressure of the liquid, so that the Young equation must be written as,

$$\gamma_L \cos \theta = \gamma_{SV} - \gamma_{SL} \tag{2}$$

for a sessile drop on a solid surface in equilibrium with the vapor of the liquid at saturation pressure. The contact angle measured under this condition will give the work of adhesion $W_A$ for the liquid and an adlayer covered solid surface. Since GGF want $W_C$, the work of adhesion of the liquid and a clean solid surface, they use the Harkins equation

$$W_C = \gamma_L(1 + \cos \theta_A) + \Pi_e \qquad (18)$$

$$2\,(\gamma_S^{\,d}\gamma_L^{\,d})^{1/2} = \gamma_L(1 + \cos \theta_A) + \Pi_e \qquad (24)$$

to convert the adlayer-covered surface work of adhesion to the clean surface work of adhesion.

### 2.4.2.1. Correct derivation of GGF approach

Since the Harkins equation, and therefore equation (24), is not correct, the GGF interaction term (and all other such terms, including acid–base formulations as well) must be constructed to equal the YDP work of adhesion. Thus

$$2\,(\gamma_{SV}^{\,d}\gamma_L^{\,d})^{1/2} = \gamma_L(1 + \cos \theta_A) \qquad (25)$$

is formally acceptable for dealing with vapor pressure. However, the quantity $\gamma_{SV}^{\,d}$ is, of course, not at all useful. It cannot be calculated from measurements of that solid against other liquids, since the vapor is different for different liquids. Also, it cannot be calculated from theoretical molecular force approaches, since the amount of vapor on the surface is not known. Furthermore, it has been stated [38] that the intermolecular force approach should only be used for a tightly packed layer. Also, there is no straightforward way to obtain a quantity $\Pi_e^{\,d}$ to convert $\gamma_{SV}^{\,d}$ to $\gamma_S^{\,d}$. What to do? One should do what has been done all along by Good and Girifalco, Fowkes, and others. Assume that adsorption, and consequently the effect of adsorption on the free energy of the solid surface, is negligible. Thus, $\gamma_S^{\,d}$ is used instead of $\gamma_{SV}^{\,d}$, and the operative equation is

$$2\,(\gamma_S^{\,d}\gamma_L^{\,d})^{1/2} = \gamma_L(1 + \cos \theta_A) \qquad (26)$$

Now, what is the actual difference from the manner in which the GGF method has been used before? Previously, $\Pi_e$ was present as a linear addition to the thermodynamic framework of the GGF equation, i.e., it was a part of the YDP equation which contained the GGF-proposed interaction term. Therefore, when the results of this approach, obtained while declaring $\Pi_e$ negligible, did not turn out to a given experimenter's satisfaction, he would sometimes conscientiously proceed to run an adsorption isotherm and calculate a $\Pi_e$ value to increase the accuracy of the result. The method, however, is not valid and the claimed improvement in the results was really the result of random changes or other factors. Furthermore, the adsorption isotherms were often in error, as a result of impurities on the surface. These often gave positive values for adsorption to low-energy surfaces when the adsorption was probably negative. When the assumption of negligible adsorption is understood to affect only the interaction approximation, however (by allowing the replacement of $\gamma_{SV}^{\,d}$ by $\gamma_S^{\,d}$ under the square root sign), there is no temptation to implement a $\Pi_e$ solution to the problem, since we are dealing with the dispersion component of the solid-surface free energy and $\Pi_e$ is a total free energy change. It is obvious that the method of choice is to keep the assumption of negligibility. However, if an attempt is going to be made to use some means or other to adjust the interaction approximation to compensate for vapor adsorption, this is not impossible since the interaction term itself is basically an empirical speculation, and the adjustment would presumably have some acceptable reasoning behind it.

Another, and no less important, offshoot of this result is the effect on intuitive qualitative expectations of the influence of vapor. The presence of $\Pi_e$ in the YDP equation itself (here also called the thermodynamic framework of the GGF equation) has helped buttress the erroneous impression that vapor adsorption onto the solid surface, when significant in magnitude, must affect the contact angle (and furthermore, according to another erroneous impression, increase it). The fact is there is no proof that vapor adsorption must affect the contact angle. There are, of course, individual cases where this has been shown to be so, but these were all instances of what has been termed autophobicity [8, 9], i.e., strong adsorption of bipolar molecules to a surface, which obviously present a different surface (such as, for example, the hydrophobic end of a long molecule with polar functional group on the adsorbing end) to the liquid. Also, not all these cases of autophobicity represent equilibrium adsorption. Many that have thus been termed involve irreversible reaction at the solid surface, such as, e.g., hydrolysis of organic esters on glass [8, 39]. The point is, that there is a thermodynamic requirement that adsorption of vapor onto the solid surface changes its free energy (although the change may be zero as well as positive or negative). However, there is no thermodynamic requirement that this change in free energy, from adsorption of vapor, must affect the contact angle, since the solid–liquid interfacial free energy also changes from adsorption of vapor, due to the change in the chosen position of the imaginary solid–liquid interface. The factor determining possible change in contact angle is the possible change in the nature of the variable representing the solid surface in the interaction term (e.g., $\gamma_S$ or $\gamma_S^d$ ). The question as to the effect on contact angle, if any, is open to judgment which may vary for each individual system. Also, from the experimental point of view, it is not clear that there is solid surface near the periphery of the sessile drop which is free of adsorbed vapor even under dry gas or vacuum [34]. In that case, there may be no reliable baseline from which to measure the effect of adsorption.

2.4.2.2. Possible qualitative effects of vapor adsorption
It is clear that the possible effect of adsorbed vapor on contact angle is a matter for speculative judgement in setting up interaction terms. Of course, the effect, if any, of high-energy molecules, such as water, on low-energy surfaces, such as Teflon-like materials, can be expected to be very small so that the considerations pointed out here are essentially theoretical. Nevertheless, these considerations can become important factors in our judgment of possible interactions in cases such as low energy molecules on low energy surfaces or high energy molecules on surfaces of intermediate energy [40, 41]. An example of the latter would be water on

the basal plane of graphite, where the contact angle is about 38° [40]. There is the substantial possibility that the discontinuity in adsorption (which occurs whenever there is a contact angle) is not due to an autophobic layer, but rather to negative adsorption, i.e., to what may be termed adsorption of holes. In this case, an experiment designed to measure the effect of adsorbed vapor could determine whether it is an adsorbed autophobic layer or the surface itself (hole adsorption) that is responsible. If removing the water vapor increases the contact angle, then it is the surface itself that is responsible for sessile drop (rather than film) formation. If removing the water vapor decreases the contact angle, then the adsorbed vapor is at least partially hydrophobic.

## 2.5. Unattached spherical drop: net free energy of adhesion

W, the YDP work of adhesion, describes the free energy necessary to detach a liquid drop from a solid surface as a function of contact angle of the drop. The value of W, however, does not describe a practical means of attachment and detachment, nor does it describe the true free energy of a complete process. What the YDP equation actually describes is the hypothetical process of removing a sessile drop from a solid surface while the drop maintains its original shape. For this process the free energy to be supplied is that required to overcome the bond of the liquid to the solid at the flat liquid–solid interface, or, the energy given off when that bond is formed.

### 2.5.1. Net Free Energy of Adhesion for a given liquid–solid interfacial area
In fact, the detachment of drop from the surface, or more important, the reverse thereof, i.e., attachment of drop to the surface, never occurs in this manner. If the drop is removed by some means and is hanging free in space, it will immediately adopt the configuration of minimum surface area, namely, a sphere [32]. Likewise, if deposited on a surface, it will generally start from an approximate spherical configuration which was modified somewhat by the utensil used to contain and transport it. Since it could be transported by a pipette, rod, needle, or other such means, all giving it a slightly different configuration which is often not reproducible, a convenient standard state, or configuration for a drop to be deposited on a solid surface, or detached from a solid surface, is the free sphere [32]. A sessile drop detached from a solid surface while artificially maintaining its original shape is depicted in Fig. 9a. In Fig. 9b we have the sessile drop detached to form the equilibrium (minimum surface area) free sphere. The YDP equation gives the free energy for the process in Fig. 9a. In Fig. 9b, the energy necessary for the process of detachment (which is the negative of the energy of depositing the spherical drop) is given by

$$\Delta F_N = \pi r^2 \gamma_L \left[ (2a / \sin \theta)^{2/3} - a \right] \tag{27}$$

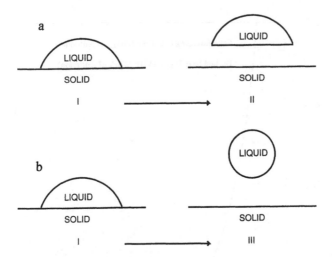

**Figure 9.** Models for the free energy of adhesion: (a) YDP work of adhesion. (b) Net Free Energy of Adhesion. Reprinted with permission from Ref. [32].

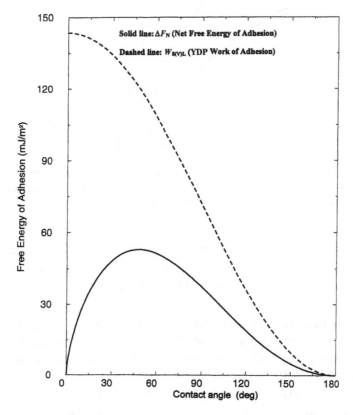

**Figure 10.** Comparison of Net Free Energy of Adhesion with YDP work of adhesion: variable contact angle with fixed solid–liquid interfacial area (unit area). Numbers on ordinate are for water as liquid. Reprinted with permission from Ref. [32].

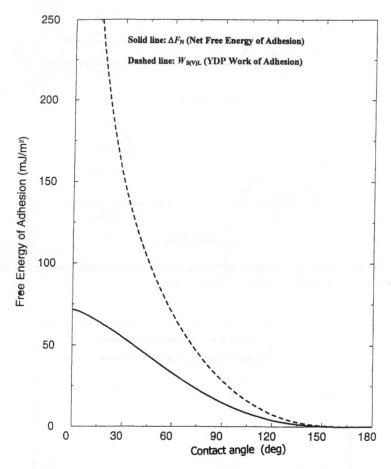

**Figure 11.** Comparison of Net Free Energy of Adhesion with YDP work of adhesion: variable contact angle with fixed liquid volume (the free liquid sphere has unit surface area). Numbers on ordinate are for water as liquid. Reprinted with permission from Ref. [32].

where $\pi r^2$ is the area of the liquid–solid interface and $a$, sometimes called the "effective area" [32, 42, 43], equals $(2/(1 + \cos \theta)) - \cos \theta$. For unit area, $\pi r^2 = 1$. The quantity $\Delta F_N$ is called the Net Free Energy of Adhesion [32].

### 2.5.2. Net Free Energy of Adhesion for a given spherical surface area

The use of spherical, or near spherical, drops of constant size to determine critical surface tensions of solids by means of Zisman plots, or $\gamma_S$ estimates from reciprocal plots, makes it convenient to express the Net Free Energy of Adhesion in terms of the radius of the free sphere. The expression then is

$$\Delta F_{N(r_0)} = 4\pi r_0^2 \gamma_L [1 - (\sin\theta)^{2/3} (a/4)^{1/3}] \tag{28}$$

where $r_0$ is the radius of the free sphere, and $\Delta F_{N(r_0)}$ is $\Delta F_N$ expressed as a function of $r_0$. This equation can then be used to calculate real thermodynamic free energies of adhesion, from the state of unattached spherical drop to attached sessile drop, on any smooth, nondeformable solid surface, from knowledge of the equilibrium contact angle and surface tension of the liquid only.

A description of the variation of Net Free Energy of Adhesion, as compared to YDP work of adhesion, with contact angle, for both constant solid–liquid interfacial area and constant drop volume, is found in Ref. [32]. Figs. 10 and 11 describe the variation.

## 2.6. Nanoscale structure

Theoretical work has been done on the shape of the edge of the sessile drop on a nanoscopic scale as determined from statistical mechanical calculations. For example, de Gennes *et al.* [44] predict a "tongue" of nanoscopic proportion protruding from the liquid drop along the solid surface. Equations for this tongue include $\theta$, the macroscopic equilibrium contact angle, as a factor. The solid surface–vapor interface should, of course, be regarded as outside the tongue [35].

# 3. DISCUSSION

## 3.1. "Comment" of reference 45

Some discussion may be useful on a "Comment" published in 1997 [45]. It was presented by its author as pertaining to papers we published in 1993 (Ref. [46]; in equations (9) and (10) in this reference, $\gamma_S$ under the square root sign should be $\gamma_{SV}$), 1995 [32] and 1996 (Ref. [47]; in this reference $\gamma_e$ should be $\pi_e$). However, the 1995 paper, which concerned the effect of the shape of the free drop on work of adhesion, is not discussed at all in the "Comment", and the 1996 full-length article on the effect of vapor pressure is hardly discussed. The "Comment" focuses almost exclusively on the 1993 introductory "Letter" on vapor-pressure effect.

### 3.1.1. Elimination of linear $\Pi_e$ from equations for cos $\theta$
The main point of both our Letter [46] and follow-up article [47] was that $\Pi_e$, the negative free energy of adsorption (integral of the Gibbs adsorption isotherm) for the vapor of the liquid on the surface of the solid, should not be used as a linear term in the GGF equation [11] nor in any of its predecessor [10] or successor [12, 13] equations in solving for cos $\theta$. At the start it was obvious that there were two essentially equivalent approaches to demonstrate this. One, derive the GG equation in the method of GG, i.e., substitute the GG interaction term into the Bangham-modified Young equation, but be careful to use vapor-saturated solid surfaces in the IMF derivation as well as in the Young equation (ultimately, this method was used in the full-length article, published in 1996). The other is to remove $\Pi_e$ directly from the Young equation at the outset, before introducing the intermolecular force derivation. This can be done since the "adhesion tension" in

the Young–Bangham equation, $\gamma_{SV} - \gamma_{SL}$, is a difference, and $\Pi_e$ can be cancelled out from both terms. This approach was used in the "Letter", with the quantity $\gamma_{SL}$ being given an artificial definition to simplify the algebra (it ultimately cancels out). The point was to simultaneously point out a limitation of the Young equation, while proving our main point about eliminating linear $\Pi_e$. The limitation of the Young equation, of course, is that any quantity algebraically added to both the solid surface free energy and the solid–liquid interfacial free energy will leave the value of the experimentally measurable $\gamma_L \cos\theta$ unchanged.

### 3.1.2. Reactions to the "Letter"
Private communications received regarding the "Letter", however, convinced us that this approach could cause considerable confusion, especially with respect to the artificial definition used for $\gamma_{SL}$. Consequently, in the follow-up article and all other publications on the subject, $\gamma_{SL}$ was used only in a conventional or other natural manner. The quantity $\gamma_{SL}$, therefore, cannot be used interchangeably between our 1993 "Letter" and any of our subsequent publications. Nevertheless, this is done throughout the "Comment" with predictably incorrect results.

A word of caution at this point. Even the conventional, or natural, definition of $\gamma_{SL}$ has a variety of versions, and delineating these versions occupies an important role in our subsequent discussions of vapor effect. They exclude, of course, the definition given in the 1993 "Letter".

### 3.1.3. Is the solid–liquid interface always "the same"?
The first conclusion in the "Comment" emphasizes that the solid–liquid interface "is the same regardless of whether it was formed from the clean or equilibrated solid surface." This misleading statement can be applied to the interfacial region, but not to the choice of interface that we use as the initial or final state of a cycle of free energy changes, e.g., detaching or attaching two different materials. The changes in free energy in the Gibbs interfacial theory depend on the concept of an imaginary two dimensional boundary [16], which is drawn through a composite, or graded, interface, at a location chosen by the investigator. In a simple graded interface, changing the location changes the numbers of the thermodynamic quantities. For example, in an interfacial region between solids, or liquids, A and B, where the composition changes gradually from nearly pure A to nearly pure B, there is an obvious difference between drawing a boundary near A or near B. It is that boundary which defines the interface. For a Dupré adhesion system, with which we are concerned in this work, this is even more clear. The imaginary boundary location determines the composition of the two surfaces on separation. They could both be almost pure A, or both be almost pure B, or somewhere in between. The quantity $\gamma_{AB}$ in a Dupré system is the tension of the interface existing at the chosen two-dimensional boundary.

For the case of an interfacial region where the choice is between an imaginary boundary that separates to a clean surface, or one that separates to a vapor equilibrated surface, the quantity $\gamma_{SL}$ is accordingly the free energy of the interface determined by the chosen imaginary boundary. The fact that it may not be possible to pinpoint the exact location of this boundary is beside the point. The virtual boundary determines the composition of the solid and liquid surfaces on separation.

### 3.1.4. Inconsistency in Young equation?
Another misconception in the "Comment" is that we have said that there "is an inconsistency in the Young equation". This is not so. Our words [46] were "an inconsistency arises in the Young equation......". On reading our page in full, it can be seen that the "inconsistency arises" when the IMF treatment which deals with a clean solid surface is substituted into the Young equation (Bangham-modified) which deals with an adlayer-covered solid surface. If an adlayer-covered solid surface free energy is inserted into the adlayer-covered solid surface free energy of the Young equation, then, of course, no inconsistency arises.

### 3.1.5. Effect of Harkins equation
There was one point of substance in the "Comment". This point had been considered by ourselves, and the complete answer is found in Ref. [35]. The point was, simply put, that the equation due to Harkins allows the conversion of the work of adhesion for a clean solid surface, as estimated, for example, by the IMF method, to the work of adhesion of the vapor equilibrated surface, and *vice versa*, by subtracting or adding $\Pi_e$. The "Comment" comes up with the equation originated by the Harkins treatment (equation (10) in "Comment"), through an invalid approach which treats the solid–liquid interface in isolation. In Ref. [35], this equation is refuted.

### 3.1.6. Correct thermodynamics
The correct thermodynamics for the effect of vapor adsorption on the work of adhesion can be found in detail in Ref. [35], the highlights of which are included in this overview. As we point out, the standard derivation of the "Harkins equation" is not correct, and furthermore this equation cannot be correct regardless of the derivation.

## 3.2. Historical note

When Bangham and Razouk [15] in 1937 first proposed their modification to the Young equation, which adds the role of the vapor of the liquid and its adsorption onto the solid surface, they also pointed out that this phenomenon required changing the meaning of the work of adhesion. They defined the real work of adhesion as that which leaves a solid surface with an adlayer after separation from the liquid. They also spoke of the commonly used definition which leaves a clean solid surface after separation from the liquid. However, despite the fact that they introduced $\Pi_e$ and showed how it can be determined from an adsorption isotherm, they did not propose converting one type of work of adhesion to the other by adding or subtracting $\Pi_e$.

## 4. CONCLUSIONS

In conclusion, we find the following:

1. $\Pi_e$, the decrease in free energy on equilibrating an evacuated solid surface with vapor at saturation, may be positive, negative or zero.

2. $\Pi_e$ does not affect cos $\theta$ directly through the change in free energy of the solid surface.

3. If the vapor adsorption does indeed affect cos $\theta$, it is through any change that might occur in $W$, the work of adhesion.

4. Since the intermolecular force interaction terms which are used as approximations for $W$, such as, for example, the GGF-term $2(\gamma_{SV}^{d}\gamma_{L}^{d})^{1/2}$, are not amenable to adjustment through the use of $\Pi_e$, it is recommended that they be used only when vapor adsorption can be ignored, so that this term becomes $2(\gamma_{S}^{d}\gamma_{L}^{d})^{1/2}$.

5. Although the GGF term has indeed always been used by its authors in the latter formulation, the reasoning was different. Consequently, their treatment provided the option, in principle, of obtaining seemingly more accurate results by determining $\Pi_e$ from an adsorption isotherm and inserting it linearly in their equation for cos $\theta$. In the present treatment there is no such option.

6. Adsorption of vapor of liquid onto an evacuated solid surface does not always raise the contact angle of the liquid, as has been commonly supposed. It may also, in principle, lower the contact angle of a liquid or have no effect on the contact angle.

*Acknowledgements*

The author thanks I. Lapides for help with the graphics.

## REFERENCES

1. M. E. Schrader, *J. Adhesion Sci. Technol.* **6**, 969 (1992).
2. B. V. Derjaguin and Z. M. Zorin, *Proc. 2nd Int. Congr. Surface Activity,* Vol. 2, p. 145. Butterworths, London (1957).
3. A. W. Adamson and I. Ling, in *Contact Angle, Wettability and Adhesion, Adv. Chem. Ser. No. 43,* p. 57, American Chemical Society, Washington, DC (1964).
4. A. W. Adamson, *J. Colloid Interface Sci.* **27**, 180 (1968).
5. P. Hu and A. W. Adamson, *J. Colloid Interface Sci.* **59**, 605 (1977).
6. M. E. Schrader and G. H. Weiss, *J. Phys. Chem.* **91**, 353 (1987).
7. P. C. Hiemenz, *Principles of Colloid and Surface Chemistry,* 2nd edition, p. 310, Marcel Dekker, New York, NY (1986).
8. W. A. Zisman, in *Contact Angle, Wettability and Adhesion, Adv. Chem. Ser. No. 43,* p. 1, American Chemical Society, Washington, DC (1964).
9. E. F. Hare and W. A. Zisman, *J. Phys. Chem.* **59**, 335 (1955).
10. R. J. Good and L. A. Girifalco, *J. Phys. Chem.* **64**, 561 (1960).
11. F. M. Fowkes, *Ind. Eng. Chem.* **56** (12), 40 (1964).

12. D. K. Owens and R. D. Wendt, *J. Appl. Polym. Sci.* **13**, 741 (1969).
13. S. Wu, *Polymer Interface and Adhesion*, Marcel Dekker, New York, NY (1982).
14. T. Young, *Philos. Trans. R. Soc. London* **95**, 65 (1805).
15. D. H. Bangham and R. I. Razouk, *Trans. Faraday Soc.* **33**, 1459 (1937).
16. J. W. Gibbs, *Collected Works* **1**, 315, Yale University Press, New Haven, CT (1928).
17. S. Glasstone, *Thermodynamics for Chemists*, p. 245, Van Nostrand, New York, NY (1947).
18. D. H. Bangham, *Trans. Faraday Soc.* **33**, 805 (1937).
19. E. A. Guggenheim and N. K. Adam, *Proc. Roy. Soc. (London)* **A 139**, 218 (1933).
20. G. Aranovich and M. Donahue, *J. Colloid Interface Sci.* **200**, 273-290 (1998).
21. G. H. Findenegg and M. Thommes, in *Physical Adsorption: Experiment, Theory and Applications*, J. Fraissard (Ed.), p. 151, Kluwer Academic, Dordrecht The Netherlands (1997).
22. W. A. Steele, in *The Solid-Gas Interface*, E. A. Flood (Ed.), Vol. 1, Chapter 10, Marcel Dekker, New York, NY (1966).
23. M. E. Schrader, *J. Phys. Chem. B* **104**, 731 (2000).
24. G. H. Findenegg, in *Theoretical Advancement in Chromatography and Related Separation Techniques*, F. Dondi and G. Guiochon (Eds.), p. 227, Kluwer, Dordrecht, The Netherlands (1992).
25. F. del Rio and A. G. Villegas, *J. Phys. Chem.* **95**, 787 (1991).
26. G. L. Aranovich and M. D. Donahue, *J. Colloid Interface Sci.* **180**, 537-541 (1996).
27. J. Specovius and G. H. Findenegg, *Ber. Bunsen-Ges. Phys. Chem.* **84**, 690 (1980).
28. D. Henderson and I. K. Snook, *J. Phys. Chem.* **87**, 2956 (1983).
29. R. J. Good, *J. Colloid Interface Sci.* **52**, 309 (1975).
30. A. Dupré, *Theorie Mechanique de la Chaleur*, pp. 369ff., Gauthier-Villars, Paris (1869).
31. N. K. Adam, *The Physics and Chemistry of Surfaces*, 3rd ed., p. 8, Oxford Press, London (1941).
32. M. E. Schrader, *Langmuir* **11**, 3585 (1995).
33. A. Pockels, *Phys. Z.* **15**, 39 (1914).
34. N. K. Adam, in *Contact Angle, Wettability and Adhesion, Adv. Chem. Ser. No. 43*, p. 52, American Chemical Society, Washington, DC (1964).
35. M. E. Schrader, *J. Colloid Interface Sci.* **213**, 602 (1999).
36. W. D. Harkins and H. K. Livingston, *J. Chem. Phys.* **10**, 342 (1942).
37. W. D. Harkins, *The Physical Chemistry of Surface Films*, p. 281. Reinhold, New York, NY (1952).
38. R. J. Good and C. J. van Oss, in *Modern Approaches to Wettability: Theory and Applications*, M. E. Schrader and G. I. Loeb (Eds.), p. 22. Plenum Press, New York, NY (1992).
39. M. E. Schrader and I. Lerner, *Nature* **204**, 759 (1964).
40. M. E. Schrader, *J. Phys. Chem.* **84**, 2774 (1980).
41. M. E. Schrader, in *Modern Approaches to Wettability: Theory and Applications*, M. E. Schrader and G. I. Loeb (Eds.), p. 67. Plenum Press, New York, NY (1992).
42. E. C. Sewell and E. W. Watson, *Bull. RILEM (Reunion Int. Lab. Essais`Rech. Mater. Constr.)* **29**, 125 (1965).
43. D. H. Everett and J. M. Haynes, *J. Colloid Interface Sci.* **38**, 125 (1972).
44. P. G. de Gennes, X. Hua and P. Levinson, *J. Fluid Mech.* **212**, 55 (1990).
45. J. J. Moura Ramos, *Langmuir* **13**, 6607 (1997).
46. M. E. Schrader, *Langmuir* **9**, 1959 (1993).
47. M. E. Schrader, *Langmuir* **12**, 3728 (1996).

# Part 2

# Contact Angle Measurements / Determination and Solid Surface Free Energy

*Contact Angle, Wettability and Adhesion*, Vol. 3, pp. 95–116
Ed. K.L. Mittal
© VSP 2003

# The molecular origin of contact angles in terms of different combining rules for intermolecular potentials

JUNFENG ZHANG and DANIEL Y. KWOK *

*Nanoscale Technology and Engineering Laboratory, Department of Mechanical Engineering, University of Alberta, Edmonton, Alberta T6G 2G8, Canada*

**Abstract**—We have formulated a combining rule for solid–liquid adhesion and intermolecular potentials using macroscopic adhesion data. The combining rule is applied successfully to determine macroscopic solid–liquid adhesion and to calculate adhesion patterns using molecular theory. We found that the results determined from our combining rule are better than those by the 9:3, Steele's and 12:6 combining rules in terms of scatters and details. Our results suggest that macroscopic data from careful contact angle and adhesion measurements can be used to infer relationships for unlike solid–fluid interactions at a molecular level.

*Keywords*: Contact angle; solid surface tension; adhesion; Lennard–Jones potential; intermolecular potential; combining rule.

## 1. INTRODUCTION

Knowledge of interfacial free energy is necessary for a better understanding and modeling of interfacial processes such as wetting, spreading and floatation. However, direct measurement of the solid–vapor ($\gamma_{sv}$) and solid–liquid ($\gamma_{sl}$) interfacial tensions is not available. Among the different indirect approaches in determining solid surface tensions, contact angle is believed to be the simplest and hence widely used approach [1, 2].

The possibility of estimating solid surface tensions from contact angles relies on a relation known as Young's equation [3]

$$\gamma_{lv} \cos \theta_Y = \gamma_{sv} - \gamma_{sl}, \tag{1}$$

where $\gamma_{lv}$ is the liquid–vapor surface tension and $\theta_Y$ is the Young contact angle, i.e., a contact angle that can be inserted into Young's equation. Within the context of this work, we assume the experimental contact angles $\theta$ to be the Young contact angle $\theta_Y$. Since Young's equation (equation (1)) contains only two measurable

---

*To whom all correspondence should be addressed: Phone: (1-780) 492-2791, Fax: (1-780) 492-2200, E-mail: daniel.y.kwok@ualberta.ca

quantities, $\gamma_{lv}$ and $\theta$, an additional expression relating $\gamma_{sv}$ and $\gamma_{sl}$ must be sought. Such an equation can be formulated using experimental adhesion or contact-angle data. Historically, the interpretation of contact angles in terms of solid surface energetics started with the pioneering work of Zisman and co-workers [4]; the key observation they made was that for a given solid the measured contact angles did not vary randomly as the liquid was varied; rather, $\cos \theta$ changed smoothly with the liquid surface tension $\gamma_{lv}$ within a band in a fashion that may suggest a straight-line relationship.

The origin of surface tensions arises from the existence of unbalanced intermolecular forces among molecules at the interface. Recently, starting from macroscopic experimental adhesion patterns, Kwok and Neumann [5] proposed a fruitful procedure in formulating a new combining rule for intermolecular potentials that is meant to better reflect solid–liquid interactions. On the other hand, using a generalized van der Waals model and a mean-field method, van Giessen *et al.* [6] presented a calculation of surface tensions from intermolecular potentials, which is, in fact, similar to that of Sullivan [7]. Their calculated surface tensions of selected liquids were in reasonably good agreement with the measured data. However, the behavior of contact angles (the curve of $\cos \theta$ *versus* $\gamma_{lv}$) deviates considerably from the experimental trend. We noticed that no combining rule was involved in the calculations of the liquid–vapor surface tension ($\gamma_{lv}$); such a relation, however, was required for the calculations of $\gamma_{sv}$ and $\gamma_{sl}$ and hence $\cos \theta$, in order to evaluate the solid–fluid intermolecular potential strength. Therefore, it is reasonable to attribute the discrepancy between the calculated and experimental $\cos \theta$ to the choice of the combining rule involved. By employing a better combining rule that represents more accurately the intermolecular attractions, we speculate that the calculation results of contact angles should be improved accordingly [8, 9].

We propose here a new formulation of solid–liquid combining rule using macroscopic adhesion data based on the original idea by Kwok and Neumann [5]. We examine the application of the derived combining rule in molecular theory by calculating the solid–liquid adhesion patterns using a van der Waals model with a mean field approximation, similar to those used in Refs. [6, 9]. The calculated adhesion and contact angle patterns by the new combining rule are compared against those from the 9:3, Steele's and 12:6 combining rules. It will be shown that the newly formulated combining rule presented here follows the experimental patterns very closely.

## 2. THEORY

### 2.1. Combining rules

In the theory of molecular interactions and the theory of mixtures, combining rules are used to evaluate the parameters for unlike-pair interactions in terms of those for the like interactions [7, 10–17]. As with many other combining rules, the Berthelot rule [18]

$$\epsilon_{ij} = \sqrt{\epsilon_{ii}\epsilon_{jj}} \tag{2}$$

is a useful approximation, but does not provide a secure basis for the understanding of unlike-pair interactions; $\epsilon_{ij}$ is the potential energy parameter (well depth) of unlike-pair interactions; $\epsilon_{ii}$ and $\epsilon_{jj}$ are for like-pair interactions.

From the London theory of dispersion forces, the attraction potential $\phi_{ij}$ between a pair of unlike molecules i and j is given by

$$\phi_{ij} = -\frac{3}{2}\frac{I_i I_j}{I_i + I_j}\frac{\alpha_i \alpha_j}{r_{ij}^6}, \tag{3}$$

where $I$ is the ionization potential, $r$ is the distance between the pair of unlike molecules and $\alpha$ the polarizability. For like molecules this term becomes

$$\phi_i = -\frac{3}{4}\frac{I_i \alpha_i^2}{r_i^6}. \tag{4}$$

The total intermolecular potential $V(r_i)$ expressed by the (12:6) Lennard–Jones potential is in the form

$$V(r_i) = 4\epsilon_{ii}\left[(\sigma_i/r_i)^{12} - (\sigma_i/r_i)^6\right], \tag{5}$$

where $\sigma$ is the collision diameter. The attractive potentials in equations (4) and (5) can be equated to give

$$\frac{3}{4}I_i \alpha_i^2 = 4\epsilon_{ii}\sigma_i^6. \tag{6}$$

Equation (6) can be used to derive $\alpha_i$ and $\alpha_j$; substituting these quantities into equation (3) yields

$$\phi_{ij} = -\frac{2\sqrt{I_i I_j}}{I_i + I_j}\frac{4\sigma_i^3 \sigma_j^3}{r_{ij}^6}\sqrt{\epsilon_{ii}\epsilon_{jj}}. \tag{7}$$

If we write $\phi_{ij}$ in the form $-4\epsilon_{ij}\sigma_{ij}^6/r_{ij}^6$ such that $\sigma_{ij} = (\sigma_i + \sigma_j)/2$, the energy parameter for two unlike molecules can be expressed as

$$\epsilon_{ij} = \frac{2\sqrt{I_i I_j}}{I_i + I_j}\left[\frac{4\sigma_i/\sigma_j}{(1 + \sigma_i/\sigma_j)^2}\right]^3 \sqrt{\epsilon_{ii}\epsilon_{jj}}. \tag{8}$$

This forms the basis of the so-called combining rules for intermolecular potential. The above expression for $\epsilon_{ij}$ can be simplified: when $I_i = I_j$, the first term of equation (8) becomes unity; when $\sigma_i = \sigma_j$ the second term becomes unity. When both conditions are met, we obtain the well-known Berthelot rule, i.e., equation (2).

For the interactions between two very dissimilar types of molecules or materials where there is an apparent difference between $\epsilon_{ii}$ and $\epsilon_{jj}$, it is clear that the Berthelot rule cannot describe the behavior adequately. It has been demonstrated [19–21] that the Berthelot geometric mean combining rule generally overestimates the strength of the unlike-pair interactions, i.e., the geometric mean value is too large an estimate. In general, the differences in the ionization potentials are not large, i.e., $I_i \approx I_j$; thus, the most serious error comes from the difference in the collision diameters $\sigma$ for unlike molecular interactions.

For solid–liquid systems, in general, the minimum of the solid–liquid interaction potential $\epsilon_{sl}$ is often expressed in the following manner [7, 11, 12]

$$\epsilon_{sl} = g(\sigma_l/\sigma_s)\sqrt{\epsilon_{ss}\epsilon_{ll}}, \tag{9}$$

where $g(\sigma_l/\sigma_s)$ is a function of $\sigma_l$ and $\sigma_s$ which are, respectively, the collision diameters for the liquid and solid molecules; $\epsilon_{ss}$ and $\epsilon_{ll}$ are, respectively, the minima in the solid–solid and liquid–liquid potentials. Several other forms for the explicit function $g(\sigma_l/\sigma_s)$ have been suggested. For example, by comparing $\epsilon_{sl}$ with the minimum in the (9:3) Lennard–Jones potential, one obtains an explicit function as

$$g(\sigma_l/\sigma_s) = \frac{1}{8}\left(1 + \frac{\sigma_l}{\sigma_s}\right)^3 \tag{10}$$

and the (9:3) combining rule becomes

$$\epsilon_{sl} = \frac{1}{8}\left(1 + \frac{\sigma_l}{\sigma_s}\right)^3 \sqrt{\epsilon_{ss}\epsilon_{ll}}. \tag{11}$$

An alternative function in the form of

$$g(\sigma_l/\sigma_s) = \frac{1}{4}\left(1 + \frac{\sigma_l}{\sigma_s}\right)^2 \tag{12}$$

has been investigated by Steele [22] and others [23], suggesting a different combining rule as

$$\epsilon_{sl} = \frac{1}{4}\left(1 + \frac{\sigma_l}{\sigma_s}\right)^2 \sqrt{\epsilon_{ss}\epsilon_{ll}}. \tag{13}$$

For comparison purpose, we label equation (13) as the Steele combining rule in this paper. Further, from the (12:6) Lennard–Jones potential, equation (8) implies an explicit function

$$g(\sigma_l/\sigma_s) = \left[\frac{4\sigma_l/\sigma_s}{(1 + \sigma_l/\sigma_s)^2}\right]^3, \tag{14}$$

resulting in a (12:6) combining rule as

$$\epsilon_{sl} = \left[\frac{4\sigma_l/\sigma_s}{(1 + \sigma_l/\sigma_s)^2}\right]^3 \sqrt{\epsilon_{ss}\epsilon_{ll}}. \tag{15}$$

These have been numerous attempts for a better representation of solid–liquid interactions from solid–solid and liquid–liquid interactions. In general, these functions are normalized such that $g(\sigma_l/\sigma_s) = 1$ when $\sigma_l = \sigma_s$; they revert to the Berthelot geometric mean combining rule equation (2) when $g(\sigma_l/\sigma_s) = 1$. Nevertheless, adequate representation of unlike solid–liquid interactions from like pairs is rare and their validity for solid–liquid systems lacks experimental support.

## 2.2. Calculation of interfacial tensions and contact angles

We employ a mean-field approximation here to calculate numerically the three interfacial tensions from molecular interactions. In our simple van der Waals model, the fluid molecules are idealized as hard spheres interacting with each other through a potential $\phi_{ff}(r)$, where $r$ is a distance between two interacting molecules. A Carnahan–Starling model [6, 13, 24] is adopted as the hard-sphere reference system. For a planar interface formed by a liquid and its vapor, each of which occupies a semi-infinite space, $z > 0$ and $z < 0$ respectively; and the surface tension is given by [6, 7]

$$\gamma_{lv} = \min_{\rho} \int_{-\infty}^{+\infty} dz \left\{ F[\rho(z)] + \frac{1}{2}\rho(z) \int_{-\infty}^{+\infty} dz' \bar{\phi}_{ff}(z' - z)[\rho(z') - \rho(z)] \right\}. \quad (16)$$

Here the minimum is taken over all possible density profiles $\rho(z)$ and $F$ is the excess free energy; $\bar{\phi}_{ff}$ represents the interaction potential that has been integrated over the whole $x'y'$ plane. For the solid–fluid (i.e., a solid–liquid or a solid–vapor) interface, the solid is modeled as a semi-infinite impenetrable wall occupying the domain of $z < 0$ and exerting an attraction potential $V(z)$ to the fluid molecule at a distance $z$ from the solid surface. The interfacial tension of such an interface can be obtained from

$$\gamma_{sf} = \gamma_s + \min_{\rho} \int_0^{+\infty} dz \left\{ F[\rho(z)] + \rho(z)V(z) \right.$$
$$+ \frac{1}{2}\rho(z) \int_0^{+\infty} dz' \bar{\phi}_{ff}(z' - z)[\rho(z') - \rho(z)]$$
$$\left. - \frac{1}{2}\rho^2(z) \int_{-\infty}^0 dz' \bar{\phi}_{ff}(z' - z) \right\}, \quad (17)$$

where $\gamma_s$ is the solid–vacuum surface tension, a constant that exists in the calculations of $\gamma_{sv}$ and $\gamma_{sl}$. This constant ($\gamma_s$) will be canceled out in the calculations of the contact angles *via* Young's equation, equation (1); it has no impact on the implication of our results since we are interested only in the difference between $\gamma_{sv}$ and $\gamma_{sl}$. Considering a solid with molecules interacting with fluid molecules through a potential $\phi_{sf}(r)$, we easily obtain the intermolecular potential $V(z)$ by integrating $\phi_{sf}(r)$ over the solid domain. We wish to point out that the above equations for $\gamma_{lv}$ and $\gamma_{sf}$ (equations (16) and (17)) are identical to those reported in Ref. [7], although the forms of equations are different.

We believe that the integral terms

$$-\frac{1}{2}\rho^2(z)\int_{-\infty}^{0} dz' \bar{\bar{\phi}}_{ff}(z'-z)$$

(18)

in equation (9) and

$$-\rho(z)\int_{-\infty}^{0} dz' \bar{\bar{\phi}}_{ff}(z'-z)$$

(19)

in equation (10) in Ref. [6] are missing. Without such integral terms, for the case of a fluid against a wall with $V(z) = 0$, $\rho(z) = \rho_f$ (the density of fluid bulk phase) would be the solution to the Euler–Lagrange equation, equation (19). Thus, we would not be finding the "drying" layer of vapor between the bulk liquid (at $z = \infty$) and the wall. In the expression of the potential exerted by the solid on the fluid $V(z)$ (equation (11) in Ref. [6]), the upper integral limit should be $-d_{sf} = -(d_s + d_f)/2$ [25]. In the expressions of all combining rules, i.e., equations (12)–(14) in Ref. [6] and throughout that paper, all of the terms $d_s/d_f$ should be corrected as $d_f/d_s$ [7]. We have employed $\sigma_f/\sigma_s$ here, rather than $d_f/d_s$.

To carry out the calculations of interfacial tensions and hence the contact angles, a given interaction potential is required. Here we assume a (12:6) Lennard–Jones potential model and consider only the attraction part. It should be pointed out that the Lennard–Jones potential function requires knowledge of two parameters: the potential strength $\epsilon$ and the collision diameter $\sigma$. The potential strength $\epsilon_{sf}$ for $\phi_{sf}(r)$ is obtained from the fluid $\epsilon_{ff}$ and solid $\epsilon_{ss}$ potential strengths via a combining rule, such as that given by equation (13).

As mentioned above, the calculation of liquid surface tension $\gamma_{lv}$ requires two parameters $\epsilon_{ff}$ and $\sigma_f$ which can be related to the critical temperature $T_c$ and pressure $P_c$ of the liquid in the following expressions for the Carnahan–Starling model [6, 13]

$$kT_c = 0.18016\alpha/\sigma_f^3,$$
$$P_c = 0.01611\alpha/\sigma_f^6,$$

(20)

where $k$ is the Boltzmann constant and $\alpha$ is the van der Waals parameter given by

$$\alpha = -\frac{1}{2}\int \phi_{ff}(r)dr.$$

(21)

The densities of the liquid $\rho_l$ and vapor $\rho_v$ were obtained by requiring the liquid and vapor to coexist at a given temperature $T$ [6, 13]. In our calculations, we have selected thirty liquids of different molecular structures and have assumed $T = 21°C$, $\sigma_s = 10$ Å and $\rho_s = 10^{27}$ molecules/m$^3$ for the solid surface.

In our theoretical model, a fluid (liquid or its vapor) is modeled by the potential parameter $\epsilon_{ll}$ and collision diameter $\sigma_l$, with the liquid density $\rho_l$ and vapor density $\rho_v$ found by requiring the liquid and vapor to coexist at a given temperature $T$. A solid is modeled by its potential parameter $\epsilon_{ss}$, density $\rho_s$ and collision diameter $\sigma_s$. The interfacial tension was obtained by considering the excess free energy and interactions of every pair of molecules in both bulk phases forming the interface with a mean-field approximation. For example, in calculation of the liquid–vapor interfacial tension at a given temperature, only $\epsilon_{ll}$ and $\sigma_l$ are needed. But for the solid–vapor and solid–liquid interfacial tension ($\gamma_{sv}$ and $\gamma_{sl}$), in addition to the parameters of fluid and solid, a combining rule $g(\sigma_l/\sigma_s)$ is also involved to determine the potential parameter $\epsilon_{sl}$ between the solid and fluid molecules from $\epsilon_{ll}$ and $\epsilon_{ss}$ by relation equation (9). Following these procedures, the three interfacial tensions $\gamma_{lv}$, $\gamma_{sv}$ and $\gamma_{sl}$ at a given temperature $T$ can then be expressed as

$$\gamma_{lv} = \gamma_{lv}(\epsilon_{ll}, \sigma_l, T)$$
$$\gamma_{sv} = \gamma_{sv}(\epsilon_{ll}, \sigma_l, \epsilon_{ss}, \rho_s, \sigma_s, g, T)$$
$$\gamma_{sl} = \gamma_{sl}(\epsilon_{ll}, \sigma_l, \epsilon_{ss}, \rho_s, \sigma_s, g, T). \tag{22}$$

According to the model, for a given solid surface at a given temperature with a selected combining rule and different fluids, $\epsilon_{ss}$, $\rho_s$, $\sigma_s$, $T$ and the function $g(\sigma_l/\sigma_s)$ are completely defined and fixed, so

$$\gamma_{lv} = \gamma_{lv}(\epsilon_{ll}, \sigma_l) \tag{23}$$
$$\gamma_{sv} = \gamma_{sv}(\epsilon_{ll}, \sigma_l) \tag{24}$$
$$\gamma_{sl} = \gamma_{sl}(\epsilon_{ll}, \sigma_l). \tag{25}$$

A first glance at the above equations might appear to be peculiar because there seems to be no relationship with solid properties and only the liquid properties $\epsilon_{ll}$ and $\sigma_l$ are involved. We shall point out that, since the solid properties have already been used to establish these functions, equations (23)–(25) indeed contain implicitly (not explicitly) the information on the solid properties. Thus, there is no guarantee that the calculated curves will always be smooth; the scatters can easily arise from the different choices of the combining rules $g(\sigma_l/\sigma_s)$ that might not truly reflect the specific solid–liquid interactions, contrary to the conclusions drawn in Ref. [6].

## 3. RESULTS AND DISCUSSION

### 3.1. Liquid–vapor surface tensions

We show in Table 1 the calculated liquid-surface tensions for the thirty liquids selected. In most cases, the difference between the calculated and experimental liquid–vapor surface tensions are less than 10–20%. The largest discrepancy is for water with a calculated $\gamma_{lv}$ value of 93 mJ/m$^2$ instead of an experimental value of 72.8 mJ/m$^2$. Considering the fact that we have only used a simple van der Waals model, the slightly larger deviations for the polar liquids are indeed expected as the Lennard–Jones potential does not reflect the complicated interactions of, e.g., water. It should be noted that the calculation of liquid–vapor surface tension does not rely on any form of a combining rule, except the van der Waals model used here.

**Table 1.**
Comparison between the calculated $\gamma_{lv}^{cal}$ and experimental $\gamma_{lv}^{exp}$ (both in mJ/m$^2$) liquid–vapor surface tensions

| Liquid | $\gamma_{lv}^{exp}$ | $\gamma_{lv}^{cal}$ |
|---|---|---|
| CH$_3$Cl | 16.20[a] | 13.72 |
| Pentane | 16.65 | 13.07 |
| Hexane | 18.13 | 15.00 |
| Methylamine | 19.89[a] | 16.98 |
| Methanol | 22.30 | 29.28 |
| Decane | 23.43 | 17.83 |
| Ethyl acetate | 23.97[a] | 18.85 |
| Acetone | 24.02[a] | 20.14 |
| Ethyl methyl ketone | 24.52[a] | 21.29 |
| Methyl acetate | 25.10 | 19.98 |
| Dodecane | 25.44 | 18.08 |
| Tetradecane | 26.55 | 16.82 |
| CCl$_4$ | 27.04[a] | 24.29 |
| Fluorobenzene | 27.26[a] | 24.59 |
| CHCl$_3$ | 27.32[a] | 25.01 |
| Hexadecane | 27.76 | 17.68 |
| CH$_2$Cl$_2$ | 27.84[a] | 24.69 |
| Benzene | 28.88[a] | 26.04 |
| trans-decalin | 29.50 | 26.15 |
| cis-decalin | 31.65 | 26.77 |
| CS$_2$ | 32.32[a] | 33.44 |
| Chlorobenzene | 33.59[a] | 30.95 |
| Bromobenzene | 35.82[a] | 34.15 |
| Iodobenzene | 39.27[a] | 38.38 |
| Aniline | 42.67[a] | 38.55 |
| Diethylene glycol | 45.04 | 36.01 |
| Ethylene glycol | 47.99 | 45.72 |
| Glycerol | 63.11 | 50.32 |
| Hydrazine | 67.60[a] | 71.81 |
| Water | 72.75 | 92.92 |

Experimental values were obtained from Ref. [1] measured at 21°C.
[a] From Ref. [26] measured at 20°C.

## 3.2. Solid–liquid adhesion patterns

Kwok and co-workers [1, 5, 27] have recently published experimental adhesion and contact-angle patterns for a large number of polar and non-polar liquids on a variety of carefully-prepared low-energy solid surfaces, including fluorocarbon FC722, hexatriacontane, cholesteryl acetate, poly(n-butyl methacrylate) (PnBMA), poly(methyl methacrylate/n-butyl methacrylate) and poly(methyl methacrylate) (PMMA). These data were obtained by the low-rate dynamic contact-angle procedures using techniques called axisymmetric drop shape analysis and capillary rise at a vertical plate. Since such angles are believed to satisfy the commonly accepted

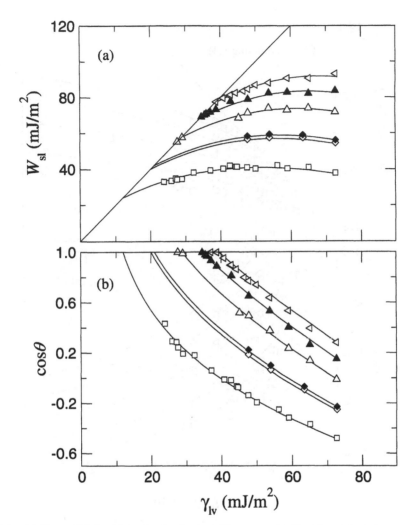

**Figure 1.** (a) The solid–liquid work of adhesion $W_{sl}$ *versus* the liquid–vapor surface tension $\gamma_{lv}$ and (b) cosine of the contact angle $\cos\theta$ *versus* the liquid–vapor surface tension $\gamma_{lv}$ for fluorocarbon FC722 ($\square$), hexatriacontane ($\diamond$), cholesteryl acetate (solid diamond), poly($n$-butyl methacrylate) ($\triangle$), poly(methyl methacrylate/$n$-butyl methacrylate) ($\blacktriangle$) and poly(methyl methacrylate) ($\triangleleft$) surfaces.

assumptions for contact angle interpretations, we reproduce these results here in Fig. 1 and compare them with our patterns calculated from intermolecular potentials. Figure 1a illustrates that, for a given solid surface, say the FC722 surface, the experimental solid–liquid work of adhesion $W_{sl}$ increases as $\gamma_{lv}$ increases up to a maximum $W_{sl}$ value identified as $W_{sl}^*$. Further increase in $\gamma_{lv}$ causes $W_{sl}$ to decrease from $W_{sl}^*$. The trend described here appears to shift systematically to the upper right for a more hydrophilic surface (such as PMMA) and to the lower left for a relatively more hydrophobic surface. There are also some indications that the location of the maximum point $W_{sl}^*$ appears to shift to the right as surface hydrophobicity decreases.

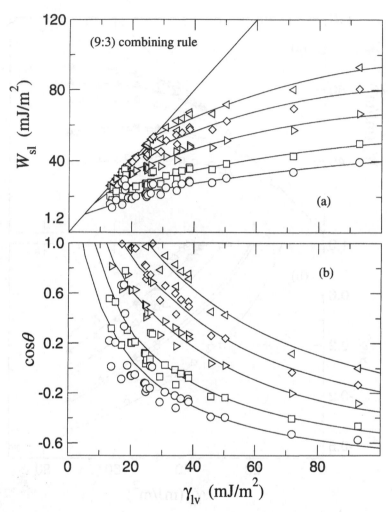

**Figure 2.** (a) The solid–liquid work of adhesion $W_{sl}$ *versus* the liquid–vapor surface tension $\gamma_{lv}$ and (b) cosine of the contact angle $\cos\theta$ *versus* the liquid–vapor surface tension $\gamma_{lv}$ calculated from the (9:3) combining rule, equation (11). The symbols are calculated data and the curves are the general trends of the data points. Each set of symbols represents a given solid surface.

Figure 1b shows the experimental contact angle patterns in $\cos\theta$ *versus* $\gamma_{lv}$. We see that, for a given solid surface, as $\gamma_{lv}$ decreases, cosine of the contact angle ($\cos\theta$) increases, intercepting at $\cos\theta = 1$ with a "limiting" $\gamma_{lv}$ value. We identify this "limiting" value as $\gamma_{lv}^c$. As $\gamma_{lv}$ decreases beyond this $\gamma_{lv}^c$ value, contact angles become more or less zero ($\cos\theta \approx 1$), representing the case of complete wetting. The trend described here appears to change systematically to the right for a more hydrophilic surface (such as PMMA) and to the left for a relatively more hydrophobic surface (such as fluorocarbon). Changing the solid surfaces in this manner change the limiting $\gamma_{lv}^c$ value, suggesting that $\gamma_{lv}^c$ might be of indicative value as a solid property.

In fact, Zisman labeled this $\gamma_{lv}^c$ value as the critical surface tension of the solid surface $\gamma_c$. It is also clear in Fig. 1b that the experimental contact angle patterns are different from what Zisman had anticipated [4].

### 3.2.1. (9:3) combining rule

The calculated adhesion and contact angle patterns for the (9:3) combining rule are shown in Fig. 2. In Fig. 2a, the (9:3) combining rule reasonably predicts the general trend of the adhesion patterns: Increasing $\gamma_{lv}$ increases $W_{sl}$ monotonically. It is not apparent, however, that there exists a maximum $W_{sl}^*$ beyond which $W_{sl}$ will

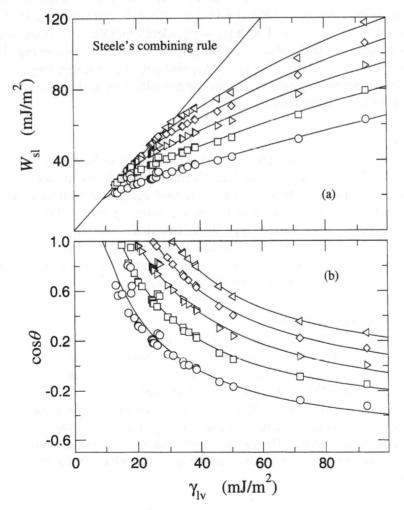

**Figure 3.** (a) The solid–liquid work of adhesion $W_{sl}$ *versus* the liquid–vapor surface tension $\gamma_{lv}$ and (b) cosine of the contact angle $\cos\theta$ *versus* the liquid–vapor surface tension $\gamma_{lv}$ calculated from the Steele combining rule, equation (13). The symbols are calculated data and the curves are the general trends of the data points. Each set of symbols represents a given solid surface.

decrease when $\gamma_{lv}$ increases, as in Fig. 1. The general adhesion pattern predicted from the (9:3) combining rule is that $W_{sl}$ increases with $\gamma_{lv}$. In order to change the hydrophilicity of the model surface and observe the change in patterns, we hypothesize that solid surface energy increases with stronger solid–solid interaction energy $\epsilon_{ss}$: increasing the solid–solid interactions increases the surface free energy required to generate a unit interfacial area. Thus, we increased the solid–solid interactions systematically to model hydrophilic surfaces and decrease the interactions for hydrophobic ones. Increasing the interactions and, hence the surface hydrophilicity, shifts the curves in Fig. 2a to the upper right; the trend is in good agreement with those from Fig. 1a. Figure 2b illustrates the calculated contact angle patterns: decreasing $\gamma_{lv}$ increases $\cos\theta$ for a given solid surface. Further decrease in $\gamma_{lv}$ causes $\cos\theta$ to intercept at $\cos\theta = 1$ with $\gamma_{lv} = \gamma_{lv}^c$, identifying the case of complete wetting. Increasing the hydrophilicity of the surface shifts the curves in Fig. 2b to the right, similar to those shown in Fig. 1b. We also note the relatively larger scatters in Fig. 2; it will become apparent later that the magnitude of such scatters depends on the the the choice of the combining rule.

### 3.2.2. Steele's combining rule

The adhesion and contact angle patterns calculated from Steele's combining rule are shown in Fig. 3. It appears that Steele's combining rule also predicts the general adhesion and contact angle patterns well, in good agreement with the experimental results shown in Fig. 1. We note that this combining rule yields results which have significantly less scatters than those calculated from the (9:3) combining rule (in Fig. 2). However, the calculated results for $W_{sl}$ in Fig. 3a still increase monotonically with increasing $\gamma_{lv}$ and did not result in a maximum $W_{sl}^*$ value that was observed in the experimental results in Fig. 1.

### 3.2.3. (12:6) combining rule

The calculated results using the (12:6) combining rule are shown in Fig. 4. In Fig. 4a, we see that the (12:6) combining rule appears to predict correctly the existence of a maximum $W_{sl}^*$ value as $\gamma_{lv}$ increases. This maximum value was not observed using the (9:3) and Steele's combining rules in Figs 2 and 3, respectively. The calculated contact angle patterns in Fig. 4b are also similar to those from the (9:3) and Steele's combining rules and in good agreement with the general patterns observed experimentally. The (12:6) combining rule, however, suffers from the same shortcoming as the (9:3) combining rule in that relatively larger scatters were apparent.

### 3.3. Reformulation of a new combining rule

With careful examination of the modification procedures in Ref. [5], we found that the assumption of $\gamma \propto \sigma^{-3}$ can be better represented by a more reasonable one: $\gamma \propto \sigma^{-2}$ [7]. Since surface tension is defined as the work required to generate

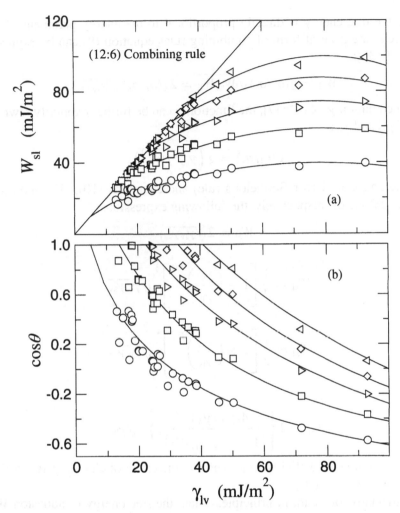

**Figure 4.** (a) The solid–liquid work of adhesion $W_{sl}$ *versus* the liquid–vapor surface tension $\gamma_{lv}$ and (b) cosine of the contact angle $\cos\theta$ *versus* the liquid–vapor surface tension $\gamma_{lv}$ calculated from the (12:6) combining rule, equation (15). The symbols are calculated data and the curves are the general trends of the data points. Each set of symbols represents a given solid surface.

a unit surface area, the relationship $\gamma \propto \sigma^{-2}$ is entirely compatible with this stipulation as $\sigma$ can be related directly to the changes in interfacial area. Therefore, we reformulate a new combining rule here and fit the expression to the experimental adhesion patterns.

According to the thermodynamic definition of the energy of adhesion $W_{sl}$, and cohesion $W_{ss}$ and $W_{ll}$ [28, 29], we have the following relations:

$$W_{sl} = \gamma_{lv} + \gamma_{sv} - \gamma_{sl}, \tag{26}$$

$$W_{ss} = 2\gamma_{sv}, \quad W_{ll} = 2\gamma_{lv}. \tag{27}$$

Because the free energy is directly proportional to the energy parameter [21, 29], i.e., $W \propto \epsilon$, the general form of combining rules equation (9) can be expressed as [18, 21, 29, 30]:

$$W_{sl} = g(\sigma_l/\sigma_s)\sqrt{W_{ss}W_{ll}} = 2g(\sigma_l/\sigma_s)\sqrt{\gamma_{sv}\gamma_{lv}}. \tag{28}$$

Due to the relation $\gamma \propto \sigma^{-2}$ [7], the function $g$ can be further explicitly rewritten in terms of $\gamma_{sv}$ and $\gamma_{lv}$:

$$g(\sigma_l/\sigma_s) = g\left(\gamma_{sv}^{1/2}/\gamma_{lv}^{1/2}\right). \tag{29}$$

Thus inserting $g = 1$ (for Berthelot's rule) and equations (10), (12) and (14) into equation (28) yields, respectively, the following expressions:

$$W_{sl} = 2\sqrt{\gamma_{lv}\gamma_{sv}}; \tag{30}$$

$$W_{sl} = \frac{1}{4}\left[1 + \left(\frac{\gamma_{sv}}{\gamma_{lv}}\right)^{1/2}\right]^3 \sqrt{\gamma_{lv}\gamma_{sv}}; \tag{31}$$

$$W_{sl} = \frac{1}{2}\left[1 + \left(\frac{\gamma_{sv}}{\gamma_{lv}}\right)^{1/2}\right]^2 \sqrt{\gamma_{lv}\gamma_{sv}}; \tag{32}$$

and

$$W_{sl} = 2\left\{\frac{4(\gamma_{sv}/\gamma_{lv})^{1/2}}{[1 + (\gamma_{sv}/\gamma_{lv})^{1/2}]^2}\right\}^3 \sqrt{\gamma_{lv}\gamma_{sv}}. \tag{33}$$

Since $W_{sl}$ now relates explicitly to $\gamma_{lv}$ and $\gamma_{sv}$, the effect of changing $\gamma_{lv}$ on $W_{sl}$ can be examined for constant $\gamma_{sv}$.

Experimentally, one can, in principle, obtain the free energy of adhesion $W_{sl}$ via contact angles through Young's equation, equation (1). Combining equation (1) with the definition of $W_{sl}$, equation (26) yields $W_{sl}$ as a function of $\gamma_{lv}$ and $\theta$:

$$W_{sl} = \gamma_{lv}(1 + \cos\theta). \tag{34}$$

Thus experimental results can be compared with those predicted from equations (30), (31), (32), and (33); i.e., contact angles of different liquids on one and the same solid surface can be employed to study the systematic effect of changing $\gamma_{lv}$ on $W_{sl}$ through $\theta$.

Figure 5 displays the free energy of adhesion $W_{sl}$ *versus* the liquid–vapor surface tension $\gamma_{lv}$ from recent experimental contact angles for polystyrene (PS) [31], poly(styrene/methyl methacrylate, 70:30) P(S/MMA, 70:30) [32] and poly(methyl methacrylate) PMMA [33]; equation (34) was used to relate $\theta$ to $W_{sl}$. The predicted patterns from equations (30), (31), (32) and (33) for a hypothetical solid surface with $\gamma_{sv} = 30$ mJ/m$^2$ are also given as solid lines. These results suggest that the above combining rules do not predict the observed adhesion patterns

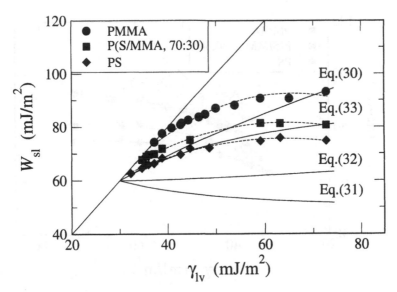

**Figure 5.** The free energy of adhesion $W_{sl}$ *versus* $\gamma_{lv}$ for polystyrene (PS) (diamonds), poly(styrene/methyl methacrylate, 70:30) P(S/MMA, 70:30) (squares) and poly(methyl methacrylate) PMMA (circles). The diagonal line is the line of zero contact angle, i.e., $W_{sl} = 2\gamma_{lv}$; other solid lines are the $W_{sl}$ values predicted by equations (30), (31), (32) and (33) using $\gamma_{sv} = 30$ mJ/m$^2$ from $\gamma_{lv} = 30$ to 75 mJ/m$^2$.

adequately. The combining rule of equation (31) (Berthelot's rule) even shows a monotonous decreasing trend of $W_{sl}$ as $\gamma_{lv}$ increases, which is clearly different from the experimental trends.

Nevertheless, closer scrutiny suggests that the form of the (12:6) combining rule in equation (33) may be useful in predicting the experimental adhesion patterns. As a step towards such an investigation, we rewrite this equation in the following generalized form:

$$W_{sl} = 2 \left\{ \frac{4(\gamma_{sv}/\gamma_{lv})^{1/2}}{\left[1 + (\gamma_{sv}/\gamma_{lv})^{1/2}\right]^2} \right\}^n \sqrt{\gamma_{lv}\gamma_{sv}}, \tag{35}$$

where $n$ is a constant to be determined. For $n = 3$, equation (35) reverts to equation (33). The question now becomes how good this equation fits the experimental data in Fig. 5. Assuming $\gamma_{sv}$ to be constant for one and the same solid surface, experimental contact angle data can be used to fit this equation using a least-squares scheme.

The best-fits of equation (35) to data are shown in Fig. 6. Clearly, equation (35) provides a good fit to the experimental data, with the fitting results summarized in the second and third columns in Table 2. Although equation (35) appears to fit the data well, there is some indication that the power term $n$ is changing with the solid surface. We noticed that $n$ decreases with solid surface tension $\gamma_{sv}$ systematically. In Table 2, we attempted to normalize such changes by expressing the fitted results as $\gamma_{sv}/n$ and $\gamma_{sv}/n^2$ in the fourth and fifth columns, respectively. We found that

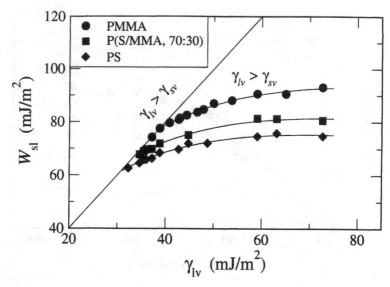

**Figure 6.** The fitted free energy of adhesion $W_{sl}$ vs. $\gamma_{lv}$ for equation (35), for the data in Fig. 5: polystyrene (PS), poly(styrene/methyl methacrylate, 70:30) P(S/MMA, 70:30) and poly(methyl methacrylate) PMMA.

**Table 2.**
Fitting results of experimental energy of adhesion from contact angles for equation (35); $\gamma_{sv}$ is expressed in mJ/m$^2$

| Solid surface | $n$ | $\gamma_{sv}$ | $\gamma_{sv}/n$ | $\gamma_{sv}/n^2$ |
|---|---|---|---|---|
| poly(methyl methacrylate), PMMA | 5.61 | 38.85 | 6.925 | 1.234 |
| poly(styrene/methyl methacrylate, 70:30), P(S/MMA, 70:30) | 5.26 | 33.59 | 6.386 | 1.214 |
| polystyrene, PS | 4.98 | 30.74 | 6.173 | 1.239 |

the values of $\gamma_{sv}/n^2$ appear to be essentially independent of the solid surface used. Averaging and weighting $\gamma_{sv}/n^2$ over the number of data points yields $\gamma_{sv}/n^2 = 1.23$ m$^2$/mJ. The relationship between the energy of adhesion $W_{sl}$ and surface tensions $\gamma_{sv}$ and $\gamma_{lv}$ can now be expressed as

$$W_{sl} = 2 \left\{ \frac{4(\gamma_{sv}/\gamma_{lv})^{1/2}}{[1 + (\gamma_{sv}/\gamma_{lv})^{1/2}]^2} \right\}^{(\tau \gamma_{sv})^{1/2}} \sqrt{\gamma_{lv}\gamma_{sv}}, \qquad (36)$$

where $\tau = 1/1.23$ m$^2$/mJ $= 0.813$ m$^2$/mJ.

### 3.3.1. Predictive power
Theoretically, combining equations (36) with (34) allows determination of the solid surface tension $\gamma_{sv}$ from a single pair of experimental data $(\gamma_{lv}, \theta)$ on this surface. Nevertheless, we fit equation (36) to the experimental $W_{sl}$ versus $\gamma_{lv}$ values on

one and the same solid surface to obtain the solid surface tension $\gamma_{sv}$. Combining equation (36) with equation (34) yields

$$\cos\theta = -1 + 2\sqrt{\frac{\gamma_{sv}}{\gamma_{lv}}}\left\{\frac{4(\gamma_{sv}/\gamma_{lv})^{1/2}}{\left[1+(\gamma_{sv}/\gamma_{lv})^{1/2}\right]^2}\right\}^{(\tau\gamma_{sv})^{1/2}}. \qquad (37)$$

With a calculated $\gamma_{sv}$ value, one can predict the contact angle from the liquid-surface tension $\gamma_{lv}$ by the above equation. To illustrate the prediction of equations (36) and (37), we selected five other surfaces which were not used in the determination of the $\tau$ value here. The five surfaces were FC722 fluorocarbon-coated surface [34], Teflon (FEP) [35], poly($n$-butyl methacrylate) (P$n$BMA) [27, 36], poly(ethyl methacrylate) (PEMA) [27, 37] and poly(propene-$alt$-$N$-($n$-propyl)maleimide) P(PPMI) [27, 38, 39]. The determined solid-surface tensions $\gamma_{sv}$ are listed in Table 3. It is apparent that the predicted $\gamma_{sv}$ value agrees well with the intuition that a fluorocarbon should have a $\gamma_{sv}$ around 12–15 mJ/m$^2$ and the $\gamma_{sv}$ for poly(ethyl methacrylate) (PEMA) should be less than that of the PMMA and should fall between 30 and 35 mJ/m$^2$.

Figure 7 displays the predicted and experimental free energy of adhesion $W_{sl}$ and contact angle, $\cos\theta$ patterns for the five surfaces in Table 3. It is clear from Fig. 7 that the predicted curves fit very well to the experimental data of $W_{sl}$ and $\cos\theta$ vs. $\gamma_{lv}$. It should be noted that these predicted values were calculated using $\tau = 0.813$ m$^2$/mJ which was not "calibrated" using these five surfaces, but with three other solid surfaces (PS, P(S/MMA, 70:30) and PMMA) described earlier. Further, the agreement between the two cases appears to be even more striking when a contact angle error of $\pm1$–2 degrees is considered.

### 3.3.2. Application to molecular theory

Since $W \propto \epsilon$ and $\gamma_{sv} \approx K/\sigma_s^2$, equation (36) can be rewritten as

$$\epsilon_{sl} = \left[\frac{4\sigma_l/\sigma_s}{(1+\sigma_l/\sigma_s)^2}\right]^{(\tau K/\sigma_s^2)^{1/2}} \sqrt{\epsilon_{ss}\epsilon_{ll}}. \qquad (38)$$

Application of this combining rule requires knowledge of the unknown constant $K$ which relates solid surface tensions to molecular collision diameters. Since $K$ is not

**Table 3.**
Fitting results of the solid surface tension $\gamma_{sv}$ using equation (37) to experimental contact angles with $\tau = 0.813$ m$^2$/mJ

| Solid surface | $\gamma_{sv}$ (mJ/m$^2$) equation (37) |
|---|---|
| FC722 | 14.90 |
| FEP | 19.65 |
| P$n$BMA | 29.95 |
| PEMA | 33.87 |
| P(PPMI) | 38.86 |

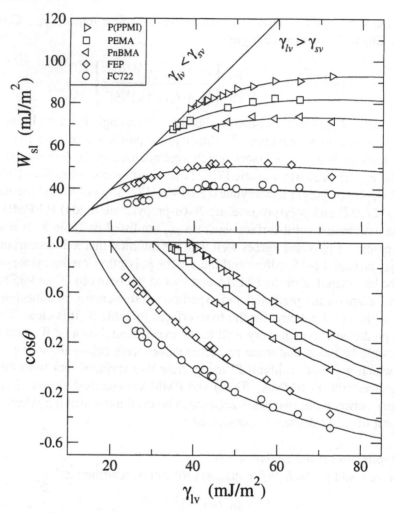

**Figure 7.** The predicted free energy of adhesion $W_{sl}$ and cosine of contact angles $\cos\theta$ vs. $\gamma_{lv}$ with $\tau = 0.813$ m$^2$/mJ for FC722 fluorocarbon-coated surface (○); Teflon (FEP) (◇); poly($n$-butyl methacrylate) (P$n$BMA) (◁); poly(ethyl methacrylate) (PEMA) (□) and poly(propene-$alt$-$N$-($n$-propyl)maleimide) P(PPMI) (▷).

readily known, here we adopted the assumption that $K \approx \gamma_{sv}\sigma_s^2 \approx \gamma_{lv}\sigma_l^2$ and hence $K/\sigma_s^2 \approx \gamma_{lv}\sigma_l^2/\sigma_s^2$, leading to

$$\epsilon_{sl} = \left[\frac{4\sigma_l/\sigma_s}{(1+\sigma_l/\sigma_s)^2}\right]^{(\tau\gamma_{lv}\sigma_l^2/\sigma_s^2)^{1/2}} \sqrt{\epsilon_{ss}\epsilon_{ll}}. \tag{39}$$

Using the theory and procedures described earlier, equation (39) was used to evaluate the intermolecular potential strength $\epsilon_{sl}$ between the solid and fluid molecules. The calculated adhesion patterns are plotted in Fig. 8. It can be seen that the newly formulated combining rule here (equation (39)) generates a much steeper trend of $\cos\theta$ versus $\gamma_{lv}$ curve similar to the experimental trend observed.

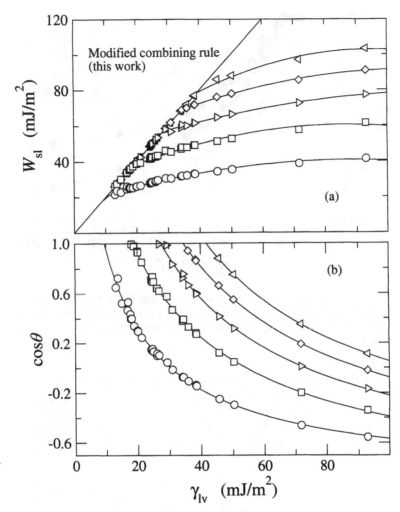

**Figure 8.** (a) The solid–liquid work of adhesion $W_{sl}$ *versus* the liquid–vapor surface tension $\gamma_{lv}$ and (b) cosine of the contact angle $\cos\theta$ *versus* the liquid–vapor surface tension $\gamma_{lv}$ calculated from the combining rule in equation (39). The symbols are calculated data and the curves are the general trends of the data points. Each set of symbols represents a given solid surface.

Other than providing a steeper trend, equation (39) also generates much smoother curves of $W_{sl}$ and $\cos\theta$ vs. $\gamma_{lv}$ than all other combining rules considered here. To elucidate our point, we superimposed our results generated from equation (39) onto Fig. 5 from Ref. [6] calculated using equation (13) and show them in Fig. 9. Here the solid intermolecular potential strength in equation (39) was adjusted to produce a curve with $\gamma_c = 18$ mJ/m$^2$ in order to be comparable with the data for a FEP surface. It can be seen in this figure that, for low surface energy liquids ($\gamma_{lv} < 50$ mJ/m$^2$), the calculated curve from equation (39) follows the experimental points almost exactly. Because of the existence of hydrogen-bonding in high surface energy liquids such as hydrazine and water, which is not considered in our theoretical model, these

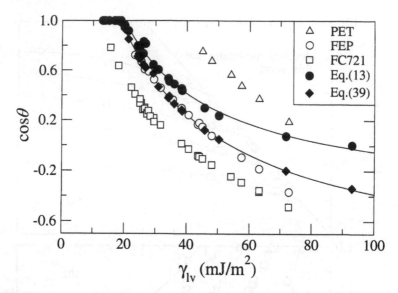

**Figure 9.** Cosine of the contact angle cos θ *versus* the liquid–vapor surface tension $\gamma_{lv}$; open symbols are experimental data; (•) and (♦) represent the calculated data from equations (13) and (39), respectively. The curves are the general trends of the calculated data points. Each set of symbols represents a given solid surface.

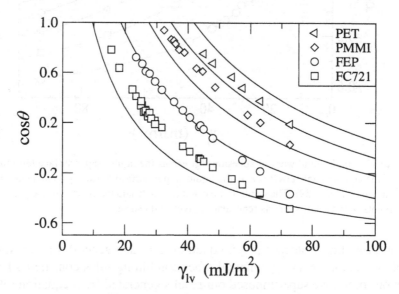

**Figure 10.** Superposition of experimental and calculated cosine of the contact angle cos θ *versus* the liquid–vapor surface tension $\gamma_{lv}$; experimental data and calculated curves are represented by open symbols and solid lines, respectively.

calculated $W_{sl}$ and cos θ tend to deviate from the experimental results. Nevertheless, if we compare the absolute values of cos θ between the calculated and experimental results for water and hydrazine, the results would be very similar: for example, the calculated $\cos \theta$ of water on the FEP is $-0.34$ while the experimental value is $-0.37$. If we plot Fig. 9 using the experimental surface tensions of water and hydrazine as 72.8 and 67.6 mJ/m$^2$, rather than the calculated ones as 92.3 and 71.8 mJ/m$^2$, respectively, our calculated curve would fall exactly on the experimental data of FEP. According to the above analysis, it is more apparent that the calculated data from our modified rule, equation (39), follows the experimental results very closely. Superposition of the calculated results onto the experimental patterns in Fig. 10 illustrates the similarity of the two patterns more clearly.

## 4. SUMMARY

We have formulated a new combining rule between $\gamma_{lv}$ and $\gamma_{sv}$ for $W_{sl}$ from experimental adhesion patterns. The obtained relation gives a good fit to experimental adhesion and contact angle data for the systems studied. By extending such a relation to the molecular level, we have derived an improved combining rule for intermolecular potentials and employed a van der Waals model using a mean field approximation for calculation of the adhesion and contact angle patterns. The calculated patterns follow the experimental results almost exactly. The remarkable agreement between the predicted and experimental data, both at macro- and microscopic levels, suggests that the combining rule we derived here is suitable to represent the relationships for solid–liquid adhesion, as well as solid–fluid intermolecular potentials.

*Acknowledgements*

This research was supported, in part, by the Alberta Ingenuity Establishment Fund, Canada Research Chair (CRC) Program, Canada Foundation for Innovation (CFI), Petro-Canada Young Innovator Research Fund, and Natural Science and Engineering Research Council of Canada (NSERC). J.Z. acknowledges a program from Dr. van Giessen and financial support from the Alberta Ingenuity Studentship Fund.

## REFERENCES

1. D. Y. Kwok and A. W. Neumann, *Adv. Colloid Interface Sci.* **81**, 167 (1999).
2. P. K. Sharma and K. H. Rao, *Adv. Colloid Interface Sci.* **98**, 341 (2002).
3. T. Young, *Philos. Trans. R. Soc. London* **95**, 65 (1805).
4. W. A. Zisman, in: *Contact Angle, Wettability, and Adhesion: Adv. Chem. Ser. No. 43*. American Chemical Society, Washington, DC (1964).
5. D. Y. Kwok and A. W. Neumann, *J. Phys. Chem. B* **104**, 741 (2000).
6. A. E. van Giessen, D. J. Bukman and B. Widom, *J. Colloid Interface Sci.* **192**, 257 (1997).
7. D. E. Sullivan, *J. Chem. Phys.* **74**, 2604 (1981).
8. J. Zhang and D. Y. Kwok, *J. Phys. Chem. B.* **106**, 12594 (2002).
9. J. Zhang and D. Y. Kwok, *Langmuir* (in press).

10. G. H. Hudson and J. C. McCoubrey, *Trans. Faraday Soc.* **56**, 761 (1960).
11. T. M. Reed, *J. Phys. Chem.* **55**, 425 (1955).
12. B. E. F. Fender and G. D. Halsey, Jr., *J. Chem. Phys.* **36**, 1881 (1962).
13. D. E. Sullivan, *Phys. Rev. B* **20**, 3991 (1979).
14. D. V. Matyushov and R. Schmid, *J. Chem. Phys.* **104**, 8627 (1996).
15. J. S. Rowlinson and F. L. Swinton, *Liquids and Liquid Mixtures*. Butterworth Scientific, London (1981).
16. K. C. Chao and R. L. Robinson, Jr., *Equation of State: Theories and Applications*. American Chemical Society, Washington, DC (1986).
17. W. A. Steele, *The Interaction of Gases with Solid Surfaces*. Pergamon Press, New York, NY (1974).
18. D. Berthelot, *Compt. Rend.* **126**, 1857 (1898).
19. J. N. Israelachvili, *Proc. R. Soc. London A.* **331**, 39 (1972).
20. J. Kestin and E. A. Mason, *AIP Conf. Proc.* **11**, 137 (1973).
21. G. C. Maitland, M. Rigby, E. B. Smith and W. A. Wakeham, *Intermolecular Forces: Their Origin and Determination*. Clarendon Press, Oxford (1981).
22. W. A. Steele, *Surface Sci.* **36**, 317 (1973).
23. J. E. Lane and T. H. Spurling, *Aust. J. Chem.* **29**, 8627 (1976).
24. N. F. Carnahan and K. E. Starling, *Phys. Rev. A* **1**, 1672 (1970).
25. H. Gouin, *J. Phys. Chem. B.* **102**, 1212 (1998).
26. J. J. Jaspers, *J. Phys. Chem. Ref. Data* **1**, 841 (1972).
27. D. Y. Kwok, H. Ng and A. W. Neumann, *J. Colloid Interface Sci.* **225**, 323 (2000).
28. A. Dupré, *Théorie Mécanique de la Chaleur*. Gauthier-Villars, Paris (1869).
29. R. J. Good and E. Elbing, *Ind. Eng. Chem.* **62**, 72 (1970).
30. L. A. Girifalco and R. J. Good, *J. Phys. Chem.* **61**, 904 (1957).
31. D. Y. Kwok, C. N. C. Lam, A. Li, K. Zhu, R. Wu and A. W. Neumann, *Polym. Eng. Sci.* **38**, 1675 (1998).
32. D. Y. Kwok, C. N. C. Lam and A. W. Neumann, *Colloid J. (USSR)* **62**, 324 (2000).
33. D. Y. Kwok, A. Leung, C. N. C. Lam, A. Li, R. Wu and A. W. Neumann, *J. Colloid Interface Sci.* **206**, 44 (1998).
34. D. Y. Kwok, R. Lin, M. Mui and A. W. Neumann, *Colloids Surfaces A* **116**, 63 (1996).
35. D. Li and A. W. Neumann, *J. Colloid Interface Sci.* **148**, 190 (1992).
36. D. Y. Kwok, A. Leung, A. Li, C. N. C. Lam, R. Wu and A. W. Neumann, *Colloid Polym. Sci.* **276**, 459 (1998).
37. D. Y. Kwok, R. Wu, A. Li and A. W. Neumann, *J. Adhesion Sci. Technol.* **14**, 719 (2000).
38. D. Y. Kwok, T. Gietzelt, K. Grundke, H.-J. Jacobasch and A. W. Neumann, *Langmuir* **13**, 2880 (1997).
39. D. Y. Kwok, C. N. C. Lam, A. Li, A. Leung and A. W. Neumann, *Langmuir* **14**, 2221 (1998).

Contact Angle, Wettability and Adhesion, Vol. 3, pp. 117–159
Ed. K.L. Mittal
© VSP 2003

# Contact angle measurements and criteria for surface energetic interpretation

DANIEL Y. KWOK[1,] * and A. WILHELM NEUMANN[2]

[1] *Department of Mechanical Engineering, University of Alberta, Edmonton, Alberta T6G 2G8, Canada*
[2] *Department of Mechanical and Industrial Engineering, University of Toronto, Toronto, Ontario M5S 3G8, Canada*

**Abstract**—We summarize here recent progress in the correlation of contact angles with solid surface tensions. We show that contact angle phenomena are complicated and extreme experimental care is required in the attempt to relate contact angles to surface energetics. We have developed experimental procedures and criteria for measuring meaningful contact angles and have showed that there exist unique experimental curves of contact angle patterns. Recent progress in contact angle hysteresis is also summarized. For a large number of liquids on different solid surfaces, the liquid–vapor surface tension times cosine of the contact angle, $\gamma_{lv} \cos \theta$, depends only on the liquid–vapor surface tension $\gamma_{lv}$ and the solid–vapor surface tension $\gamma_{sv}$ when appropriate experimental techniques and procedures are used.

*Keywords*: Contact angle; dynamic contact angle; solid surface tension; equation of state for interfacial tensions; contact angle complexity; surface tension of solids; surface tension component approaches; solid–liquid adhesion; intermolecular potentials.

## 1. INTRODUCTION

Solid surface tension is an important thermodynamic quantity governing many technological processes. However, because of the absence of surface mobility, a solid phase is very different from a liquid phase; hence, one cannot measure the surface tension of a solid phase directly as is the case for a liquid phase. For example, the process of particle adhesion to a substrate can be modeled by the net free energy change, $\Delta F^{\mathrm{adh}}$, of the system during the adhesion process, which depends explicitly on the solid (particle) surface tensions: the net free energy change per unit surface area for the adhesion process of a square particle is given by

$$\Delta F^{\mathrm{adh}} = \gamma_{pv} - \gamma_{pl} - \gamma_{sl} \tag{1}$$

---

*To whom all correspondence should be addressed. Phone: (1-780) 492-2791, Fax: (1-780) 492-2200, E-mail: daniel.y.kwok@ualberta.ca where $\gamma_{pv}$, $\gamma_{pl}$ and $\gamma_{sl}$ are, respectively, the particle–vapor, particle–liquid, and solid–liquid interfacial tensions. If $\Delta F^{adh} < 0$, the adhesion process is thermodynamically favorable. The free energy of adhesion, $W^{adh}$, is simply

$$W^{adh} = -\Delta F^{adh} \tag{2}$$

Because of the difficulties involved in measuring directly the surface tension involving a solid phase, indirect approaches are called for. Several independent approaches have been used to estimate solid surface tensions, including direct force measurements [1–9], contact angles [10–17], capillary penetration into columns of particle powder [18–21], sedimentation of particles [22–25], solidification front interaction with particles [26–33], film flotation [34–38], gradient theory [39–42], Lifshitz theory of van der Waals forces [42–45] and theory of molecular interactions [46–49]. Among these methods, contact angle measurements are believed to be the simplest.

Contact angle measurement is easily performed by establishing the tangent (angle) of a liquid drop with a solid surface at the base. The attractiveness of using contact angles $\theta$ to estimate the solid–vapor and solid–liquid interfacial tensions is due to the relative ease with which contact angles can be measured on suitably prepared solid surfaces. It will become apparent later that this seeming simplicity can be misleading.

The possibility of estimating solid surface tensions from contact angles relies on a relation which was recognized by Young [50] in 1805. The contact angle of a liquid drop on an ideal solid surface is defined by the mechanical equilibrium of the drop under the action of three interfacial tensions: solid–vapor, $\gamma_{sv}$, solid–liquid, $\gamma_{sl}$, and liquid–vapor, $\gamma_{lv}$. This equilibrium relation is known as Young's equation:

$$\gamma_{lv} \cos \theta_Y = \gamma_{sv} - \gamma_{sl} \tag{3}$$

where $\theta_Y$ is the Young contact angle, i.e., a contact angle which can be inserted into Young's equation. It will become apparent later that the experimentally accessible contact angles may or may not be equal to $\theta_Y$: there exist many metastable contact angles which are not equal to $\theta_Y$.

Young's equation (3) contains only two measurable quantities: the contact angle $\theta$ and the liquid–vapor surface tension, $\gamma_{lv}$. In order to determine $\gamma_{sv}$ and $\gamma_{sl}$, an additional relation relating these quantities must be sought. Nevertheless, equation (3) suggests that the observation of the equilibrium contact angles of liquids on solids may be a starting point for investigating the solid surface tensions, $\gamma_{sv}$ and $\gamma_{sl}$. This has inspired many studies in an attempt to develop methodologies for determining solid surface tensions. A common feature of these approaches is the assumption that contact angle measurement is a trivial task.

Since $\gamma_{lv}$, $\gamma_{sv}$ and $\gamma_{sl}$ are thermodynamic properties of the liquid and solid, equation (3) implies a single, unique contact angle; in practice, however, contact angle phenomena are complicated [51–53]. In particular, the contact angle made by an advancing liquid ($\theta_a$) and that made by a receding liquid ($\theta_r$) are not identical;

nearly all solid surfaces exhibit contact angle hysteresis, $H$ (the difference between $\theta_a$ and $\theta_r$):

$$H = \theta_a - \theta_r \tag{4}$$

Contact angle hysteresis can be due to roughness and/or heterogeneity of a solid surface. If roughness is the primary cause, then the measured contact angles are meaningless in terms of Young's equation. On very rough surfaces, experimental contact angles are found to be larger than those on chemically identical, smooth surfaces [21]. Obviously, interpreting such angles in terms of equation (3) would lead to erroneous results because the contact angle would inevitably reflect surface topography, rather than exclusively surface energetics.

Because of these various complexities, models have been employed to gain a deeper understanding of the thermodynamic status of contact angles. In general, it has been found that the experimentally observed apparent contact angle $\theta$ may or may not be equal to the Young contact angle, $\theta_Y$ [51, 52], because of the reasons given below.

(1) On ideal solid surfaces, there is no contact angle hysteresis and the experimentally observed contact angle is equal to $\theta_Y$.

(2) On smooth, but chemically heterogeneous solid surfaces, the experimentally observed $\theta$ is not necessarily equal to the thermodynamic equilibrium angle. Nevertheless, the experimental advancing contact angle $\theta_a$ can be expected to be a good approximation of $\theta_Y$. This has been illustrated using a model of a vertical surface consisting of heterogeneous and smooth strips [51, 52]. Therefore, care must be exercised to ensure that the experimental apparent contact angle $\theta$ is the advancing contact angle in order to be inserted into the Young equation. While the receding contact angle on a heterogeneous and smooth surface can also be a Young angle, it is usually found to be non-reproducible because of sorption of the liquid into the solid and swelling of the solid by the liquid [54].

(3) On rough solid surfaces, no such equality between $\theta_a$ and $\theta_Y$ exists: all contact angles on rough surfaces are meaningless in terms of Young's equation [51, 52].

The thermodynamic equilibrium angles on rough and heterogeneous surfaces are the so-called Wenzel [55] and Cassie and Baxter [56–58] angles, respectively. They are not equal to $\theta_Y$ [51, 52]; furthermore, they are not experimentally accessible quantities.

There are as yet no general guidelines to answer the question of how smooth a solid surface has to be for surface roughness not to have an effect on the contact angle. This and similar problems may be linked to line tension, which has its own complexities [59]. It is, therefore, of utmost importance to prepare solid surfaces as smooth as possible so that the experimental advancing angles can be a good approximation of $\theta_Y$. In addition to these complexities, penetration of the liquid into the solid, swelling of the solid by the liquid and physical/chemical reactions can all play a role. For example, swelling of a solid by a liquid [54] can change the chemistry of the solid in an unknown manner and hence affect the values of $\gamma_{sl}$ and/or $\gamma_{sv}$. Therefore, it is also important to ensure that the solid surfaces are as inert as possible in order to minimize such effects, by appropriate choice of the liquid.

Several contact angle approaches [10–17], of current interest, were largely inspired by the idea of using Young's equation for the determination of surface energetics. While these approaches are, logically and conceptually, mutually exclusive, they share, nevertheless, the following basic assumptions:

(1) All approaches rely on the validity and applicability of Young's equation for surface energetics from experimental contact angles.

(2) Pure liquids are always used; surfactant solutions or mixtures of liquids should not be used, since they would introduce complications due to preferential adsorption.

(3) The values of $\gamma_{lv}$, $\gamma_{sv}$ (and $\gamma_{sl}$) are assumed to be constant during the experiment, i.e., there should be no physical/chemical reaction between the solid and the liquid.

(4) The surface tensions of the test liquids should be higher than the anticipated solid surface tension.

(5) The values of $\gamma_{sv}$ in going from liquid to liquid are also assumed to be constant, i.e., independent of the liquids used.

With respect to the first assumption, one requires the solid surfaces to be rigid, smooth and homogeneous so that Young's equation is the appropriate equilibrium condition; the experimentally observed contact angles should also be advancing contact angles. However, many attempts have been made in the literature to interpret surface tensions of solid surfaces, which are not rigid (e.g., gels [60]) and not smooth (e.g., biological surfaces [61]), in conjunction with Young's equation. Clearly, these results are open to question, since Young's equation may not be valid or applicable in these situations. With respect to the other assumptions, the solid surfaces should be as inert as possible so that effects such as swelling and physical/chemical reactions are minimized.

In order to assure that the experimentally measured contact angles do not violate any of the above assumptions, one requires careful experimentation and suitable methodology. However, contact angles are typically measured simply by depositing a drop of liquid on a given solid surface and manually placing a tangent to the drop at its base using a so-called goniometer sessile drop technique [62, 63]. Apart from the subjectivity of the technique, it normally yields contact angle accuracy of no better than ±2°. More important, the technique cannot be expected to reflect the complexities of solid–liquid interactions. It will become apparent in this paper that much of the controversy with respect to the interpretation of contact angles in terms of surface energetics lies in the fact that not enough attention is paid to the above generally accepted assumptions.

## 2. CONTACT ANGLE MEASUREMENTS

### 2.1. Experimental contact angle patterns

On carefully prepared solid surfaces, Li and co-workers [64, 65] have performed static (advancing) contact angle experiments using an automated drop shape analysis technique: axisymmetric drop shape analysis-profile (ADSA-P). ADSA-P [66] is a technique to determine liquid–fluid interfacial tensions and contact angles from the shape of axisymmetric menisci, i.e., from sessile as well as pendant drops. Assuming that the experimental drop is Laplacian and axisymmetric, ADSA-P finds the theoretical profile that best matches the drop profile extracted from the image of a real drop, from which the surface tension, contact angle, drop volume, surface area and three-phase contact radius can be computed. The strategy employed is to fit the shape of an experimental drop to a theoretical drop profile according to the Laplace equation of capillarity, using surface/interfacial tension as an adjustable parameter. The best fit identifies the correct surface/interfacial tension from which the contact angle can be determined by a numerical integration of the Laplace equation. Details can be found elsewhere [67–69]. It has been found [64, 65] that a contact angle accuracy of better than $\pm 0.3°$ can be obtained on well-prepared solid surfaces.

In the experiments of Li and co-workers [64, 65], static (advancing) contact angles were measured by supplying test liquids from below the surface into the sessile drop, using a motor-driven syringe device. A hole of about 1 mm in the center of each solid surface was required to facilitate such procedures. Liquid was pumped slowly into the drop from below the surface until the three-phase contact radius was about 0.4 cm. After the motor was stopped, the sessile drop was allowed to relax for approx. 30 s to reach equilibrium. Then three pictures of this sessile drop were taken successively at intervals of 30 s. More liquid was then pumped into the drop until it reached another desired size, and the above procedure was repeated [64]. These procedures ensure that the measured static contact angles are indeed the advancing contact angles.

The three carefully prepared solid surfaces were FC-721-coated mica, Teflon FEP heat-pressed against quartz glass slides, and poly(ethylene terephthalate) PET. FC-721 is a 3M "Fluorad" brand antimigration coating designed to prevent the creep of lubricating oils out of bearings. The FC-721-coated mica was prepared by a dip-coating technique. Teflon FEP surfaces were prepared by a heat-pressing method. The material was cut to $2 \times 4$ cm$^2$, sandwiched between two glass slides, and heat-pressed by a jig in an oven. Poly(ethylene terephthalate) PET is the condensation product of ethylene glycol and terephthalic acid. The surfaces of PET films were exceedingly smooth as received and were cleaned by ethanol before measurement. Details of the solid surface preparation can be found elsewhere [64].

Figure 1 shows these contact angle results in a plot of $\gamma_{lv} \cos \theta$ *vs.* $\gamma_{lv}$ for a large number of pure liquids with different molecular properties: they are, in the order of increasing surface tension, pentane, hexane, methanol, decane, methyl acetate, dodecane, tetradecane, hexadecane, *trans*-decalin, *cis*-decalin, dimethylfor-

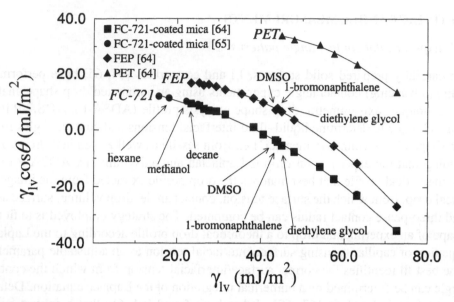

**Figure 1.** $\gamma_{lv} \cos \theta$ vs. $\gamma_{lv}$ for three well-prepared inert solid surfaces: FC-721-coated mica, heat-pressed Teflon FEP and poly(ethylene terephthalate) PET [64, 65]. The smoothness of curves suggests that $\gamma_{lv} \cos \theta$ depends only on $\gamma_{lv}$ and $\gamma_{sv}$.

mamide, ethyl cinnamate, dibenzylamine, dimethyl sulfoxide, 1-bromonaphthalene, diethylene glycol, ethylene glycol, 2,2′-thiodiethanol, formamide, glycol and water. The curves are the best-fits of experimental data to a simple quadratic equation. The choice of plotting $\gamma_{lv} \cos \theta$ in Fig. 1 instead of $\cos \theta$ reflects our intention to follow Young's equation as closely as possible. It can be seen that, for a given solid surface, $\gamma_{lv} \cos \theta$ changes smoothly and systematically with $\gamma_{lv}$. Since the surface tension, $\gamma_{sv}$, of a given solid surface is expected to be constant, i.e., independent of the choice of the test liquid used, Fig. 1 implies that $\gamma_{lv} \cos \theta$ depends only on $\gamma_{lv}$ at constant $\gamma_{sv}$. Replacing the solid surface from the hydrophobic FC-721 surface by the hydrophilic PET surface (i.e., increasing the $\gamma_{sv}$) shifts the curve in a very regular manner. These results suggest that the values of $\gamma_{lv} \cos \theta$ depend only on $\gamma_{lv}$ and $\gamma_{sv}$, independent of any specific intermolecular forces of the liquids and solid surfaces [64, 65, 70], i.e.,

$$\gamma_{lv} \cos \theta = f(\gamma_{lv}, \gamma_{sv}) \tag{5}$$

where $f$ is as yet an unknown function. Because of Young's equation, the experimental contact angle patterns in Fig. 1 imply, in light of equation (5), that the solid–liquid interfacial tension $\gamma_{sl}$ depends only on the liquid–vapor ($\gamma_{lv}$) and solid–vapor ($\gamma_{sv}$) surface tensions. Combining Young's equation with equation (5) yields

$$\gamma_{sv} - \gamma_{sl} = f(\gamma_{lv}, \gamma_{sv}) \tag{6}$$

and hence

$$\gamma_{sl} = \gamma_{sv} - f(\gamma_{lv}, \gamma_{sv}) = F(\gamma_{lv}, \gamma_{sv}) \tag{7}$$

where $F$ is as yet another unknown function. This is in agreement with thermodynamics [71] and the thermodynamic phase rule for capillary systems [72–75] which states that there are only two degrees of freedom for such solid–liquid systems. Thus, the contact angle can be changed simply by changing either $\gamma_{lv}$ or $\gamma_{sv}$. While the specific intermolecular forces (e.g., dipole–dipole moments and hydrogen bondings) determine the primary surface tensions of liquids and solids, they do not appear to have any additional and independent effects on the contact angles, in the context of Young's equation. In principle, a plot of $\cos\theta$ *vs.* $\gamma_{lv}$, rather than $\gamma_{lv}\cos\theta$ *vs.* $\gamma_{lv}$, can also be used to deduce the functional dependence of equation (7).

We shall point out that the above findings have been controversial [70, 76–82]. It is to be noted that the experimental contact angle patterns shown in Fig. 1 do not always appear in the literature: curves far less smoother or no unique curves at all are very frequently reported (see later). Such patterns can have a variety of causes. Accurate contact angle measurements require extreme experimental care. Even very minor vibrations can cause advancing contact angles to decrease, resulting in errors of several degrees. Surface roughness can affect the contact angles and make Young's equation inapplicable. Swelling of a solid by a liquid [54] can change the chemistry of the solid and hence the values of $\gamma_{sl}$ and $\theta$ in an unpredictable manner. Non-constancy of $\gamma_{lv}$, $\gamma_{sv}$ and $\gamma_{sl}$ during the experiment, and non-constancy of $\gamma_{sv}$ from liquid to liquid can produce scatter in plots of the type of Fig. 1 easily.

It will be illustrated in subsequent sections that disagreement with respect to the experimental contact angle patterns arises from the fact that contact angle phenomena are often complex, and cannot be unraveled by the simple goniometer techniques. In addition to the solid surface tensions given by Young's equation (equation (3)), experimental contact angles often contain a variety of other information about a given solid surface, such as molecular orientation at the surface or surface topography. Substantial experimental efforts are required to measure contact angles which represent solid surface tensions exclusively as given by Young's equation and which do not violate the widely accepted assumptions for the interpretation of contact angles. An example of the experimental patterns, which are very different from those of Fig. 1, will be illustrated, simply by measuring advancing contact angles by a goniometer technique.

## 2.2. Experimental contact angle patterns from a goniometer

In order to compare the experimental contact angle patterns obtained by a goniometer with those by an automated ADSA-P technique shown in Fig. 1, Kwok *et al.* [83] selected two well-defined copolymers: poly(propene-*alt*-N-(n-propyl)maleimide) and poly(propene-*alt*-N-(n-hexyl)maleimide). These copolymers were selected

purposely to be not as inert as the FC-721-coated mica, Teflon FEP and PET surfaces used by Li and co-workers [64, 65]. For each copolymer, a 2% solution
was prepared using tetrahydrofuran as the solvent. Silicon wafers <100> were
selected as the substrate for the copolymer coating. The wafers were obtained as
circular discs of about 10 cm diameter and were cut into rectangular shapes of
about $2 \times 3$ cm$^2$. Each surface was first cleaned with ethanol, acetone, and then
soaked in chromic acid for at least 24 h. The cleaned surfaces were rinsed with
doubly-distilled water, and dried under a heat lamp before the copolymer coating.
A few drops of the 2% copolymer/tetrahydrofuran solution were deposited on the
dried wafers inside petri dishes overnight; the solution spread and a thin layer of
the copolymer formed on the wafer surface after tetrahydrofuran evaporated. This
preparation produced good quality coated surfaces, as manifested by light fringes,
due to refraction at these surfaces, suggesting that roughness was on the order of
nanometers or less. It should be noted that if preparation of the solid surfaces and
liquids for contact angle measurements is less careful, impurities can easily conta-
minate the test liquids and solids. This would inevitably result in the contact angle
measurements of contaminated liquids and solids, rather than the presumed pure
liquids and solids.

Thirteen liquids were chosen in this study. Selection of these liquids was based
on the following criteria: (1) liquids should include a wide range of intermolecular
forces; (2) liquids should be non-toxic; and (3) the liquid surface tension should be
higher than the anticipated solid surface tension [10, 13, 21]. They are, in the order of increasing surface tension, *cis*-decalin, 2,5-dichlorotoluene, ethyl cinnamate,
dibenzylamine, dimethyl sulfoxide (DMSO), 1-bromonaphthalene, diethylene glycol, ethylene glycol, diiodomethane, 2,2′-thiodiethanol, formamide, glycerol and
water.

The procedures to measure the advancing contact angles using a goniometer
sessile drop technique were the same as those typically used in the literature: a
sessile drop of about 0.4–0.5 cm radius was formed from above through a Teflon
capillary. The three-phase contact line of the drop was then slowly advanced by
supplying more liquid from above through the capillary which was always kept in
contact with the drop. The maximum (advancing) contact angles were measured
carefully from the left and right sides of the drop and subsequently averaged. The
above procedures were repeated for five drops on five new surfaces. All readings
were then averaged.

Figure 2 shows the contact angle results for the two copolymers from the
goniometer sessile drop measurements in a plot of $\gamma_{lv} \cos \theta$ *vs.* $\gamma_{lv}$. In contrast
to the contact angle patterns shown in Fig. 1, considerable scatter is apparent. On
a given solid, say the poly(propene-*alt*-*N*-(*n*-hexyl)maleimide) copolymer, $\gamma_{sv}$ is
expected to be constant; since the values of $\gamma_{lv} \cos \theta$ here do not appear to give a
smooth and systematic change with $\gamma_{lv}$, one might argue that the contact angle (or
$\gamma_{lv} \cos \theta$) cannot be a simple function $g$ of only $\gamma_{lv}$ and $\gamma_{sv}$, but has to depend also
on the various specific intermolecular forces (such as polarities) of the liquids and

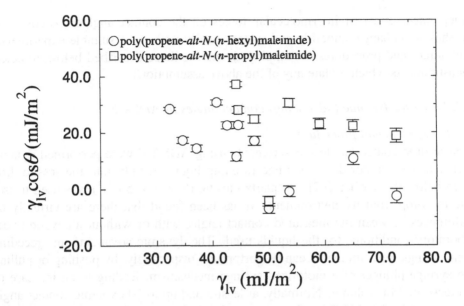

**Figure 2.** $\gamma_{lv} \cos \theta$ *vs.* $\gamma_{lv}$ for poly(propene-*alt*-*N*-(*n*-propyl)maleimide) and poly(propene-*alt*-*N*-(*n*-hexyl)maleimide) copolymers, from advancing contact angles measured by a conventional goniometer technique [83]. Due to the scatter, one might argue that $\gamma_{lv} \cos \theta$ cannot be a simple function of only $\gamma_{lv}$ and $\gamma_{sv}$.

solids, i.e.,

$$\gamma_{lv} \cos \theta = g(\gamma_{lv}, \gamma_{sv}, \text{dipole–dipole, hydrogen bonding, etc.}) \tag{8}$$

Because of Young's equation, the apparent scatter would seem to favor the stipulation of the surface tension component approaches [11, 15–17, 84–86] that $\gamma_{sl}$ depends on various surface-tension components due to the specific intermolecular forces, in addition to $\gamma_{lv}$ and $\gamma_{sv}$, i.e.,

$$\gamma_{sv} - \gamma_{sl} = g(\gamma_{lv}, \gamma_{sv}, \text{dipole–dipole, hydrogen bonding, etc.}) \tag{9}$$

and hence

$$\gamma_{sl} = \gamma_{sv} - g(\gamma_{lv}, \gamma_{sv}, \text{dipole–dipole, hydrogen bonding, etc.}) \tag{10}$$

or

$$\gamma_{sl} = G(\gamma_{lv}, \gamma_{sv}, \text{dipole–dipole, hydrogen bonding, etc.}) \tag{11}$$

where $G$ is an unknown function. While the contact angle patterns in Fig. 2 can be easily found in the literature, they do not really support the above stipulation of the surface tension component approaches [11, 15–17, 84–86]. In the next section, it will be shown that this scatter indeed comes from the fact that many of the experimental contact angles in Fig. 2 have violated some (or all) of the assumptions usually made by all present contact angle approaches. Thus, the apparent additional degrees of freedom (inferred from the scatter) do not come from the putative independent effect of intermolecular forces on the contact angles: $\gamma_{lv} \cos \theta$ can be shown to change smoothly and systematically with $\gamma_{lv}$ if suitable experimental techniques and procedures are employed, such as those described below, to delete measurements which violate any of the above assumptions.

## 2.3. Low-rate dynamic (advancing) contact angles by ADSA-P

### 2.3.1. Experimental procedures

Sessile drop contact angle measurements using ADSA-P were performed dynamically, using a motor-driven syringe to pump liquid steadily into the sessile drop from below the surface [87]. A quartz cuvette ($5 \times 5 \times 5$ cm$^3$) was used to isolate the drop from its environment. It has been found that there are virtually no differences between the measured contact angles with or without a cuvette under laboratory conditions, for the liquids used. The dynamic advancing and receding contact angle measurements can be performed, respectively, by pushing or pulling the syringe plunger of a motorized syringe mechanism, leading to an increase or decrease in drop volume. Normally, at least 5 and up to 10 dynamic contact angle measurements were performed on a new solid surface each time, at velocities of the three-phase contact line in the range from 0.1 to 1.5 mm/min. It will become apparent later that low-rate dynamic contact angles in this velocity range are essentially identical to the static contact angles for relatively smooth surfaces.

In these dynamic procedures, liquid is supplied into the sessile drop from below the solid surface using a motorized-syringe device [87]. In order to facilitate such an experimental procedure, a hole of about 1 mm diameter in the solid surface is required. The strategy of pumping liquid from below the surface was pioneered by Oliver and co-workers [88, 89] because of its potential for avoiding drop vibrations and for measuring true advancing contact angles without disturbing the drop profile. In order to avoid leakage between the stainless steel needle and the hole (in the surface), Teflon tape was wrapped around the end of the needle before insertion into the hole. In the literature, it is customary to first deposit a drop of liquid on a given solid surface using a syringe or a Teflon needle; the drop is then made to advance by supplying more liquid from above using a syringe or a needle in contact with the drop. Such experimental procedures cannot be used for ADSA-P, since ADSA-P determines the contact angles and surface tensions based on the complete and undisturbed drop profile.

In actual experiments, an initial liquid drop of about 0.3 cm radius was carefully deposited from above using a gas-tight Hamilton syringe with a stainless steel needle, covering the hole in the surface. This is to ensure that the drop will increase axisymmetrically in the center of the image field when liquid is supplied from the bottom of the surface and will not hinge on the lip of the hole. The motor in the motorized-syringe mechanism was then set to a specific speed, by adjusting the voltage from a voltage controller. Such a syringe mechanism pushes the syringe plunger, leading to an increase in drop volume and hence the three-phase contact radius. A sequence of pictures of the growing drop was then recorded by the computer typically at a rate of 1 picture every 2 s, until the three-phase contact radius was about 0.5 cm or larger. For each low-rate dynamic contact angle experiment, at least 50 and up to 500 images were normally taken. Since ADSA-P determines the contact angle and the three-phase contact radius simultaneously for each image, the advancing dynamic contact angles as a function of the three-phase contact radius (i.e., location on the surface) can be obtained. In addition, the change in the contact angle, drop volume, drop surface area and the three-phase contact radius can also be studied as a function of time. The actual rate of advancing can be determined by linear regression, by plotting the three-phase contact radius over time. For each liquid, different rates of advancing were studied, by adjusting the speed of the pumping mechanism [87].

Measuring contact angles as a function of the three-phase contact radius has an additional advantage: the quality of the surface is observed indirectly in the measured contact angles. If a solid surface is not smooth, irregular and inconsistent contact angle values will be seen as a function of the three-phase contact radius. When the measured contact angles are essentially constant irrespective of surface location, the mean contact angle for a specific rate of advancing can be obtained by averaging the contact angles, after the three-phase contact radius reaches 0.3 to 0.5 cm (see later). The purpose of choosing these relatively large drops is to avoid any possible line tension effects on the measured contact angles [53, 90, 91].

### 2.3.2. Inert (non-polar) surface: FC-722-coated mica surface

Mica dip-coated with a fluorocarbon, FC-722, was chosen as the substrate for the dynamic contact angle experiments. FC-722 is a fluorochemical coating available from 3M and is chemically very similar to the FC-721 used in earlier studies. The samples were prepared by the same dip-coating procedures described elsewhere [64, 65], using freshly cleaved mica surfaces, originally received as sheets.

Before dip-coating, the mica surfaces were prepared using the following procedures: (1) mica sheets were cut into small mica plates of about $2 \times 3$ cm$^2$; (2) a hole of about 1 mm in diameter was made, by drilling, in the center of each surface; (3) each mica surface was then cleaved with a sharp knife, cleaned with ethanol, and acetone, and dried in air before dip-coating.

Seventeen liquids were chosen for the contact angle measurements [87]. The surface tensions of these liquids were determined independently by applying ADSA-P to pendant drops at room temperature, $23.0 \pm 0.5°$C. Selection of these liquids was based on the criteria described in Section 2.2.

Figure 3 shows a typical example of a low-rate dynamic contact angle experiment: water on a FC-722 surface. As can be seen in Fig. 3, increasing the drop volume $V$ linearly from 0.18 cm$^3$ to 0.22 cm$^3$, by the motorized-syringe mechanism, increases the contact angle $\theta$ from about 108° to 119° at essentially constant three-phase contact radius $R$. This is due to the fact that even carefully placing an initial drop from above on a solid surface can result in a contact angle somewhere between advancing and receding. Therefore, it takes a certain amount of additional liquid

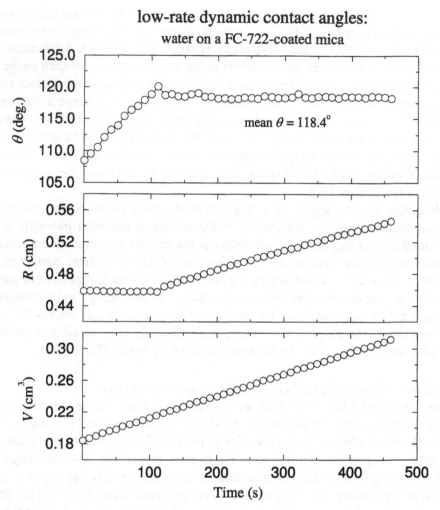

**Figure 3.** Low-rate dynamic contact angles $\theta$ of water on a FC-722-coated mica [87]. $R$ and $V$ are radius and volume of the sessile drop, respectively.

for the initial drop front to start advancing. Further increase in the drop volume causes the three-phase contact line to advance, with $\theta$ essentially constant as $R$ increases. Increasing the drop volume in this manner ensures the measured $\theta$ to be an advancing contact angle. The rate of advancing for this experiment can be determined by linear regression of the three-phase contact radius over time: the drop periphery was advanced at a rate of 0.14 mm/min., in the specific example given in Fig. 3.

In Fig. 3, the measured contact angles are essentially constant as $R$ increases, indicating good surface quality of the solid used. It turns out that averaging the measured contact angles after $R$ reaches 0.50 cm is convenient, since the drop is guaranteed to be in the advancing mode and that possible line tension effects are negligible [59, 90, 91]. Averaging the measured contact angles, after $R$ reaches

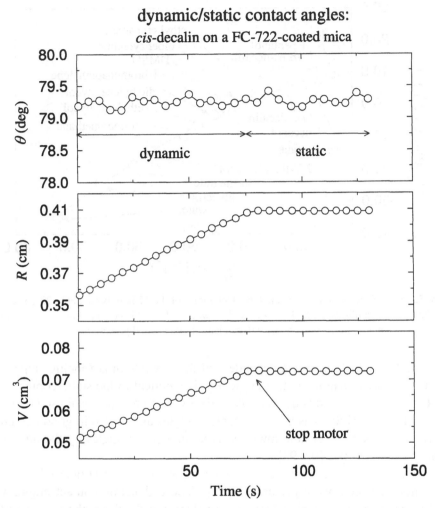

**Figure 4.** Dynamic and static contact angles of *cis*-decalin on a FC-722-coated mica [87]. This result suggests that the low-rate dynamic contact angle is identical to the static angle. *R* and *V* are radius and volume of the sessile drop, respectively.

0.50 cm, yields a mean contact angle of 118.4° for water. While a three-phase contact radius of 0.50 cm may seem to be an arbitrary value, it turns out that there is virtually no difference between averaging $\theta$ for $R$ larger than 0.48 cm or 0.54 cm; the contact angles are essentially constant after $R = 0.50$ cm. Similar results were also obtained for other liquids [87].

Dynamic/static contact angle experiments have also been performed: A liquid drop is first selected to advance at a specific rate of advancing and then stopped, while a sequence of images is recorded. A typical experiment is shown in Fig. 4 for *cis*-decalin. As drop volume increases from 0.05 to 0.07 cm³, the three-phase contact line advances from about 0.36 to 0.41 cm at a rate of 0.41 mm/min. A sequence of drop images was acquired after the motor was stopped at $R = 0.41$ cm.

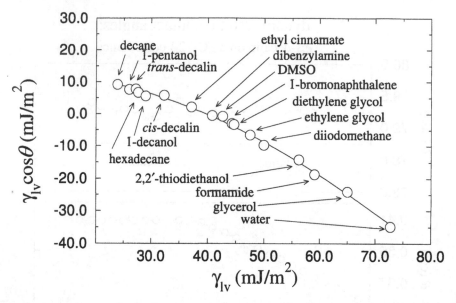

**Figure 5.** $\gamma_{lv} \cos \theta$ vs. $\gamma_{lv}$ for the dynamic contact angles of the 17 liquids on a FC-722-coated mica [87]. This result reconfirms the finding of Li and co-workers [64, 65] shown in Fig. 1 that at constant $\gamma_{sv}$, $\gamma_{lv} \cos \theta$ changes smoothly with $\gamma_{lv}$, independent of other liquid properties.

In Fig. 4, the contact angle is independent of the low rate of advancing, suggesting that the low-rate dynamic contact angle $\theta_{dyn}$ is identical to the static contact angle $\theta_{stat}$. This result re-confirms the validity of the experimental protocol used by Li and co-workers [64, 65] to measure static contact angles and is also in good agreement with recent work to determine low-rate dynamic contact angles by the automated capillary rise technique [92, 93].

With each liquid, ten different measurements (i.e., ten different rates of advancing on ten new surfaces) were performed [87]. These dynamic contact angles were found to be essentially independent of the velocity of the three-phase contact line, as is, in principle, obvious from Fig. 4. Since the low-rate dynamic contact angles are essentially independent of the velocity of the three-phase contact line, a mean dynamic contact angle for each pure liquid was determined by averaging the contact angles at the ten different rates of advancing.

Figure 5 shows all experimental contact angle results at very slow motion of the three-phase contact line in a plot of $\gamma_{lv} \cos \theta$ vs. $\gamma_{lv}$. This result reconfirms the finding of Li and co-workers [64, 65] shown in Fig. 1 that, at constant $\gamma_{sv}$, $\gamma_{lv} \cos \theta$ changes smoothly with $\gamma_{lv}$, independent of the liquid properties. Clearly, if there were any additional and independent effects of intermolecular forces on the contact angles, the data points for the polar liquids (1-pentanol, 1-decanol, DMSO, 2,2'-thiodiethanol, diethylene glycol, ethylene glycol, glycerol and water) in Fig. 5 would not fall completely on a smooth curve along with the non-polar liquids. It can be seen that the data point for 1-pentanol is slightly below the curve; 1-decanol is again slightly below; DMSO is above and ethylene glycol is on the curve: there is no evidence of any systematic deviation of the polar liquids away from such a curve. This question will also be addressed in Section 2.4.1.

In these low-rate dynamic contact angle experiments, images of an advancing drop (and hence information such as surface tension and contact angle) are recorded continuously as drop volume is steadily increased from below the surface. The procedures used here are different from those by Li and co-workers [64, 65] in that the contact angles measured by Li and co-workers were static angles, i.e., contact angles at zero velocity of the three-phase contact line.

### 2.3.3. Non-inert (polar) surfaces: poly(propene-alt-N-(n-alkyl)-maleimide)

Experience has shown that non-polar surfaces, such as Teflon and fluorocarbons, often are quite inert with respect to many liquids; however, polar surfaces often are less inert and hence may show different contact angle patterns, due to such causes as physical/chemical reaction and/or swelling and dissolution of the solid by the liquid. Since low-rate dynamic contact angle experiments when interpreted by ADSA-P have many advantages over the conventional way of manually placing tangents to the sessile drops, ADSA-P is employed here to measure low-rate dynamic contact angles on the copolymers, poly(propene-*alt*-*N*-(*n*-propyl)maleimide) and poly(propene-*alt*-*N*-(*n*-hexyl)maleimide), in order to elucidate the discrepancies between the results in Figs 1 and 5 on the one hand, and those in Fig. 2 by a goniometer on the other. It will become apparent that a goniometer technique is liable to produce a mixture of meaningful and meaningless angles, with no criterion to distinguish between the two. If one disregards the meaningless angles to be identified by dynamic ADSA-P measurements, the results are in harmony with the patterns shown in Figs 1 and 5.

Eight liquids were selected for the contact angle measurements on poly(propene-*alt*-*N*-(*n*-propyl)maleimide) and thirteen liquids on poly(propene-*alt*-*N*-(*n*-hexyl)-maleimide). The chemical structure of these copolymers has been described elsewhere [83]. Silicon wafers <100> were again selected as the substrate. They were obtained as circular discs of about 10 cm diameter and were cut into rectangular size of about $2 \times 3$ cm$^2$. Before soaking the rectangular wafer surfaces into chromic acid, a hole of about 1 mm diameter was made by drilling, using a diamond drill bit (SMS-01027) from Lunzer (Saddle Brook, NJ, USA). The general procedures to prepare the polymer-coated solid surfaces are similar to those described in Section 2.2.

Figure 6 shows the contact angle results for glycerol on the poly(propene-*alt*-*N*-(*n*-propyl)maleimide) surface. Since the contact angles are essentially constant for all experiments, they can be averaged: in this specific example, averaging yields a mean contact angle of $70.8 \pm 0.1°$. Similar results were also obtained for water, 2,2'-thiodiethanol and 1-bromonaphthalene [83]. For each liquid, at least 5 and up to 10 different experiments (for different rates of advancing) were performed, each on a newly prepared surface.

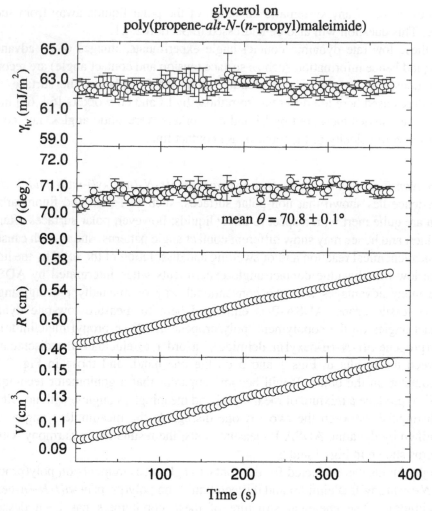

**Figure 6.** Low-rate dynamic contact angles of glycerol on a poly(propene-*alt*-N-(*n*-propyl)maleimide) copolymer [83]. $\gamma_{lv}$, R and V are liquid–vapor surface tension, radius and volume of the sessile drop, respectively.

Unfortunately, not all liquids yield constant advancing contact angles. Figure 7 shows the results of formamide on the same copolymer. It can be seen that as drop volume increases initially, contact angle increases from 60° to 63° at essentially constant three-phase radius. As V continues to increase, $\theta$ suddenly decreases to 60° and the three-phase contact line starts to move. The contact angle then decreases slowly from 60° to 54°. The surface tension–time plot indicates that the surface tension of formamide decreases with time. This suggests that dissolution of the copolymer occurs, causing $\gamma_{lv}$ to change from that of the pure liquid. Similar behavior was also observed in other experiments [83]. It is an important question to ask which contact angles one should use for the interpretation of surface

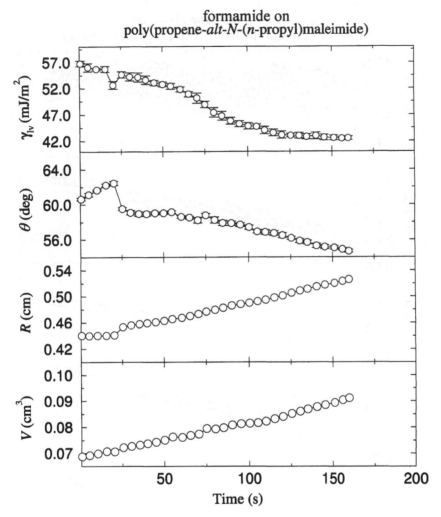

**Figure 7.**   Low-rate dynamic contact angles of formamide on a poly(propene-*alt*-*N*-(*n*-propyl)maleimide) copolymer [83]. $\gamma_{lv}$, $R$ and $V$ are liquid–vapor surface tension, radius and volume of the sessile drop, respectively.

energetics. Since physical/chemical reactions, such as polymer dissolution, alter the liquid–vapor, solid–liquid and solid–vapor interfaces (interfacial tensions) in an unknown manner, the contact angle data in Fig. 7 should be disregarded for the interpretation in terms of surface energetics. The criteria for rejecting contact angles for calculations of solid surface tensions will be discussed in detail in Section 3. Obviously, it is virtually impossible for goniometer measurements to detect the complexities shown in Fig. 7, e.g., the decrease in $\gamma_{lv}$. Thus, the contact angles obtained from a goniometer for this and similar solid–liquid systems cannot be meaningful.

The results of ethylene glycol are shown in Fig. 8. It can be seen that the contact angle increases slowly from 54° to 57° as the three-phase contact line advances from

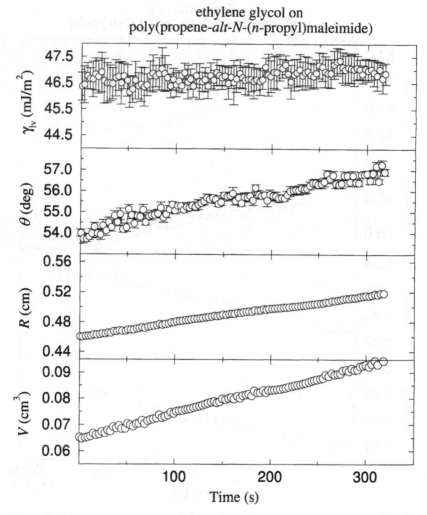

**Figure 8.** Low-rate dynamic contact angles of ethylene glycol on a poly(propene-*alt*-*N*-(*n*-propyl)maleimide) copolymer [83]. $\gamma_{lv}$, $R$ and $V$ are liquid–vapor surface tension, radius and volume of the sessile drop, respectively.

0.46 to 0.54 cm. While the cause of this increase in the contact angle is unclear, it is suspected that the operative solid–liquid interfacial tension is changed slowly due to physico-chemical reaction. According to Young's equation, if the values of $\gamma_{lv}$ and $\gamma_{sv}$ are constant, a change in the contact angle must be a consequence of a change in $\gamma_{sl}$. The observed trends in the contact angle may well start immediately after drop formation, not only after the measurement procedure was set in motion. Also, there is no reason to suspect that the change of contact angle with time (or radius) would cease once the measurement was terminated. More likely, such trends would continue. Because there is no unique apparent contact angle and it is unclear whether or not $\gamma_{sl}$ will remain constant and whether Young's equation is applicable, these angles should be excluded from the interpretation in terms of surface energetics. However, one might consider averaging the contact angles for $R$ larger than, e.g., 0.48 cm, since $\gamma_{lv}$ seems to be constant and since the contact angle error after averaging would not be very large, i.e., $\theta = 55.7 \pm 0.6°$. Such averaging is not allowed because apart from the experimental reasons given above, the question of whether averaging over some data is allowed or not has firm quantitative answers based on the laws of statistics. The main fact is that the statistical analysis rejects "averaging" of the above contact angle data over time by a very large margin [83].

While the contact angle data of formamide, diiodomethane, ethylene glycol and diethylene glycol for poly(propene-*alt-N*-(*n*-propyl)maleimide) should be disregarded, the contact angles of water, glycerol, 2,2′-thiodiethanol and 1-bromo-naphthalene can be used for the interpretation in terms of surface energetics.

A similar study was conducted for the second copolymer, poly(propene-*alt-N*-(*n*-hexyl)maleimide). It was found that only the advancing contact angles of water, glycerol, diethylene glycol and *cis*-decalin were essentially constant [83]. The remaining liquids, formamide, 2,2′-thiodiethanol, diiodomethane, ethylene glycol, 1-bromonaphthalene, dimethyl sulfoxide (DMSO), dibenzylamine, ethyl cinnamate and 2,5-dichlorotoluene all showed very complex contact angle patterns which had to be excluded from the interpretation in terms of surface energetics and testing of approaches for interfacial tensions.

A different contact angle experiment is shown in Fig. 9 for diiodomethane. Initially the apparent drop volume, as perceived by ADSA-P, increases linearly, and the contact angle increases from 88° to 96° at essentially constant three-phase radius. Suddenly, the drop front jumps to a new location as more liquid is supplied into the sessile drop. The resulting contact angle decreases sharply from 96° to 88°. The contact angle increases again as more liquid is supplied into the drop. It should be emphasized that such slip/stick contact angles occurred even for $R$ moved by only 0.2 mm (200 $\mu$m). Obviously, this mechanism cannot be easily detected by the usual goniometer technique and procedures. Such slip/stick behavior could be due to non-inertness of the surface. Phenomenologically, an energy barrier for the drop front exists, resulting in sticking, which causes $\theta$ to increase at constant $R$. However, as more liquid is supplied into the sessile drop, the drop front possesses enough energy to overcome the energy barrier, resulting in slippage, which causes $\theta$ to decrease suddenly. It should be noted that as the drop front jumps from one location to the next, it is unlikely that the drop will remain axisymmetric. Such a non-axisymmetric drop will obviously not meet the basic assumptions underlying ADSA-P, causing possible errors, e.g., in the apparent surface tension and drop volume. This can be seen from the discontinuity of the apparent drop volume and apparent surface tension with time as the drop front sticks and slips. Similar behavior was also observed in other experiments [83, 94]. Obviously, the observed contact angles cannot all be Young contact angles: since $\gamma_{lv}$, $\gamma_{sv}$ (and $\gamma_{sl}$) are constants, then, because of Young's equation, $\theta$ ought to be a constant. In addition, it is difficult to decide unambiguously at this moment whether or not Young's equation

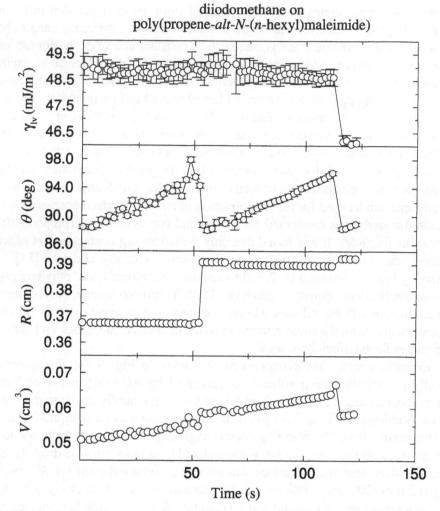

**Figure 9.** Low-rate dynamic contact angles of diiodomethane on a poly(propene-*alt*-*N*-(*n*-hexyl)maleimide) copolymer [83]. $\gamma_{lv}$, $R$ and $V$ are liquid–vapor surface tension, radius and volume of the sessile drop, respectively.

is applicable at all. Therefore, these contact angles should not be used for the interpretation in terms of surface energetics.

While pronounced cases of slip/stick behavior can indeed be observed by the goniometer, it is virtually impossible to record the entire slip/stick behavior manually. In this case, the goniometer contact angle can be very subjective, depending on the skill of the experimentalist. It is expected that a contact angle thus recorded by the goniometer would agree with the maximum angle obtained by ADSA-P. Indeed, a contact angle of 98° was observed, in reasonable agreement with the maxima in the entire slip/stick pattern of the ADSA-P results ($\theta \approx 96°$ in Fig. 9). In cases where the liquid–vapor surface tension of the sessile drop decreases due to, e.g., dissolu-

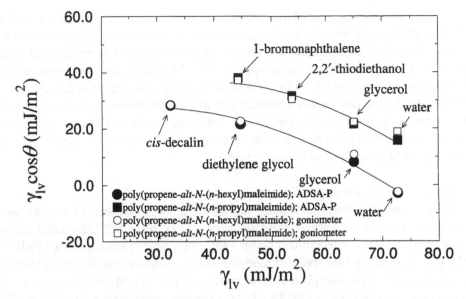

**Figure 10.** $\gamma_{lv} \cos\theta$ *vs.* $\gamma_{lv}$ for the poly(propene-*alt*-*N*-(*n*-propyl)maleimide) and poly(propene-*alt*-*N*-(*n*-hexyl)maleimide) copolymers [83]. Figure 2 changes drastically upon elimination of the angles shown to be meaningless using the dynamic contact angle procedures by ADSA-P.

tion of the surface, only an automated strategy such as ADSA-P can detect changes in the liquid–vapor surface tension. The distinctions and differentiations made by ADSA-P are not possible in a goniometer study. Thus, circumspection is necessary in the decision whether or not experimental contact angles can be used in conjunction with Young's equation. Contact angles from a conventional goniometer-sessile drop technique may produce contact angles which do not fulfill the basic assumptions made in all surface energetic approaches [10–17], e.g., constancy of $\gamma_{lv}$ and applicability of Young's equation, already mentioned in the Introduction. These various assumptions will be discussed in detail in Section 3.

A comparison of the angles measured here and those by the goniometer technique was made [83]. It was found that the goniometer angles corresponded very well with those from ADSA-P only in situations where the contact angles were constant; in cases where complexities of contact angles arose, the goniometer contact angles corresponded only to the maxima of the angles from ADSA-P. The picture emerging in Fig. 2 changes drastically upon elimination of the angles shown to be meaningless in the ADSA-P study, see Fig. 10. The curves in Fig. 10 are in harmony with the results obtained for more inert polar and non-polar surfaces in Figs 1 and 5. Again, it can be concluded that $\gamma_{lv} \cos\theta$ changes smoothly and systematically with $\gamma_{lv}$ at constant $\gamma_{sv}$. Because of Young's equation, a relation of the form of equation (7) can be deduced:

$$\gamma_{sl} = F(\gamma_{lv}, \gamma_{sv})$$

Thus, the experimental procedures and techniques used are crucial in the collection of contact angle data for the determination of surface energetics; conventional goniometer techniques may produce contact angles which violate the basic assumptions made in all surface energetic approaches. In other words, the most serious shortcoming of goniometer studies is not subjectivity and lack of accuracy, but its inability to distinguish between meaningful and meaningless contact angle measurements. A recent study [95] has shown that using the above dynamic procedures with a simple polynomial fit to sessile drop images may also allow one to identify meaningless and meaningful contact angles.

### 2.3.4. Other non-polar and polar surfaces

Similarly, low-rate dynamic contact angles of a large number of liquids were studied extensively on various solid surfaces. They were FC-722-coated silicon wafer [95], FC-725-coated silicon wafer [96], poly($n$-butyl methacrylate) P$n$BMA [97], polystyrene PS [98], poly(styrene-(hexyl/10-carboxydecyl 90:10)maleimide) [99], poly(methyl methacrylate/$n$-butyl methacrylate) P(MMA/$n$BMA) [100] and poly(methyl methacrylate) PMMA [101]. Details with respect to solid surface preparation and experimental results have been given elsewhere [95–101].

Again, it was found that not all contact angles could be used for surface energetics purposes. For example, slip/stick of the three-phase contact line, $\gamma_{lv}$ decrease, $\theta$ decrease or increase as the drop front advances and dissolution of the polymers by the liquids were observed. The meaningful angles for these polymers (copolymers) are plotted in Fig. 11, in plots of $\gamma_{lv} \cos \theta$ *vs.* $\gamma_{lv}$. These curves are in harmony with the patterns shown in Figs 1, 5 and 10. The universality of these contact angle patterns will be discussed below.

### 2.4. Low-rate dynamic (advancing and receding) contact angles by ADSA-P

Although contact angle hysteresis has been studied extensively in the past several decades, the underlying causes and their origins are not completely understood. We shall discuss in the following sections why receding angles are typically not used in conjunction with Young's equation in surface energetic calculations.

Contact angle hysteresis has been attributed to surface roughness [89, 102–106] and heterogeneity [107–114], as well as metastable surface energetic states [110, 114–117]. Some found that contact angle hysteresis decreased with increasing molecular volume of the liquid on monolayers [118, 119]. In more recent studies, contact angle hysteresis was found to be related to molecular mobility and packing of the surface, liquid penetration [120–122] and surface swelling [54, 123]. Previous studies [124, 125] showed that contact angle hysteresis was strongly dependent on the liquid molecular size and solid–liquid contact time. These findings lead to the presumption that liquid sorption and liquid retention are causes of contact angle hysteresis. In order to verify these causes further, we present the following dynamic cycling measurements which are basically a modification of the low-rate dynamic contact angle procedures described above.

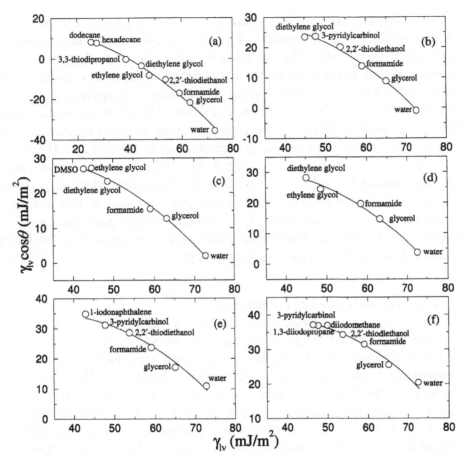

**Figure 11.** $\gamma_{lv} \cos\theta$ *vs.* $\gamma_{lv}$ for (a) FC-725-coated silicon wafer [96], (b) poly($n$-butyl methacrylate) P$n$BMA [97], (c) polystyrene PS [98], (d) poly(styrene-(hexyl/10-carboxydecyl 90:10)maleimide) [99], (e) poly(methyl methacrylate/$n$-butyl methacrylate) P(MMA/$n$BMA) [100] and (f) poly(methyl methacrylate) PMMA [101].

### 2.4.1. Experimental procedures

ADSA-P was used as the technique for contact angle measurements. All procedures were similar to those described before, except the adjustment of drop volume in advancing and receding experiments. A motorized syringe mechanism was employed to push and pull the syringe plunger for advancing and receding measurements, respectively. The growing and shrinking drops were recorded in a sequence of images by a computer, with approximately one image every 2 s. By adjusting the speed of the pumping mechanism, different rates of advancing and receding can be studied.

### 2.4.2. Dynamic one-cycle contact angle experiment

Dynamic one-cycle contact angle measurements were conducted with different liquids on different surfaces. Systems which exhibited constant advancing contact angles were chosen for the study. Typical contact angle results obtained from a dynamic one-cycle contact angle experiment on a poly(methyl methacrylate/$n$-butyl methacrylate) surface are illustrated in Fig. 12. For convenience, these results are divided into three domains. The first domain ranges from the beginning of the experiment to time $t_0$ when the motor was switched to the reverse mode and the liquid started to flow back into the syringe. This domain is characterized by constant $\theta$, and increases of both the drop volume and radius. The advancing contact angle, which is commonly used for surface characterization, is found by averaging the constant contact angles with increasing contact radius in this domain. The second domain ranges from $t_0$ to time $t_r = 0$; $t_0$ is the time where the contact angle starts to decrease at constant three-phase contact radius; $t_r = 0$ is the time where the periphery starts to recede. Among the liquid/solid systems that were studied, four patterns of receding contact angles were obtained: (1) receding contact angle decreases with time (as in Fig. 12); (2) receding contact angle is nearly constant; (3) a "stick/slip" recession pattern; and (4) no receding contact angle.

### 2.4.3. Time-dependent receding contact angle

Time-dependent receding contact angles were observed in most liquid/solid systems. Figure 12 illustrates a typical result of time dependence of the receding contact angle as obtained from the dynamic contact angle experiment with formamide on poly(methyl methacrylate/$n$-butyl methacrylate). In this case, the third domain is characterized by decreasing contact radius, and a decrease of the receding contact angle. Unlike the advancing contact angle, the receding contact angle changes with time in most cases, so that a mean receding contact angle cannot be obtained. Details of the time-dependent receding contact angle will be discussed below.

### 2.4.4. Constant receding contact angle

Constant receding contact angles have been observed in some biodegradable surface/liquid systems, such as poly(lactic acid). Figure 13 illustrates a typical result with constant receding contact angle as obtained from a dynamic one-cycle contact angle experiment with water on poly(lactic acid)-coated silicon wafer. It can be seen clearly that contact angles obtained from the third domain were essentially constant. For such systems, mean receding contact angles can be found by averaging the contact angles beyond $t_r = 0$. Presumably, any molecular mechanism/process will eventually approach a steady state. A steady state might be achieved when the solid is saturated by the liquid. If the existence of the molecular mechanism does contribute to the behavior of time-dependent receding contact angle, one could interpret the constant receding angle in two ways. Liquid sorption could have reached its saturation point in a very short period of time, such that the process had finished prior to the measurement of receding angles. On the other hand, liquid sorption could have been very slow, so that the process would not have really started when the measurements of the receding angles (third domain) were finished.

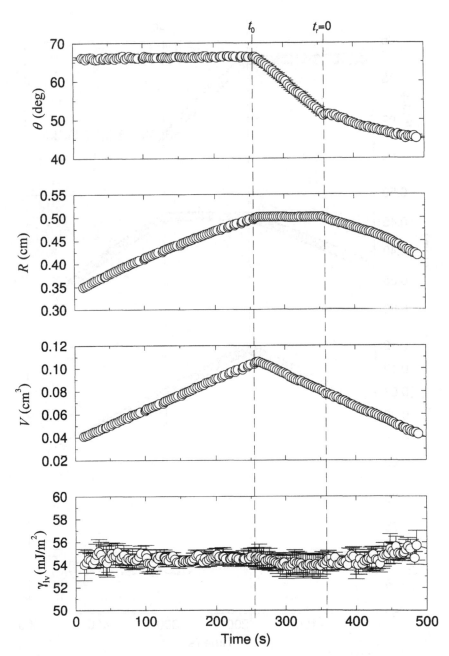

**Figure 12.** Low-rate dynamic one-cycle contact angle measurements of formamide on poly(methyl methacrylate/*n*-butyl methacrylate)-coated silicon wafer, illustrating the time dependence of receding angle. $t_0$ is the time where the contact angle starts to decrease at constant three-phase contact radius. $t_r = 0$ is the time where the periphery starts to recede.

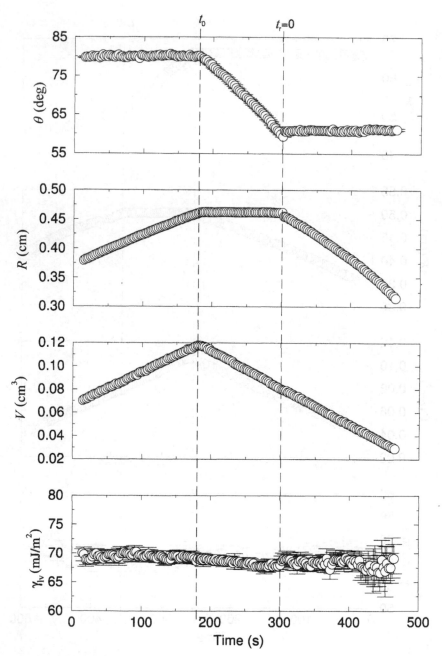

**Figure 13.** Low-rate dynamic one-cycle contact angle measurements of water on poly(lactic acid)-coated silicon wafer, illustrating the constant receding contact angle. $t_0$ is the time where the contact angle starts to decrease at constant three-phase contact radius. $t_r = 0$ is the time where the periphery starts to recede.

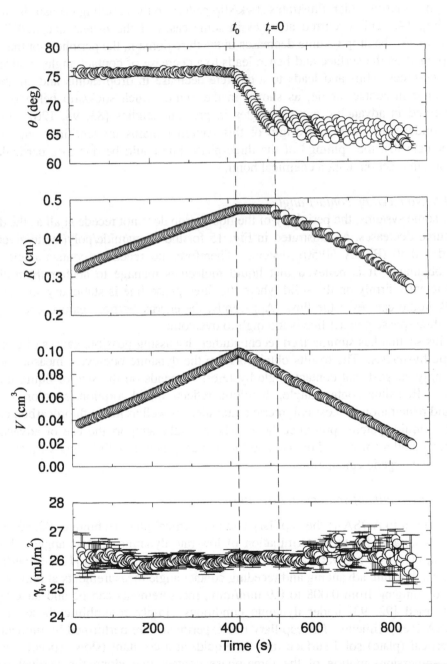

**Figure 14.** Low-rate dynamic one-cycle contact angle measurements of *n*-pentadecane on FC-732-coated silicon wafer, illustrating the slip/stick pattern. $t_0$ is the time where the contact angle starts to decrease at constant three-phase contact radius. $t_r = 0$ is the time where the periphery starts to recede.

### 2.4.5. Stick/slip pattern

A typical result which illustrates stick/slip pattern in the receding domain is shown in Fig. 14. This occurred at times in some cases of the n-pentadecane/FC-732 system. As the drop volume decreases in the third domain, the periphery of the drop is pinned on the surface and hence leads to a decrease of contact angle. Suddenly, the periphery slips and leads to a discrete decrease in drop radius and an abrupt increase in contact angle, as shown in the figure. Such stick/slip behavior was observed in advancing contact angles in previous studies [83, 96, 126]. While the underlying cause and origin of this pattern remains unidentified, one might speculate that the "pining" of the three-phase line could be due to a particularly strong interaction, e.g., a chemical bond.

### 2.4.6. No receding contact angle

For some systems, the periphery of the liquid drop does not recede at all as the drop volume decreases, as illustrated in Fig. 15 for the formamide/poly(lactic/glycolic acid) 50:50 (PLGA 50/50) system. Therefore, no receding contact angle can be obtained. It is believed that liquid molecules manage to anchor themselves sufficiently firmly on the solid where the three-phase line is stationary so that the bulk liquid can resist the flow. Apparently, the energy barrier against receding of the three-phase contact line is too high to overcome.

This section has summarized recent studies discussing possible causes of contact angle hysteresis. The results obtained from the dynamic one-cycle measurements strongly suggest that contact angle hysteresis depends on the size of liquid molecules. Receding contact angle, therefore, reflects liquid sorption/retention and/or liquid penetration in most polymeric materials – as well as in the fluorocarbon coating – liquid systems presented here. These results support the use of advancing contact angles (measured on a dry surface) in conjunction with Young's equation in surface energetic calculations.

### 2.5. Universality of contact angle patterns

In addition to ADSA-P, the capillary rise at a vertical plate technique [92, 93, 127] is also suitable for the determination of low-rate dynamic contact angles. Since the capillary rise at a vertical plate technique has been automated [127] to perform various dynamic advancing and receding contact angle measurements at immersion speeds ranging from 0.008 to 0.9 mm/min., measurements can be and have been performed [92, 93] under dynamic conditions closely resembling those in the ADSA-P experiments. The capillary rise experiments are performed by immersing a vertical (plate) solid surface into the liquid, at a constant (slow) speed; during the immersion, motion of the three-phase contact line along the vertical solid surface is always monitored. The task of measuring a contact angle is reduced to the measurement of a length (capillary rise), which can be performed optically with a high degree of accuracy. In several instances, the capillary rise technique has been employed on relatively inert and well-prepared surfaces: FC-721-coated

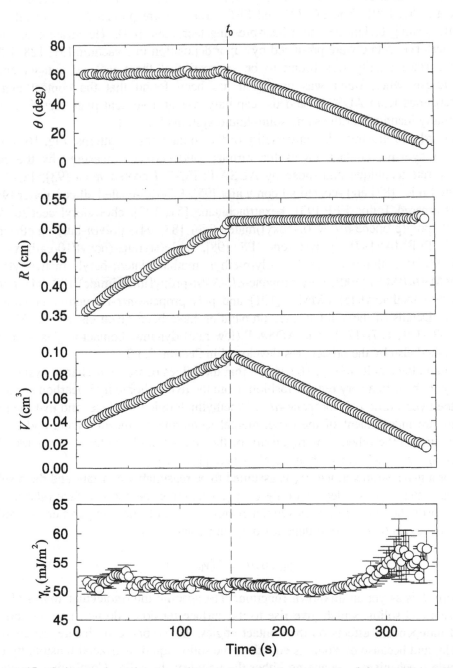

**Figure 15.** Low-rate dynamic one-cycle contact angle measurements of formamide on 50:50 poly(lactic/glycolic acid)-coated silicon wafer, illustrating the absence of receding contact angle. $t_0$ is the time where the contact angle starts to decrease at constant $R$.

mica [93], heat-pressed Teflon FEP [93], hexatriacontane [51, 128], and cholesteryl acetate [51, 129]. The FC-721 and FEP surfaces were prepared, respectively, by a dip-coating technique and a heat-pressing technique [64]. Hexatriacontane and cholesteryl acetate were produced by vapor deposition in a vacuum [51, 128, 129]; the surface quality was found to be so good that there was no contact angle hysteresis when water was used. It has been found that the contact angles determined from ADSA-P and the capillary rise at a vertical plate technique are virtually identical for the same solid–liquid systems [92, 93].

In order to illustrate the universality of the contact angle patterns, Fig. 16 shows recent dynamic contact angles for various solid surfaces measured by the capillary rise technique and those by ADSA-P: FC-721-coated mica [93], FC-722-coated mica [87] and -coated silicon wafer [95], FC-725-coated silicon wafer [96], heat-pressed Teflon FEP [93], hexatriacontane [51, 128], cholesteryl acetate [51, 129], poly(propene-*alt*-*N*-(*n*-hexyl)maleimide) [83, 94], poly(*n*-butyl methacrylate) P*n*BMA [97], polystyrene PS [98], poly(styrene-(hexyl/10-carboxyde-cyl 90:10)-maleimide) [99], poly(methyl methacrylate/*n*-butyl methacrylate) P(MMA/*n*BMA) [100], poly(propene-*alt*-*N*-(*n*-propyl)maleimide) [83, 94], poly(methyl methacrylate) PMMA [101] and poly(propene-*alt*-*N*-methylmaleimide) [95]. Details of the solid surface preparation have been given elsewhere [51, 83, 87, 93–101, 127–129]. The ADSA-P (low-rate) dynamic contact angles were established also by the procedures developed in Section 2.3.3.

Again, the results in Fig. 16 suggest that the values of $\gamma_{lv} \cos \theta$ change systematically with $\gamma_{lv}$ in a very regular fashion, from the hydrophobic hydrocarbon surfaces to the hydrophilic poly(propene-*alt*-*N*-methylmaleimide) surface, and that the patterns are independent of the experimental technique on the one hand and liquid structure on the other. The regularity of the contact angle patterns is remarkable. Similar results were also obtained elsewhere [130–133].

For a given solid surface, $\gamma_{sv}$ is assumed to be reasonably constant and the results suggest that $\gamma_{lv} \cos \theta$ depends only on $\gamma_{lv}$. Changing the solid surface (and hence $\gamma_{sv}$) shifts the curve in a very regular manner. Overall, these experimental results suggest again that $\gamma_{lv} \cos \theta$ depends only on $\gamma_{lv}$ and $\gamma_{sv}$, i.e.,

$$\gamma_{lv} \cos \theta = f(\gamma_{lv}, \gamma_{sv})$$

where $f$ is as yet an unknown function. The specific intermolecular forces of the liquids and solids, which give rise to the surface tensions, do not have additional and independent effects on the contact angles. Thus, one can change the contact angle, and because of Young's equation, the solid–liquid interfacial tension, by the simple mechanism of changing either the liquid or the solid. Combining equation (5) with Young's equation, we can express $\gamma_{sl}$ as another unknown function $F$ of only $\gamma_{lv}$ and $\gamma_{sv}$, allowing a search for an equation (equation (7))in the form

$$\gamma_{sl} = \gamma_{sv} - f(\gamma_{lv}, \gamma_{sv}) = F(\gamma_{lv}, \gamma_{sv})$$

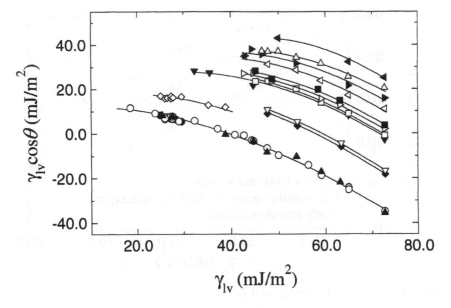

**Figure 16.** $\gamma_{lv}\cos\theta$ *vs.* $\gamma_{lv}$ for various solid surfaces: FC-721-coated mica [93], FC-722-coated mica [87] and -coated silicon wafer [95], FC-725-coated silicon wafer [96], heat-pressed Teflon FEP [93], hexatriacontane [51, 128], cholesteryl acetate [51, 129], poly(propene-*alt*-*N*-(*n*-hexyl)maleimide) [83, 94], poly(*n*-butyl methacrylate) P*n*BMA [97], polystyrene PS [98], poly(styrene-(hexyl/10-carboxydecyl 90:10)-maleimide) [99], poly(methyl methacrylate/*n*-butyl methacrylate) P(MMA/*n*BMA) [100], poly(propene-*alt*-*N*-(*n*-propyl)maleimide) [83, 94], poly(methyl methacrylate) PMMA [101] and poly(propene-*alt*-*N*-methylmaleimide) [95].

This is called an equation-of-state relation, a relation involving intensive properties [134]. Indeed, equations of this form have been in the literature for a long time, e.g., Antonow's [135] and Berthelot's [136] rules.

### 2.5.1. Reasons for deviation from smoothness

As Fig. 16 stands, on one and the same solid surface, there are minor deviations of some contact angle data from the curves and one might wish to argue that

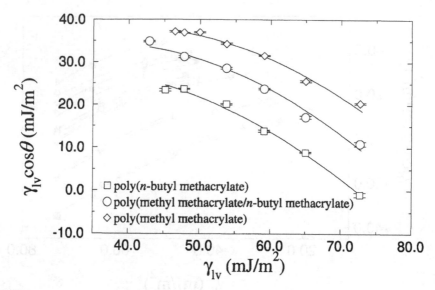

**Figure 17.** $\gamma_{lv} \cos \theta$ vs. $\gamma_{lv}$ for various solid surfaces.

intermolecular forces (e.g., polarity or hydrogen bonding) could still have some independent effects on the contact angles. This question will be addressed here, by focusing on the contact angle data of three chemically similar methacrylate polymers.

Figure 17 shows $\gamma_{lv} \cos \theta$ vs. $\gamma_{lv}$ curves for the PnBMA, P(MMA/nBMA) and PMMA polymers. As can be seen in Fig. 17, there are slight deviations of the data points from the curves; however, it is not apparent that these deviations are systematic and are caused by any independent effects of intermolecular forces on the contact angles. For example, in the case of 3-pyridylcarbinol, with $\gamma_{lv}$ of about 48.0 mJ/m$^2$, the data point on the PMMA curve appears to be slightly higher, while the point on the P(MMA/nBMA) is slightly lower and the one for PnBMA slightly higher again. There is no evidence for any systematic variation, for this and other liquids. Clearly, if there were a deviation due to specific intermolecular forces, one would, at least, expect a monotonous deviation when going from PMMA to P(MMA/nBMA) to PnBMA. Obviously, this does not mean that intermolecular forces are irrelevant; they determine the primary surface tensions of the liquids and solids.

To keep matters in perspective, it has to be realized that the curves in Fig. 17 would have to be considered completely smooth if a conventional goniometer technique with ±2° contact angle accuracy had been used and the unsuitable systems, as described above, could have been eliminated in some fashion. Further, the fact that an equilibrium spreading pressure $\pi_e$

$$\pi_e = \gamma_s - \gamma_{sv} \tag{12}$$

of as low as 1 mJ/m$^2$ would easily contribute to such a variation should not be overlooked; $\gamma_s$ is the surface tension of a solid in a vacuum. On one and the same solid surface, $\gamma_{sv}$ is expected to be constant when vapor adsorption is negligible; however, if vapor adsorption of liquids does play a role, then $\gamma_{sv}$ can be different from liquid to liquid, even on one and the same type of solid surface. The answer to the question of whether or not vapor adsorption has an effect on the contact angle patterns in Fig. 17 (and similar figures) would require general criteria to distinguish this from all other effects, such as swelling of a solid surface. As Fig. 17 stands, it does not appear that vapor adsorption plays a significant role in the contact angles presented here; otherwise, one would expect such an effect to manifest itself as random variations of the contact angles, rather than the remarkable regularity. Nevertheless, minor adsorption which causes the equilibrium spreading pressure $\pi_e$ to vary by as much as 1 mJ/m$^2$ has been considered possible for low-energy solid surfaces [137]. A recent study has examined the effect of adsorption on the contact angles for some high-energy ceramic/metal systems [138].

Although the liquids and solids (polymers) selected are of high purity, e.g., > 99%, minute impurities present in either the liquids or the solids are unavoidable, and finding liquids and solids (polymers) which contain absolutely no impurities is unrealistic. There is no guarantee that such matters could not have caused the apparent minor deviations of the points in Fig. 17 (and similar figures). Even very minor swelling of the polymer [54] or creeping of the liquid could easily introduce a slight deviation of some points from such curves. Therefore, only after considering all such possibilities would the need or justification arise for explanations in terms of direct effects of intermolecular forces [45]. There is no reason to suppose that intermolecular forces have any additional and independent effects on the contact angles beyond the fact that intermolecular forces determine the primary interfacial tensions, $\gamma_{lv}$ and $\gamma_{sv}$; the two interfacial tensions determine $\gamma_{sl}$ through equation (7); and the three interfacial tensions then determine the contact angle, as given by Young's equation. This fact is most easily understood in terms of the experimental contact angle patterns shown in Figures 1, 5, 10, 11, 16 and 17. Recent studies of dynamic contact angles on various copolymer/polymer surfaces revealed similar findings [139–141].

## 3. CRITERIA FOR SURFACE ENERGETIC CALCULATIONS

In Section 2.3.3 we have shown that there exist a large number of contact angle complexities which prevent use of the measurements for surface energetic calculations. To arrive at the curves shown in Fig. 16, several assumptions already mentioned before were used to eliminate meaningless contact angles. However, there remains one last assumption which has not been used and discussed: the constancy of $\gamma_{sv}$ in going from liquid to liquid. It will be apparent in this section that such an assumption is indeed needed for surface energetic calculations; the importance of the various underlying assumptions in the determination of solid surface tensions will be discussed in detail.

### 3.1. Accepted assumptions for calculations of surface energetics

(1) It has to be realized that Young's equation (3)

$$\gamma_{lv} \cos \theta_Y = \gamma_{sv} - \gamma_{sl}$$

has to be used to interrelate the three interfacial tensions, $\gamma_{lv}$, $\gamma_{sv}$ and $\gamma_{sl}$, with the experimental contact angle $\theta$. Therefore, the expectation of the applicability of Young's equation has to be fulfilled. As in the case of slip/stick of the three-phase contact line shown in Fig. 9, Young's equation cannot be applicable: since $\gamma_{lv}$, $\gamma_{sv}$ and $\gamma_{sl}$ are material properties and are expected to be constant, Young's equation implies a unique contact angle. Thus, contact angles from slip/stick of the three-phase contact line have to be discarded.

(2) Obviously, contact angle interpretation of surfactant solutions or mixtures of liquids is expected to be more complicated than that of a pure liquid. It has been found that if one measures contact angles of mixed liquids or solutions on one and the same solid surface, scatter, or patterns different from those in plots of the type as shown in Fig. 16 would arise [142]. Thermodynamically, such systems have three degrees of freedom [72–75]. The additional degree of freedom comes from the effect of an additional liquid component. While Young's equation may still be applicable in this case, no contact angle approach as yet allows the determination of solid surface tensions from such angles. Therefore, pure liquids should always be used in contact angle measurements. However, even if this is the case, one has to ensure that $\gamma_{lv}$ remains constant during the experiment. In an example shown earlier in Fig. 7, polymer dissolution occurs, causing the liquid surface tension to differ from that of the pure liquid. This reflects the fact that the presumed pure liquid has been changed to a mixture of liquid/polymer solution. If a conventional goniometer technique had been used instead, the change in the operative $\gamma_{lv}$ might not have been detected. Such goniometer measurements would inevitably reflect a solid–liquid system with a changed $\gamma_{lv}$, rather than the $\gamma_{lv}$ of the presumed pure liquid. Thus, deductions and conclusions from such angles would be in error. A recent study has attempted to interpret contact angles of binary liquids in terms of surface energetics; the results obtained from a thermodynamic approach similar to that used in the equation-of-state approach appear to be promising [143].

(3) If a chemical/physical reaction takes place, any of $\gamma_{lv}$, $\gamma_{sv}$ and $\gamma_{sl}$ could change during the experiment, and because of Young's equation (equation (3)), the contact angle $\theta$ would also change. Therefore, changes in $\theta$ suggest that at least one of $\gamma_{lv}$, $\gamma_{sv}$ and $\gamma_{sl}$ is changing. In an example shown in Fig. 8, the contact angle increases as the drop front advances with essentially constant $\gamma_{lv}$. While there is no reason to question the applicability of Young's equation, $\gamma_{sl}$

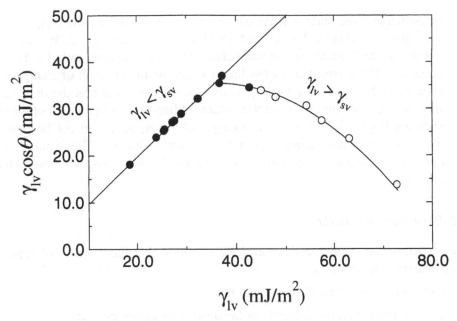

**Figure 18.** $\gamma_{lv} \cos \theta$ *vs.* $\gamma_{lv}$ for a PET surface: o [64]; • [144]. Contact angle data for $\gamma_{lv} < \gamma_{sv}$ and $\gamma_{lv} > \gamma_{sv}$ are shown.

is suspected to change. From Young's equation, if $\gamma_{lv}$ and $\gamma_{sv}$ are constant, changes in $\theta$ must be a consequence of a change in $\gamma_{sl}$. While such cases might be very interesting in a different context, they have to be discarded because our goal is to deduce solid surface tension from the simplest possible situations.

(4) If $\gamma_{lv} \approx \gamma_{sv}$ or $\gamma_{lv} < \gamma_{sv}$, complete wetting occurs; a change in $\gamma_{sv}$ induced by the solid/liquid contact (autophobicity) may also occur. In order to illustrate this, a plot of $\gamma_{lv} \cos \theta$ *vs.* $\gamma_{lv}$ is shown in Fig. 18 for a PET surface. In Fig. 18, liquids having surface tensions less than the anticipated PET surface tension are included, i.e., $\gamma_{lv} < \gamma_{sv}$. Starting from high surface tension (e.g., 72 mJ/m$^2$), the values of $\gamma_{lv} \cos \theta$ increase as $\gamma_{lv}$ decreases, reaching a global maximum. Further decrease in $\gamma_{lv}$ causes the data points to fall on a 45° straight line where the contact angles are zero. Thus, the liquid surface tensions of the test liquids should be higher than that of the anticipated solid surface tension, by the appropriate choice of the liquids. Another possible effect of $\gamma_{lv} < \gamma_{sv}$ is liquid adsorption, which could cause $\gamma_{sv}$ to be different from liquid to liquid. Therefore, the test liquids used in this study were selected to fulfill the condition $\gamma_{lv} > \gamma_{sv}$. A recent contact angle study has reconfirmed that contact angles should be measured for $\gamma_{lv} > \gamma_{sv}$ and that the measurements with $\gamma_{lv} < \gamma_{sv}$ contain no information about $\gamma_{sv}$ [144].

(5) One last assumption which has not been used to identify meaningful contact angles in Section 2.3 is the assumption of the constancy of $\gamma_{sv}$ in going from liquid to liquid. This assumption is needed for reasons of procedures

in deducing solid surface tensions. Between the three variables, $\gamma_{lv}$, $\gamma_{sv}$ and $\gamma_{sl}$, only $\gamma_{lv}$ is measurable; an additional measurable quantity is, however, the contact angle $\theta$ which can be interrelated to these quantities only by Young's equation. Therefore, it is impossible to study the direct effect of changing $\gamma_{lv}$ on a second (non-measurable) quantity, $\gamma_{sl}$, through $\theta$, unless the third (non-measurable) quantity, $\gamma_{sv}$, is kept constant. Otherwise, interpretation of the data showing implicitly the effect of changes of $\gamma_{lv}$ on $\gamma_{sl}$ would not be possible by any of the contact angle approaches of current interest. The assumption of constant $\gamma_{sv}$ will be used below to deduce solid surface tensions from contact angles.

## 3.2. Experimental criteria

The experimental results in Section 2.3.3 illustrate that there are three apparent contact angle complexities:

(1) slip/stick of the three-phase contact line;

(2) contact angle increase/decrease as the drop front advances; and

(3) liquid-surface tension changes as the drop front advances.

Such contact angles should not be used for the determination of solid surface tensions. With respect to the first point, slip/stick of the three-phase contact line indicates that Young's equation is not applicable. Increase/decrease in the contact angle and change in the liquid surface tension as the drop front advances violate the expectation of no physical/chemical reaction. Therefore, when the experimental contact angles and liquid-surface tensions are not constant, they should be disregarded. However, the question arises as to whether the reverse is true, i.e., whether constancy of $\theta$ and $\gamma_{lv}$ in the dynamic measurements described in Section 2.3 will always guarantee that the above contact angle assumptions (in Section 3.1) are fulfilled. This question will be explored below.

In practice, many solids are not truly inert with respect to many liquids; even swelling of fluorocarbon surfaces by alkanes has been reported [54]. For example, if swelling of a solid surface occurs quickly upon the contact with a liquid as the drop front advances, $\theta$ will change to reflect the changed energetics. If the time scale of such an effect is much shorter than that of the contact angle measurements, $\theta$ could reflect the changed $\gamma_{sl}$. This might result in a rather constant contact angle, although the energetics would have changed. If this mechanism would not affect the liquid, $\gamma_{lv}$ would remain constant, i.e., will be independent of time. However, such constant contact angles would reflect only the energetics of the already swollen solid surface in contact with the liquid. Further, one and the same solid surface could swell differently by different test liquids in an unknown manner. This effect would manifest itself in scatter in $\gamma_{lv} \cos \theta$ vs. $\gamma_{lv}$ plots because

$$\gamma_{sl} \neq F(\gamma_{lv}, \gamma_{sv}) \tag{13}$$

In this case, a systematic study of the effect of surface energetics is not easily possible, since the energetics has changed in an unknown, more complex manner. Obviously, such contact angles would have to be disregarded, even if the measured liquid-surface tensions and contact angles are constant, because the assumption of no physical/chemical reaction would have been violated. Unfortunately, the strategies employed here will not allow identification of such matters, and it remains a possibility that such effects contribute to the minor deviations of some data points from a smooth curve.

An example for this more complex pattern is self-assembled monolayers (SAMs), which have been widely used to produce monolayer surfaces of different chemical compositions and wettabilities [145–148]. For this type of surfaces, it is expected that penetration of liquids into the SAMs is inevitable; different liquids would have different effects on such surfaces. Because penetration of liquid is expected to occur almost instantly as the liquid contacts the solid, the observed advancing angle might reflect a $\gamma_{sl}$ value caused by a modification of the solid surface.

Low-rate dynamic contact angles of water on an octadecanethiol ($HS(CH_2)_{17}CH_3$) SAM on an evaporated gold substrate are shown in Fig. 19. It can be seen that the advancing contact angles and liquid surface tension are quite constant. A plot of $\gamma_{lv}\cos\theta$ vs. $\gamma_{lv}$ for various liquids on this surface is shown in Fig. 20 (solid symbols). These contact angles were determined dynamically by ADSA-P [91]. Contrary to the patterns shown in Figs 1, 5, 10, 11, 16 and 17, no smooth curve results in Fig. 20. The reason for the difference in pattern is believed to be the effect of liquid penetration into the SAMs. Such mechanism is believed to be absent or negligible for the relatively thick polymer films used in Section 2.

It is instructive to compare this contact angle pattern with that obtained from the hexatriacontane surface [51, 128, 129, 149] (also plotted in Fig. 20). Since both surfaces are expected to consist entirely of $CH_3$ groups, their solid surface tensions should be similar; one would then expect the two contact angle patterns to be essentially the same. However, the hexatriacontane data follow a smooth curve. From the point of view of surface energetics, the only difference for the two surfaces is that the SAM is a monolayer, whereas the hexatriacontane is a relatively thick crystalline layer. Obviously, it is the effect of liquid penetration and possible contact with the SAM substrate which causes $\gamma_{sl}$ to change in a pattern different from that which would prevail in the absence of liquid penetration. In this case, there is no reason to question the validity and applicability of Young's equation; however, different liquids are expected to penetrate differently even on one and the same solid surface (in an unknown manner) and hence the surface energetics could be very different from liquid to liquid. Since nothing is known about the changed energetics, use of these contact angles in any contact angle approach naively could be meaningless [150]. Although the interpretation of contact angles on SAMs in terms of solid surface tensions could be misleading, wettability studies of SAMs can be very interesting [148, 151]. A more detailed explanation of contact angles and SAMs in terms of surface energetics can be found elsewhere [152].

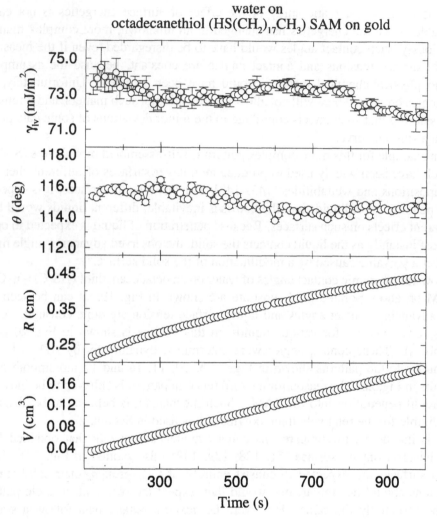

**Figure 19.** Low-rate dynamic contact angles of water on an octadecanethiol $(HS(CH_2)_{17}CH_3)$ SAM on a gold substrate [91].

## 4. SUMMARY

This paper illustrates that obtaining meaningful contact angles for the determination of solid surface tensions depends heavily on how contact angles are measured and whether or not the widely made assumptions have been violated. Experimental procedures and criteria for measuring and interpreting meaningful contact angles have been developed. A large amount of meaningful contact angle data was also generated on various solid surfaces. Unique experimental curves of $\gamma_{lv} \cos \theta$ vs. $\gamma_{lv}$ were obtained for a large number of solid surfaces. The new contact angle data established here provide an experimental foundation for future development of contact angle approaches for the determination of solid surface tensions.

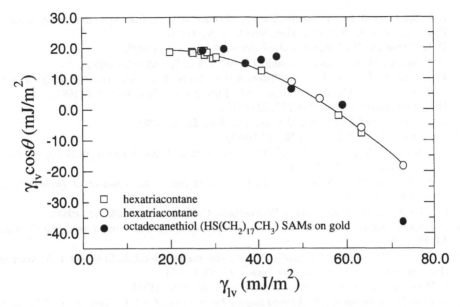

**Figure 20.** $\gamma_{lv} \cos \theta$ *vs.* $\gamma_{lv}$ for an octadecanethiol $(HS(CH_2)_{17}CH_3)$ SAM on a gold substrate, and for a hexatriacontane surface; o [51, 128]; □ [149]; • [91].

## Acknowledgements

We gratefully acknowledge financial support from the Natural Science and Engineering Research Council (NSERC) of Canada, University of Alberta Innovation & Science Fund, Canada Research Chair Program (CRCP) and Canada Foundation for Innovation (CFI).

## REFERENCES

1. B. V. Derjaguin, V. M. Muller and Yu. P. Toporov, *J. Colloid Interface Sci.* **73**, 293 (1980).
2. K. L. Johnson, K. Kendall and A. D. Roberts, *Proc. Roy. Soc. (London)* **A324**, 301 (1971).
3. V. M. Muller, V. S. Yushchenko and B. V. Derjaguin, *J. Colloid Interface Sci.* **92**, 92 (1983).
4. A. Fogden and L. R. White, *J. Colloid Interface Sci.* **138**, 414 (1990).
5. R. M. Pashley, P. M. McGuiggan, R. G. Horn and B. W. Ninham, *J. Colloid Interface Sci.* **126**, 569 (1988).
6. H. K. Christenson, *J. Phys. Chem.* **90**, 4 (1986).
7. P. M. Claesson, C. E. Blom, P. C. Horn and B. W. Ninham, *J. Colloid Interface Sci.* **114**, 234 (1986).
8. P. M. Pashley, P. M. McGuiggan and R. M. Pashley, *Colloids Surfaces* **27**, 277 (1987).
9. R. M. Pashley, P. M. McGuiggan, B. W. Ninham and D. F. Evans, *Science* **229**, 1088 (1985).
10. W. A. Zisman, in: *Contact Angle, Wettability and Adhesion, Adv. Chem. Ser., No. 43*. American Chemical Society, Washington, DC (1964).
11. F. M. Fowkes, *Ind. Eng. Chem.* **56**, 40 (1964).
12. O. Driedger, A. W. Neumann and P. J. Sell, *Kolloid - Z. Z. Polym.* **201**, 52 (1965).
13. A. W. Neumann, R. J. Good, C. J. Hope and M. Sejpal, *J. Colloid Interface Sci.* **49**, 291 (1974).

14. J. K. Spelt and D. Li, in: *Applied Surface Thermodynamics*, J. K. Spelt and A. W. Neumann (Eds), pp. 239–292. Marcel Dekker, New York, NY (1996).
15. D. K. Owens and R. C. Wendt, *J. Appl. Polym. Sci.* **13**, 1741 (1969).
16. C. J. van Oss, M. K. Chaudhury and R. J. Good, *Chem. Rev.* **88**, 927 (1988).
17. R. J. Good and C. J. van Oss, in: *Modern Approaches to Wettability: Theory and Applications*, M. E. Schrader and G. Loeb (Eds), pp. 1–27. Plenum Press, New York, NY (1992).
18. H. G. Bruil, *Colloid Polym. Sci.* **252**, 32 (1974).
19. G. D. Cheever and J. C. Ulcing, *J. Coating Technol.* **55**, 53 (1983).
20. H. W. Kilau, *Colloids Surfaces* **26**, 217 (1983).
21. K. Grundke, T. Bogumil, T. Gietzelt, H.-J. Jacobasch, D. Y. Kwok and A. W. Neumann, *Progr. Colloid Polym. Sci.* **101**, 58 (1996).
22. E. I. Vargha-Butler, T. K. Zubovits, D. R. Absolom and A. W. Neumann, *J. Dispersion Sci. Technol.* **6**, 357 (1985).
23. E. I. Vargha-Butler, E. Moy and A. W. Neumann, *Colloids Surfaces* **24**, 315 (1987).
24. E. I. Vargha-Butler, T. K. Zubovits, D. R. Absolom and A. W. Neumann, *Chem. Eng. Commun.* **33**, 255 (1985).
25. D. Li and A. W. Neumann, in: *Applied Surface Thermodynamics*, J. K. Spelt and A. W. Neumann (Eds), pp. 509–556. Marcel Dekker, New York, NY (1996).
26. S. N. Omenyi and A. W. Neumann, *J. Appl. Phys.* **47**, 3956 (1976).
27. D. Li and A. W. Neumann, in: *Applied Surface Thermodynamics*, J. K. Spelt and A. W. Neumann (Eds), pp. 557–628. Marcel Dekker, New York, NY (1996).
28. A. E. Corte, *J. Geophys. Res.* **67**, 1085 (1962).
29. P. Hoekstra and R. D. Miller, *J. Colloid Interface Sci.* **25**, 166 (1967).
30. J. Cissé and G. F. Bolling, *J. Crystal Growth* **10**, 67 (1971).
31. J. Cissé and G. F. Bolling, *J. Crystal Growth* **11**, 25 (1971).
32. A. M. Zubko, V. G. Lobonov and V. V. Nikonova, *Sov. Phys. Crystallogr.* **18**, 239 (1973).
33. K. H. Chen and W. R. Wilcox, *J. Crystal Growth* **40**, 214 (1977).
34. D. W. Fuerstenau and M. C. Williams, *Colloids Surfaces* **22**, 87 (1987).
35. D. W. Fuerstenau and M. C. Williams, *Particle Characterization* **4**, 7 (1987).
36. D. W. Fuerstenau and M. C. Williams, *Int. J. Miner. Process.* **20**, 153 (1987).
37. D. W. Fuerstenau, M. C. Williams, K. S. Narayanan, J. L. Diao and R. H. Urbina, *Energy Fuels* **2**, 237 (1988).
38. D. W. Fuerstenau, J. Diao and J. Hanson, *Energy Fuels* **4**, 34 (1990).
39. S. J. Hemingway, J. R. Henderson and J. R. Rowlinson, *Faraday Symp. Chem. Soc.* **16**, 33 (1981).
40. R. Guermeur, F. Biquard and C. Jacolin, *J. Chem. Phys.* **82**, 2040 (1985).
41. B. S. Carey, L. E. Scriven and H. T. Davis, *AIChE J.* **26**, 705 (1980).
42. E. Moy and A. W. Neumann, in: *Applied Surface Thermodynamics*, J. K. Spelt and A. W. Neumann (Eds), pp. 333–378. Marcel Dekker, New York, NY (1996).
43. H. C. Hamaker, *Physica* **4**, 1058 (1937).
44. J. N. Israelachvili, *Proc. R. Soc. London A.* **331**, 39 (1972).
45. A. E. van Giessen, D. J. Bukman and B. Widom, *J. Colloid Interface Sci.* **192**, 257 (1997).
46. T. M. Reed, *J. Phys. Chem.* **55**, 425 (1955).
47. B. E. F. Fender and G. D. Halsey, Jr., *J. Chem. Phys.* **36**, 1881 (1962).
48. D. E. Sullivan, *J. Chem. Phys.* **74**, 2604 (1981).
49. D. V. Matyushov and R. Schmid, *J. Chem. Phys.* **104**, 8627 (1996).
50. T. Young, *Philos. Trans. R. Soc. London* **95**, 65 (1805).
51. A. W. Neumann, *Adv. Colloid Interface Sci.* **4**, 105 (1974).
52. D. Li and A. W. Neumann, in: *Applied Surface Thermodynamics*, J. K. Spelt and A. W. Neumann (Eds), pp. 109–168. Marcel Dekker, New York, NY (1996).
53. A. Marmur, *Colloids Surfaces A* **116**, 25 (1996).

54. R. V. Sedev, J. G. Petrov and A. W. Neumann, *J. Colloid Interface Sci.* **180**, 36 (1996).
55. R. N. Wenzel, *Ind. Eng. Chem.* **28**, 988 (1936).
56. A. B. D. Cassie, *Discuss. Faraday Soc.* **3**, 11 (1948).
57. S. Baxer and A. B. D. Cassie, *J. Textile Inst.* **36**, 67 (1945).
58. A. B. D. Cassie and S. Baxter, *Trans. Faraday Soc.* **40**, 546 (1944).
59. J. Gaydos and A. W. Neumann, in: *Applied Surface Thermodynamics*, J. K. Spelt and A. W. Neumann (Eds), pp. 169–238. Marcel Dekker, New York, NY (1996).
60. C. J. van Oss, L. Ju, M. K. Chaudhury and R. J. Good, *J. Colloid Interface Sci.* **128**, 313 (1989).
61. C. J. van Oss and R. J. Good, *J. Macromol. Sci-Chem.* **A26**, 1183 (1989).
62. A. W. Neumann and R. J. Good, in: *Surface and Colloid Surface*, R. J. Good and R. R. Stomberg (Eds), Vol. 11, pp. 31–91. Plenum Press, New York, NY (1979).
63. J. K. Spelt and E. I. Vargha-Butler, in: *Applied Surface Thermodynamics*, J. K. Spelt and A. W. Neumann (Eds), pp. 379–411. Marcel Dekker, New York, NY (1996).
64. D. Li and A. W. Neumann, *J. Colloid Interface Sci.* **148**, 190 (1992).
65. D. Li, M. Xie and A. W. Neumann, *Colloid Polym. Sci.* **271**, 573 (1993).
66. Y. Rotenberg, L. Boruvka and A. W. Neumann, *J. Colloid Interface Sci.* **93**, 169 (1983).
67. P. Cheng, D. Li, L. Boruvka, Y. Rotenberg and A. W. Neumann, *Colloids Surfaces* **93**, 169 (1983).
68. P. Cheng, Ph.D. thesis, University of Toronto (1990).
69. S. Lahooti, O. I. del Río, P. Cheng and A. W. Neumann, in: *Applied Surface Thermodynamics*, J. K. Spelt and A. W. Neumann (Eds), pp. 441–507. Marcel Dekker, New York, NY (1996).
70. D. Y. Kwok, D. Li and A. W. Neumann, *Colloids Surfaces A* **89**, 181 (1994).
71. C. A. Ward and A. W. Neumann, *J. Colloid Interface Sci.* **49**, 286 (1974).
72. R. Defay, *Etude Thermodynamique de la Tension Superficielle*. Gauthier Villars, Paris (1934).
73. R. Defay and I. Prigogine, *Surface Tension and Adsorption*. Longmans, London (1954).
74. D. Li, J. Gaydos and A. W. Neumann, *Langmuir* **5**, 293 (1989).
75. D. Li and A. W. Neumann, *Adv. Colloid Interface Sci.* **49**, 147 (1994).
76. R. E. Johnson, Jr. and R. H. Dettre, *Langmuir* **5**, 293 (1989).
77. I. D. Morrison, *Langmuir* **5**, 540 (1989).
78. I. D. Morrison, *Langmuir* **7**, 1833 (1991).
79. D. Li, E. Moy and A. W. Neumann, *Langmuir* **6**, 885 (1990).
80. J. Gaydos, E. Moy and A. W. Neumann, *Langmuir* **6**, 888 (1990).
81. J. Gaydos and A. W. Neumann, *Langmuir* **9**, 3327 (1993).
82. D. Li and A. W. Neumann, *Langmuir* **9**, 3728 (1993).
83. D. Y. Kwok, T. Gietzelt, K. Grundke, H.-J. Jacobasch and A. W. Neumann, *Langmuir* **13**, 2880 (1997).
84. D. H. Kaelble, *J. Adhesion* **2**, 66 (1970).
85. S. Wu, in: *Adhesion and Adsorption of Polymers*, L. H. Lee (Ed.), pp. 53–65. Plenum, New York, NY (1980).
86. C. J. van Oss, R. J. Good and M. K. Chaudhury, *Langmuir* **4**, 884 (1988).
87. D. Y. Kwok, R. Lin, M. Mui and A. W. Neumann, *Colloids Surfaces A* **116**, 63 (1996).
88. J. F. Oliver, C. Huh and S. G. Mason, *J. Colloid Interface Sci.* **93**, 169 (1983).
89. J. F. Oliver, C. Huh and S. G. Mason, *Colloids Surfaces* **1**, 79 (1980).
90. D. Duncan, D. Li, J. Gaydos and A. W. Neumann, *J. Colloid Interface Sci.* **169**, 256 (1995).
91. A. Amirfazli, D. Y. Kwok, J. Gaydos and A. W. Neumann, *J. Colloid Interface Sci.* **205**, 1 (1998).
92. D. Y. Kwok, D. Li and A. W. Neumann, in: *Applied Surface Thermodynamics*, J. K. Spelt and A. W. Neumann (Eds), pp. 413–440. Marcel Dekker, New York, NY (1996).
93. D. Y. Kwok, C. J. Budziak and A. W. Neumann, *J. Colloid Interface Sci.* **173**, 143 (1995).
94. D. Y. Kwok, C. N. C. Lam, A. Li, A. Leung and A. W. Neumann, *Langmuir* **14**, 2221 (1998).
95. O. I. del Río, D. Y. Kwok, R. Wu, J. M. Alvarez and A. W. Neumann, *Colloids Surfaces A* **143**, 197 (1998).

96. D. Y. Kwok, C. N. C. Lam, A. Li, A. Leung, R. Wu, E. Mok and A. W. Neumann, *Colloids Surfaces A* **142**, 219 (1998).

97. D. Y. Kwok, A. Leung, A. Li, C. N. C. Lam, R. Wu and A. W. Neumann, *Colloid Polym. Sci.* **276**, 459 (1998).

98. D. Y. Kwok, C. N. C. Lam, A. Li, K. Zhu, R. Wu and A. W. Neumann, *Polym. Eng. Sci.* **38**, 1675 (1998).

99. D. Y. Kwok, A. Li, C. N. C. Lam, R. Wu, S. Zschoche, K. Pöschel, T. Gietzelt, K. Grundke, H.-J. Jacobasch and A. W. Neumann, *Macromol. Chem. Phys.* **200**, 1121 (1999).

100. D. Y. Kwok, C. N. C. Lam, A. Li and A. W. Neumann, *J. Adhesion* **68**, 229 (1998).

101. D. Y. Kwok, A. Leung, C. N. C. Lam, A. Li, R. Wu and A. W. Neumann, *J. Colloid Interface Sci.* **206**, 44 (1998).

102. F. E. Bartell and J. W. Shepard, *J. Phys. Chem.* **57**, 211 (1953).

103. R. E. Johnson, Jr. and R. H. Dettre, in: *Contact Angle, Wettability and Adhesion, Adv. Chem. Ser.*, Vol. 43, p. 112. American Chemical Society, Washington, DC (1964).

104. J. D. Eick, R. J. Good and A. W. Neumann, *J. Colloid Interface Sci.* **53**, 235 (1975).

105. J. F. Oliver, C. Huh and S. G. Mason, *J. Adhesion* **8**, 223 (1977).

106. J. F. Oliver and S. G. Mason, *J. Mater. Sci.* **15**, 431 (1980).

107. R. J. Good, *J. Am. Chem. Soc.* **74**, 5041 (1952).

108. R. E. Johnson, Jr. and R. H. Dettre, *J. Phys. Chem.* **68**, 1744 (1964).

109. R. H. Dettre and R. E. Johnson, Jr., *J. Phys. Chem.* **69**, 1507 (1965).

110. A. W. Neumann and R. J. Good, *J. Colloid Interface Sci.* **38**, 341 (1972).

111. L. W. Schwartz and S. Garoff, *Langmuir* **1**, 219 (1985).

112. E. L. Decker and S. Garoff, *Langmuir* **12**, 2100 (1996).

113. E. L. Decker and S. Garoff, *Langmuir* **13**, 6321 (1997).

114. A. Marmur, *J. Colloid Interface Sci.* **168**, 40 (1994).

115. B. V. Derjaguin, *C. R. Acad. Sci. USSR* **51**, 361 (1946).

116. R. E. Johnson, Jr. and R. H. Dettre, in: *Surface and Colloid Science*, E. Matijevic (Ed.), pp. 85–153. Wiley Interscience, New York, NY (1969).

117. S. Brandon and A. Marmur, *J. Colloid Interface Sci.* **183**, 351 (1996).

118. C. O. Timmons and W. A. Zisman, *J. Colloid Interface Sci.* **22**, 165 (1966).

119. A. Y. Fadeev and T. J. McCarthy, *Langmuir* **15**, 7238 (1999).

120. W. Chen and T. J. McCarthy, *Macromolecules* **30**, 78 (1997).

121. A. Y. Fadeev and T. J. McCarthy, *Langmuir* **15**, 3759 (1999).

122. J. P. Youngblood and T. J. McCarthy, *Macromolecules* **32**, 6800 (1999).

123. R. V. Sedev, C. J. Budziak, J. G. Petrov and A. W. Neumann, *J. Colloid Interface Sci.* **159**, 392 (1993).

124. C. N. C. Lam, N. Kim, D. Hui, D. Y. Kwok, M. L. Hair and A. W. Neumann, *Colloids Surfaces A.* **189**, 265 (2001).

125. C. N. C. Lam, L. H. Y. Ko, L. M. Y. Yu, A. Ng, D. Li, M. L. Hair and A. W. Neumann, *J. Colloid Interface Sci.* **243**, 208 (2001).

126. D. Y. Kwok and A. W. Neumann, *Adv. Colloid Interface Sci.* **81**, 167 (1999).

127. C. J. Budziak and A. W. Neumann, *Colloids Surfaces A* **43**, 279 (1990).

128. G. H. E. Hellwig and A. W. Neumann, *Proc. Fifth International Congress of Surface Activity, Section B*, p. 727, Barcelona (1968).

129. G. H. E. Hellwig and A. W. Neumann, *Kolloid-Z. Z. Polym.* **40**, 229 (1969).

130. A. Augsburg, K. Grundke, K. Pöschel, H.-J. Jacobasch and A. W. Neumann, *Acta Polym.* **49**, 417 (1998).

131. M. Wulf, K. Grundke, D. Y. Kwok and A. W. Neumann, *J. Appl. Polym Sci.* **77**, 2493 (2000).

132. C. N. C. Lam, R. Wu, D. Li, M. L. Hair and A. W. Neumann, *Adv. Colloid Interface Sci.* **96**, 169 (2002).

133. K. Grundke, S. Zschoche, K. Pöschel, T. Gietzelt, S. Michel, P. Friedel, D. Jehnichen and A. W. Neumann, *Macromolecules* **49**, 417 (1998).
134. H. B. Callen, *Thermodynamics and an Introduction to Thermostatistics,* 2nd ed. John Wiley, New York, NY (1985).
135. G. Antonow, *J. Chim. Phys.* **5**, 372 (1907).
136. D. Berthelot, *Compt. Rend.* **126**, 1857 (1898).
137. R. J. Good, in: *Adsorption at Interfaces,* K. L. Mittal (Ed.), Symp. Ser. No. 8, pp. 28–47. American Chemical Society, Washington, DC (1975).
138. R. Asthana, *J. Colloid Interface Sci.* **165**, 256 (1994).
139. D. Y. Kwok, A. Li and A. W. Neumann, *J. Polym. Sci. B.* **37**, 2039 (1999).
140. D. Y. Kwok, R. Wu, A. Li and A. W. Neumann, *J. Adhesion Sci. Technol.* **14**, 719 (1999).
141. D. Y. Kwok, A. Li and A. W. Neumann, *Colloid J. (USSR)* **62**, 324 (2000).
142. D. Li, C. Ng and A. W. Neuman, *J. Adhesion Sci. Technol.* **6**, 601 (1992).
143. E. Tronel-Peyroz and A. Lhassani, *J. Colloid Interface Sci.* **171**, 552 (1995).
144. D. Y. Kwok, H. Ng and A. W. Neumann, *J. Colloid Interface Sci.* **225**, 323 (2000).
145. C. D. Bain, E. B. Troughton, Y. Tao, J. Eval, G. M. Whitesides and R. G. Nuzzo, *J. Am. Chem. Soc.* **111**, 321 (1989).
146. C. D. Bain and G. M. Whitesides, *Angew. Chem. Int. Ed. (Engl.)* **28**, 506 (1989).
147. P. E. Laibinis and G. M. Whitesides, *J. Am. Chem. Soc.* **113**, 7152 (1991).
148. P. E. Laibinis, Ph.D. thesis, Harvard University (1991).
149. H. W. Fox and W. A. Zisman, *J. Colloid Sci.* **7**, 428 (1952).
150. J. Drelich and J. D. Miller, *J. Colloid Interface Sci.* **167**, 217 (1994).
151. G. S. Ferguson and G. M. Whitesides, in: *Modern Approaches to Wettability: Theory and Applications,* M. Schrader and G. Loeb (Eds), pp. 143–177. Plenum Press, New York, NY (1992).
152. J. Yang, J. Han, K. Isaacson and D. Y. Kwok, in: *Contact Angle, Wettability and Adhesion, Volume 3,* K. L. Mittal (Ed.), pp. 319–338. VSP, Utrecht (2003).

*Contact Angle, Wettability and Adhesion*, Vol. 3, pp. 161–173
Ed. K.L. Mittal
© VSP 2003

# Advancing, receding and vibrated contact angles on rough hydrophobic surfaces

M. FABRETTO, R. SEDEV* and J. RALSTON

*Ian Wark Research Institute, University of South Australia, Mawson Lakes, SA 5095, Australia*

**Abstract**—Hydrophobic solid surfaces with controlled roughness were prepared by coating glass slides with an amorphous fluoropolymer (Teflon AF1600) containing varying amounts of 50 μm silica spheres. The roughness of the surfaces was estimated through the surface concentration of particles (assessed from SEM images). Quasi-static advancing and receding contact angles were measured with the Wilhelmy plate technique. The contact angle hysteresis was small but significant. By subjecting the system to acoustic vibrations the hysteresis was practically eliminated. At low particle concentration the vibrated contact angle, $\theta_V$, is roughness independent. At higher particle concentration the vibrated contact angle changes in general agreement with Wenzel equation: $\theta_V$ increases when its value on a smooth surface, $\theta_V^0$, is obtuse and decreases when $\theta_V^0$ is acute. The transition, however, occurs at about 80° rather than exactly at 90°. The contact angle obtained by averaging the cosines of the advancing and receding angles is a good approximation for the vibrated one provided that surface roughness is small or the angles relatively large.

*Keywords*: Contact angle; surface roughness; Teflon AF1600; acoustic vibrations; Wenzel equation.

## 1. INTRODUCTION

The contact angle is the most important macroscopic characteristic of a three-phase contact line, i.e. the line where the liquid/vapour interface crosses the solid surface. The equilibrium condition at the contact line is given by the Young equation [1]

$$\cos\theta_0 = \frac{\gamma_{SV} - \gamma_{SL}}{\gamma} \tag{1}$$

It relates the equilibrium contact angle, $\theta_0$ (i.e. the inclination of the liquid/gas interface at the solid surface), to the solid/vapour, $\gamma_{SV}$, solid/liquid, $\gamma_{SL}$, and liquid/vapour, $\gamma$, interfacial tensions. Surface roughness significantly alters the con-

---

*To whom all correspondence should be addressed. Phone: +61 8 8302 3225,
Fax: +61 8 8302 3683, E-mail: rossen.sedev@unisa.edu.au

tact angle and the relation between the angle on a smooth surface, $\theta_0$, and the angle on a rough surface, $\theta$, is given by the Wenzel equation [1]:

$$\cos\theta = \frac{A}{A_0}\cos\theta_0 = r\cos\theta_0 \tag{2}$$

where $r$ is the ratio of the actual, $A$, to the projected, $A_0$, solid surface area.

Equations (1) and (2) can be derived using strict thermodynamic arguments (see e.g. [2]) and therefore their validity can hardly be questioned [3, 4]. In practice, however, any contact angle measurement is confronted with complications arising from the contact angle hysteresis. For almost any real system a whole range of contact angles can be achieved without actually moving the contact line. The advancing, $\theta_A$, and receding, $\theta_R$, contact angles are the upper and lower limits of this interval and hysteresis is defined as $\theta_A - \theta_R$. Contact angle hysteresis is usually seen as an irreversible phenomenon [2-4], hence the use of experimental contact angle values in conjunction with Eqns. (1) and (2) is questionable. There is no agreement on how to treat contact angles measured on rough solid surfaces. Kwok and Neumann [5] assert that: "all contact angles on rough surfaces are meaningless in terms of the Young equation". At the other extreme Good et al. [6] used both advancing and receding contact angles in conjunction with the van Oss-Chaudhury-Good theory of contact angles. A practical solution of this problem could be the development of a method for measuring single-valued contact angles. It has been known that mechanical vibrations reduce hysteresis [7-9] and it is tempting to expand this possibility into a systematic approach.

In this study hydrophobic surfaces with increasing roughness were produced by coating glass slides with an amorphous fluoropolymer (Teflon AF1600) containing 50 μm silica spheres. Advancing and receding contact angles were measured by the conventional Wilhelmy plate technique. The contact angle hysteresis was small but significant. By subjecting the system to acoustic vibrations the hysteresis was effectively eliminated. The vibrated contact angle follows the Wenzel equation. The contact angle obtained by averaging the cosines of the advancing and receding angles is a good approximation for the vibrated one provided that surface roughness is rather small.

## 2. MATERIALS AND METHODS

Glass microscope slides ($22 \times 50 \times 0.15$ mm³, Deckglasser, Menzel-Gläser, Germany) were used as substrates. Gross contaminants were removed by sonicating (Soniclean Model 160HT, Australia) the slides in warm ethanol (Aldrich, spectrophotometric grade) for 80 min. The slides were then removed and rinsed with copious amounts of ultra-pure water (resistivity > 18 MΩ·cm; UHQPS, Elgastat, UK). The wet slides were transferred without delay into a plasma cleaner (Harrick

Model PDC-32G, Ossining, NY) and treated for 5 min on medium power setting (4 min to evaporate water + 1 min actual plasma treatment).

The clean microscope slides were immediately coated with a layer of an amorphous fluoropolymer (Teflon AF1600, DuPont). A 1.5% by weight solution of AF1600 in a mixture of perfluorinated cyclic ethers (Fluorinert FC-75, 3M) was used to dip-coat the slides at a constant withdrawal speed of 10 mm/s. Surfaces with different roughnesses were produced by adding varying amounts of silica spheres (30-50 µm, GelTech, Orlando, FL) into the coating mixture (the bare glass slides had an RMS roughness of about 5 nm which is negligible in comparison to the particle induced roughness). The hydrophobic coatings were then heat-treated to remove residual solvent and increase the bond strength to the substrate: 7 min at 112°C, 5 min at 165°C and 15 min at 330°C [10].

Several test liquids were used to probe the wettability of the coated (both smooth and rough) surfaces. All liquids were used as received. Their surface tension was checked by the drop weight method and was always within 2% of reference values (Table 1).

SEM images of the samples were taken at 10° tilt from the perpendicular (CamScan 44 FE, electron energy 10-19 keV). The samples were carbon coated (20-50 nm) to avoid surface charging. A typical image is shown in Fig. 1 and illustrates the random nature of the particle deposition.

The number of particles per unit area, $n$, was determined from 5 separate images of each sample. The size distribution of the particles on the surface (Fig. 2) was obtained from images taken at higher magnification. The mean diameter of the silica spheres was $d = 48$ µm.

Contact angles were determined by the Wilhelmy plate method [11-13]. The mechanical apparatus consisted of an electrobalance (Cahn C-2000, USA) and a screw-feed platform (Time & Precision, UK) controlled by a micro-step controller (Compumotor AX series, Parker, Petaluma, CA). A 5-inch audio speaker was

**Table 1.**
Surface tension of the liquids used

| Liquid | Surface tension [mJ/m²] | Supplier/Quality |
|---|---|---|
| Water | 72.8 | Elgastat (> 18 MΩ·cm) |
| Formamide | 58.2 | Aldrich 98% |
| Ethylene glycol | 47.7 | Sigma 99+% |
| Dimethyl sulfoxide | 42.7 | BDH Chemicals 99% |
| Methylnaphthalene | 38.3 | Aldrich 95% |
| Dimethylformamide | 35.2 | Sigma 100% |
| Hexadecane | 27.6 | Aldrich 99% |
| Octane | 21.8 | Aldrich 99% |
| Heptane | 20.1 | BDH Chemicals 99.5% |
| Hexane | 18.4 | EM Science 98.5% |

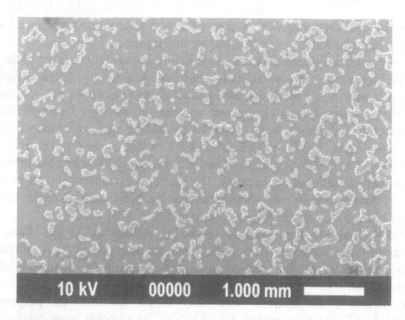

**Figure 1.** SEM image of a glass slide coated with a layer of amorphous Teflon (AF1600). The thickness of the fluoropolymer coating is about 50 nm and 50 μm silica spheres are embedded into the coating.

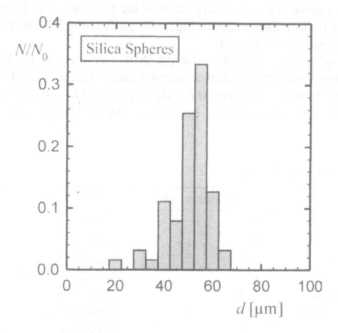

**Figure 2.** Size distribution of the silica spheres embedded into the fluoropolymer coating (see Fig. 1). The number of particles within a certain size range, $N$, is normalized by the total number of particles, $N_0$. The mean particle diameter is $d = 48$ μm.

mounted on the platform and fitted with an in-house made cradle for holding the beaker with the test liquid. The setup was housed in an environmental chamber at $22 \pm 1°C$. Data acquisition was performed on a PC via an in-house designed system at a rate of 10 points per second.

The microscope slide was suspended from the balance via a 100 mm long steel piano wire (0.7 mm diameter) with a polyethylene clasp. The wire was slightly bent to align the lower edge of the microscope slide to the liquid surface. The sensitivity of the balance was set to 0.001 mN. As soon as the microscope slide had touched the liquid surface, the immersion depth, time and force were set to zero and continuous data recording commenced.

All contact angles were measured at a constant pull rate of 50 µm/s. It was independently checked that the values obtained were not significantly affected by liquid motion, i.e. all reported angles are quasi-static. Advancing and receding contact angles were recorded during immersion and withdrawal of the sample, respectively. At certain depth of immersion the plate was halted and the following procedure was used to obtain a vibrated contact angle. Initially the triple line was allowed to relax for 10 s and then the audio speaker was turned on. A sinusoidal voltage was applied (nominal amplitude 6 V, frequency 50-110 Hz depending on the liquid) for 60 seconds after which the amplitude was slowly reduced to zero over a 10 s period. Another 10 s were allowed, during which the vibrated system came to rest, and at the end a force measurement (corresponding to the vibrated contact angle) was taken.

All experiments were carried out in a dust-free environment (Cleanroom class 1000) at ambient temperature of 22°C.

## 3. RESULTS AND DISCUSSION

Samples with an increasing number of particles (and, therefore, roughness) were produced. For the calculation of the roughness ratio the following simple model was adopted: $N$ individual particles are randomly scattered on the surface (Fig. 3a). The Wenzel ratio is given by

$$r = 1 + \frac{3\pi}{4} d^2 n \tag{3}$$

where $d$ is the particle diameter and $n$ ($= N/A_0$) is the number concentration of particles. The thickness of the Teflon layer (~ 50 nm) is much smaller than the particle size and thus is neglected. The numerical factor in (3) corresponds to a cylinder of diameter $d$ and height $d/2$ capped with a hemisphere of diameter $d$, i.e. it is assumed that the space below the overhanging sphere is filled with Teflon as revealed by SEM imaging (Fig. 3b).

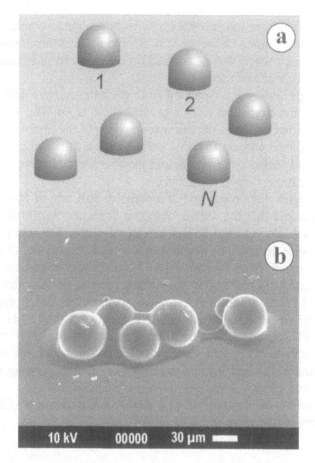

**Figure 3.** (a) Model: $N$ randomly scattered cylinders capped with hemispheres; (b) high magnification SEM image.

Fig. 3b also shows that it is necessary to account for particle agglomeration (this can also be seen in Fig. 1). This effectively reduces the additional area due to the presence of particles and the Wenzel ratio, $r$, is calculated as

$$r = 1 + \frac{3\pi}{4} d^2 \frac{n}{\sqrt{M}} \qquad (4)$$

where $M$ is the number of particles in a cluster. Eqn. (4) is an approximation (it would be exact in the case of $\sqrt{M} \times \sqrt{M}$ clusters of cubic particles of size $d$). The distribution of the cluster size is shown in Fig. 4 and the mean cluster size is $M = 2.2$.

The samples used in this study are listed in Table 2.

**Table 2.**
Surface concentration of silica spheres, $n$ (= $N/A_0$), and Wenzel ratio, $r$, (Eqn. (4)) for the samples used

| Sample | $N/A_0$ [1/mm²] | $r$ |
|---|---|---|
| 0 | 0 | 1.00 |
| 1 | 4.4 | 1.02 |
| 2 | 8.0 | 1.03 |
| 3 | 44.3 | 1.16 |
| 4 | 93.7 | 1.35 |

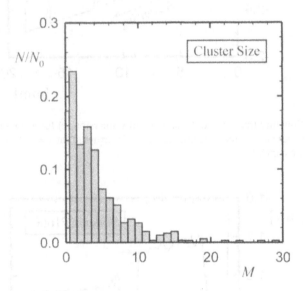

**Figure 4.** Cluster size distribution of the silica sphere aggregates embedded into the fluoropolymer coating (see Figs. 1 and 3b). The mean cluster size is $M = 2.2$.

A typical wetting cycle for smooth Teflon AF1600 (Sample 0) in water is shown in Fig. 5. The only difference from a conventional scan is that at several locations along the plate ($x$ = 4, 8, 12 and 16 mm) vibration energy was supplied as described in the previous section. The obvious effect of the acoustic treatment is that it decreases the advancing and increases the receding contact angle.

Fig. 6 shows the time evolution of the measured force during the same run. For clarity only a portion of the advancing stage is shown. As acoustic vibration is commenced (vertical arrows) the force trace immediately spikes upwards indicating a decrease in the contact angle. It then oscillates as the amplitude is slowly ramped down to zero allowing the system to find a vibrated-equilibrium (horizontal arrows).

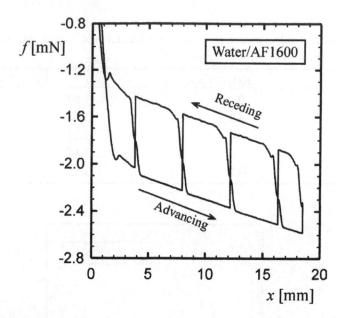

**Figure 5.** Typical Wilhelmy trace (force, $f$, vs. depth of immersion, $x$) for a smooth Teflon AF1600 surface in water. The advancing and receding traces are interrupted (at $x = 4$, 8, 12 and 16 mm) and acoustic vibrations are applied.

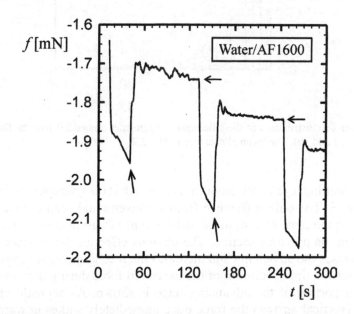

**Figure 6.** Measured force, $f$, as a function of time, $t$ (data are from the run shown in Fig. 5). The vertical arrows show the moment when acoustic vibrations are started. The horizontal arrows mark the force reading used for obtaining the vibrated contact angles.

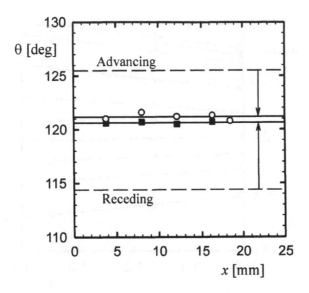

**Figure 7.** Advancing, receding and vibrated (O - starting from an advanced position; ■ - starting from a receded position) contact angles obtained during the run shown in Fig. 5. The arrows show how hysteresis (~ 11°) was mitigated to practically zero (~ 0.5°).

The advancing, receding and vibrated contact angles thus obtained are shown in Fig. 7. The hysteresis is small (~ 11°) but significant. However, the vibrated contact angles obtained from an advanced position practically coincide with those obtained from a receding one, i.e. hysteresis is practically eliminated.

Thus by subjecting the three-phase system to acoustic vibrations we have effectively achieved mechanical equilibrium [7-9]. Since our system is chemically inert we can now relate the vibrated contact angle, $\theta_V$, to the Wenzel equation (2) with some confidence.

All the contact angles measured with water are listed in Table 3. Even on the slightly rough surfaces (samples 1 and 2) the advancing and receding angles depart from their values on a smooth surface. The vibrated contact angle, however, is not affected until the roughness has exceeded some threshold value ($r \approx 1.1$).

**Table 3.**
Advancing, $\theta_A$, receding, $\theta_R$, and vibrated $\theta_V$, contact angles of water on smooth and increasingly rough surfaces

| Sample | $\theta_A$ [deg] | $\theta_R$ [deg] | $\theta_V$ [deg] |
|--------|------------------|------------------|------------------|
| 0 | 125.5 | 114.4 | 121.0 |
| 1 | 127.9 | 118.1 | 121.1 |
| 2 | 129.5 | 111.0 | 120.3 |
| 3 | 145.5 | 116.6 | 126.8 |
| 4 | 165.2 | 121.9 | 132.1 |

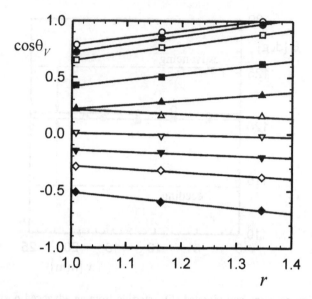

**Figure 8.** Vibrated contact angle, $\theta_V$, vs. Wenzel ratio, $r$ ($\bigcirc$ - hexane, $\bullet$ - heptane, $\square$ - octane, $\blacksquare$ - hexadecane, $\triangle$ - dimethylformamide, $\blacktriangle$ - methylnaphthalene, $\triangledown$ - dimethylsulphoxide, $\blacktriangledown$ - ethylene glycol, $\diamondsuit$ - formamide, $\blacklozenge$ - water).

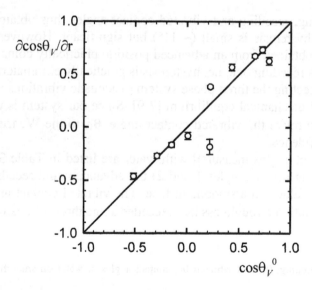

**Figure 9.** Comparison of the Wenzel slope, $\partial\cos\theta_V/\partial r$, determined from Fig. 8 ($\bigcirc$) and the theoretical value, $\cos\theta_V^0$, predicted by Eqn. (2) (solid line).

Thus two different regimes are found: (*i*) at low particle concentration (i.e. low roughness) the vibrated contact line is still entirely located on smooth portions of the surface and the vibrated contact angle is essentially equal to that measured on a smooth sample; (*ii*) at higher particle concentration (i.e. at higher roughness) such an obstacle-free position is not possible and the vibrated contact angle becomes roughness-dependent.

The vibrated contact angles for all the test liquids on the rougher surfaces (samples 3 and 4) are shown in Fig. 8, in the format prescribed by Wenzel equation (2).

The dependence of the $\cos\theta_V$ on the roughness ratio is apparently linear and the slope $\partial\cos\theta_V/\partial r$ changes from negative to positive as the angle measured on a smooth surface, $\theta_V^0$, changes from obtuse to acute. The slopes of the lines drawn in Fig. 8 are quantitatively compared with Wenzel's prediction in Fig. 9.

The solid line corresponds to Eqn. (2) and in spite of some scatter, the experimental results closely follow the theoretical trend. Thus Wenzel equation adequately accounts for the influence of the solid surface roughness on the vibrated contact angle. One single point (obtained with dimethylformamide) clearly deviates from the general trend in Fig. 9. It is also apparent in Fig. 8 that the Wenzel slope changes sign at about $\cos\theta = 0.2$ (i.e. $\theta \approx 78°$) rather than exactly at 90° as implied by the Wenzel equation. There is no clear explanation for this departure but this fact should be further investigated as Busscher *et al.* [14] have also reported that the Wenzel slope changed between 60 and 86°.

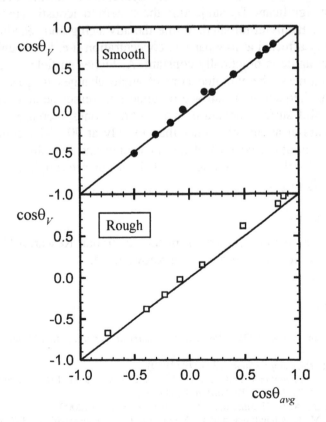

**Figure 10.** Vibrated contact angle, $\theta_V$, vs. averaged contact angle, $\theta_{avg}$, on smooth ($\bullet$, sample 0) and rough ($\square$, sample 4) surfaces.

The vibrated contact angle is rarely measured [7-9] and it is, therefore, appropriate to compare it with the routinely reported advancing, $\theta_A$, and receding, $\theta_R$, contact angles. The most popular way of averaging the two angles is the one suggested by Adam and Jessop [15], i.e.

$$\cos \theta_{avg} = \frac{1}{2} \left( \cos \theta_A + \cos \theta_R \right) \tag{5}$$

In Fig. 10 the vibrated contact angle is compared with the averaged one.

While for the smooth surface the agreement is excellent, for the rough surface it is less convincing, particularly at lower contact angles. A similar discrepancy has been reported by Andrieu *et al.* [7] for substrates which were both chemically and physically heterogeneous. In the present case the deviation is clearly caused by the roughness of the solid surface.

## 4. CONCLUSION

The contact angle hysteresis on a smooth hydrophobic surface (Teflon AF1600) was small but significant. By subjecting the system to acoustic vibrations the hysteresis was practically eliminated. While the advancing and receding contact angles are already affected at low particle concentration (i.e. low roughness), the vibrated contact angle is practically constant. At higher particle concentration (i.e. at higher roughness) the vibrated contact angle changes in general accordance with the Wenzel equation: as surface roughness increases acute contact angles on the smooth solid surface decrease, while obtuse ones increase. The transition, however, occurs at about 80° rather than exactly at 90°. The contact angle obtained by averaging the cosines of the advancing and receding angles is a good approximation for the vibrated one provided that surface roughness is small or the angles relatively large.

*Acknowledgements*

Financial support for this project from the Australian Research Council Special Research Centre Scheme is gratefully acknowledged.

## REFERENCES

1. A. W. Adamson and A. P. Gast, *Physical Chemistry of Surfaces*, 6th edition, Wiley, New York (1997).
2. D. H. Everett, *Pure Appl. Chem.* **52**, 1279 (1980).
3. T. D. Blake, in: *Surfactants*, Th.F. Tadros (Ed.), Ch. 10, Academic Press, London (1984).
4. R. J. Good, *J. Adhesion Sci. Technol.* **6**, 1269 (1992).
5. D. Y. Kwok and A. W. Neumann, *J. Phys. Chem. B* **104**, 741 (2000).
6. R. J. Good, M. K. Chaudhury and C. Yeung in: *Mittal Festschrift on Adhesion Science and Technology*, W. J. van Ooij and H. R. Anderson, Jr. (Eds.) p. 181, VSP, Utrecht (1998).

7. C. Andrieu, C. Sykes and F. Brochard, *Langmuir* **10**, 2077 (1994).
8. E. L. Decker and S. Garoff, *Langmuir* **12**, 2100 (1996).
9. C. Della Volpe, D. Maniglio, S. Siboni and M. Morra, *Oil Gas Sci. Technol. Rev. IFP* **56**, 9 (2001).
10. DuPont Product Information Sheet, *Teflon AF*, Cat. No. H-44015-3 (1998).
11. R. Hayes and J. Ralston, *Colloids Surfaces A* **80**, 137 (1993).
12. R. Hayes and J. Ralston, *J. Colloid Interface Sci.* **159**, 429 (1993).
13. R. Hayes and J. Ralston, *Colloids Surfaces A* **93**, 15 (1994).
14. H. J. Busscher, A. W. J. van Pelt, P. de Boer, H. P. de Jong and J. Arends, *Colloids Surfaces* **9**, 319 (1984).
15. N. K. Adam and G. Jessop, *J. Chem. Soc.* **127**, 1863 (1925).

*Contact Angle, Wettability and Adhesion*, Vol. 3, pp. 175–190
Ed. K.L. Mittal
© VSP 2003

# Dynamic evolution of contact angle on solid substrates during evaporation

S.K. BARTHWAL,[*,1] AMRISH K. PANWAR[1] and S. RAY[2]

[1]*Department of Physics, Indian Institute of Technology Roorkee, Roorkee - 247667, India*
[2]*Department of Metallurgy and Material Engineering, Indian Institute of Technology Roorkee, Roorkee - 247667, India*

**Abstract**—The interfacial forces which determine the interaction between a liquid and solid surface have been investigated under dynamic conditions of evaporation. The evaporation characteristics of probe liquids and their influence on the droplet of the liquid on a solid substrate have been investigated. The changes in mass, contact angle, solid-liquid contact radius during evaporation of droplets (3-90 mg) of water on glass, polycarbonate and PTFE substrates and droplets of methyl alcohol on polycarbonate, polypropylene, PTFE and high density polyethylene substrates have been examined. The evolution of contact angle and contact radius with the progress of evaporation has been investigated for each droplet-substrate system in order to identify the common trend. In the systems of water droplets on polycarbonate and glass, the contact radius remained constant with the progress of evaporation but such behavior was not observed in the case of methyl alcohol on polycarbonate, polypropylene, PTFE and high density polyethylene.

*Keywords*: Contact angle; sessile drop; evaporation of sessile drop; surface energy.

## 1. INTRODUCTION

The phenomena of wetting and adhesion between a liquid and a solid phase are very important because of their technological importance. Wetting is characterized by the contact angle formed when a liquid drop rests on a smooth, homogeneous, non-deformable and inert surface of a solid. Considering the equilibrium of various interfacial forces acting on the sessile drop, the contact angle, $\theta$, is given by Young's equation as

$$\gamma_{sv} = \gamma_{sl} + \gamma_{lv} \cos \theta \tag{1}$$

where, $\gamma_{sl}, \gamma_{sv}$ and $\gamma_{lv}$ are solid-liquid, solid-vapor and liquid-vapor interfacial energies. The wetting characteristic of a solid surface can be controlled by altering the surface energy [1-5] of the solid.

*To whom all correspondence should be addressed. Phone: +91-1332-285750,
Fax: +91-1332-273560, E-mail: skbarfph@iitr.ernet.in

Experimentally, the surface energy of a solid is probed by forming a sessile drop of a test liquid on the solid surface. The test liquids are so chosen that their polar and dispersive components of liquid-vapor interface energy are known. One may then calculate the surface energy components of the solid by measuring contact angles, $\theta$, for two liquids of known polar and dispersive energy components [6, 7]. Water has been commonly chosen as a probe liquid since its properties are well established. It has been observed that evaporation of liquid affects the measured contact angle profoundly, thus raising the question of water as a suitable test liquid in a sessile drop experiment. This paper discusses the effect of evaporation of liquids with widely different evaporation rates on the contact angle and contact radius of sessile drops on different substrate materials.

The evaporation of a small droplet of a volatile liquid under various conditions has been described by many workers [8-12]. Birdi and Vu [9] have discussed the evaporation kinetics of water and other liquids on PTFE and glass substrates and concluded that for substrate on which $\theta < 90°$, the contact radius of the drop remains constant with time during evaporation; but for substrates which show $\theta > 90°$, the contact angle of the drop remains constant and the contact radius decreases with time during evaporation. They explained their observations by considering evaporation as diffusion through the boundary layer of vapour around the droplet. However, no satisfactory explanation was put forward for constant contact angle mode of evaporation by these authors. The evaporation kinetics of water droplets on poly (methyl methacrylate) (PMMA) has been described in details by Rowan et al. [13]. Considering the diffusion process, the authors found that $\theta$ varied linearly with time for larger values of contact angles but deviations from linearity were observed for smaller values of contact angles. Shanahan and coworkers [14, 15] investigated the evaporation of water and n-decane droplets on polyethylene, epoxy resin and PTFE and observed that the contact radius and contact angle behavior depended upon the vapour concentration around the droplets.

All the calculations and experiments discussed above have been carried out on small drops having spherical cap geometry. Calculations on drops with non-spherical cap geometry have been performed by Erbil and coworkers [11, 16, 17]. These authors concluded that experimental results were better described for the drops with pseudo-spherical cap geometry as compared to spherical cap geometry.

In all the studies mentioned above, the role of the substrate was not investigated and the effect of substrate entered indirectly through contact angle only. In the present study, the influence of evaporation on the contact angle and contact radius of sessile drops on different substrate materials has been investigated. We have chosen water and methyl alcohol as liquids since their evaporation rates are widely different. In order to investigate the role of substrate glass, polycarbonate, high density polyethylene (HDPE), polypropylene and PTFE were chosen as substrate materials in the present studies.

## 2. EXPERIMENTAL DETAILS

The substrates used in this study were flat and smooth samples of glass (microscope slides) and polymer sheets of size 2 cm x 2 cm and thickness in the range 1.5 to 3 mm. The glass, polypropylene (PP), high density polyethylene (HDPE) and poly(tetrafluroethylene) (PTFE) substrates were cleaned ultrasonically in analytical grade acetone and the polycarbonate substrates were cleaned in petroleum ether. The substrates were dried subsequently by hot air. The surface roughness was measured by Sloan profilometer model Dektak IIA. The average surface roughness of glass, polycarbonate, polypropylene and HDPE was found to be about 0.15 μm, 0.04 μm, 0.28 μm and 0.04 μm, respectively. Apart from the PTFE substrates which show a large value of roughness (about 2.0 μm), all other substrates can be assumed fairly smooth. All roughness measurements were done over 6 to 10 mm length of the substrates. For measurement of mass of sessile drop the substrates were placed on a Mettler microbalance Model AJ100. A drop of probe liquid was then carefully placed on the substrate using a specially designed syringe. Entire assembly was enclosed inside a chamber such that there was no vapor built up with time nor was draft of air over the drops so as to affect the evaporation rates. The drops were illuminated using a quartz halogen lamp light source. The shadow photographs of the drops were taken using Leica Stereo zoom microscope Model M3Z fitted with a Nikon E995 digital camera. The intensity of the source was kept low enough such that the exposure time of the order of 1/8 s yielded satisfactory photographs. Further, a mechanical shutter was also placed between the light source and drops and opened during exposure only, thereby reducing the heating due to light source. No appreciable increase in temperature was found due to light source. The temperature and humidity of the ambient were continuously monitored.

The digital images were then transferred to a personal computer for image processing and contact angle measurements. The microscope was also fitted with a goniometer and could be moved horizontally. The horizontal motion of the microscope was measured by a vernier with an accuracy of 0.0001 cm. The repeatability in the contact angle measurements was about ±2°. Although, the contact angle and the drop base radius were evaluated from the digital images, the data were also verified by manually measuring the radius and the contact angle using vernier and goniometer.

## 3. RESULTS AND DISCUSSION

The humidity and temperature of the ambient are given in Table 1. It is well known that the evaporation rate is strongly dependent on the humidity and temperature. We could not measure these parameters very close to the drop surface, so the data presented here were obtained in the chamber surrounding the drop. Since each evaporation experiment lasted only for about an hour, there was a variation of only about 1°C in room temperature and about 1% in relative humidity.

The variations of the mass of the sessile drop of water on polycarbonate substrate with time due to evaporation are shown in Fig. 1 (a) for different initial masses. Figure 2 (a) shows the variations of the mass of the sessile drop of methyl alcohol on polycarbonate substrate as a function of time. As can be seen from Fig. 1 (a), the mass of the water drop on polycarbonate decreases linearly with time, indicating a steady state evaporation. The rate of change of mass i.e. the evaporation rate ($dm/dt$) increases with increasing initial mass of the drop. A similar behavior was also observed for water drops on the glass substrate [18] and for small drops by Birdi *et al.* [8]. Although the data are not presented here, in general, a similar behavior was also seen for other water-substrate systems. However, the linearity of drop mass as a function of time is not so good for HDPE and PTFE substrates. On the other hand, drops of methyl alcohol on polycarbonate show a non-linear behavior Fig. 2 (a). The evaporation is much faster during initial period and then proceeds slowly in a linear fashion indicating the onset of equilibrium. A similar behavior, in general, is seen for methyl alcohol on HDPE, polypropylene, and PTFE substrates. The data for methyl alcohol on glass could not be taken as the drops of methyl alcohol spread over the glass substrate. This faster evaporation during the initial period is perhaps due to higher volatility of methyl alcohol as compared to water. The subsequent slower evaporation could be attributed to increased concentration of alcohol vapours around the drops. The alcohol vapour concentration could not be measured due to lack of any facility for such measurements in our laboratory. Further, it is possible that such fast evaporation of methyl alcohol may change the temperature of the drop significantly, affecting the evaporation rates. As mentioned earlier, it was not possible for us to measure the temperature of the drop directly, however, no change in the ambient temperature was observed.

The variation of contact radius, $r_b$, of drops of water on polycarbonate as a function of time as evaporation proceeds is shown in Fig. 1 (b). The contact radius remains constant until the drop size becomes so small as to make an accurate drop profile measurement difficult. A similar behavior was also seen for water

**Table 1.**

Temperature and humidity of the ambient during evaporation time

| System | Water on glass | Water on PC | Water on HDPE | Water on PTFE | Methyl alcohol on glass | Methyl alcohol on PC | Methyl alcohol on PP | Methyl alcohol on HDPE | Methyl alcohol on PTFE |
|---|---|---|---|---|---|---|---|---|---|
| Temperature in °C | 27-28 | 25-26 | 25-26 | 24-25 | 22-23 | 21-22 | 22 | 21 | 22 |
| Relative humidity in % | 61-63 | 69-71 | 68 | 60-63 | 65-66 | 65-68 | 61-62 | 62 | 59-60 |

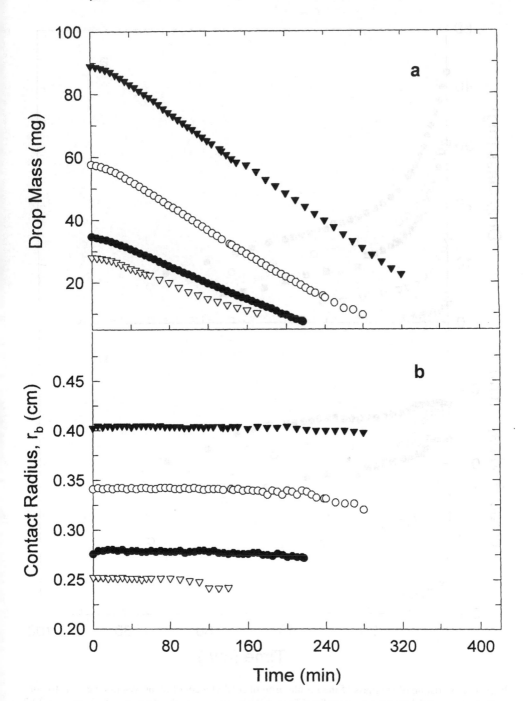

**Figure 1.** Variation of (a) mass of the sessile drop of water on polycarbonate and (b) contact radius, $r_b$, with evaporation time for different initial masses of the drop: ▼ - 89 mg, ○ - 57.6 mg, ● - 34.5 mg, ▽ - 28.1 mg.

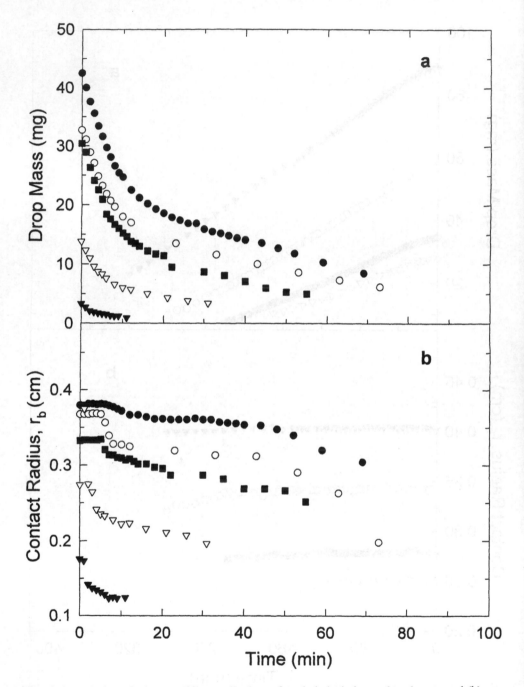

**Figure 2.** Variation of (a) mass of the sessile drop of methyl alcohol on polycarbonate and (b) contact radius, $r_b$ with evaporation time for different initial masses of the drop: ● - 42.5 mg, ○ - 32.7 mg, ■ - 30.4 mg, ▽ - 13.8 mg, ▼ - 3.3 mg.

drops on glass substrate [18]. On the other hand, the variation of contact radius of water drops on PTFE and HDPE is much more complex as shown in Fig. 3 (a). The contact radius in case of water on HDPE continuously decreases with time. On the other hand, in case of PTFE, it remains constant initially and then shows a slow decrease with time. It has been observed [13, 18] that in systems where drop mass changes linearly with evaporation time, the contact radius remains constant during evaporation. This bears out the fact that evaporation rate is proportional to the contact radius of the drop, as has also been suggested by other researchers [8]. Figure 4 shows the variation of evaporation rate as a function of contact radius of the drops of water on three different substrates. Figure 5 shows the variation of drop mass and contact radius as a function of evaporation time for methyl alcohol on polypropylene. The contact radius remains constant during the initial phase of evaporation then it decreases rapidly and again becomes constant thereafter. It has been observed that the rapid decrease corresponds to time when the evaporation rate reaches the steady state value in mass versus time plots shown in Fig. 5 (a). If one approximates the variation of mass as a function of time by two different evaporation rates (straight lines), the kink in contact radius versus time plot corresponds to time when these two straight lines intersect each other. This initial constancy and subsequent sudden decrease in the contact radius has also been observed for methyl alcohol on polycarbonate, PTFE and HDPE as shown in Figs. 6 (a) and 7 (a); however, the contact radius does not remain constant after the sudden decrease. One may, therefore, conclude that evaporation rate is closely related to the contact radius or vice versa. A closer look at diffusion equation may give a clearer picture for the relationship between evaporation rate and contact radius.

The variation of contact angle as a function of evaporation time is more complex. It must be pointed out here that the contact angles were measured directly and were not calculated from height and contact radius data of the drops; therefore, the cap geometry is not important as for angle measurement. The variation of contact angle, $\theta$, for water on glass as a function of time is given in Fig. 8 (a), which shows that the contact angle decreases almost linearly with time. The slope of the contact angle versus time plots $(d\theta/dt)$ is greater for drops with higher initial mass. Figure 8 (b) shows the variation of contact angle with time for drops of water on polycarbonate substrate. While the variation of contact angle for drops of water on glass is linear, this is not the case for polycarbonate. For both cases of water on glass and water on polycarbonate, the contact angle decreases more rapidly for smaller drops. The variation of contact angle for water on PTFE as a function of time is shown in Fig. 3 (b). As can be seen, the contact angle first decreases linearly with increasing time and then shows a non-linear behavior. If one correlates the behavior of contact angle and contact radius as a function of evaporation time, it can be seen that a constant contact radius implies a linear variation of contact angle and the deviation from constancy of contact radius results in a non linear behavior of contact angle. It must be pointed out here that the contact angle of water on PTFE is more than 90° whereas in all other systems investi-

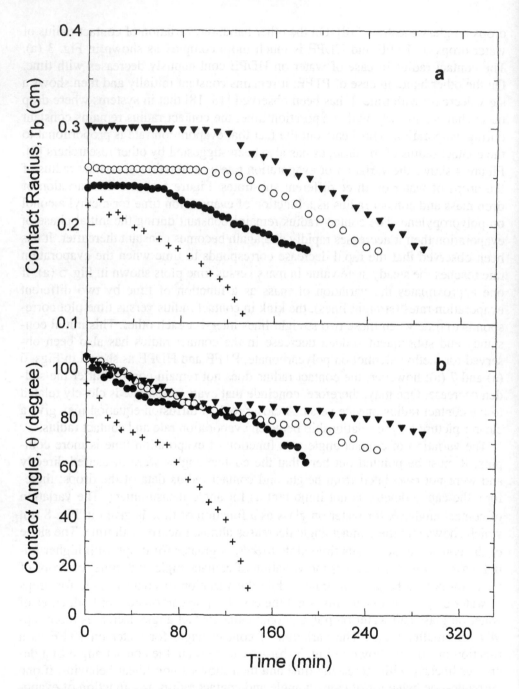

**Figure 3.** Variation of (a) contact radius, $r_b$ of the sessile drop of water on PTFE and high density polyethylene (HDPE) and (b) contact angle, $\theta$ with evaporation time for different initial masses of the drop: on PTFE: ▼ - 73.5 mg, ○ - 50.2 mg, ● - 37.5 mg, and on HDPE: + 30 mg.

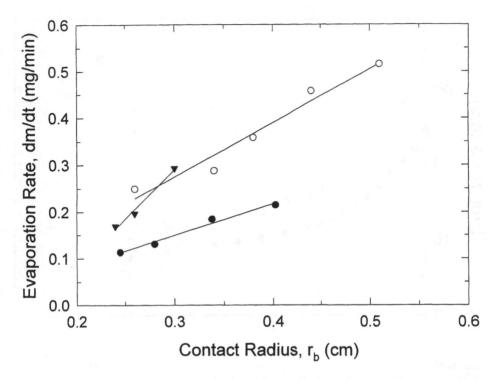

**Figure 4.** Variation of evaporation rate (dm/dt) of the sessile drop of water with contact radius, $r_b$ for different substrates: ○ - glass, ▼ - PTFE and ● - polycarbonate.

gated, the contact angles are smaller than 90°. The variation of contact angle with evaporation time for methyl alcohol on PTFE is shown in Fig. 6 (b). The contact angle first decreases linearly with time corresponding to a constant contact radius, then shows a hump corresponding to the rapid change in contact radius and finally decreases non-linearly. A similar behavior was also observed for methyl alcohol on polypropylene as shown in Fig. 9 (a). However the hump is not so prominent in this case. This seems to be more or less true for methyl alcohol on polycarbonate Fig. 9 (b) and HDPE Fig. 7 (b). We have not observed any case in which the contact angle remains constant and contact radius decreases as evaporation proceeds.

It is thus clear that the contact angle and contact radius are related to each other during evaporation. A relationship between these could be sought in terms of free energy changes for the drop during evaporation. The changes in the free energy are determined by the changes in the volume free energy of the drop due to evaporation and those resulting from the various interfaces. The total free energy of the drop is the sum of interfacial energies of all the three interfaces, i.e., liquid-vapour ($A_{lv}\gamma_{lv}$), solid-liquid ($A_{sl}\gamma_{sl}$) and solid-vapour ($A_{sv}\gamma_{sv}$), where $\gamma_{sl}$, $\gamma_{sv}$ and $\gamma_{lv}$ are solid-liquid, solid-vapor and liquid-vapor interfacial energies, respectively and $A_{sl}$, $A_{sv}$ and $A_{lv}$ are the corresponding interface areas. When the drop size

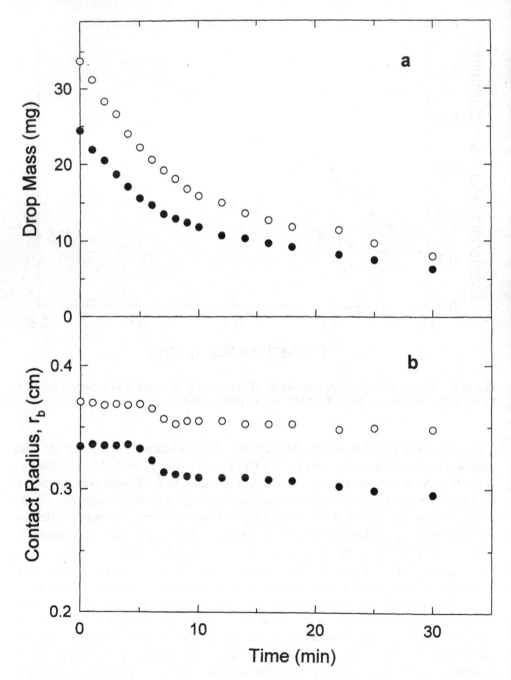

**Figure 5.** Variation of (a) mass of the sessile drop of methyl alcohol on polypropylene and (b) contact radius, $r_b$ with evaporation time for different initial masses of the drop: ○ - 33.7 mg and ● - 24 mg.

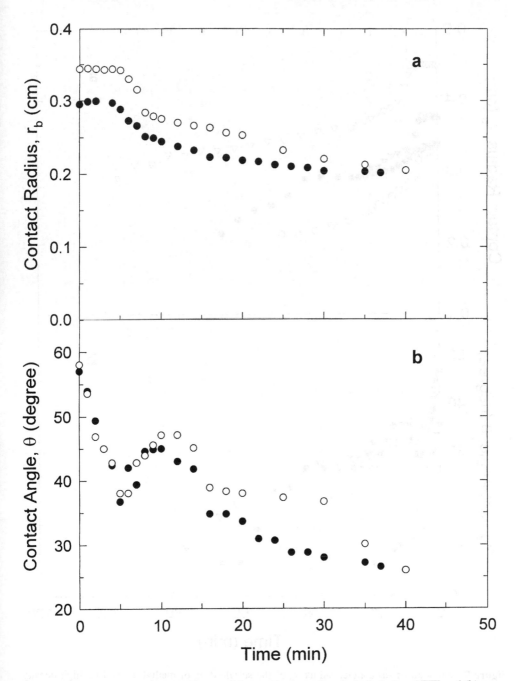

**Figure 6.** Variation of (a) contact radius, $r_b$ of the sessile drop of methyl alcohol on PTFE and (b) contact angle, $\theta$ with evaporation time for different initial masses of the drop: $\circ$ - 34.8 mg and $\bullet$ - 25 mg.

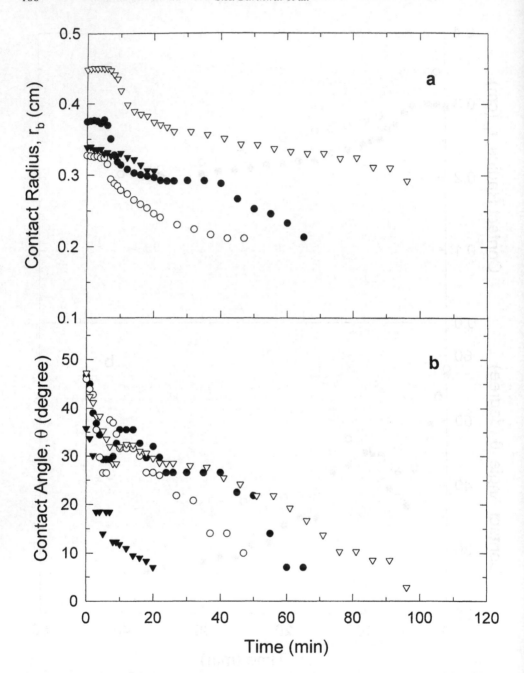

**Figure 7.** Variation of (a) contact radius, $r_b$ of the sessile drop of methyl alcohol on high density polyethylene and (b) contact angle, $\theta$ with evaporation time for different initial masses of the drop: $\nabla$ - 51.5 mg, ● - 34.5 mg, ▼- 25.6 mg and ○ - 21.2 mg.

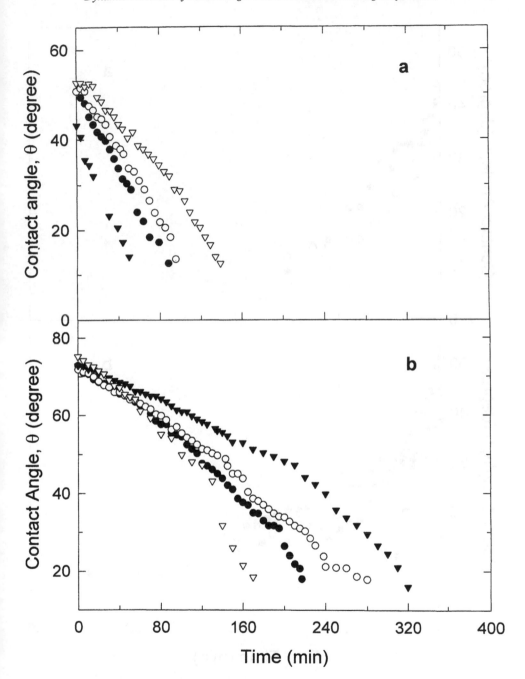

**Figure 8.** (a) Variation of contact angle, θ with evaporation time for sessile drop of water on glass for different initial masses: ∇ - 73.5 mg, ○ - 45 mg, ● - 32.4 mg and ▼ - 15.4 mg. (b) Variation of contact angle, θ with evaporation time for sessile drop of water on polycarbonate for different initial masses: ▼ - 89 mg, ○ - 57.6 mg, ● - 34.5 mg and ∇ - 28.1 mg.

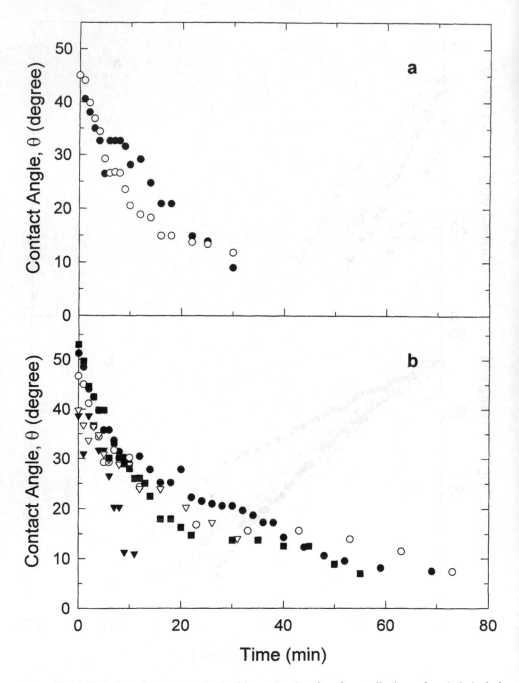

**Figure 9.** (a) Variation of contact angle, θ with evaporation time for sessile drop of methyl alcohol on polypropylene for different initial masses: ○ - 33.7 mg and ● - 24 mg. (b) Variation of contact angle, θ with evaporation time for sessile drop of methyl alcohol on polycarbonate for different initial masses: ● - 42.5 mg, ○ - 32.7 mg, ■ - 30.4 mg, ▽ - 13.8 mg, ▼ - 3.3 mg.

changes due to evaporation, the geometrical parameters of the drop will adjust in such a way as to minimize the free energy. The change in total free energy is given by

$$\Delta G = \Delta A_{lv}\gamma_{lv} + \Delta A_{sl}\gamma_{sl} + \Delta A_{sv}\gamma_{sv} + \Delta V \Delta G_v^e \qquad (2)$$

where, $\Delta A$ is the change in interfacial area indicated by the subscripts $l$, $v$, $s$ denoting liquid, vapour and the solid substrate, respectively. $\Delta V$ is the change in the volume of the drop and $\Delta G_v^e$ is the change in free energy per unit volume of the liquid due to evaporation at the ambient temperature. The mode of evaporation would depend on the change in free energy along either constant radius or constant contact angle route. This problem is under investigation and the results will be published elsewhere.


## 4. CONCLUSIONS

The behaviors of contact radius and contact angle of sessile drops of water and methyl alcohol on glass, polycarbonate, high density polyethylene, polypropylene and PTFE substrates with the progress of evaporation have been studied by direct measurements of contact angles. It has been shown that a constant contact radius corresponds to a constant evaporation rate leading to a linear variation of contact angle with evaporation time. Further, the substrate does play a significant role in determining the behavior of the contact angle.

*Acknowledgements*

The authors thankfully acknowledge the financial support provided by All India Council for Technical Education, Government of India.


## REFERENCES

1. K. L. Mittal (Ed.), *Polymer Surface Modification; Relevance to Adhesion,* VSP, Utrecht (1996).
2. K. L. Mittal (Ed.), *Polymer Surface Modification; Relevance to Adhesion,* Vol. 2, VSP, Utrecht (2000).
3. E. M. Liston, L. Martinu and M. R. Wertheimer, *J. Adhesion Sci. Technol.* **7**, 1091-1127 (1993).
4. S. O. Kell, T. Henshaw, G. Farrow, M. Aindow and C. Jones, *Surface Interface Anal.* **23**, 319-327 (1995).
5. F. Poncin-Epaillard, B. Chevet and J.-C. Brosse, *J. Adhesion Sci. Technol.* **8**, 455-468 (1994).
6. J. Kinloch, *J. Mater. Sci.* **15**, 2141-2166 (1980).
7. S. Bhomick, P. K. Ghosh, S. Ray and S. K. Barthwal, *J. Adhesion Sci. Technol.* **12**, 1181-1204 (1998).
8. K. S. Birdi, D. T. Vu and A. J. Winter, *J. Phys. Chem.* **93**, 3702-3703 (1989).
9. K. S. Birdi and D. T. Vu, in: *Contact Angle, Wettability and Adhesion,* K. L. Mittal (Ed.), pp. 285-293, VSP, Utrecht, The Netherlands (1993).
10. R. A. Meric and H. Y. Erbil, *Langmuir,* **14**, 1915-1920 (1998).

11. H. Y. Erbil, *J. Adhesion Sci. Technol.* **13**, 1405-1413 (1999).
12. G. McHale, S. M. Rowan, M. I. Newton and M. K. Banerjee, *J. Phys. Chem. B*, **102**, 1964-1967 (1998).
13. S. M. Rowan, M. I. Newton and G. McHale, *J. Phys. Chem.* **99**, 13268-13271 (1995).
14. M. E. R. Shanahan and C. Bourges, *Int. J. Adhesion Adhesives*, **14**, 201 (1994).
15. C. Bourges-Monnier and M. E. R. Shanahan, *Langmuir*, **11**, 2820 (1995).
16. H. Y. Erbil and R. A. Meric, *J. Phys. Chem. B*, **101**, 6867-6873 (1997).
17. H. Y. Erbil, G. McHale, S. M. Rowan and M. I. Newton, *J. Adhesion Sci. Technol.* **13**, 1375-1391 (1999).
18. A. K. Panwar, S. K. Barthwal and S. Ray, *J. Adhesion Sci. Technol.* (communicated).

*Contact Angle, Wettability and Adhesion*, Vol. 3, pp. 191–210
Ed. K.L. Mittal
© VSP 2003

# The concept, characterization, concerns and consequences of contact angles in solid-liquid-liquid systems

DANDINA N. RAO*

*The Craft and Hawkins Department of Petroleum Engineering, Louisiana State University, 3516 CEBA Bldg., Baton Rouge, LA 70803*

**Abstract**—An aura of skepticism surrounds the use of the concept of contact angles to characterize wettability of solid-liquid-liquid (S-L-L) systems such as those consisting of petroleum reservoir rock, crude oil and brine. This review article analyzes the various conventional techniques used in contact angle measurements and their inability to account for the effect of the adhesion interaction at the S-L-L interface on measured contact angles that has led to the prevailing skepticism. The recent development of the dual-drop dual-crystal (DDDC) technique that overcomes these limitations and yields highly reproducible dynamic contact angles is presented with example case studies. The important role of the liquid-liquid interfacial tension on the spreading behavior in solid-liquid-liquid systems is discussed. The paper attempts also to address the commonly encountered concerns of practicing engineers in the petroleum industry regarding contact angles and their relation to multiphase flow in porous media.

*Keywords*: Adhesion; dynamic contact angles; interfacial tension; solid-fluids interactions; spreading; wettability.

## 1. INTRODUCTION

The production of petroleum fluids from underground reservoirs involves the movement of fluid-fluid interfaces through the tortuous paths within the porous rocks. The movement of oil, brine and gas phases relative to one another depends on the nature of distribution of these fluids in the rock pores as well as on the strength of interaction between the fluids and the rock surface. While the distribution of the phases is largely governed by the spreading coefficient, which involves the interfacial tensions at the three interfaces (gas/oil/brine), the strength of rock-fluids interactions depends on the dynamic (advancing and receding) contact angles subtended by the moving fluid-fluid interface with the rock surface. These interfacial phenomena of fluid-fluid spreading and the rock-fluids interactions have

---

*Phone: 225-578-6037, Fax: 225-578-6039, E-mail: dnrao@lsu.edu

generally been lumped into one parameter called wettability in petroleum engineering literature. Although yielding the advantage of a single global term for all the interactions, the term wettability has masked our understanding of the surface and interfacial phenomena, e.g., adhesion, spreading, and their influence on the dynamics of fluids in petroleum reservoirs. The situation is further complicated by the fact that the conventional techniques used to measure dynamic contact angles in solid-liquid-vapor (S-L-V) systems have failed to yield meaningful results when applied to solid-liquid-liquid (S-L-L) systems such as the rock-oil-brine system of interest in petroleum engineering. This has resulted in concerns and skepticism surrounding the applicability of the contact angle concept to S-L-L systems to characterize their wetting tendency. The presence of a thin wetting film of an immiscible liquid on the solid surface and the parameters governing the stability of the wetting film are factors that control the preferential wetting behavior in S-L-L systems. In addition to addressing these concerns, this paper highlights the recent developments related to techniques for making meaningful measurements of dynamic contact angles in crude oil-brine-rock systems at actual reservoir conditions of pressure and temperature in addition to utilizing these data to understand the role of interfacial interactions involved in adhesion, spreading and wettability in S-L-L systems.

## 2. CONCEPTS AND DEFINITIONS

The year 2005 will mark the 200th anniversary of Young's equation that introduced the concept of contact angle to describe the extent of interaction between fluids and solids. On one hand, questions have been raised in the published literature regarding the limitations of Young's equation based on the grounds that it ignores the normal force component at the three-phase contact line, and hence cannot describe the whole equilibrium relationship. On the other hand, lengthy thermodynamic proofs of Young's equation have also been published. Experimental verification of Young's equation had not been possible mainly due to the difficulty in measuring the interfacial tensions involving solid surfaces. However, only in 1981 a single study was reported where all four terms in the Young's equation were independently measured and the equation was thus verified to be correct [1]. A single experimental verification of an equation that has been in use for nearly two centuries can hardly be considered adequate. However, with the advent of successful experimental techniques to measure the interaction forces (to within about $10^{-8}$ N) between flat surfaces separated in a third medium by very short distances (down to 0.1 nm), more verifications could be expected in the not too distant future.

Irrespective of the arguments about its limitations, the concept of contact angle, introduced by Young's equation, was valuable because it gave a definition to the notion of wettability and led to the development of the concept of dynamic contact angles to describe moving interfaces on solid surfaces. Much of these devel-

**Table 1.**
The Concept of Contact Angles in S-L-V and S-L-L Systems

| S-L-V System | S-L-L System |
|---|---|
| 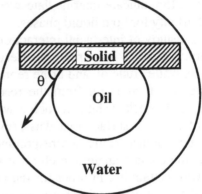  | |

| S-L-V System | S-L-L System |
|---|---|
| 1. Contact angle measured in denser phase (i.e: liquid) | 1. Contact angle measured in denser phase (i.e: water) |
| 2. Denser fluid (liquid) advances on the solid surface as drop is placed on it. Hence, measured angle corresponds to advancing angle | 2. Denser phase (water) recedes from the solid surface as oil drop is placed on it. Hence, measured angle is the water-receding contact angle |
| 3. Smaller advancing angles indicate spreading of the liquid drop displacing vapor | 3. Large water-receding angles indicate spreading of the oil drop displacing water |
| 4. Spreading indicates wetting of the solid surface by liquid | 4. Spreading is not necessarily an indication of wetting preference. Oil can spread on a film of water covering the surface |
| 5. Spreading, wetting and adhesion are almost synonymous | 5. Extent of adhesion of oil on surface determines wettability. Hence spreading, wetting and adhesion are not synonymous |

opments dealt with solid-liquid-vapor (S-L-V) systems, which were then extended to solid-liquid-liquid (S-L-L) systems. However, for sake of clarity, it is important to recognize the differences in the definitions of the concepts of wettability and dynamic contact angles between these systems. Some of these differences are listed in Table 1, which indicate that while we have adopted the same convention of measuring contact angles in the denser phase in both S-L-V and S-L-L systems, the definition of wettability is quite different in the two systems. Wettability has similar meaning as spreading or adhesion of the denser fluid in S-L-V systems and is characterized by decreasing advancing angle (measured in the denser phase). In S-L-L systems, however, spreading and adhesion are two different phenomena. While spreading of the less dense fluid in S-L-L systems correlates well with increasing receding angle (measured in the denser fluid), wettability of the system is governed by the advancing angle (as explained below) which reflects the extent of adhesion of the less dense fluid to the solid surface. Advancing angles less than 70° (measured in the denser phase) indicate the preference of the solid surface to be wetted by the denser phase; advancing angles larger than 115° indicate the less dense fluid to be the wetting phase; advancing angles in the range of 70°–115° indicate intermediate wettability where the solid surface has nearly equal affinity for both liquid phases.

In the study of interfacial interactions, namely in spreading, adhesion and wettability, occurring in petroleum reservoirs, the porous rock represents the solid surface, while crude oil and brine represent the two immiscible liquid phases. The production of crude oil from the reservoir requires that the oil-water interface move on the rock surface in such a way that water (i.e., brine) advances over an area of the rock surface that was previously occupied by the produced oil. The dynamic contact angle, corresponding to this flow scenario, is the water-advancing contact angle; therefore, it has become well accepted as a measure of reservoir wettability. In order for the oil-water interface to move on the rock surface, the force of adhesion between the oil and the rock surface needs to be overcome by the viscous forces that are responsible for mobilizing the oil in the rock pores. Thus, adhesion is reflected by the water-advancing contact angle, while spreading behavior is quantified by the water-receding contact angle. Since wettability is defined in terms of the advancing angle, it refers more to the adhesion aspect of the rock-fluids interactions than spreading of oil. An assumption that is commonly made in S-L-V systems is that if the liquid spreads on the surface, it must also adhere to it. This is not necessarily true in S-L-L systems, since oil can spread on a film of water that is wetting the solid surface. On the contrary, oil adhering to the surface need not spread on it. Both spreading and adhesion phenomena are important and affect the dynamics of oil-water flow in the porous rock medium. Hence wettability of a reservoir rock-oil-brine system, which is meant to be a catchall term to represent all of the rock-fluids interactions, should ideally incorporate both the adhesion and spreading phenomena, represented by the advancing and receding contact angles, respectively. Do the techniques of measurements commonly used allow us to accomplish this objective? An attempt to address this question is made in the following section.

## 3. CHARACTERIZATION OF DYNAMIC CONTACT ANGLES IN S-L-L SYSTEMS

### 3.1. Sessile Drop Technique

The commonly used techniques, (such as the drop volume expansion and contraction, lateral shift of the base of the sessile drop resting on the solid surface while anchoring the tip of the drop to the needle used for drop placement, and tilting of the solid surface), to measure dynamic (advancing and receding) contact angles in S-L-V systems, seldom work in S-L-L systems. The development of adhesion be-

| **Water-Wet** | **Intermediately-Wet** | **Oil-Wet** |
|:---:|:---:|:---:|

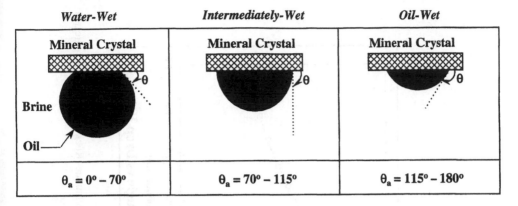

| $\theta_a = 0° - 70°$ | $\theta_a = 70° - 115°$ | $\theta_a = 115° - 180°$ |
|:---:|:---:|:---:|

**Figure 1.** The conventional sessile drop contact angle technique.

tween the oil drop and the solid surface and the consequent pinning of the contact line prevents the movement of the contact line required to satisfy the definition of advancing and receding contact angles. The lack of such movement of the contact line appears to be the main reason for the poor reproducibility of dynamic angle measurements in S-L-L systems encountered by other researchers [2] using the sessile drop (changing volume) technique illustrated in Figure 1. In this technique, a capillary tube is used to place a crude oil drop on a solid surface immersed in reservoir brine. As the oil drop is placed on the surface immersed in water, water recedes to make room for the oil. Therefore, Figure 1 depicts in actuality the water-receding contact angle ($\theta_r$), while wettability refers to the water-advancing contact angle ($\theta_a$). It can be further observed in Figure 1 that as the receding angle (measured in water, the denser phase) increases, the oil drop spreads on the solid surface. Therefore, the spreading behavior is related to the water-receding contact angle ($\theta_r$).

In order to measure the advancing angle, $\theta_a$, the oil drop needs to be moved on the solid surface so that water can advance over an area previously occupied by oil. To accomplish this, attempts were made to withdraw the oil drop back into the capillary tube used for placing the drop at the beginning of the experiment. This technique repeatedly failed to advance water in experiments in which the crude oil developed even slight adhesion with the solid surface. An example of this is shown in Figure 2, where drop profiles obtained from image analysis of the photographs of a crude oil drop resting on a smooth quartz surface immersed in deionized water are shown in part A and the measured contact angles in part B. The drop profiles clearly demonstrate the pinning of the contact line, which resulted in the monotonically increasing (when measured in the water phase) contact angle as shown in part B of Figure 2. Such results have led to the currently prevailing skepticism that contact angles are meaningless and of no use in characterizing wettability.

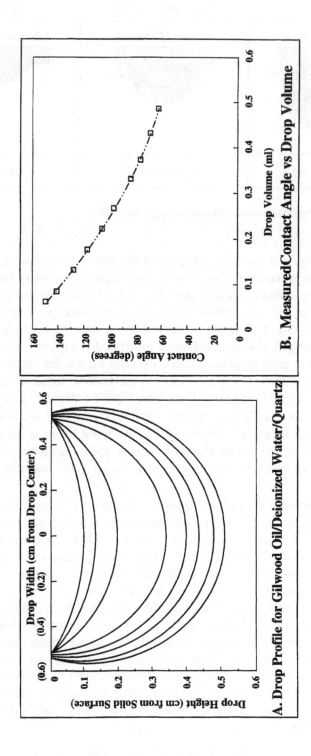

**Figure 2.** Effect of Rock-Oil Adhesion on Drop Shape (A) and Measured Contact Angle (B) in the Sessile Drop Technique (Rock-oil adhesion is indicated by the pinning of the contact line shown by the converging drop profiles at the two ends in part A).

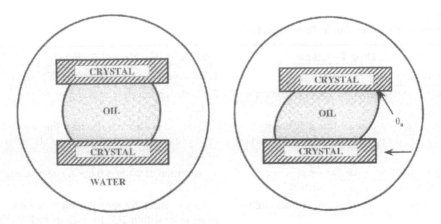

A. Depiction of the Modified Sessile Drop Technique. (The picture on the left shows the initial shape of the oil drop while that on the right depicts the expected shape when the lower crystal position is shifted)

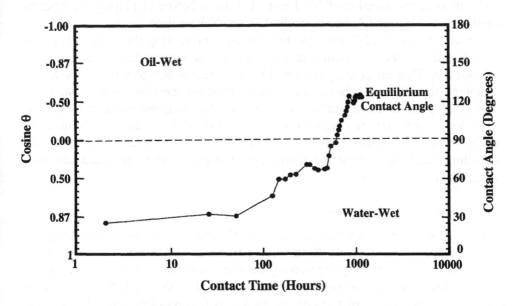

B. Example Results for the Lindbergh Crude Oil-Brine-Rock System

**Figure 3.** Modified Sessile Drop Technique for S-L-L Systems.

## 3.2. Modified Sessile Drop Technique

This technique involves placing a drop of crude oil between two solid surfaces immersed in water, aging the drop and then shifting the lower surface laterally, as shown in Figure 3 - part A. Part B shows the results obtained in an experimental study using this technique. It can be seen that the advancing angle increased

**Table 2.**
Comparison of Contact Angle Techniques for S-L-L Systems

| Modified Sessile Drop Technique | Dual-Drop Dual-Crystal Technique |
|---|---|
| 1. Long test durations (30-60 days) | 1. Relatively short test durations (3-5 days) |
| 2. Different rates of approach to equilibrium on two surfaces | 2. Similar approach to equilibrium on both surfaces |
| 3. Uncertainty of contact-line movement within the oil-exposed area of the solid surface | 3. Concrete evidence for contact-line movement within the oil-exposed area of the solid surface |
| 4. Doubt whether the definition of water-advancing contact angle is satisfied | 4. Definition of water-advancing contact angle is satisfied |
| 5. No opportunity to check for reproducibility | 5. Easy to check for reproducibility and it is found to be within 2-3° for various rock-liquid systems |

gradually over the duration of 1000 hours (41 days) before displaying an apparent stability at 124° around 1176 hours (49 days) of contact time.

The end result of 124° was questionable as representing the water-advancing angle because of the uncertainty in movement of the contact line on either of the two surfaces. This uncertainty is caused by the effect of net gravity (or buoyancy) in making the oil drop behave like a sessile drop on the upper surface and as a pendent drop on the lower surface. Furthermore, the long test duration of 49 days precluded a repeat test to check for reproducibility of the results. Such uncertain results from the modified sessile drop technique have served to strengthen the prevailing skepticism about the usefulness of contact angles in evaluating wettability of S-L-L systems.

### 3.3. Wilhelmy Plate Technique

This technique, although widely used in S-L-V systems, has seen limited applications in S-L-L systems [3, 4]. Both water-advancing and receding contact angles can be measured by raising or lowering the rock surface through the oil-water interface. This technique measures adhesion tension (i.e., the product of the interfacial tension and cosine of the contact angle), which requires separate measurements of interfacial tension in order to obtain the value of contact angle. The experimental results for S-L-L systems obtained in our laboratory using the Wilhelmy technique are presented by Muhammad [4]. The only limitation of this apparatus for reservoir wettability studies is that it operates at ambient conditions, but optical cells to measure contact angles can be designed and operated at reservoir pressures and temperatures.

## 3.4. The Dual-Drop Dual-Crystal (DDDC) Contact Angle Technique

The problems of the lack of reproducibility and long test durations associated with the sessile drop and modified sessile drop techniques, discussed earlier, have led to the development of this technique. The main reason behind these problems with the sessile and modified sessile drop contact angle techniques was that they did not allow the contact line to move and enable the water to advance over a previously oil-occupied area of the surface. Hence they failed to satisfy the definition of water-advancing contact angle and, consequently, the measured angles did not represent the wettability of the S-L-L system. The DDDC technique overcomes this limitation by allowing both surfaces of the solid to possess a similar history of exposure to crude oil under the influence of buoyancy. The development, testing and several applications of the DDDC technique have been presented elsewhere [5-7]. This technique involves initiating the experiment by placing two separate drops of crude oil on two parallel crystal surfaces immersed in formation brine at reservoir conditions of temperature (up to 200°C) and pressure (up to 70 MPa). The two-drops/crystal/brine system is then aged for 24 hours to attain equilibrium. Longer aging periods of 2 days, 7 days and 60 days have all been tested with results being similar to those from 24 hours of aging. The force of buoyancy under which the drops are aged appears to be the reason for rapid establishment of rock-fluids interactions in various oil-brine-rock systems tested so far. After the initial aging period, the lower crystal is turned over, when one of three following events occurs: (i) due to strong oil-rock adhesion the entire drop remains attached to the surface, (ii) due to weaker adhesion a small portion of the oil drop remains on the surface while the rest floats away due to buoyancy, and (iii) due to negligible oil-rock adhesion all of the oil drop detaches from the lower surface. Then the upper surface is lowered to mingle the two drops into one, or to let the upper drop contact the same area of the lower surface where the drop of oil was initially placed, and to squeeze the resulting drop between the two surfaces. The lower crystal is then shifted sideways in small steps until the three-phase contact line (TPCL) begins to move on it, thus enabling the water to advance onto a previously oil occupied area of the lower surface. The peak angle measured in water at which the movement of the contact line is initiated is the water-advancing contact angle. The angle on the other side of the oil drop where water recedes and oil advances is the water-receding contact angle. The technique also enables accurate monitoring of the movement of the contact line, which enables checking for reproducibility of results by repeating the test, and ending the test when the contact line moves past the oil-exposed area of the lower surface. Each DDDC test, including thorough cleaning of the apparatus, the initial aging time, several measurements of the advancing and receding angles, and repeatability checks, can be completed within 3-5 days. This is a significant reduction compared to 30-60 days reported for the modified sessile drop technique without repeatability checks. The reproducibility of water-advancing contact angles has been within 2-3°, as shown by

**Table 3.**
Comparison of Wettability (expressed as contact angles in degrees) from DDDC and Modified
Sessile Drop Technique

| Case | DDDC Technique | | | | Modified Sessile Drop Technique |
|---|---|---|---|---|---|
| | Trial #1 | Trial #2 | Trial #3 | Trial #4 | |
| Water-Wet (24.13 MPa, 96.5°C) | 50 | 53 | 52 | 50 | 40 |
| Intermediate-Wet (17.24 MPa, 60°C) | 110 | 110 | 108 | 108 | 38 |
| Oil-Wet (24.13 MPa, 96.5°C) | 147 | 144 | 144 | 146 | 25 |

the three example cases in Table 3, for all measurements made using the DDDC technique for various rock-fluids systems at reservoir conditions over a period of ten years. The salient differences between the DDDC and the modified sessile drop techniques are summarized in Table 2. Thus the long standing concerns of poor reproducibility and long test duration that have plagued the applicability of the concept of contact angle to characterize reservoir wettability appear to be well addressed by the DDDC technique. However, there remain some other concerns, which are discussed briefly in the next section.

## 4. CONCERNS IN RESERVOIR WETTABILITY CHARACTERIZATION THROUGH CONTACT ANGLES

1. *All contact angle methods, including the DDDC technique, suffer from the basic limitation that they use smooth crystal surfaces in place of reservoir rocks that are heterogeneous in their mineralogical and roughness characteristics.*
   The reservoir rock has four different characteristics of interest to us - mineralogy, surface roughness, porosity and permeability. Heterogeneity refers to variation in the reservoir of these four basic characteristics. The effects of porosity and permeability can be largely captured in multiphase flow dynamics in porous rocks through relative permeability measurements using reservoir cores. The rock characteristics that affect its wettability are mineralogy and surface roughness - both of which can be easily studied using our DDDC technique. The effects of mineralogy are well evaluated in an earlier paper [5] in the results of Beaverhill Lake crude oil on both quartz and calcite surfaces, each showing a different wettability. Our ongoing work on the effect of surface roughness on dynamic contact angles [8], which is published in these proceedings, appears to refute the general notion of increasing hysteresis with surface roughness. Although the current work attempts to address the concerns regarding relevance of contact angles to reservoir wettability, questions do remain on the nature of relationship between microscopic and macroscopic (or apparent) contact angles. The point of this discussion is that it is not sufficient to say that "rock affects wettability"; but it is important

to identify which particular characteristics of the rock affect wettability, and then to look for ways to include them in the measurement techniques used to characterize wettability. Dynamic contact angles do not suffer from this limitation; instead, they offer a convenient way to address the concerns.

2. *Since there is no evidence to show that light ends in the crude oil affect wettabiliy, the Amott and United States Bureau of Mines (USBM) tests, using reservoir core sample and stocktank oil at ambient conditions, may be sufficient for characterizing reservoir wettability.*

How could there be any such evidence when the widely used USBM and Amott methods have always omitted the light ends from the tests by using dead crude oils, and the conventional contact angle techniques - that did include the light ends - could not yield reproducible results? For example, the Gilwood live and stocktank oils differed markedly in their wettability behavior on quartz. The live oil displayed an intermediate wettability state with a water-advancing contact angle of 110° while the stocktank oil showed an oil-wet behavior with an advancing angle of 150°. Anderson [9] notes, "the use of dead crude at ambient or reservoir pressure may change the wettability because the properties of the crude are altered", and "...the effects of pressure are not known at present".

3. *All contact angle techniques, including the DDDC technique, are not suited to evaluate spotted and mixed-wettability systems.*

The spotted and mixed-wettability situations are special types of fluids distribution that no technique at present is capable of confidently characterizing. Spotted wettability is a heterogeneous state of wettability and is believed to be caused by heterogeneity in mineralogy and clay types that make up the rock structure. Let us suppose, for the sake of this discussion, that the Beaverhill Lake rock contains both quartz and calcite minerals in a checker-board arrangement. The results contained in our paper would immediately indicate a spotted wettability state - meaning that quartz squares would be water-wet and the calcite squares oil-wet. Thus spotted wettability could be interpreted from the knowledge of reservoir mineralogy and the true water-advancing contact angles on each of these mineral components. The DDDC technique enables such measurements.

Mixed-wettability is another form of heterogeneous wettability that is "interpreted" from oil/water flow behavior observed in flow experiments using reservoir cores [10]. This wettability state has been conjectured to explain abnormally high oil recoveries in waterflood experiments. A careful examination of core flood data would be essential to infer this state of wettability. It would be dangerous to rely on a simple USBM or Amott test, with stocktank crude oil at pressures and temperatures far from those in the reservoir, to conclude regarding the absence or existence of mixed-wettability. Any technique that cannot conclusively determine the existence of continuous oil films in larger pores of the rock cannot be claimed as indicating mixed-wet conditions. Considerable work in this area remains to be done to correlate different methods of wettability determination to the location and distribution of oil and water in rock pores, as examined by such techniques as Magnetic Resonance Imaging (MRI). Such progress will, no doubt, be made in the future. Until such time, corefloods, oil-water relative permeabilities and their ratios are the only recourse for identifying special events such as mixed-wettability. It can be expected that the DDDC technique would probably identify the mixed-wet systems as being oil-wet with large advancing angles to support the existence of oil films in larger pores. Thus, it appears possible to identify mixed-wet systems by conducting waterflood experiments using representative reservoir cores and analyzing the data in conjunction with the DDDC results.

4.  *Instead of waiting for the buoyancy force to create the initial oil occupied area on the crystal surfaces and later monitoring contact line movements within this oil exposed area during the DDDC tests, why not start with two crystals that are first soaked in crude oil before being mounted in the contact angle cell and immersed in brine?*

Such a procedure would be an unnatural representation of the crude oil reservoirs where oil migrates into initially brine filled formations. Thus, crude oil, being the latecomer, could interact with the rock only in the presence of formation brine. There exists an abundance of eminent published literature [11, 12] concerning the role of stability of thin wetting films of water in establishing the equilibrium wettability of rock-oil-brine systems. In fact, it is important to establish the equilibrium of interactions between the crystal surface and the formation brine at reservoir conditions before exposing it to the crude oil in the contact angle cell. It is for this reason that we normally perform overnight aging of the crystals in formation brines at reservoir conditions before placing the two crude oil drops on the crystal surfaces. It is also for the reason that the three-phase (oil-water-rock) equilibrium is not the same as the two-phase (oil-rock) equilibrium, that the above-suggested option was considered and eliminated early in the development of the DDDC technique.

5.  *Is there any correlation of DDDC wettability with that inferred from multiphase flow behavior observations in reservoir rocks?*

This question has been discussed in detail in a recent publication [13] in which wettability inferred from oil-water relative permeabilities (derived using dynamic displacement tests and their computer simulations) and wettability from DDDC contact angles have been compared for ten different rock-fluids systems at the respective reservoir conditions. For eight of the ten systems considered, the wettabilities from corefloods and contact angles appeared to correlate well with each other. For the remaining two systems, however, the two techniques yielded contradictory wettabilities. Out of these two cases where disagreement occurred, one was a sandstone reservoir and the other a carbonate formation. The wettability disagreement observed in two of the ten cases appeared to be related to the effect of core-scale heterogeneities and the level of pore interconnectivity on measured oil-water relative permeabilities. The pore size distribution in reservoir rocks, as characterized by capillary pressure measurements [14], appears to be a factor that may negate the wettability inferences from relative permeability rules. Moreover, the broad rules-of-thumb commonly used for inferring wettability from relative permeability data may work in a majority of cases, but may completely fail in others because of the dependence of relative permeability on factors other than wettability. This emphasizes the need to have some other independent measure of wettability in addition to that inferred from relative permeability data.

Both water-advancing and receding contact angles obtained using the DDDC technique were reproducible within 3°, yielding confidence in the results. The same level of confidence cannot be claimed for the wettability inferred from oil-water relative permeabilities due mainly to the dependence of relative permeabilities on factors other than wettability and also because the rules-of-thumb contradict each other in some cases. Thus the DDDC technique appears to merit consideration as an independent measure of reservoir wettability since it not only yields reproducible dynamic contact angles using live fluids at actual reservoir pressure and temperature, but also it is significantly more rapid than conventional sessile and modified sessile drop techniques.

6. *If the water-advancing contact angle correlates well with wettability and accounts for the influence of adhesion on wettability, what is the role of the water-receding contact angle in S-L-L systems?*

As noted earlier, the receding contact angles in S-L-L systems correlate well with spreading of oil on the solid surface immersed in brine. Such spreading of oil against brine in the reservoir has serious implications to oil recovery due to its influence on the nature of distribution of oil and water in rock pores and, consequently, on the flow dynamics of the two phases in the rock. The Zisman-plot [15] of cosine of contact angle versus surface tension in S-L-V systems - to determine the critical solid surface tension below which the liquid spreads on the solid against the vapor phase - is familiar to those interested in surface phenomena. Can we correlate the spreading in S-L-L systems also in a similar manner by plotting the water-receding contact angle (since it correlates with spreading) against the oil-water interfacial tension? One affirmative answer to this was discovered while studying the effect of temperature on spreading and wettability in heavy oil reservoirs [6]. In that study, both the oil-water interfacial tension and water-receding contact angle decreased with increasing temperature and a Zisman-type plot yielded a critical spreading tension of about 12 mN/m. When the temperature was increased further, and the interfacial tension fell below this critical value of 12 mN/m, the crude oil drop was seen to gradually spread on the solid surface against brine. This gradual spreading in S-L-L systems, as opposed to the spontaneous spreading observed by Zisman in S-L-V systems, appears to be reasonable considering the fact that for the oil to spread, it has to displace brine of similar density and viscosity, whereas the liquid spreads spontaneously in S-L-V systems by displacing a vapor phase of much lower density and viscosity.

Recently we have encountered another example of oil spreading on a solid surface against brine - but in a totally different S-L-L system - as a result of brine composition change at a constant ambient temperature. In order to study the effect

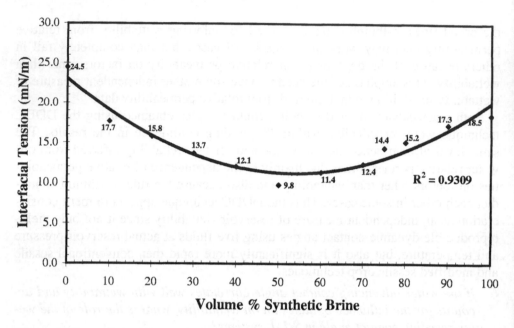

**Figure 4.** Effect of Brine Strength on Interfacial Tension of Yates Crude Oil against Yates Synthetic Brine.

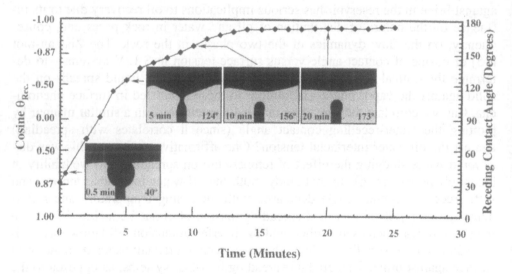

**Figure 5.** Receding Contact Angle of Spreading Yates Crude Oil Drop against Brine (75% Synthetic Brine and 25% DIW) on Smooth Dolomite Surface.

of brine composition on wettability, synthetic brine was prepared which had the same composition as that of reservoir brine. It was then filtered through 0.2 μm filter paper under vacuum. Various mixtures of synthetic brine-deionized water were prepared and the interfacial tensions of the Yates crude oil against these

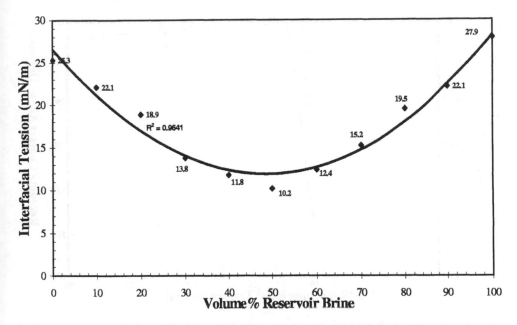

**Figure 6.** Effect of Brine Strength on Interfacial Tension of Yates Crude Oil against Yates Reservoir Brine.

mixtures were determined using Computerized Drop Shape Analysis technique. The DDDC test procedure for contact angles determination was described earlier.

There was a noticeable trend in the interfacial tensions which can be observed in Figure 4 indicating an initial decrease and a later increase in interfacial tension with the amount of synthetic brine in the mixture, having a minimum interfacial tension of about 9.8 mN/m at the 50-50 mixture composition.

To examine the effect of changing interfacial tension with brine composition on dynamic contact angles, a DDDC test was begun with a mixture containing 75% synthetic brine and 25% deionized water (DIW) using two smooth dolomite substrates. As soon as the Yates crude oil drop was placed on the lower dolomite surface it began to spread on the surface yielding a pancake-like appearance with large receding angles. Figure 5 shows the drop profiles captured by the digital video system at short time intervals along with the plot of receding contact angle against time of drop spreading. As can be seen, after about 15 minutes of placing the drop on the dolomite surface, the oil drop spread completely as indicated by a receding contact angle of about 173 degrees.

Similar mixtures were prepared using actual reservoir brine to study if it had the same effect as the synthetic brine. Interfacial tension of Yates crude oil was measured against these mixtures and a very similar trend to that of the synthetic brine was observed, as shown in Figure 6. Experiments conducted with various mixtures of actual reservoir brine and DIW with varying levels of total dissolved solids also indicated similar spreading behavior of Yates crude oil on dolomite

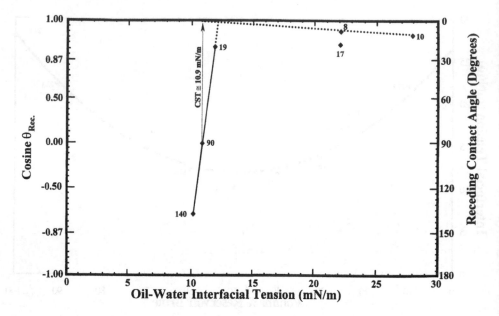

**Figure 7.** Zisman-type Spreading Correlation for the Yates Brine-Yates Crude Oil-Dolomite Rock System.

surfaces. As noted earlier, such spreading behavior in crude oil-brine-rock systems has been previously correlated with receding contact angles [6] while studying the effect of temperature on spreading and wettability in heavy oil reservoirs. Hence an attempt was made here to correlate the spreading behavior observed as an effect of changing brine dilution against the oil-brine interfacial tension. The resulting plot is shown in Figure 7, which has a similar appearance to the one found by the author earlier [6] in a totally different system (Lindbergh heavy oil against reservoir brine at elevated temperatures). The interfacial tension value of 10.9 mN/m corresponding to the point of 90-degree receding angle (as shown by the upward arrow in Figure 7) has been termed the Critical Spreading Tension (CST) for solid-liquid-liquid systems. It is hypothesized [6] that if the interfacial tension between the fluid pairs falls below the CST, then the drop-phase will spread on the solid surface with a large water-receding contact angle. In order to check this hypothesis, another test was conducted using a brine containing 50% each of reservoir brine and DIW because this 50-50 mixture had an interfacial tension of 10.2 mN/m, which is below the CST of 10.9 mN/m. This test resulted in the drop spreading with a receding angle of 140° (also plotted in Figure 7) - further confirming the hypothesis based on the critical spreading tension. These findings have serious implications because the use of surfactants to decrease interfacial tension could certainly bring the rock-fluids system into the oil spreading region. The spreading aspect, observed in the course of the study of brine composition effects, needs further detailed examination because of its potentially strong impact on oil recovery and production strategies in the future.

## 5. CONSEQUENCES OF CONTACT ANGLES IN S-L-L SYSTEMS

Dynamic contact angles along with interfacial tension data are a good measure of interfacial interactions involved in spreading, adhesion and wettability, in solid-liquid-liquid systems. The early applications of the concept of contact angles and the various techniques of measuring them originated from solid-liquid-vapor systems. However, their applicability to solid-liquid-liquid systems have met with skepticism because of poor reproducibility and long test durations. The poor reproducibility appears to have been mainly due to adhesion of the drop phase to the solid (even in the presence of another immiscible continuous liquid phase), which prevents contact line movement. The long test durations can generally be attributed to the difference in initial exposure of solid surfaces to the less dense liquid used in contact angle tests. The DDDC contact angle technique, developed specifically for S-L-L systems, resolves both these problems effectively, and hence it has seen several applications to quantify reservoir wettability at actual pressures and temperatures encountered in petroleum reservoirs. Therefore, continuing skepticism about the applicability of contact angles in S-L-L systems appears unjustified.

In S-L-L systems contained within a porous rock medium (such as the rock-oil-brine systems), the multiphase flow behavior is of importance. Both the spreading and adhesion aspects of wettability influence the flow behavior. While the spreading behavior can be quantified by the water-receding contact angle, the adhesion aspect is accounted for by the advancing angle. The difference between the advancing and the receding contact angles, which is generally referred to as hysteresis, has even more significance to the multiphase flow dynamics in porous media since it enables a preliminary estimation of the role of adhesion forces relative to capillary forces in rock-fluids systems of different wettabilities. This is further discussed below.

## 6. ADHESION VERSUS CAPILLARY FORCES IN S-L-L SYSTEMS

Several theories have appeared in the literature for estimating the adhesion forces in solid-liquid-vapor systems [16-20]. The relationships proposed by these authors for the adhesion force in solid-liquid-vapor systems are quite similar to the form in equation (1) below, except that they differ only in the value of the constant of proportionality.

$$F/R = K \gamma_{LV} (Cos\ \theta_r - Cos\ \theta_a) \tag{1}$$

where F is the adhesion force, R is the radius of liquid-solid contact area, K is the constant of proportionality, $\gamma_{LV}$ is the surface tension of the liquid, and $\theta_a$ and $\theta_r$ are advancing and receding contact angles, respectively. Wolfram and Faust [18] proposed K to be equal to the value of $\pi$, Furmidge [17] and Dussan and Chow [19] proposed K to be equal to unity, and Extrand and Gent [20] proposed a value

$4/\pi$ for K. In all these studies, a common factor appears in the equations, which is the difference between the cosines of advancing and receding contact angles. These relationships are valid for solid-liquid-air (or vacuum) systems. However, they seem equally applicable to systems comprising of a solid and two immiscible liquid phases, such as the rock-oil-brine systems of our interest, if both the liquid-liquid interfacial tension and the true values of advancing and receding contact angles can be measured under realistic reservoir conditions. The measurements of oil-water interfacial tensions and their precision at elevated pressures and temperatures using the computerized axisymmetric drop shape analysis have been reported elsewhere [21]. These measurements have been made at pressures up to 70 MPa (10,000 psi) and temperatures up to 200°C (nearly 400° F) with standard deviations in the range of 0.1% to 0.8% of the mean. The DDDC technique has yielded highly reproducible advancing and receding contact angles, also at elevated pressures and temperatures encountered in the petroleum reservoirs. Therefore, by combining the results of the interfacial tension measurements with those from the DDDC contact angle measurements, an attempt has been made here to estimate the adhesion forces using the form of the equation suggested by Extrand and Gent [20], and a value of 26 mN/m for the oil-water interfacial tension which was measured in one of the projects. The results are presented in Table 4. The results of Table 4 show the increasing trend in the estimated adhesion force per unit drop radius as the wettability shifts towards oil-wet behavior.

In order to understand the physical significance of the adhesion forces, it appears reasonable to define an Adhesion Number as the ratio of the adhesion force to the capillary force. Capillary force is proportional to the product of interfacial tension and cosine of the equilibrium or static contact angle (which is usually low, making the cosine of the angle close to unity). Therefore, as a first approximation, the Adhesion Number, $N_a$, can be defined as:

$$N_a \cong \{\gamma_{ow} (\text{Cos } \theta_r - \text{Cos } \theta_a)\} / \gamma_{ow} = (\text{Cos } \theta_r - \text{Cos } \theta_a) \qquad (2)$$

Thus, an indication of the extent of adhesion with respect to capillary force, is given by the difference between the cosines of receding and advancing angles.

**Table 4.**
Evaluation of interfacial adhesion forces at reservoir conditions

|  | Water-Wet Case[1] | Intermediate-Wet Case[2] | Oil-Wet Case[3] |
|---|---|---|---|
| Water-Advancing Contact Angle (degrees) | 53 | 110 | 147 |
| Water-Receding Contact Angle (degrees) | 15 | 31 | 10 |
| (Cos $\theta_r$ - Cos $\theta_a$) | 0.361 | 1.20 | 1.823 |
| Adhesion Force per Unit Length (F/R in mN/m) | 12.05 | 39.70 | 60.37 |

[1] Beaverhill Lake live oil - reservoir brine - quartz system at 25 MPa and 96°C
[2] Gilwood live oil - reservoir brine - quartz system at 17.5 MPa and 61°C
[3] Beaverhill Lake live oil - reservoir brine - calcite system at 25 MPa and 96°C

This difference is also given in Table 4, from which it can be seen that adhesion forces play a significant role compared to capillary forces in retaining crude oil within the formation. While the adhesion force was about a third of the capillary force in the water-wet case, it increased to 1.2 times the capillary force in the intermediate-wet case and was almost double the capillary force in the oil-wet case.

It is also interesting to note that the term, $(\text{Cos } \theta_r - \text{Cos } \theta_a)$, has been referred to as hysteresis in our efforts to understand the differences in flow behaviors in porous media during imbibition (increasing wetting phase saturation in porous medium) and drainage (decreasing wetting phase saturation). In the past we have attributed hysteresis to surface roughness and impurities in the system. The above definition of the simplified Adhesion Number points out that hysteresis could indeed be due, partly or entirely, to the adhesion phenomenon at the solid-liquid-liquid interface. Thus the traditional approach of attributing trapped oil to capillary forces may be justifiable in water-wet reservoirs, while adhesion forces may be controlling the trapping process in non-water-wet reservoirs, which happen to be the majority. Hence enhanced recovery processes aspiring to produce trapped oil should incorporate mechanisms to overcome the adhesion forces in rock-oil-brine systems. Precise measurements of liquid-liquid interfacial tension and solid-fluids interactions through dynamic contact angles are basic requirements in understanding and overcoming these forces. This paper has presented techniques that enable such precise measurements.

## 7. SUMMARY

The classical works in the field of surface science involving solid-liquid-vapor systems have been of significant help in the development of theories and techniques for solid-liquid-liquid systems. However, significant differences between the S-L-V and S-L-L systems need to be accounted for in measurement techniques and interpretation of results. As to which of the two immiscible liquid phases would wet the solid surface in S-L-L systems depends on spreading and adhesion characteristics involving interfacial interactions. While the receding contact angle correlates with the spreading of one liquid against the other on the solid surface, the adhesion phenomenon is reflected in the advancing contact angle. The dual-drop dual-crystal technique has enabled reproducible measurements of both advancing and receding contact angles in several S-L-L systems thereby yielding information on spreading and adhesion aspects of wettability. There appears to be evidence for the existence of a Zisman-type critical spreading tension in S-L-L systems as indicated by two very different rock-oil-brine systems. Further work is needed in order to generalize such a Zisman-type relationship between receding contact angle and liquid-liquid interfacial tension in S-L-L systems. The preliminary estimates of adhesion forces based on experimental measurements of oil-brine interfacial tension and advancing and receding contact angles indicate that the phenomenon of adhesion plays a significant role compared to capillary forces in trapping crude oil in reservoir rocks. The development of enhanced recovery processes aspiring to recover the trapped oil should be based on understanding and overcoming the adhesion forces. This paper has presented recent developments in experimental techniques to make precise measurements of liquid-liquid interfacial tension and solid-liquid interactions through dynamic contact angles, which would enable the development of processes to overcome the adhesion forces.

*Acknowledgements*

This work has been made possible by the financial support of the Louisiana Board of Regents Support Fund (Contract No. LEQSF(2000-2003)-RD-B-06), and Marathon Oil Company. Sincere thanks to Chandrasekhar Vijapurapu for carefully conducting the spreading tests and Rajesh Pillai for his help with the manuscript. The author thanks MST Conference organizers for the invitation to present this work at the 3rd International Symposium on Contact Angle, Wettability and Adhesion.

**REFERENCES**

1. R.M. Pashley and J.N. Israelachvili, *Colloids Surfaces*, **2**, 169 (1981).
2. O.S. Hjelmeland, Thesis, The Norwegian Institute of Technology, The University of Trondheim, Trondheim, Norway (1984).
3. M.A. Andersen, D.C. Thomas and D.C. Teeters, *Paper No. SPE/DOE 17368, Proc. 1988 SPE/DOE Enhanced Oil Recovery Symposium*, Tulsa (1988).
4. Z. Muhammad, "Compositional Dependence of Reservoir Wettability", MS Thesis, Louisiana State University, Baton Rouge, LA, pp. 5-23 (May 2001).
5. D.N. Rao and M.G. Girard, *J. Can. Petroleum. Technol.*, **35** (1), 31 (1996).
6. D.N. Rao, *SPE Reservoir Evaluation Eng.*, **2** (**5**), 420 (1999).
7. D.N. Rao and R.S. Karyampudi, *J. Adhesion Sci. Technol.*, **16**, 579 (2002).
8. C.S. Vijapurapu and D.N. Rao, *These proceedings*.
9. W.G. Anderson, *J. Petroleum Technol.*, **38**, 1125 (1986).
10. D.N. Rao, M. Girard and S.G. Sayegh, *SPE Reservoir Eng.*, **7**, 204 (1992).
11. M.P. Aronson, M.F. Petko and H.M. Princen, *J. Colloid. Interface Sci.* **65**, 296 (1978).
12. C. Maldarelli and R. Jain, in *Thin Liquid Films: Fundamentals and Applications*, I.B. Ivanov (Ed.), Marcel Dekker, New York (1988).
13. D.N. Rao, *J. Can. Petroleum Technol.*, **41** (**7**), 31 (2002).
14. D. Tiab and E.C. Donaldson, *Petrophysics*, p. 294, Gulf Publishing Co., Houston, TX (1996).
15. W. Zisman, in *Contact Angle, Wettability and Adhesion*, Adv. Chem. Ser., No. 43, pp. 1-51, American Chemical Society, Washington, D.C. (1964).
16. K.L. Mittal (Ed.), *Contact Angle, Wettability and Adhesion*, Vol. 2, VSP, Utrecht (2002).
17. C.G.L. Furmidge, *J. Colloid Sci.*, **17**, 309 (1962).
18. E. Wolfram and R. Foust, in *Wetting, Spreading and Adhesion*, J.F. Padday (Ed.), Academic Press, New York (1978).
19. E.B. Dussan V and R.T.P. Chow, *J. Fluid Mech.*, **137**, 1 (1983).
20. C.W. Extrand and A.N. Gent, *J. Colloid Interface Sci.*, **138**, 431 (1990).
21. D.N. Rao, *Fluid Phase Equilibria*, **139**, 311 (1997).

*Contact Angle, Wettability and Adhesion*, Vol. 3, pp. 211–218
Ed. K.L. Mittal
© VSP 2003

# A thermodynamic model for wetting free energies of solids from contact angles

C.W. EXTRAND*

*Entegris, 3500 Lyman Blvd., Chaska, MN 55318, USA*

**Abstract**—A simple thermodynamic model to determine the wetting free energies of a small sessile drop spreading on a solid surface has been derived by assuming an adsorption mechanism. According to the model, wetting free energies are expected to be small for large contact angles and to increase exponentially as the contact angles tend toward zero. Wetting free energies calculated from contact angle data taken from the literature for various liquid-solid pairs agreed reasonably well with the interaction strengths measured by other experimental techniques.

*Keywords*: Contact angles; wetting free energies; polymers; adsorption.

## 1. INTRODUCTION

Wetting is important in many industrial processes, such as cleaning, drying, painting, coatings, adhesion, heat transfer, and pesticide application. The most common method of evaluating wetting is contact angle measurements. Figure 1 shows the case of a sessile drop measurement, where a small liquid drop is deposited on a solid surface. As contact is made, molecular interactions between the liquid and solid pull at the drop, advancing the contact line. When forces at the contact line reach equilibrium, the drop exhibits an advancing contact angle, $\theta_a$. The stronger the interaction, the greater the wetting. If $\theta_a$ is substantially greater than zero, then the liquid is referred to as partial wetting. On the other hand, a zero or near-zero value is considered to be complete wetting.

In this work, a simple model has been developed which assumes that wetting is adsorption of a contact liquid onto a solid substrate. This leads to the determination of wetting free energies directly from contact angle data, without knowledge of gas/liquid/solid interfacial tensions. Wetting free energies were calculated for a wide range of contact angles and their significance discussed. Wetting free energies also were computed from experimental data taken from the literature and then compared with values from other techniques.

---

*Phone: (952) 556-8619, Fax: (952) 556-8023, E-mail: chuck_extrand@entegris.com

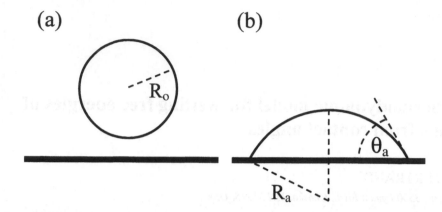

**Figure 1.** Side view of a small, liquid drop of volume, V. (a) Before deposition, the drop has an initial radius of curvature, $R_o$. (b) After deposition, the drop exhibits an advancing contact angle, $\theta_a$, and has a radius of curvature of $R_a$.

## 2. ANALYSIS

A number of attempts have been made to link wetting to adsorption by examining the interplay between the solid, the contact liquid, and its vapor [1-8]. Here, we assume that wetting can be described as adsorption of a contact liquid onto a solid [8] and employ the Gibbs adsorption equation relating the change in surface free energy of the solid due to wetting, g, to its chemical potential, $\mu_s$ [9],

$$dg = - (1/A)d\mu_s, \tag{1}$$

where A is the molar surface area of the solid. For a smooth polymer surface, A can be calculated from its density, $\rho$, its monomer weight, M, and Avogadro's number, N,

$$A = (M/\rho)^{2/3}N^{1/3}. \tag{2}$$

Consider the small sessile drop on a solid surface shown in Figure 1. The drop is sufficiently small such that the influence of gravity is negligible and it retains spherical proportions. The gas phase that surrounds the drop is dry and inert, *i.e.*, it does not contain vapor from the contact liquid. Since our sessile drop exhibits an equilibrium value of $\theta_a$ [10], the chemical potential of the solid surface, $\mu_s$, is equal to that of the liquid drop, $\mu_l$,

$$\mu_s = \mu_l, \tag{3}$$

and $\mu_l$ can be related to the pressure within the liquid drop, p, by

$$\mu_l = \mu_{l,o} + RT \cdot \ln(p), \tag{4}$$

where $\mu_{l,o}$ is the standard-state chemical potential of the liquid, R is the ideal gas constant and T is absolute temperature. Combining equations (1), (3), and (4) gives

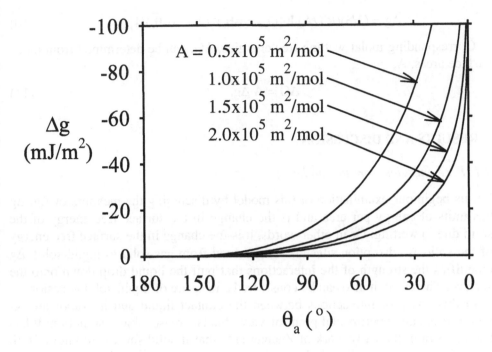

**Figure 2.** Surface wetting free energy, $\Delta g$, *versus* advancing contact angle ($\theta_a$) for various molar surface areas, A, at room temperature (T = 298 K), eq. (10).

$$dg = - (RT/A)\cdot d(\ln p). \tag{5}$$

Integrating equation (5) between the equilibrium and initial pressures, $\Delta p_a$ and $\Delta p_o$, one obtains the change in the surface free energy of the solid, $\Delta g$, due to wetting, *i.e.*, surface wetting free energy (See Figure 2)

$$\Delta g = (RT/A)\cdot \ln(\Delta p_a/\Delta p_o). \tag{6}$$

The pressure differential across the air-liquid interface of the drop, $\Delta p_i$, depends upon its radius of curvature, $R_i$, and the surface tension of the contact liquid, $\gamma$ [11, 12],

$$\Delta p_i = 2\gamma/R_i, \tag{7}$$

where i = o and i = a represent the drop before and after deposition. Before deposition, the initial radius of curvature of the drop, $R_o$, is related to its volume, V, by

$$R_o = (3V/4\pi)^{1/3}. \tag{8}$$

After deposition, the radius of curvature of the sessile drop, $R_a$, is related to V and $\theta_a$ by [13, 14],

$$R_a = [3V/\pi (1 - \cos\theta_a)^2 (2 + \cos\theta_a)]^{1/3}. \tag{9}$$

Ultimately, $\Delta g$ can be written in terms of $\theta_a$ by combining equations (6-9),

$$\Delta g = (1/3)(RT/A)\cdot\ln[(1 - \cos\theta_a)^2(2 + \cos\theta_a)/4].$$  (10)

Corresponding molar wetting free energies, $\Delta G$, can be determined from molar surface areas, $A$,

$$\Delta G = A\cdot\Delta g.$$  (11)

## 3. RESULTS AND DISCUSSION

### 3.1. Free energies from the model

Let us begin our examination of this model by discussing the meaning of $\Delta g$. $\Delta g$ has units of energy per area and is the change in the surface free energy of the solid due to wetting. Or in other words, it is the change in the surface free energy of the solid as adsorption sites are transformed from gas-solid to liquid-solid. $\Delta g$ quantifies the strength of the interactions that pull the liquid drop down onto the surface. Therefore, it also can be considered a measure of liquid-solid adhesion.

In this analysis, interactions between the contact liquid and its vapor are ignored. From an experimental point-of-view this is a reasonable assumption. It has been shown in the early work of Zisman [15] that if solid surfaces do not adsorb vapor from the contact liquid, the partial pressure of the contact liquid, if it exists, has no effect on $\theta_a$.

It should be noted that the model does not depend on the drop volume, $V$. As $V$ is constant during deposition and spreading, it cancels out of equation (10). This theoretical finding agrees with experimental observation – contact angles are independent of $V$, except for very small drops where line tension is influential [16-18]. Also note that if the surface energy of the liquid, $\gamma_l$, is constant during spreading, then it also cancels out of equation (10). Subsequently, the model does not depend explicitly on $\gamma_l$. Similar to the solubility parameter, $\gamma_l$ is a gauge of intramolecular interactions within the contact liquid, but is not necessarily a measure of the intermolecular interactions at the contact line.

Figure 2 shows the surface wetting free energy, $\Delta g$, versus advancing contact angle, $\theta_a$, at room temperature. $\Delta g$ is negative for all values of $\theta_a$, as expected for a spontaneous process. (However, from the perspective of the liquid drop, it is coerced into spreading.) Prior to being placed on a surface, the drop has minimized its surface free energy by minimizing its area. Interactions occurring at the contact line drag the drop downward, flattening it onto the solid surface. Without these contact line interactions, drops do not spread [19].

Wetting free energies generated by this model appear to be of the correct magnitude. For the largest angles, $\theta_a > 130°$, the molar wetting free energies approached zero. This finding agrees with practice, as very large contact angles are difficult to achieve solely by chemical modification [20]. Interactions of liquids on smooth surfaces rarely produce $\theta_a > 130°$. Larger $\theta_a$ values usually require both lyophobicity and roughness [21-27].

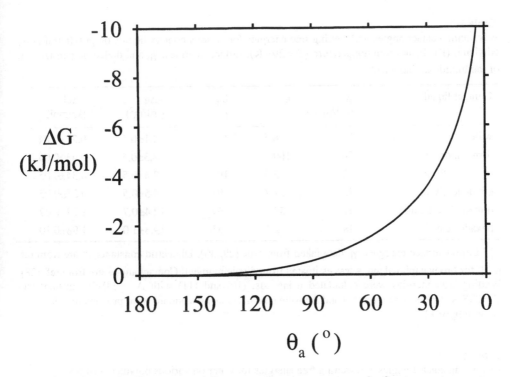

**Figure 3.** Molar wetting free energy, $\Delta G$, *versus* advancing contact angle, $\theta_a$, at room temperature (T = 298 K), according to eqs. (10) and (11).

Progressively lower contact angles gave negative $\Delta g$ values of greater magnitude. $\theta_a$ values on the order of 100° yielded $\Delta g$ values of −10 mJ/m² or less. With further decrease in $\theta_a$, the wetting free energies increased exponentially. As contact angles tended toward zero, $\Delta g$ values exceeded −50 mJ/m². For a given $\theta_a$, smaller A values gave larger absolute values of $\Delta g$.

Figure 3 shows the corresponding relation between molar wetting free energy, $\Delta G$, and advancing contact angle, $\theta_a$. Similar to the surface wetting free energies, the absolute magnitude of $\Delta G$ values was small for large contact angles, but increased dramatically as $\theta_a$ declined – for $\theta_a < 5°$, |−$\Delta G$| > 10 kJ/mol.

### 3.2. Comparison with experimental data

Experimental contact angle data from the literature were used to calculate surface and molar wetting free energies. Table 1 shows contact angles and wetting free energies for various liquids on poly(tetrafluoroethylene) (PTFE) [28]. The liquids have a broad span of polarities, as demonstrated by their surface energies, $\gamma_l$, and dielectric constants, $\varepsilon$ [29, 30]. PTFE is an inert, homogeneous polymer, comprised solely of low surface energy fluoromethylene (CF$_2$) groups. Contact angles ranged from 113° to 51°. Molar values of the wetting free energies for PTFE and

**Table 1.**
Advancing contact angles and wetting free energies for various contact liquids on poly(tetrafluoro-ethylene) (PTFE) at room temperature (T = 298 K); surface energies, $\gamma_l$, and dielectric constants, $\varepsilon$, of the liquids are included[a]

| Contact liquid | $\gamma_l$ (mJ/m$^2$) | $\varepsilon$ | $\theta_a$ (°) | $-\Delta g$ (mJ/m$^2$) | $-\Delta G$ (kJ/mol) |
|---|---|---|---|---|---|
| water | 73 | 78.5 | 113 | 2.1±0.4 | 0.21±0.04 |
| formamide | 56 | 109 | 97 | 4.3±0.5 | 0.43±0.05 |
| diiodomethane | 51 | 5.3 | 103 | 3.4±0.5 | 0.34±0.05 |
| ethylene glycol | 48 | 37.7 | 91 | 5.5±0.5 | 0.55±0.05 |
| dimethyl acetamide | 37 | 59 | 67 | 12.4±0.7 | 1.24±0.07 |
| hexadecane | 28 | 2.1 | 51 | 19.8±1.0 | 1.98±0.10 |

[a] Liquid surface energies, $\gamma_l$, data taken from refs. [29, 30]. Dielectric constants, $\varepsilon$, are from ref. [30]; for bromonaphthalene, $\varepsilon$ was estimated from dipole moment. Contact angles are from ref. [28]. Wetting free energies were calculated using eqs. (10) and (11) with A = $1.0 \times 10^5$ m$^2$/mol and T = 298 K. Errors shown for $\Delta G$ and $\Delta g$ were calculated using standard error propagation assuming error in $\theta_a$ of ±2°.

**Table 2.**
Advancing contact angles and wetting free energies for water on various polymer surfaces[a]

| Surface | $\gamma_s$ (mJ/m$^2$) | A (m$^2$/mol) | $\theta_a$ (°) | $-\Delta g$ (mJ/m$^2$) | $-\Delta G$ (kJ/mol) |
|---|---|---|---|---|---|
| PTFE | 19 | $1.0 \times 10^5$ | 113 | 2.1±0.4 | 0.21±0.04 |
| PP | 32 | $1.1 \times 10^5$ | 104 | 2.9±0.4 | 0.32±0.05 |
| PS | 33 | $1.8 \times 10^5$ | 92 | 2.9±0.3 | 0.53±0.05 |
| PVC | 39 | $1.0 \times 10^5$ | 90 | 5.7±0.5 | 0.57±0.05 |
| PA6 | 42 | $1.8 \times 10^5$ | 73 | 5.7±0.4 | 1.03±0.07 |
| PC | 45 | $2.9 \times 10^5$ | 66 | 4.4±0.3 | 1.27±0.07 |

[a] The polymers are: poly(tetrafluoroethylene) (PTFE), polypropylene (PP), polystyrene (PS), poly(vinyl chloride) (PVC), polyamide6 (PA6), and polycarbonate (PC). Surface energies, $\gamma_s$, were taken from ref. [29], except PC, from [33]. Molar surface areas were taken from ref. [8]. Contact angles are room temperature values from ref. [28]. Wetting free energies were calculated using eqs. (10) and (11) for T = 298 K. Errors shown for $\Delta G$ and $\Delta g$ were calculated using standard error propagation assuming error in $\theta_a$ of ±2°.

the various liquids ranged from –0.2 kJ/mol to –2.0 kJ/mol and were of the magnitude expected from weak to moderate dispersive interactions [31-33]. Where the polarity of the contact liquid was less (or the dispersive component of the liquid was larger), interaction with the PTFE was greater, as demonstrated by smaller contact angles and larger wetting free energies. The more similar the contact liquid and the solid, the greater the interaction and the lower the contact angle.

Table 2 lists water contact angles and associated wetting free energies for various polymer surfaces [28]. The two non-polar polymers, PTFE and PP [29, 33], interact weakly with water, as demonstrated by large $\theta_a$ values and small wetting free energies. For the polar polymers (PS, PVC, PA6, & PC), water contact angles were lower and wetting free energies were higher. While liquid/solid functionalities capable of hydrogen bonding can produce $\Delta G$ values of $-10$ kJ/mol or more [34, 35], these surfaces contained a mixture of non-polar and polar functionalities and thus displayed more moderate interactions with strengths of only a few kJ/mol [8].

## 4. CONCLUSIONS

The wetting of small, sessile drops can be modeled as adsorption. Wetting free energies from very large contact angles are effectively zero. Wetting free energies increase exponentially as contact angles decrease. Large angles (about 100°) produce wetting free energies of the magnitude expected from weak interactions, while near-zero angles give large wetting free energies indicative of much stronger interactions. Molar wetting free energies calculated from contact angle data for several liquid-solid pairs generally agreed with interaction strengths measured by other techniques.

*Acknowledgments*

The author thanks Entegris management for supporting this work and allowing publication.

## REFERENCES

1. M.E. Tadros, P. Hu and A.W. Adamson, *J. Colloid Interface Sci.*, **49**, 184 (1974).
2. J. Tse and A.W. Adamson, *J. Colloid Interface Sci.*, **72**, 515 (1979).
3. P.R. Chidambaram, G.R. Edwards and D.L. Olson, *Metall. Trans.*, **23B**, 215 (1992).
4. M.E. Schrader and G.H. Weiss, *J. Phys.Chem.*, **91**, 353 (1987).
5. M.E. Schrader, in *Contact Angle, Wettability and Adhesion*, K.L. Mittal (Ed.), pp. 109-121, VSP, Utrecht, The Netherlands (1993).
6. M.E. Schrader, *Langmuir*, **11**, 3585 (1995).
7. M. Turmine and P. Letellier, *J. Colloid Interface Sci.*, **227**, 71 (2000).
8. C.W. Extrand, *J. Colloid Interface Sci.*, **207**, 11 (1998).
9. J.W. Gibbs, *The Scientific Papers*, Vol. 1, Dover, New York (1961); P.C. Hiemenz, *Principles of Colloid and Surface Chemistry*, 2nd ed., Marcel Dekker, New York (1986).
10. The definition of equilibrium for contact angles has been contentious. An advancing angle certainly constitutes a mechanical equilibrium and in many cases meets the requirements for thermodynamic equilibrium. After spreading, the temperatures of the liquid and solid are equal. Contact angles are also path independent – different test geometries (sessile drop, captive bubble, Wilhelmy plate, *etc.*) usually produce the same value of the contact angle.
11. P.S. Laplace, *Méchanique Céleste*; t. 4, Supplément au X livre; Courier, Paris (1805).
12. A.W. Adamson, *Physical Chemistry of Surfaces*, 5th ed., Wiley, New York (1990).

13. W.H. Beyer (Ed.), *Standard Mathematical Tables*, 27th ed., CRC Press, Boca Raton, FL (1984).
14. C.W. Extrand, *J. Colloid Interface Sci.*, **157**, 72 (1993).
15. H.W. Fox and W.A. Zisman, *J. Colloid Sci.*, **5**, 514 (1950).
16. R.J. Good and M.N. Koo, *J. Colloid Interface Sci.*, **71**, 283 (1979).
17. M. Yekta-Fard and A.B. Ponter, *J. Colloid Interface Sci.*, **126**, 134 (1988).
18. F.Y.H. Lin, D. Li and A.W. Neumann, *J. Colloid Interface Sci.*, **159**, 86 (1993).
19. P. Aussillous and D. Quéré, *Nature*, **411**, 924 (2001).
20. L. Mahadevan, *Nature*, **411**, 895 (2001).
21. J.W. Gibbs, *The Collected Works of J. Willard Gibbs*, Vol. 1, Yale University Press, New Haven (1961).
22. A.B.D. Cassie and S. Baxter, *Trans. Faraday Soc.*, **40**, 546 (1944).
23. R. Shuttleworth and G.L.J. Bailey, *Disc. Faraday Soc.*, **3**, 16 (1948).
24. F.E. Bartell and J.W. Shepard, *J. Phys. Chem.*, **57**, 211 (1953).
25. J.F. Oliver, C. Huh and S.G. Mason, *J. Colloid Interface Sci.*, **59**, 568 (1977).
26. D. Öner and T.M. McCarthy, *Langmuir*, **16**, 7777 (2000).
27. C.W. Extrand, *Langmuir*, **18**, 7991 (2002).
28. L.S. Penn and B. Miller, *J. Colloid Interface Sci.*, **78**, 238 (1980).
29. S. Wu, *Polymer Interface and Adhesion*, Marcel Dekker, New York (1982).
30. R.C. Weast (Ed.), *Handbook of Chemistry and Physics*, 73rd ed., CRC Press, Boca Raton, FL (1992).
31. J.N. Israelachvili, *Intermolecular and Surface Forces*, 2nd ed., Academic Press, New York (1992).
32. J.C. Bolger and A.S. Michaels, in *Interface Conversion*, P. Weiss and G.D. Cheevers (Eds.), Elsevier, New York (1969); J.C. Berg, in *Wettability*, J.C. Berg (Ed.), Marcel Dekker, New York (1993).
33. S. Wu, in *Polymer Handbook*, 3rd ed., J. Brandrup and E.H. Immergut (Eds.), Wiley, New York (1989).
34. M.D. Joesten and L.J. Schaad, *Hydrogen Bonding*, Marcel Dekker, New York (1974).
35. G.A. Jeffrey, *An Introduction to Hydrogen Bonding*, Oxford University Press, New York (1997).

*Contact Angle, Wettability and Adhesion*, Vol. 3, pp. 219–264
Ed. K.L. Mittal
© VSP 2003

# Characterization of surface free energies and surface chemistry of solids

FRANK M. ETZLER*

*Boehringer-Ingelheim Pharmaceuticals, Inc., 900 Ridgebury Road, P.O. Box 368, Ridgefield, CT 06877-0368, USA*

**Abstract**—The surface chemistry and surface energetics of materials are important to the performance of many products and processes – sometimes in as yet unrecognized ways. This review is written for the researcher interested in exploring the nature of surfaces and their relation to processes involving spreading, wetting, liquid penetration and adhesion. Researchers concerned with many types of products including pharmaceuticals, printing and the making of composite materials should have interest in this topic. More specifically, this work is a review of the literature concerning the surface free energy of solids. Both theoretical approaches for understanding the surface free energy of solids are explored and contrasted, as are experimental methods for measuring surface free energy of solids. Experimental methods that offer insight into the chemical nature of surfaces but do not measure surface free energy are also discussed as these two subjects are intertwined.

*Keywords*: Surface free energy; surface free energy of solids; surface tension; contact angle; atomic force microscopy; ESCA; inverse gas chromatography; adhesion; wetting.

## 1. INTRODUCTION

The surface chemical and surface energetic nature of materials used in the formulation of commercial products or used in the manufacture of these products is often important to the final quality of the product. Despite the importance of surface chemistry to the ultimate performance of the product, not as much recognition as is deserved has been given to the characterization of surface chemistry and the effects of its variation on product performance. Difficulties with both theoretical descriptions of interfaces and the measurement of surface chemical characteristics make the incorporation of material surface chemical specifications for the manufacture of many products challenging. Both theoretical developments related to surface characterization and experimental methods used to characterize materials are discussed in this paper.

*Phone: (1-203) 798-5445, Fax: (1-203) 791-6197,
E-mail: fetzler@rdg.boehringer-ingelheim.com

Surface chemistry is important to processes involving spreading, wetting, liquid penetration and adhesion [1]. Such processes might include printing, drug formulation, painting, and gluing. The formulation of composite materials used for construction materials, tires, gaskets and pharmaceutical capsules also rely on the surface chemical nature of component materials. The relation of processes such as spreading, wetting, liquid penetration and adhesion to surface chemical properties is also discussed.

## 2. OVERVIEW OF SURFACE THERMODYNAMICS

### 2.1. Surface thermodynamic quantities of pure materials

The fundamental principles of surface chemistry have been discussed by Adamson [1], among others.

The creation of surface area, $A$, results in an increase in the Gibbs free energy of a material. For a process at constant temperature and pressure the resulting free energy change is given by:

$$dG = \gamma \, dA \tag{1}$$

Thus

$$G^s = \gamma = \left(\frac{\partial G}{\partial A}\right)_{T,P} \tag{2}$$

where $G^s$ is referred to as the surface free energy and $\gamma$ is the surface tension. From thermodynamics it is known that

$$\left(\frac{\partial G}{\partial T}\right)_P = -S \tag{3}$$

thus

$$\left(\frac{\partial G^s}{\partial T}\right)_P = \left(\frac{\partial \gamma}{\partial T}\right)_P = -S^s \tag{4}$$

where $S^s$ is the surface entropy. The surface energy, $E^s$, can be determined via the usual thermodynamic relations and is given by the following expression:

$$E^s \approx H^s = G^s + T S^s \tag{5}$$

$S^s$ is positive and thus indicates increased entropy or disorder at the interface. The interfacial disorder suggested by the positive surface entropy is an equilibrium process and is not related to non-equilibrium amorphous states that may exist near the surfaces of some solids and which are sometimes referred to as Beilby layers [1, 2]. $E^s$ may be determined via Eq. 5 and can sometimes be calculated from fundamental theoretical calculations (see Section 4).

For liquids the temperature dependence of the surface free energy is nearly linear and for many purposes a linear function can be used to describe the temperature dependence of surface free energy [1]. More exact equations have been proposed by Eötvös [3] and by van der Waals and Guggenheim [4].

The orientation of molecules near surfaces may differ from those in the interior. Langmuir [5] proposed the principle of independent action to account for the surface tension of liquids. According to Langmuir's sound statistical thermodynamic reasoning individual chemical groups within molecules could be assigned surface tension values. Langmuir reasoned using this principle that the surface tension of ethanol was more similar to those of hydrocarbon liquids as ethyl groups pointed towards the exterior. Langmuir's principle of independent action must be considered in any estimation of surface tension or surface free energy values. Because preferred molecular orientations may exist and surface free energies are determined by nature of the first 1-2 atomic layers, it is important to note that surface atomic composition can be quite different from that of the bulk.

Spreading of a liquid across a surface is controlled by the interfacial properties of the materials under study. Let material A be the substrate and material B the liquid spreading across the surface. At constant temperature and pressure the free energy changes are described via using the exact differential equation below.

$$dG = \left(\frac{\partial G}{\partial A_A}\right)_{T,P} dA_A + \left(\frac{\partial G}{\partial A_{AB}}\right)_{T,P} dA_{AB} + \left(\frac{\partial G}{\partial A_B}\right)_{T,P} dA_B \tag{6}$$

and

$$dA_B = dA_{AB} = -dA_A$$

The coefficient

$$-\left(\frac{\partial G}{\partial A_B}\right)_{Area} \equiv S_{B/A} = \gamma_A - \gamma_B - \gamma_{AB} \tag{7}$$

can be used to describe the spreading of B over A. $S_{B/A}$ is referred to as the spreading coefficient and is positive for spontaneous processes. Here

$$\gamma_{AB} = \left(\frac{\partial G}{\partial A_{AB}}\right)_{T,P} \tag{8}$$

and is the interfacial free energy at the A/B interface.

In most instances the liquids do not spread indefinitely across solid surfaces. Instead a drop with an angle of contact between the solid surface and the upper surface of the drop remains on the solid surface. The angle thus formed is referred to as the contact angle, $\theta$. Young's equation [6] describes the relation between the interfacial tensions or free energies of the solid and liquid and contact angle.

$$\gamma_{SV} - \gamma_{SL} = \gamma_{LV}\cos(\theta) \tag{9}$$

Here the subscripts SV, SL and LV refer to the solid–vapor, solid–liquid and liquid–vapor interfaces, respectively. The subscripts recognize that contact angle results from an equilibrium of all three phases.

The definition for work of adhesion, $W_A = -\Delta G_A$, follows directly from the definition of interfacial free energy and is expressed using the following relation.

$$W_A = \gamma_A + \gamma_B - \gamma_{AB} \tag{10}$$

Similarly the work of cohesion, $W_C$, can be defined as:

$$W_C = 2\gamma_A \tag{11}$$

Combining Eqns. 9 and 10 the relation between contact angle and surface free energies can be now be written

$$W_A = \gamma_{LV}[1 + \cos(\theta)] \tag{12}$$

Equation 12 is referred to as the Young–Dupre equation [1, 7].

The penetration of liquids into porous media is of practical importance in such applications as water repellency of fabrics and printing. The Young–Laplace equation [1, 6, 8] describes the pressure differential, $\Delta P$, across a curved spherical interface and its relation to surface free energy and the radius of curvature.

$$\Delta P = \frac{2\gamma}{r} \tag{13}$$

The Young–Laplace equation can be used to describe the rise of a liquid in a capillary and indeed forms the basis of the capillary rise method for determining the surface tension of liquids. For a liquid which imperfectly wets the walls of a capillary ($\theta > 0$) Eqn. 13 becomes:

$$\Delta P = \frac{2\gamma\cos(\theta)}{r} \tag{14}$$

Washburn [9] has described the rate of penetration, $v$, of liquid into cylindrical capillaries. Washburn's equation is written as follows:

$$v = \frac{dl}{dt} = \frac{r\gamma_L\cos(\theta)}{4\eta l} \tag{15}$$

where $\eta$ is the liquid viscosity and $l$ the depth of penetration. For $\theta < 90°$, $v$ is positive indicating that penetration is spontaneous while for $\theta > 90°$ $v$ is negative indicating liquid withdrawal from capillaries is spontaneous.

The above equations represent the principal surface thermodynamic relations that are likely to form the scientific basis for a large number of development problems associated with product performance or product manufacture. The ability to design, control and measure surface free energies would allow for easier development of products which require knowledge of wetting, adhesion, spreading and liquid penetration for their use or manufacture. Knowledge of the relevant interfacial and surface free energies is critical to the understanding of each of the above discussed phenomena.

The measurement of surface tensions of liquids is a relatively easy task and several suitable techniques exist. Mulqueen and Huibers have given a recent review of the measurement of liquid surface tensions [10]. Measurement of liquid–liquid interfacial tensions are more difficult due to the mutual solubility of the liquids under study but many of techniques used for liquid surface tensions are available to conduct these measurements. Kwok *et al.* have discussed the measurement of liquid–liquid interfacial tensions in light of the models discussed below [11]. The measurement of solid surface free energies and liquid–solid or solid–solid interfacial free energies is considerably more difficult and is the subject of this work (see also the recent review by Butt and Raiteri [12]). Here strategies and methods to estimate solid surface free energies critical to an understanding of the above-discussed phenomena are elaborated.

## 3. THERMODYNAMIC APPROACHES FOR THE ESTIMATION OF SURFACE FREE ENERGY OF SOLIDS

### 3.1. Zisman critical surface tension

Following the influence of Thompson [13, 14] and Gibbs [15], Zisman and co-workers [16] made pioneering investigations of the thermodynamics of wetting and adhesion.

Let us recall that Eqns. 9 and 10 are the basic thermodynamic equations describing the relation between surface free energy and wetting. These equations might be rewritten to clarify the present argument in the following way.

$$\gamma_{SV^0} - \gamma_{SL} = \gamma_{LV^0} \cos(\theta) \tag{16}$$

$$W_a = \gamma_{S^0} + \gamma_{LV^0} - \gamma_{SL} \tag{17}$$

where the subscript $LV^0$ refers to liquid–saturated vapor interface, $SV^0$ is the solid saturated with vapor interface and $S^0$ is the solid interface in absence of liquid vapor. From Eqns. 16 and 17 it follows that:

$$W_a = \left(\gamma_{S^0} - \gamma_{SV^0}\right) + \gamma_{LV^0} \cos(\theta) \tag{18}$$

where

$$\gamma_{S^0} - \gamma_{SV^0} = RT \int_0^{p^0} \Gamma \, d\ln p = \pi_e \tag{19}$$

here $\Gamma$ is the surface excess (amount adsorbed) of the liquid vapor and $P^0$ the saturation vapor pressure of liquid phase. The vapor is assumed to be ideal. Also note that Eqn. 18 reduces to Eqn. 12 when the film pressure, $\pi_e$, approaches zero. For low surface free energy solids $\pi_e \approx 0$.

Zisman and co-workers also noted that plots of $\cos(\theta)$ vs. $\gamma_{LV^0}$ were nearly linear, particularly when homologous series of liquids were used. It thus follows that:

$$\cos(\theta) = 1 + \beta\left(\gamma_c - \gamma_{LV^0}\right) \tag{20}$$

where $\gamma_c$ is referred to as the Zisman critical surface tension. A plot of $\cos(\theta)$ vs. $\gamma_{LV^0}$ will intersect $\cos(\theta) = 1$ when $\gamma_{LV^0} = \gamma_c$ (see Fig. 1). Zisman's relation is empirical.

From Eqn. 16 it can be seen that:

$$\gamma_{s^0} - \gamma_{LV^0} \approx \gamma_{sv^0} - \gamma_{LV^0} = \gamma_c \tag{21}$$

Despite the fact that $\gamma_c$ is not the solid surface free energy, the critical surface tension has been shown to correlate with the known surface chemistry of several solids. Determination of $\gamma_c$ is an adequate measure of solid surface free energy for many practical problems. Little information on the underlying surface is, however, learned.

Banerjee and Etzler [17] have explored the relation between surface chemistry and contact angle by combining Zisman's early ideas with the UNIFAC models advanced by Prausnitz and co-workers [18].

Eqns. 16 and 20 can be combined to give the following expression:

$$\cos(\theta) = 1 + \beta W_a - 2\beta\gamma_{LV^0} \tag{22}$$

The UNIFAC model is used to calculate the residual activity for the liquid interacting with the monomer of the polymer surface at infinite dilution. This activity coefficient measures the chemical interaction between the two materials in contact and is thus related to, but not numerically equal, to the work of adhesion. The UNIFAC model is a group contribution model so that the interactions of many materials may be estimated from the UNIFAC library of group interactions.

Using an equation of the form of Eqn. 22 and the UNIFAC model, Banerjee and Etzler [17] determined the semi-empirical relation:

$$\cos(\theta) = 1.42 - 0.102\ln(\gamma_r) - 0.115\,\gamma_{LV^0} \tag{23}$$

where $\gamma_r$ is the UNIFAC residual activity coefficient. The coefficients in Eqn. 23 were determined using 54 polymer–liquid pairs. The equation is able to predict contact angles within $\pm 7°$.

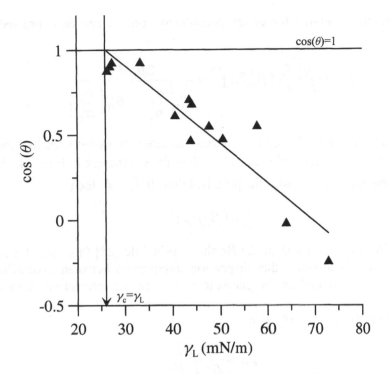

**Figure 1.** A Zisman plot. The critical surface tension is found where the linear fit to the data intersects $\cos(\theta) = 1$ and is about 26 mN/m in this instance.

### 3.2. The van Oss, Chaudhury and Good approach

The Zisman model for estimating surface free energies provides a number of insights into the relation between surface chemistry and contact angle. Furthermore, the Zisman's model can be of importance to practical applications including, for instance, printing. Several investigators, including Fowkes [19-23], Good [24, 25], van Oss [26] and Chang [27], have constructed statistical thermodynamic models for wetting and adhesion. These models may be considered as statistical thermodynamic in the sense that they offer molecular interpretations for origins of wetting and adhesion. In this section, the basic principles common to these statistical thermodynamic theories are explored and in particular the van Oss, Chaudhury and Good [24, 26] model is discussed.

Intermolecular forces between molecules result from interaction between their corresponding electron orbitals. The principal non-bonding interactions result from induced dipole–induced dipole (London), dipole–induced dipole (Debye) and dipole–dipole (Keesom) interactions. The intermolecular potential energy function for each of these three types of interactions is of the same form.

$$U = \frac{-\beta_{12}}{r^6} \qquad (24)$$

If London dispersion forces are considered Eqn. 24 can be expressed as follows:

$$\beta_{12}^d = \frac{2\sqrt{I_1 I_2}}{I_1 + I_2}\left(\beta_{11}^d \beta_{22}^d\right)^{1/2} = \frac{2\beta_{11}^d \beta_{22}^d}{\beta_{11}^d\left(\dfrac{\alpha_2}{\alpha_1}\right) + \beta_{22}^d\left(\dfrac{\alpha_1}{\alpha_2}\right)} \tag{25}$$

Here the subscripts 11, 22 and 12 refer to interactions between like molecules (11, 22) and dissimilar molecules (12). $\beta$ is the coefficient in Eqn. 24. Also, $I$ is the ionization potential and $\alpha$ the polarizability. If $I_1 \approx I_2$ then

$$\beta_{12}^d \cong \left(\beta_{11}^d \beta_{22}^d\right)^{1/2} \tag{26}$$

Eqn. 26 forms the basis of the Berthelot principle [28] (also see, for instance, Chang [29]) which states that dispersion interactions between dissimilar molecules can be estimated as the geometric mean of the interactions between like molecules.

Alternatively, if $\alpha_1 \approx \alpha_2$ then

$$\beta_{12}^d = \frac{2\beta_{11}^d \beta_{22}^d}{\beta_{11}^d + \beta_{22}^d} \tag{27}$$

The harmonic mean estimation expressed in Eqn. 27 has been used less often and is frequently numerically similar to the geometric mean approximation.

The interaction potentials between molecules have been used to determine the interactions between macroscopic bodies. For a column of material 1 interacting with material 2, the free energy, $\Delta G_{12}^d$, to move the plates of material from distance, $d$, to infinity is:

$$\Delta G_{12}^d = -W_{12}^d = \int_{d_{12}}^{-\infty} \frac{-\pi \beta_{12}^d N_2}{6x^3} N_1 \; dx = -\frac{A_{12}}{12\pi d_{12}^2} \tag{28}$$

Here $A_{12}$ is the Hamaker constant (see, for instance, Refs. [1, 29]) and

$$A_{12} = \pi^2 \beta_{12} N_1 N_2 \tag{29}$$

$N_1$ and $N_2$ are the numbers of molecules of type 1 and 2, respectively.

Recalling Berthelot's principle [28],

$$A_{12} \cong \left(A_{11} A_{22}\right)^{1/2} \tag{30}$$

and further assuming that the intermolecular distance, $d_{12}$, can be approximated as the geometric mean of the intermolecular distances found in the pure components ($d_{11}$ and $d_{12}$).

$$d_{12} = (d_{11}d_{22})^{1/2} \tag{31}$$

Thus the work of adhesion, $W_{12}^d$, resulting from London dispersion forces is

$$W_{12}^d \cong \frac{(A_{11}A_{22})^{1/2}}{12\pi \, d_{11}d_{22}} = \left(W_{11}^d W_{22}^d\right)^{1/2} = 2\left(\gamma_1^d \gamma_2^d\right)^{1/2} \tag{32}$$

Fowkes [22, 23] suggested that the surface free energy of materials could be considered to be a sum of components resulting from each class of intermolecular interaction; thus,

$$\gamma = \sum_i \gamma_i \tag{33}$$

Eqn. 33 might be expanded as

$$\gamma = \gamma^d + \gamma^p + \gamma^{AB} + ... \tag{34}$$

Here the superscript d refers to dispersion forces, p refers to dipole–dipole (Debye) interactions and AB refers to Lewis acid–base interactions. London, Kessom and Debye interactions are non-bonding orbital interactions and Lewis acid–base interactions which by definition involve electron acceptance and donation.

Van Oss, Chaudhury and Good choose to express surface free energy in terms of two principal components, Lifshitz–van der Waals (LW) and Lewis acid–base components (AB).

The Lifshitz–van der Waals term is composed of the interactions covered by Eqn. 24. The work of adhesion due to Lifshitz–van der Waals interactions is estimated using the geometric mean rule discussed above. Thus

$$W_a^{LW} = 2(\gamma_1^{LW} \gamma_2^{LW})^{1/2} \tag{35}$$

The use of the geometric mean approximation with regard to Lifshitz–van der Waals interactions is not unique to the van Oss, Chaudhury and Good approach and is used in the models to be discussed later by Chang–Chen and by Fowkes. The harmonic mean approximation (Eqn. 27) has also sometimes been used although the results of the two calculations are often nearly identical. The use of the geometric mean approximation is not a subject of current controversy. The relative merits of the geometric and harmonic mean approximations have been discussed in the literature [21, 30–34].

According to the van Oss, Chaudhury and Good model [24, 26] the Lewis acid–base parameter is modeled as follows:

$$\gamma_i^{AB} = 2\left(\gamma_i^+ \gamma_i^-\right)^{1/2} \tag{36}$$

where $\gamma^+$ is the Lewis acid parameter and $\gamma^-$ the Lewis base parameter. Van Oss, Chaudhury and Good further choose

$$\gamma_i^+ = \gamma_i^- \equiv 0 \tag{37}$$

for alkanes, methylene iodide and $\alpha$-bromonaphathalene which presumably interact only through Lifshitz–van der Waals interactions. For water

$$\gamma_{H_2O}^+ = \gamma_{H_2O}^- \equiv 25.5 \ \text{mJ/m}^2 \tag{38}$$

Based on these above numerical choices $\gamma^+$ and $\gamma^-$ have been experimentally determined for a variety of liquids. Van Oss [26] has compiled and reviewed the determination of these values (Table 1). Della Volpe and co-workers [35-37] have argued that the choice of $\gamma^+$ and $\gamma^-$ for water is inappropriate and van Oss [38] has argued that this numerical choice is not scientifically significant. Kwok [39], for instance, has criticized the use of surface free energy components.

Earlier, Owens and Wendt [31] described surface free energy in terms of two components which were called dispersive $\gamma^d$ and polar $\gamma^p$. Thus

$$\gamma = \gamma^d + \gamma^p \tag{39}$$

While it is generally recognized that $\gamma^d \approx \gamma^{LW}$ the meaning of $\gamma^p$ is perhaps hopelessly confused in the literature. According to Fowkes [22, 23], $\gamma^p$ should refer to dipole–dipole (Debye) interactions. In the van Oss, Chaudhury and Good model such interactions are incorporated into $\gamma^{LW}$. Good [40] no longer recommends the use of $\gamma^p$. Good's argument follows in the next paragraph.

Eqn. 36 reminds us that for monopolar materials ($\gamma^+$ or $\gamma^- = 0$) $\gamma^{AB} = 0$. On the other hand, for two dipolar ($\gamma^+, \gamma^- \neq 0$) materials interacting $\gamma^p > 0$; thus $\gamma^{AB} \neq \gamma^p$. For example, the surface tensions of carbon tetrachloride and chloroform are nearly identical, yet their interfacial tensions with water are 45.0 mJ/m$^2$ and 31.6 mJ/m$^2$, respectively. Because chloroform is a monopolar acid ($\gamma^+ \neq 0, \quad \gamma^- = 0 \rightarrow \gamma^{AB} = 0$) a descriptor such as $\gamma^p$ is inadequate to describe the difference in the observed interfacial tensions as $\gamma^{AB} = 0$ for these two substances. Instead Eqn. 36 is a better descriptor.

From a practical point of view, reported values of $\gamma^d$ and $\gamma^p$ can be regarded as $\gamma^{LW}$ and $\gamma^{AB}$, respectively. The values, however, should be interpreted in terms of the van Oss, Chaudhury and Good model.

Recalling Eqns. 10, 12 and 36 together with the relation,

$$\gamma = \gamma^{LW} + \gamma^{AB} \tag{40}$$

it follows that

$$W_a = \gamma_1 \left[1 + \cos(\theta)\right] = 2(\gamma_1^{LW} \gamma_s^{LW})^{1/2} + 2(\gamma_1^+ \gamma_s^-)^{1/2} + 2(\gamma_1^- \gamma_s^+)^{1/2} \tag{41}$$

**Table 1.**
Van Oss, Chaudhury and Good surface tension parameters for various liquids at 20°C in mJ/m$^2$

| Liquid | $\gamma$ | $\gamma^{LW}$ | $\gamma^{AB}$ | $\gamma^+$ | $\gamma^-$ |
|---|---|---|---|---|---|
| Methanol | 22.5 | 18.2 | 4.3 | ~ 0.06 | ~ 77 |
| Ethanol | 22.4 | 18.8 | 2.6 | ~ 0.019 | ~ 68 |
| Methyl ethyl ketone | 24.6 | 24.6 | 0 | 0 | 24.0 |
| Tetrahydrofuran | 27.4 | 27.4 | 0 | 0 | 15.0 |
| Ethylene glycol | 48 | 29 | 19 | 1.92 | 47.0 |
| Glycerol | 64 | 34 | 30 | 3.92 | 57.4 |
| Formamide | 58 | 39 | 19 | 2.28 | 39.6 |
| Dimethylsulfoxide | 44 | 36 | 8 | 0.5 | 32 |
| Water | 72.8 | 21.8 | 51 | 25.5 | 25.5 |
| Chlorobenzene | 33.6 | 32.1 | 1.5 | 0.9 | 0.61 |
| Nitrobenzene | 43.9 | 41.3 | 2.6 | 0.26 | 6.6 |
| Diiodomethane | 50.8 | 50.8 | 0 | 0 | 0 |
| α-Bromonaphthalene | 44.4 | 44.4 | 0 | 0 | 0 |
| Pentane | 16.05 | 16.05 | 0 | 0 | 0 |
| Hexane | 18.40 | 18.40 | 0 | 0 | 0 |
| Heptane | 20.14 | 20.14 | 0 | 0 | 0 |
| Octane | 21.62 | 21.62 | 0 | 0 | 0 |
| Nonane | 22.85 | 22.85 | 0 | 0 | 0 |
| Decane | 23.83 | 23.83 | 0 | 0 | 0 |
| Undecane | 24.66 | 24.66 | 0 | 0 | 0 |
| Dodecane | 25.35 | 25.35 | 0 | 0 | 0 |
| Tridecane | 25.99 | 25.99 | 0 | 0 | 0 |
| Tetradecane | 26.56 | 26.56 | 0 | 0 | 0 |
| Pentadecane | 27.07 | 27.07 | 0 | 0 | 0 |
| Hexadecane | 27.47 | 27.47 | 0 | 0 | 0 |
| Nonadecane | 28.59 | 28.59 | 0 | 0 | 0 |
| Eicosane | 28.87 | 28.87 | 0 | 0 | 0 |
| Cyclohexane | 25.24 | 25.24 | 0 | 0 | 0 |

If the van Oss, Chaudhury and Good parameters are known for at least three liquids and the contact angles of these liquids on a solid are measured, then Eqn. 41 can be used to determine the van Oss, Chaudhury and Good parameters for the surface free energy of the solid. Van Oss [26] has reviewed the numerous publications which have reported the determination of the van Oss, Chaudhury and Good parameters for various liquids.

The simultaneous solution of Eqn. 41 for a set of several liquids would seem straightforward, at first glance. Gardner *et al.* [41], however, have pointed out that different calculation approaches may give numerically different results and that a particular author's choice may not be readily apparent to the reader.

Dalal [42] has discussed the choice of liquid sets used to determine surface free energy parameters. While Dalal's discussion addresses the older Owens–Wendt model [31] much of the discussion applies directly to the van Oss, Chaudhury and Good model as well. Because the Owens–Wendt model has only two parameters, it is only necessary to measure contact angles for two liquids. Dalal noted that the calculated values for the surface free energy components depended upon the choice of liquids. The use of dissimilar liquid pairs (e.g., water, methylene iodide) minimized the dependence of the calculated results upon the precise choice of probe liquids. Dalal recommended that many liquids be used and that the contact angle results from this overdetermined set of liquids be used to find the best fit surface free energy components. Etzler *et al.* [43] concur with Dalal's conclusions but realize that many authors have not heeded this advice.

For the solution of Eqn. 41, most authors choose to measure the contact angle with at least one non-polar liquid ($\gamma^+, \gamma^- = 0$) such as methylene iodide and two polar liquids one of which is usually water as water is most dissimilar to other liquids.

If a non-polar liquid is chosen then Eqn. 37 reduces to

$$W_a = \gamma_1 \left[ 1 + \cos(\theta) \right] = 2(\gamma_1^{LW} \gamma_s^{LW})^{1/2} \tag{42}$$

The Lifshitz–van der Waals component may thus be calculated from a single contact angle measurement (subject, of course, to Dalal's warnings).

The parameters $\gamma_s^j$, ($j = LW, +, -$) in Eqn. 41 have been interpreted in two ways for fitting purposes. The first method involves determination of $\gamma_s^j$ directly from Eqn. 41. In this first case one will find $\gamma_s^j \geq 0$ and $W_a \geq 0$. In the second case investigators let $c_s^j = \sqrt{\gamma_s^j}$. The second choice allows $c_s^j$ to become negative during the fitting process and, thus, $W_a$ may also be negative. Again, Gardner *et al.* [41] have explored the consequences of theses two choices. The first choice is the correct van Oss, Chaudhury and Good model [24, 26].

The fitting to Eqn. 41 of contact angle data can be accomplished in at least three ways. The first way involves direct fitting to Eqn. 41 using data from the entire pool of chosen liquids. This method works best when an overdetermined set (more than three liquids) of data is available. Care should be taken to select a well-conditioned set of liquids, as Dalal [42] suggested earlier. If this first method is chosen when data for only a few liquids are available, it possible to obtain an unrealistic result. The second method involves the use of Eqn. 42 to fit the data from measurements made with non-polar liquids (alkanes, methylene iodide, α-bromonaphthalene, etc.) to determine $\gamma_s^{LW}$. Data from the polar liquids and Eqn. 41 are then used to determine $\gamma_s^-$ and $\gamma_s^+$. When data from many liquids are available both the first and second method will yield comparable results. When only a few liquids have been chosen the second method is preferable. A third method would be to borrow from Owens and Wendt the expression

$$W_a = \gamma_1 \left[ 1 + \cos(\theta) \right] = 2(\gamma_1^{LW} \gamma_s^{LW})^{1/2} + 2(\gamma_1^{AB} \gamma_s^{AB})^{1/2} \tag{43}$$

Eqn. 43 and data from measurements using polar and nonpolar liquids can be used to determine $\gamma_s^{LW}$ and $\gamma_s^{AB}$. Equation 41 can then be used to calculate $\gamma_s^+$ and $\gamma_s^-$ subject to the constraints imposed by Eqn. 43.

It is important to the reader of the literature to realize that various authors have calculated so-called van Oss, Chaudhury and Good parameters using various numerical methods which may influence the calculated result significantly. The van Oss, Chaudhury and Good model expressed in Eqn. 41 requires $W_a > 0$ and $(\gamma_s^j \gamma_1^k) \geq 0$. Furthermore, $\gamma_1^k \geq 0$ for all investigated liquids ($j, k = +, -$). Again, tabulated values for van Oss, Chaudhury and Good parameters have been compiled by van Oss, for instance [26] (see Table 1). The best estimates of the van Oss, Chaudhury and Good components require the use of an overdetermined set of probe liquids and carefully and properly measured contact angles. The set of chosen liquids should contain liquids that are non-polar as well as liquids that are polar.

### 3.3. The Chang–Chen model

The Chang–Chen model [27, 29] for interfacial free energy is largely based upon the same principles which govern the van Oss, Chaudhury and Good model. Both models treat Lifshitz–van der Waals interactions in the same way. Calculation of the surface free energy components requires the knowledge of the same experimental data. The two models, however, differ in the way that Lewis acid–base interactions are modeled.

Recall that

$$W_a = W_a^{LW} + W_a^{AB} \tag{44}$$

and

$$\gamma = \gamma^{LW} + \gamma^{AB}. \tag{45}$$

The Chang–Chen model uses the same geometric mean approximation for $W_a^{LW}$ as does the van Oss, Chaudhury and Good model. Thus

$$W_a^{LW} = W_a^L = 2\left( \gamma_1^{LW} \gamma_2^{LW} \right)^{1/2} = P_1^L P_2^L \tag{46}$$

where

$$P_i^L = \left( 2\, \gamma_i^L \right)^{1/2} \tag{47}$$

$P_i^L$ is the dispersion parameter. The superscript L is equivalent to LW.

Like the van Oss, Chaudhury and Good model the acid–base interaction is modeled using two parameters. These parameters, $P_i^a$ and $P_i^b$, are referred to as principal values. The acid–base work of adhesion can be represented using the following relation:

$$W_a^{AB} = -\Delta G_a^{AB} = -(P_1^a P_2^b + P_1^b P_2^a) \tag{48}$$

The surface free energy of the material is thus

$$\gamma = \gamma^{LW} + \gamma^{AB} = \frac{1}{2}\left(P^L\right)^2 - P^a P^b \tag{49}$$

Tabulated $P_i^a$ and $P_i^b$ values (Table 2) are substituted into Eqn. 48 such that the work of adhesion is maximized and the free energy of adhesion is minimized.

The acid–base character of a material is characterized by the sign of $P_i^a$ and $P_i^b$. If $P_i^a = P_i^b = 0$, then the material is neutral (or non-polar). If $P_i^a$ and $P_i^b$ are both positive, then the material is monopolar acidic and if both are negative, then the material is monopolar basic. If $P_i^a$ and $P_i^b$ are of opposite sign, then the material is amphoteric.

Despite some similarities to the van Oss, Chaudhury and Good model, the Chang–Chen model differs from the former model in a number of ways. The Chang–Chen model only applies the geometric mean rule to Lifshitz–van der Waals interactions. In determining values for $P_i^a$ and $P_i^b$ interactions involving only n-alkanes are assumed to exclusively result from Lifshitz–van der Waals interactions. The van Oss, Chaudhury and Good model, for instance, assumes that both methylene and α-bromonaphthalene also interact exclusively by Lifshitz–van der Waals interactions. A major difference is that the Chang–Chen model allows for both attractive and repulsive interactions. In other words, $-\infty \leq W_a^{AB} \leq \infty$, whereas in the van Oss, Chaudhury and Good model $W_a^{AB} \geq 0$. The Lewis acid–base concept is general enough to include traditional ion–ion and dipole–dipole repulsions and, thus, it may not be unreasonable to suggest the existence of repulsive interactions [29]. Furthermore, entropic effects may contribute to the overall repulsion.

## 3.4. The Fowkes approach

As discussed above, Fowkes [22, 23] first suggested that surface free energy could be considered as a sum of components resulting from different classes of intermolecular interactions. Both, the van Oss, Chaudhury and Good and Chang–Chen models draw upon the idea of Fowkes and as such all use the geometric mean approximation to model Lifshitz–van der Waals interactions.

However, Fowkes [19, 22, 34] has suggested a different approach to evaluating the acid base character of surfaces (also see Ref. [44]). Fowkes has criticized the

**Table 2.**
Chang–Chen parameters for various liquids at 20°C

| Liquid | $\gamma^{LW}$(calc) mJ/m$^2$ | $P_L^{LW}$ (mJ/m$^2$)$^{\frac{1}{2}}$ | $P_L^a$ (mJ/m$^2$)$^{\frac{1}{2}}$ | $P_L^b$ (mJ/m$^2$)$^{\frac{1}{2}}$ |
|---|---|---|---|---|
| Pentane | 15.8 | 5.62 | 0 | 0 |
| Hexane | 18.4 | 6.07 | 0 | 0 |
| Heptane | 20.1 | 6.34 | 0 | 0 |
| Decane | 23.8 | 6.90 | 0 | 0 |
| Dodecane | 25.4 | 7.13 | 0 | 0 |
| Hexadecane | 27.5 | 7.42 | 0 | 0 |
| Cyclohexane | 25.6 | 7.15 | 0 | 0 |
| Water | 72.75 | 6.60±0.01 | 6.88±0.01 | -7.40±0.01 |
| Glycerol | 63.64 | 8.00±0.01 | 3.40±0.02 | -9.30±0.08 |
| Formamide | 58.76 | 7.30±0.04 | 6.92±0.11 | -4.64±0.09 |
| Methylene iodide | 50.35 | 11.60±0.04 | -4.11±0.07 | -4.12±0.07 |
| α-Bromonaphthalene | 44.93 | 10.50±0.03 | -2.67±0.07 | -3.82±0.01 |
| Tricresyl phosphate | 41.10 | 9.00±0.02 | 0.11±0.04 | -5.45±0.46 |
| Ethylene glycol | 48.2 | 7.50±0.06 | 3.69±0.16 | -5.44±0.21 |
| Chloroform | 27.1 | 7.10±0.01 | 2.04±0.01 | -0.92±0.01 |
| Sym-tetrachloroethane | 36.5 | 8.30±0.05 | 2.69±0.31 | -0.77±0.17 |
| Acetone | 26.0 | 6.90±0.01 | 0.32±0.01 | -6.95±0.14 |
| Propyl ether | 17.8 | 6.00±0.01 | -0.05±0.02 | -4.70±0.19 |
| Ethyl ether | 16.9 | 6.10±0.01 | -0.28±0.01 | -5.94±0.01 |
| Tetra-butylnaphthalene | 34.0 | 8.20±0.01 | 0.14±0.01 | -2.83±0.08 |
| Trichlorobiphenyl | 45.0 | 10.20±0.04 | -2.15±0.12 | -3.26±0.20 |
| Tetrachlorobiphenyl | 44.2 | 10.00±0.01 | -2.35±0.02 | -2.45±0.02 |
| Benzyl phenylundecanoate | 37.6 | 9.50±0.01 | -1.76±0.03 | -4.30±0.07 |
| Benzene | 28.6 | 7.40±0.03 | 1.00±0.09 | -1.26±0.22 |
| Toluene | 28.3 | 7.40±0.02 | 1.90±0.05 | -0.50±0.04 |
| Carbon tetrachloride | 26.8 | 7.30±0.01 | 0.73±0.01 | -0.17±0.01 |
| Tetrachloroethylene | 33.1 | 8.20±0.01 | -0.70±0.02 | -0.71±0.02 |
| Carbon disulfide | 31.3 | 7.90±0.03 | 0.51±0.12 | -0.11±0.14 |

use of contact angles for measurement of interfacial properties [21]. His approach is, for experimental reasons, more applicable to powdered samples. As stated previously,

$$W_a = W_a^{LW} + W_a^{AB} \qquad (50)$$

$W^{AB}$ is then, according to Fowkes, expressed by the following relation

$$W_a^{AB} = -f \cdot N \cdot \Delta H_a^{AB} \qquad (51)$$

where $N$ is the number of sites per unit area and

$$f = \left[1 - \frac{\partial \ln W_a^{AB}}{\partial \ln T}\right]^{-1} \tag{52}$$

and

$$f \approx 0.2 \ldots 1.0 \tag{53}$$

When using the Fowkes approach some authors have taken $f$ as unity although this does not seem to be a good approximation [44]. Because $f$ and $N$ are generally not known, direct calculations of the work of adhesion are often not made. Determination of $\Delta H^{AB}$ for multiple probe liquids on a given solid together with models by Drago [45] or Gutmann [46, 47] can be used to assess the acid–base nature of the surface (also see Ref. [48]).

Lewis acid–base interactions encompass hydrogen bonding, electron donor–acceptor and organic nucleophile–electrophile interactions. The Mulliken theory [49] is based on a valence-bond model for the ground state energy of the donor–acceptor complex. This model considers the complex in terms of two resonance forms – non-bonded and ionic (electrons fully transferred).

This notion of covalent and Coulombic contributions forms the basis for the Drago parameters (also see, for instance, Ref. [50]).

According to Drago

$$-\Delta H^{AB} = E_A \cdot E_B + C_A \cdot C_B \tag{54}$$

where $E_i$'s are the electrostatic susceptibility parameters and $C_i$'s are the covalent susceptibility parameters. Drago parameters allow one to easily consider the hardness of the acid–base interaction. Hardness is related to polarizability. Substances with low polarizability are referred to as hard and those with high polarizability are said to be soft. In Drago's model, $E/C$ is a measure of hardness. Drago's equation shows the preference for interaction with molecules of like hardness. While useful, Drago's model does not allow one to consider easily the amphoteric nature of many materials (for a discussion of hard and soft acids, see Refs. [51, 52]).

Many adsorbates are amphoteric; therefore, it becomes necessary to recognize the dual polarity of many compounds. Gutmann [46, 47] introduced the notion of electron donor numbers (DN) and electron acceptor numbers (AN). These parameters are similar to the van Oss, Chaudhury and Good surface tension parameters as they both describe the same molecular parameters but from different points of view. In 1966 Gutmann [46] introduced the donor number based on the interaction with $SbCl_5$. DN has units corresponding to enthalpy (e.g., kJ/mol). In 1975, Mayer et al. [53] introduced the acceptor number based on the relative [31]P shift induced by triethylphosphine oxide. AN has arbitrary units. In 1990 Riddle and Fowkes [48] removed the dispersive component from AN. The corrected AN*

**Table 3.**
Gutmann donor numbers in kcal/mol [47]

| Liquid | DN | Liquid | DN |
|---|---|---|---|
| 1,2-Dichloroethane (DCE) | – | Ethylene carbonate (EC) | 16.4 |
| Benzene | 0.1 | Phenylphosphonic difluoride | 16.4 |
| Thionyl chloride | 0.4 | Methyl acetate | 16.5 |
| Acetyl chloride | 0.7 | n-Butyronitrile | 16.6 |
| Tetrachloroethylene carbonate | 0.8 | Acetone (AC) | 17.0 |
| (TCEC) | | Ethylacetate | 17.1 |
| Benzoyl fluoride (BF) | 2.3 | Water | 18.0 |
| Benzoyl chloride | 2.3 | Phenylphosphoric dichloride | 18.5 |
| Nitromethane (NM) | 2.7 | Diethyl ether | 19.2 |
| Nitrobenzene (NB) | 4.4 | Tetrahydrofuran (THF) | 20.0 |
| Acetic anhydride | 10.5 | Diphenylphosphoric chloride | 22.4 |
| Phosphorus oxychloride | 11.7 | Trimethyl phosphate (TMP) | 23.0 |
| Benzonitrile (BN) | 11.9 | Tributyl phosphate (TBP) | 23.7 |
| Selenium oxychloride | 12.2 | Dimethyl formamide (DMF) | 26.6 |
| Acetonitrile | 14.1 | n-Methyl pyrrolidinone (NMP) | 27.3 |
| Tetramethylenesulfone (TMS) | 14.8 | n-Dimethyl acetamide (DMA) | 27.8 |
| Dioxane | 14.8 | Dimethyl sulfoxide (DMSO) | 29.8 |
| Propandiol-(1 ,2)-carbonate | 15.1 | n-Diethyl formamide (DEF) | 30.9 |
| (PDC) | | n-Diethyl acetamide (DEA) | 32.2 |
| Benzyl cyanide | 15.1 | Pyridine (PY) | 33.1 |
| Ethylene sulphite (ES) | 15.3 | Hexamethylphosphoric triamide (HMPA) | 38.8 |
| iso-Butyronitrile | 15.4 | | |
| Propionitrile | 16.1 | | |

values have the usual units of enthalpy. According to Gutmann, water is somewhat more nucleophilic. Chang [27, 29] also finds that $|P_{H_2O}^b| > P_{H_2O}^a$ which suggests that water is somewhat more nucleophilic than it is electrophilic. These conclusions contrast with van Oss' assumption that water is equally electrophilic and nucleophilic [26] (again, the question of the electrophile/nucleophile balance for water remains an open issue as does the exact relation between each of the various acidity scales proposed in the literature [35-38]). Table 3 lists known values of DN and Table 4 lists known values of AN and AN*.

According to Gutmann's theory

$$\Delta H^{AB} = K_a \, DN + K_d AN^*  \tag{55}$$

where $K_a$ and $K_d$ reflect the acceptor and donor characteristics of a solid. Gutmann's model works best with hard (low polarizability) atoms.

**Table 4.**
Van der Waals and acid–base contributions to Gutmann acceptor numbers

| Liquid | AN | $AN^d$ | $AN-AN^d$ | $AN^*$ (kcal/mol) |
|---|---|---|---|---|
| Dioxane | 10.8 | 10.9 | 0.0 | 0.0 |
| Hexane | 0.0 | 0.0 | 0.0 | 0.0 |
| Benzonitrile | 15.5 | 15.3 | 0.2 | 0.06 |
| Pyridine | 14.2 | 13.7 | 0.5 | 0.14 |
| Benzene | 8.2 | 7.6 | 0.6 | 0.17 |
| Tetrahydrofuran | 8.0 | 6.1 | 1.9 | 0.5 |
| Carbon tetrachloride | 8.6 | 6.3 | 2.3 | 0.7 |
| Diethyl ether | 3.9 | -1.0 | 4.9 | 1.4 |
| Ethyl acetate | 9.3 | 4.0 | 5.3 | 1.5 |
| Methyl acetate | 10.7 | 5.0 | 5.7 | 1.6 |
| n,n-dimethylacetamide | 13.6 | 7.9 | 5.7 | 1.6 |
| 1,2-dichloroethane | 16.7 | 10.3 | 6.4 | 1.8 |
| n,n-dimethylacetamide | 16.0 | 9.4 | 6.6 | 1.9 |
| Acetone | 12.5 | 3.8 | 8.7 | 2.5 |
| Dimethyl sulfoxide | 19.3 | 8.5 | 10.8 | 3.1 |
| Dichloromethane | 20.4 | 6.9 | 13.5 | 3.9 |
| Nitromethane | 20.5 | 5.7 | 14.8 | 4.3 |
| Acetonitrile | 18.9 | 2.6 | 16.3 | 4.7 |
| Chloroform | 25.1 | 6.4 | 18.7 | 5.4 |
| Tert-butyl alcohol | 27.1 | 0.6 | 26.5 | 7.6 |
| 2-propanol | 33.6 | 2.1 | 31.5 | 9.1 |
| N-butyl alcohol | 36.8 | 5.1 | 31.7 | 9.1 |
| Formamide | 39.8 | 7.6. | 32.2 | 9.3 |
| Ethanol | 37.9 | 2.0 | 35.9 | 10.3 |
| Methanol | 41.5 | -0.2 | 41.7 | 12.0 |
| Acetic acid | 52.9 | 3.6 | 49.3 | 14.2 |
| Water | 54.8 | 2.4 | 52.4 | 15.1 |
| Trifluoroethanol | 53.3 | -2.9 | 56.2 | 16.2 |
| Hexafluoro-2-propanol | 61.6 | -4.7 | 66.3 | 19.1 |
| Trifluoroacetic acid | 105.3 | 5.7 | 111.0 | 31.9 |

N.B. $AN = AN^d + c\, AN^*$. AN and $AN^d$ are dimensionless. $AN^d$ is Lifshitz–van der Waals component. See for details, Riddle and Fowkes [48].

## 3.5. Neumann's approach

Neumann and co-workers [54-64] have discussed the surface tension of solids from a purely thermodynamic point of view. Their view contrasts with the statistical thermodynamic approaches used by van Oss, Chaudhury and Good, Chang–Chen and Fowkes (see earlier sections for references). Because the Neumann's approach does not consider the molecular origins of surface tension no statistical mechanical insight is gained.

Kwok and Neumann [54] correctly remind us that that contact angle measurements can be difficult. Measured contact angles can deviate from the true Young's contact angle which satisfies Eqn. 9. Both surface topography and surface chemical heterogeneity can affect the measured value for contact angle [65-68]. Grundke *et al.* [69] have recently discussed wetting on rough surfaces.

All of the thermodynamic approaches to calculate surface free energy which are explored in this paper assume that the following conditions apply:

1.  Young's equation is valid.
2.  Probe liquids are pure compounds. Mixtures are likely to exhibit selective surface adsorption (see, e.g., Adamson [1]).
3.  The surface tensions of materials are independent of the test conditions at fixed $T$ and $P$ (i.e., the surface tension of a solid is not modified by the probe liquid).
4.  $\gamma_l > \gamma_s$ according to Neumann or more generally $\cos(\theta) > 0$.
5.  $\gamma_{sv}$ is not influenced by liquids.

According to Kwok and Neumann [54], the contact angle can be expressed as a function of $\gamma_{LV}$ and $\gamma_{SV}$ only. Thus,

$$\gamma_{LV} \cos(\theta) = f(\gamma_{LV}, \gamma_{SV}) \tag{56}$$

$$\gamma_{SV} - \gamma_{SL} = f(\gamma_{LV}, \gamma_{SV}) \tag{57}$$

$$\gamma_{SL} = \gamma_{SV} - f(\gamma_{LV}, \gamma_{SV}) = F(\gamma_{LV}, \gamma_{SV}) \tag{58}$$

where $f$ and $F$ are suitable functions. Kwok and Neumann have observed smooth monotonic dependence of $\gamma_{LV}\cos(\theta)$ with $\gamma_{LV}$ consistent with Eqn. 56 when liquid–solid pairs conform closely to the assumptions listed above. For arbitrary solid–liquid pairs such a plot may show considerable scatter because the measured contact angles deviate significantly from the true Young's contact angle. Stick/slip behavior in advancing contact angles, time-dependent contact angles and liquid–surface tension changes during the course of the experiments result from physico-chemical interactions not consistent with the use of Young's equation for determining surface free energies.

The function, $F$, in Eqn. 58 has been historically modeled in several ways. Antonow [70] stated that:

$$\gamma_{SL} = |\gamma_{LV} - \gamma_{SV}| \tag{59}$$

or combining with Young's equation

$$\cos(\theta) = -1 + 2\frac{\gamma_{SV}}{\gamma_{LV}} \tag{60}$$

Alternatively, Berthelot's rule [28] (recall Eqn. 35, for instance) has been used such that

$$\gamma_{SL} = \gamma_{LV} + \gamma_{SV} - 2\left(\gamma_{LV}\gamma_{SV}\right)^{1/2} = \left[\left(\gamma_{LV}\right)^{1/2} - \left(\gamma_{SV}\right)^{1/2}\right]^2 \tag{61}$$

and again combining with Young's equation

$$\cos(\theta) = -1 + 2\left(\frac{\gamma_{SV}}{\gamma_{LV}}\right) \tag{62}$$

Eqn. 58 would be identical to that used in the van Oss, Chaudhury and Good model if the solid and liquid had only Lifshitz–van der Waals interactions such that $\gamma_{LV}^{LW} = \gamma_{LV}$. If Eqns. 56 and 58 were adequate descriptors, then calculated values for $\gamma_{SV}$ would be independent of the choice of probe liquid used. This is unfortunately not the case. Li and Neumann [64] have considered a modified Berthelot equation such that

$$\gamma_{SL} = \gamma_{LV} + \gamma_{SV} - 2\left(\gamma_{SV}\gamma_{LV}\right)^{1/2} e^{-\beta(\gamma_{LV} - \gamma_{SV})^2} \tag{63}$$

and

$$\cos(\theta) = -1 + 2\left(\frac{\gamma_{SV}}{\gamma_{LV}}\right)^{1/2} e^{-\beta(\gamma_{LV} - \gamma_{SV})^2} \tag{64}$$

Empirically it has been shown that

$$\beta \approx 0.0001247 \tag{65}$$

and that the measured solid surface free energy using this choice for $\beta$ is nearly independent of the choice of liquid. Eqn. 64 can be used in two ways. First, $\beta$ can be chosen from Eqn. 65 and a suitable contact angle can be used to determine $\gamma_{SV}$. Second, $\beta$ and $\gamma_{SV}$ can be treated as adjustable parameters. Least-squares analysis using contact angles measured for several liquids is then used to determine the best fit values for $\beta$ and $\gamma_{SV}$. The second approach would seem to be preferable. An alternate combining rule has been suggested recently by Kwok and Neumann [71].

Kwok and Neumann [54] have criticized the van Oss, Chaudhury and Good approach for several reasons. The reasons include at least the following:

1.  The Neumann model uses fewer adjustable parameters; two versus six
2.  Surface free energies measured using the van Oss, Chaudhury and Good may appear to depend on the choice of liquids
3.  The assigned values for the van Oss, Chaudhury and Good parameters are assigned in ways not fully satisfactory to Neumann and co-workers.

It is my opinion that Kwok and Neumann have properly identified experimental factors that result in significant deviations of measured contact angles from the true Young's contact angle. Application of any of the above models requires that Young's equation be valid. Undoubtedly, contact angles reported in the literature sometimes deviate significantly from the Young's angle for a variety of reasons.

The apparent variation of surface free energy with the choice of probe liquid sets undoubtedly, in part, results from physico-chemical interactions between solid and liquid that cause Young's equation to be invalid for the present purposes. On the other hand, Kwok and Neumann have not always reported results from liquid sets containing diverse liquids. The mathematics of least squares fitting requires the use of a diverse set of liquids to achieve reliable values for the fitted parameters. Kwok, Li and Neumann [39, 54, 72] have also chosen a fitting procedure that allows $\sqrt{\gamma_i^j}$ to be negative ( $j = +, -$ ) which appears contrary to van Oss' statements (see Section 3.2). It is not clear, at this time, that the model proposed by Neumann and the model by van Oss, Chaudhury and Good are mutually exclusive as Neumann and co-workers suggest.

## 3.6. Overview of various thermodynamic approaches for determining surface free energy

In the preceding sections several models for calculating the surface free energies of solids have been discussed.

The models by Zisman [16] and Kwok and Neumann [54] yield total surface free energies. As the models are thermodynamic (classical) in nature no direct and detailed information is given on the molecular origins of the observed contact angles. The Zisman's approach is operationally simple and can be adequate for some product quality applications. Because Neumann's approach is theoretically more rigorous, it would seem more suitable, particularly when the contact angles from several liquids are used to determine the surface free energy of the solid under study. These models use a small number of adjustable parameters.

Banerjee and Etzler [17] have offered a model which allows for some molecular insight into the intermolecular interactions responsible for observed contact angles. If the molecular structure of the surface is known and suitable UNIFAC coefficients are known, the model can be a good predictor of observed contact angles.

The models advanced by van Oss, Chaudhury and Good [26], Chang–Chen [29] and Fowkes (see, for instance, Ref. [44]) are of statistical thermodynamic origin and as such seek to identify some of the molecular characteristics of the surface which might add insight into the origin of experimental observations. All of these models consider intermolecular interactions to be divided into classes Lifshitz–van der Waals and Lewis acid–base. The theoretical treatment of Lewis acid–base interactions differs between models as do the reference points used to construct numerical acidity and basicity scales. Perhaps not surprisingly, Morra [73] and Etzler *et al.* [43] have observed that the acid–base character of a given material is not always assessed to be the same by the various models. Because each of the statistical thermodynamic models hopes to extract more information (acid–base character), the models contain more adjustable parameters than the models addressed in previous paragraphs of this section.

The van Oss, Chaudhury and Good model has been most widely applied and in many respects it is theoretically the most pleasing; furthermore, it is similar to the earlier ideas of Owens and Wendt [31]. The assumptions made in determining the published tabular values of the van Oss, Chaudhury and Good parameters for various probe liquids have been criticized by Della Volpe and Siboni [35] and Kwok and Neumann [54]. At present, however, the relative merits of the arguments made by van Oss, Chaudhury and Good, and Chang–Chen are not settled. The van Oss, Chaudhury and Good, and Chang–Chen models are best applied to systems where contact angle data can be collected.

The Fowkes model is mathematically similar to the van Oss, Chaudhury and Good model when it is coupled with Gutmann's donor–acceptor model [47, 48]. Fowkes' model, however, is more adaptable to experimental techniques like inverse gas chromatography and flow calorimetry, which can be easily applied to powders.

## 4. THEORETICAL CALCULATION OF SURFACE FREE ENERGY

While this paper focuses on thermodynamic and experimental approaches for the estimation of surface free energy, it would be remiss not to mention that theoretical calculation of surface energies is possible. Such calculations do, however, present some challenges.

The calculation of surface energy of solids requires a good knowledge of the crystal structure of the solid and the nature of the intermolecular potentials. This knowledge, along with knowledge of the mathematical methods of molecular simulation, allows for the estimation of surface energies. Such methods have been discussed by a number authors [74-80] .

Each crystal plane has a unique surface energy. The most common surfaces found on crystal surfaces are those of low Miller index. Stable interfaces will be electrically neutral and will have no net dipole moment perpendicular to the surface.

The Wulff theorem [81-86],

$$\gamma_{\text{cryst}} = \sum_i \gamma_i A_i , \tag{66}$$

provides theoretical guidance for the equilibrium morphology of the crystal. Here $\gamma_i$ is the surface free energy of a particular crystal plane and $A_i$ is the specific surface area of that plane. $A_i$'s are adjusted to minimize the overall surface free energy.

Simulations must recognize structural differences between surface planes and similar planes found in the interior. Relaxation is caused by the asymmetry of the intermolecular force field surrounding surface atoms relative to the force field surrounding similar interior atoms. Such structural relaxation in $\alpha$-$Al_2O_3$, for instance, accounts for a 2-fold reduction in surface free energy [77]. Diamonds and some metals, for instance, have surfaces covered with oxide, thus their chemical composition differs from that of the bulk material. Surface chemical composition must be accounted for in estimates of surface free energy.

For simple materials like halides it is possible to determine surface free energies which closely approximate those determined by experiment. Stoneham [76], for instance, has shown good agreement between experiment and theory for alkaline earth fluorides.

Clearly, the theoretical calculation of surface energies is best suited to rather simple surfaces. Calculations can assist, however, in the understanding of experimentally determined values. The reader is directed to published reviews for more comprehensive information regarding the calculations discussed here [74-80].

## 5. EXPERIMENTAL DETERMINATION OF SURFACE FREE ENERGY

### 5.1. Contact angle

Measurement of the contact angle at the solid–liquid–vapor interface has been used extensively for the study of the surface properties of both solids and liquids. Many different techniques have been developed for the measurement of contact angle. The most widely used methods are discussed below.

The contact angles measured for probe liquids on solids of interest are used to calculate surface free energies of solids based on one of the above described models.

As described earlier, Young's equation (Eqn. 9) relates the equilibrium contact angle of a liquid drop on a solid surface with the three relevant interfacial free energies, solid–vapor, solid–liquid and liquid–vapor, present in the system. The use of contact angle measurements to determine surface free energy relies on the use of Young' equation. In order for contact angle measurements to comply with Young's equation certain conditions must apply [54] (also, see Section 3.5). Again, equilibrium contact angle measurements which satisfy this relationship can be made on ideal solid surfaces that are homogeneous and flat, and have properties which do not change appreciably due to interactions with the liquid or vapor phases. In practice, the measured contact angle may not fully satisfy the assumptions implicit in Eqn. 9. For instance, significant contact angle hysteresis is frequently observed between the advancing and receding angles. The subject of contact angle measurements has been reviewed extensively by several authors including Neumann and Good [87].

## 5.1.1. Sessile drops

The most commonly used method for measuring contact angle as described by Zisman [88] employs an eyepiece goniometer to directly measure the angle of interest. A small liquid drop is formed on the surface of the solid, usually from the tip of a small syringe. The contact angle may be obtained directly by measuring the angle between the tangent to the drop surface at the point of contact with the solid and the horizontal solid surface. An identical measurement may be taken from a bubble trapped at a solid surface submerged in the liquid of interest. Advancing and receding angles can be determined by repeatedly adding or removing small quantities of the liquid (or gas) to the drop (or bubble). This represents the simplest and most direct method for obtaining the contact angle, but is less accurate than determinations from drop shape described below.

Contact angles can be calculated from drop dimensions. Drop shape is determined from the combination of the forces of surface tension and gravity. Several methods are available, including the axisymmetric drop shape analysis (ADSA) techniques developed by Neumann and co-workers [89]. These methods are based on Eqn. 13, the Young–Laplace equation.

Drop profiles can be determined automatically though digital images and computer analysis. Surface tension, contact angle, as well as drop characteristics such as volume and surface area, are calculated from a nonlinear regression of the measured drop profile to the Young–Laplace equation (Eqn. 13). ADSA measurements produce higher accuracy and less subjectivity compared to direct measurements using a traditional goniometer. As with measurement using a goniometer, the contact angle is determined at the point where the three-phase line intersects the largest drop diameter and may not reflect any variations along the three-phase line due to heterogeneity or surface roughness.

Using liquids of known surface tension, profiles of the contact diameter ($\theta < 90°$) or maximum equatorial diameter ($\theta > 90°$) determined using the above apparatus can provide average contact angles occurring on rough or heterogeneous surfaces though these values are not easily related to Young's contact angle and surface free energies [54, 65-68].

Contact angles on powders can be made using compressed tablets or on single particles. It is customary to use either equlibrium or advancing contact angles for surface free energy analysis. It is possible for the surface chemistry of the powder to be altered by the compression process.

## 5.1.2. Wilhelmy plate and dynamic contact angle analysis

Wilhelmy [90] reported the use of a plate (such a microscope slide or thin piece of Pt foil) to measure surface tension of liquids. Such Wilhelmy plates have been employed in several ways to measure surface tensions and contact angles.

A plate can be used to measure the detachment force similar to the more familiar du Noüy method [1]. Correction factors are not needed when the Wilhelmy plate is used instead of the du Noüy ring [91, 92].

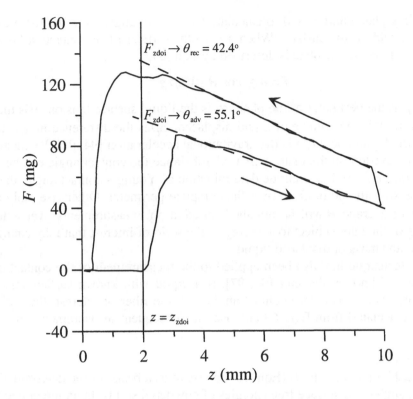

**Figure 2.** Force versus depth of immersion for Wilhelmy plate experiment in dynamic mode. Contact angle is calculated from the force extrapolated to the zero depth of immersion (zdoi).

A Wilhelmy plate can be suspended from a balance and the liquid raised until the plate just touches the liquid. The increase in weight, $\Delta W$, is related to the surface tension and contact angle through the following relation.

$$\gamma \cos(\theta) = \frac{\Delta W}{p} \tag{67}$$

where $p$ is the perimeter of the plate. If necessary, it is possible to determine both liquid surface tension and contact angle in the same experiment.

The capillary rise, $h$, at the surface of the plate can also be measured. Contact angle is then calculated through Eqn. 68.

$$\sin\theta = 1 - \frac{\Delta \rho g h^2}{2\gamma_{lv}} \tag{68}$$

Neumann and Tanner [93] provide an example of the use of this method.

A Wilhelmy plate may be used in dynamic mode. In this mode, the stage that holds the sample is raised or lowered at a fixed and controlled rate to allow the meniscus to pass over the surface. Using a liquid of known surface tension, the

Wilhelmy plate can be used to calculate the contact angle from the force exerted by the liquid on the surface. When a smooth, vertical plate contacts a liquid the resulting force on the plate is described by Eqn. 69,

$$f = p\ \gamma_{lv} \cos\ \theta - V\Delta\rho\ g \tag{69}$$

where $p$ is the perimeter of the plate, $\gamma_{lv}$ is the liquid surface tension, $\theta$ is the contact angle, $V$ is the volume of liquid displaced, $\Delta\rho$ is the difference in density between the liquid and air, $g$ is the gravitational acceleration [94, 95]. By measuring the force exerted on the plate with a microbalance the contact angle can be determined (see Fig. 2). For accurate determination of Young's contact angle, the plate must be smooth and uniform over the complete perimeter. Otherwise, only an effective or average $\theta$ will be obtained. In addition, measurements over extended periods of time are subject to swelling of the solid of interest that may change the volume and mass of displaced liquid.

This technique has also been applied to the measurement of the contact angles on fibers of known diameter [96, 97]. If a liquid with known surface tension is available which has a zero contact angle with the fiber of interest, the perimeter can be calculated from Eqn. 69 and used in subsequent measurements of contact angles.

### 5.1.3. Wicking

The Washburn equation [9] (Eqn. 15) can serve as a basis for the determination of contact angles and surface free energies of powdered solids. In an integrated form, the Washburn equation is expressed as:

$$h^2 = \frac{t\ r\ \gamma_L \cos(\theta)}{2\eta} \tag{70}$$

where $h$ is the penetration depth of the liquid, $t$ the elapsed time and $\eta$ the liquid viscosity. Eqn. 70 has two variables which are unknown – $r$ and $\theta$. $r$ may be determined using a liquid which completely wets $(\cos(\theta) = 1)$ the solid such as an alkane. Once $r$ has been determined using the wetting liquid, the contact angles of other liquids can be calculated using the Washburn equation. Wicking experiments cannot be done for liquids which have contact angles greater than 90° or which have high viscosities. Wicking experiments, furthermore, are not suitable for materials that exhibit significant solubility in the usual probe liquids. Because pharmaceuticals are soluble in many of the usual probe liquids the wicking method is usually not suitable for these materials.

At least, two different methods for measuring the rate of wicking have been employed. These methods are column and thin-layer wicking.

In column wicking capillary tubes are filled with powder. One end of the tube is plugged with cotton to prevent loss of the powder. The time to reach several heights along the capillary are recorded. In order to use column wicking, powder packing must not be disturbed or the observed wetting line will be irregular (non-uniform height across the tube).

In thin-layer wicking, a microscope cover slide is coated with a powder by evaporation of a dilute suspension of particles. This method requires that an even coating of particles with good adhesion to the substrate is achieved. As in column wicking the plate is immersed slighly into the probe liquid and the times to reach several heights are recorded.

From the data collected, plots of $h^2$ vs. $t$ are constructed. For wetting liquids $(\cos(\theta) = 1)$ the slope of the plot will be $\frac{r\gamma_L}{2\eta}$ . $r$ can then be calculated from this slope. $\gamma_s^{LW}$ can then be determined using the calculated value of $r$ and the slope of the plot of $h^2$ vs. $t$ for the liquid with finite contact angle but interacting exclusively through Lifshitz–van der Waals interactions such as methylene iodide. Similar data using polar liquids could then be used to estimate the acid–base contributions to surface free energy.

Alternatively one may plot $\dfrac{2\eta h^2}{t}$ vs. $\gamma_L$ for a series of liquids which include those with zero contact angle and those with finite angle. Such a plot will exhibit two slopes. For a low surface tension liquid where $\cos(\theta) = 1$, the slope will be positive and equal to the capillary radius. When the liquid reaches a surface tension where the contact angle becomes finite the slope will be negative. The negative slope region is a Zisman plot; thus the solid surface free energy can be estimated from the value of the liquid surface tension corresponding to the point of intersection between the positive and negative sloped regions. A plot of penetration time versus liquid surface tension will provide an estimate of the Zisman critical surface free energy. Presumably, this method could be applied to a bed of powder.

The use of wicking has been discussed in the literature by van Oss and co-workers [98, 99]. The work of Ku *et al.* is also of interest [100].

A much less sophisticated method to test the wetting of porous materials such as paper has been used by printers and it is sometimes called the "dyne solution method". This method is also based on the Washburn equation. A series of aqueous solutions containing various concentrations of isopropanol or other suitable solvent are prepared and their surface tension is measured. A small drop of fixed size is placed on top of the substrate and the time for the drop to penetrate completely into the surface is recorded. For high surface tensions where the contact angle is greater than 90° the drop will not penetrate in a reasonable length of time. In the region where the contact angle is between 90 and 0° the time for penetration drops rapidly with surface tension. The "dyne solution method" should be regarded as a crude test as it does not, for instance, account for selective adsorption of co-solvents on the solid surface.

*5.1.4. Contact angle from enthalpy of immersion*

Enthalpies of immersion can be used to determine contact angles. The enthalpy of immersion, $h_i$, can be expressed in terms of the following relation.

$$h_i = \gamma_{SL} - \gamma_S - T\left(\frac{\partial \gamma_{SL}}{\partial T} - \frac{\partial \gamma_S}{\partial T}\right) \tag{71}$$

and recalling and substituting Young's equation

$$h_i = -\gamma_{LV}\cos\theta - (\gamma_S - \gamma_{SV}) - T\left(\frac{\partial \gamma_{SL}}{\partial T} - \frac{\partial \gamma_S}{\partial T}\right). \tag{72}$$

Furthermore, it is often found that

$$(\gamma_S - \gamma_{SV}) \approx 0 \qquad \frac{\partial \gamma_S}{\partial T} \approx 0 \tag{73}$$

and

$$\frac{\partial \gamma_{SL}}{\partial T} = -0.07 \text{ mJ m}^{-2}\text{K}^{-1}, \tag{74}$$

thus,

$$\cos(\theta) = \frac{-0.07T - h_i}{\gamma_{LV}}. \tag{75}$$

Preadsorbed materials like water often affect the measured heats of immersion and thus could influence conclusions made from contact angles calculated from enthalpies of immpersion. A number of authors have described these measurements in the literature [101-107].

*5.2. Inverse gas chromotography*

Inverse gas chromatography (IGC) can be used to study the interfacial characteristics of solids and is particularly well suited to powders. IGC is identical to conventional GC except for the fact that the solid phase is the subject of study. It appears that the technique was first mentioned by Kiselev [108]. Lloyd, Ward and Schreiber have published a comprehensive review [109]. The subject has also been reviewed on several other occasions [110-116].

IGC is a useful technique to study a number of phenomena. Here we discuss the use of IGC to study surface acid–base and Lifshitz–van der Waals interactions and at a later stage (in Section 6.5) its use in the study of surface heterogeneity.

When studying the acid–base properties of a material, IGC is used with small vapor concentrations where Henry's Law holds and sorbate-sorbate interactions are negligible.

The retention volume, $V_n$, is defined as the amount of carrier gas required to elute the injected volume of probe molecules from the column. The retention volume is determined by its interaction energy with the solid. $V_n$ can be expressed in terms of the sorption equilibrium constant, $K_s$. Thus,

$$V_n = K_s \, mA \tag{76}$$

where $m$ is the mass of the adsorbent and $A$ its specific surface area.

Assuming the vapor behaves as an ideal gas, then

$$K_s = \frac{\Gamma RT}{p}, \tag{77}$$

where $\Gamma$ is the surface excess (surface concentration) and $p$ the partial pressure of the adsorbate. $RT$ has its usual meaning. At infinite dilution

$$\Gamma = \left( \frac{p}{RT} \right) \left( \frac{\pi_e}{p} \right), \tag{78}$$

where $\pi_e$ is the surface pressure (recall Eqn. 19).

From the above two equations it follows that

$$K_s = \frac{\pi_e}{p}. \tag{79}$$

The standard free energy change for adsorption can be expressed by the usual relation,

$$\Delta G_{ads}^{\circ} = -RT \ln \left( \frac{p_g}{p_s} \right), \tag{80}$$

where

$$p_s = \left( \frac{\pi_e}{K_s} \right) \tag{81}$$

and $p_g$ refers to the Henry's law standard state for the gas phase. Kembell and Rideal [117], as well as DeBoer and Druyer [118], have discussed the choice of adsorption standard states.

From Eqn. 80 it follows that

$$\Delta G_{ads}^{\circ} = -RT \ln \left( \frac{V_n \, p_g}{\pi \, m A} \right) = -RT \ln \left( V_n \right) + C \tag{82}$$

Using van 't Hoff's relations and data collected at different experimental temperatures $\Delta H$ and $\Delta S$ can be calculated.

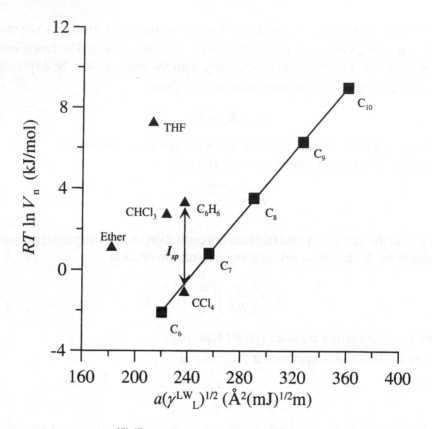

**Figure 3.** $RT \ln(V_n)$ *vs.* $a(\gamma_L^{LW})^{1/2}$. Squares, alkane liquids; triangles, other liquids. Slope of linear fit to alkane liquids points gives $\gamma_s^{LW}$ and vertical displacement of triangular points from linear fit gives $I_{sp}$. $I_{sp}$ ($\Delta G^{ab}$) is the free energy resulting from acid–base interactions.

The standard free energy of adsorption can be related to the work of adhesion of gas molecules to the surface. Here

$$-\Delta G_{ads}^{\circ} = N_A\, a\, W_a \tag{83}$$

where $a$ is the surface area occupied by an adsorbed molecule. Recalling Berthelot's rule for Lifshitz –van der Waals interactions and Eqn. 82 and 83

$$RT \ln\left(V_n\right) = 2\, N_A a (\gamma_S^{LW} \gamma_L^{LW})^{1/2} + C\,. \tag{84}$$

A graph of $RT \ln\left(V_n\right)$ *vs.* $a(\gamma_L^{LW})^{1/2}$ for a series of n-alkane probes will be linear and will allow calculation of $\gamma_S^{LW}$ from its slope (see Fig. 3).

Dorris and Gray [119] proposed a different method for calculation of $\gamma_S^{LW}$. In this method the area occupied by each methylene group is taken to be constant for a series of n-alkane probes. Thus,

$$\gamma_S^{LW} = \frac{RT \ln \left( V_{n_{C_{n+1}H_{2n+4}}} / V_{n_{C_{2n+2}}} \right)}{4 N_A^2 a_{CH_2} \gamma_{CH_2}}$$ (85)

where $a_{CH_2}$ is the surface area occupied by a methylene group (6 $\text{Å}^2$) and the surface tension of a methylene group, $\gamma_{CH_2}$ is given by the relation below.

$$\gamma_{CH_2} = 35.6 + 0.058(20 - t)$$ (86)

where the result has units of $mJ/m^2$ and $t$ is the Celcius temperature. Both methods for determination of the Lifshitz–van der Waals component of surface free energy give comparable results [120, 121]. The correct choice for the molecular area is, however, important to the calculations and is the largest source of error.

It is worthwhile to mention that the literature on the whole typically finds that $\gamma_S^{LW}$ values determined via IGC are larger than those found using contact angle data. It is the general belief that the larger values result from the fact that in IGC adsorption occurs at low levels of surface coverage thus only the most energetic surface sites are occupied by adsorbed molecules.

If absolute values for the Lifshitz–van der Waals components are not required plots of $RT \ln (V_n)$ vs. $\ln(p^0)$ (vapor pressure) or boiling point are also linear when a series of n-alkane liquids is used. A more complete description of the molecular descriptors suitable for use with IGC has been published [122].

As discussed in preceding sections, the work of adhesion can be considered as being composed of two components; one resulting from Lifshitz–van der Waals interactions and the other from Lewis acid–base interactions.

The acid–base contribution to the surface free energy can be estimated from a plot of $RT \ln (V_n)$ vs. $a(\gamma_L^{LW})^{1/2}$ (see Fig. 3). One of the previously mentioned variables (e.g., boiling point) may also be substituted for the x-axis variable in such a plot if information on $\gamma_S^{LW}$ is not needed. The plot should contain points representing data collected using a series of n-alkane probes and data collected using probes having Lewis acid–base interactions with the surface. As stated before, the n-alkane points will lie on a single straight line. Points representing probes which interact, in part, through Lewis acid–base interactions will lie above the line connecting the n-alkane points. The vertical distance between the line and the point representing the adsorption of the probe interacting, in part, through Lewis–acid base interactions is the free energy of adsorption due to Lewis acid–base interactions, $\Delta G_{ads}^{AB}$ or $I_{sp}$. $\Delta H_{ads}^{AB}$ may be determined by measuring $\Delta G_{ads}^{AB}$ at several temperatures and using the van 't Hoff relation.

Eqns. 54 and 55 describe relations between $\Delta H_{ads}^{AB}$ and acid–base properties of the solid (usually unknown) and the acid–base properties of various probes according to the theories of Drago [45] and Gutmann [46, 47]. If $\Delta H_{ads}^{AB}$ is known for

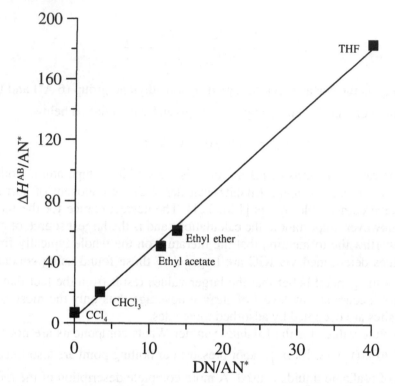

**Figure 4** . $\Delta H^{AB}/AN^*$ *vs.* $DN/AN^*$. Slope is $K_d$ and intercept $K_a$.

several probes which have known acid–base properties, then the acid–base properties of the solid can be calculated.

If Gutmann's model (Eqn. 55) is chosen, a plot of $\dfrac{\Delta H^{AB}_{ads}}{DN}$ *vs.* $\dfrac{AN^*}{DN}$ will be linear with a slope equal to $K_d$ and and intercept of $K_a$ (see Fig. 4) (N.B.: For statistical reasons it is better to extract the parameters, $K_a$ and $K_d$, using a least squares analysis which does not rely on coordinates which are ratios of experimental quantities).

## 6. COMPLEMENTARY EXPERIMENTAL METHODS

### 6.1. Overview

In previous sections of this paper, the theoretical basis for determination of the surface free energy solids is addressed. Experimental methods which yield numerical values for surface free energy components are also discussed. In the following section methods which do not yield numerical values for surface free energy but which nonetheless can shed considerable insight into the nature of the surface are addressed. The methods discussed below should complement measurements made by the methods discussed above.

## 6.2. Flow microcalorimetry

Flow microcalorimetry for use in surface chemistry has been pioneered by Groszek and co-workers [123-130]. Fowkes and co-workers have also contributed significantly to the study of the acid–base properties of surfaces via flow microcalorimetry [20, 131-133]. The reader is also directed to the work of other groups [134-136]. The apparatus consists of a pair of syringe pumps, a sensitive calorimeter and a downstream detector to monitor the concentration of the adsorbate. The downstream detector monitors refractive index, ultraviolet absorbance or other property sensitive to concentration. The apparatus measures the heat of adsorption and the amount of material adsorbed. From the two measured quantities the molar heat of adsorption can be determined. Commercial instrumentation is manufactured by Microscal (UK).

Fowkes and co-workers have in particular addressed the use of Flow microcalorimetry to assess the acid-base properties of surfaces [20, 131-133]. Inverse gas chromatography and flow microcalorimetry are similar in that both techniques can be used to determine $\Delta H^{AB}$. Thus Fowkes' approach combined with either Drago's or Gutmann's model could be used to determine the acid-base character of materials. Inverse gas chromatography and flow microcalorimetry are complementary with respect to determining the acid–base character of materials. IGC uses the van 't Hoff approach for calculating enthalpies of adsorption while microcalorimetry measures the quantity directly.

In practice, the heat of adsorption of an acid–base probe from a neutral solvent such as iso-octane is measured. This quantity is essentially $\Delta H^{AB}$ as the heat of adsorption of a neutral solvent from another neutral solvent is negligible. Fowkes and others have used probes for which Drago parameters are available to assess the acid–base properties of materials. Such probes usually yield heats of adsorption which are large enough to be measured with good accuracy.

Flow microcalorimetry has not been used to determine Lifshitz–van der Waals contributions to surface free energy.

## 6.3. Electron spectroscopy for chemical analysis (ESCA)

One technique for monitoring the chemical nature of surfaces, which is applicable to a wide variety of materials is electron spectroscopy for chemical analysis (ESCA). ESCA can be used to determine the elemental composition and oxidation state composition of surfaces. ESCA is also known in the literature as X-ray photoelectron spectroscopy (XPS). The method has been used extensively in the literature to study adhesion and other surface chemical problems. ESCA/XPS has been reviewed in the literature by Barr [137]. Here we merely highlight the potential for its use to complement surface free energy calculations.

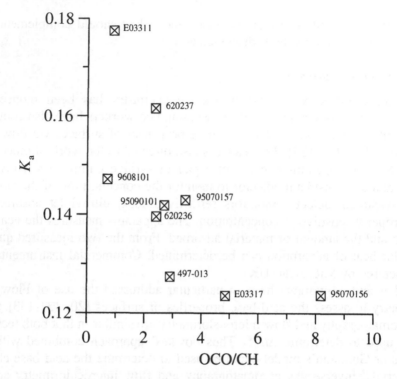

**Figure 5.** Correlation between $K_a$ and carbons in OCO state versus aliphatic carbons on Pharmatose 200M (see Etzler *et al.* for details [138]) ESCA can be used to better understand the relation between surface free energy and surface chemical changes. Numbers next to points are lot numbers of Pharmatose 200M. Aliphatic carbons indicate the presence of surface contaminants.

ESCA is based on the photoelectric effect. When X-ray photons strike a surface, electrons are emitted. The ejected electrons come from inner electron orbitals (1s, 2s, etc.). The kinetic energy of the ejected electrons can be related to the electron orbital energy via the following expression.

$$hv = \frac{1}{2} \cdot mv_e^2 + E_b + q\Phi \tag{87}$$

The binding energies, $E_b$, are determined from the kinetic energy of the emitted electrons. The electron binding energy for a given electron is characteristic for the element and the particular electron orbital belonging to that element. $q\Phi$ is the workfunction and is usually characteristic of the apparatus. It represents the extra work to extract electrons from condensed media. *In vacuo* $q\Phi = 0$.

In addition to being characteristic for a given element, $E_b$ is also dependent on the electronegativity of the surrounding atoms; thus, both an elemental analysis and some information about the chemical environment of each atom under study can be obtained from ESCA.

The X-ray source typically illuminates a spot on the order of 1-5 mm. For powders, it is not possible to obtain information about single particles. Electrons emitted from atoms located within approx. 5 nm of the surface are detectable using ESCA. Because the electrons come from atoms located near the surface, ESCA is usually regarded as a surface sensitive technique.

The surface abundance of certain elements or variations in the distribution of oxidation states of surface carbons may, for instance, correlate with surface free energy measurements. Fig. 5 shows the correlation between Gutmann's $K_a$ parameter and the ratio of OCO to aliphatic carbons. OCO carbons are carbon atoms bonded to two oxygen atoms.

## 6.4. Scanning probe microscopy

The ability to study the characteristics of a surface down to atomic dimensions allows one to gain considerable insight into interfacial phenomenon. Scanning probe microscopy is a powerful technique for studying surfaces down to atomic dimensions [139, 140].

A scanning probe microscope (SPM) consists of a needle-like probe attached to a cantilever spring that can be moved along the surface [139]. Surface chemically modified tips or particles of interest can be substituted for the usual SPM tip. The practice of substituting particles for the tip is referred to as the colloidal probe technique. The lateral $(x, y)$ position of the probe and deflection of the cantilever spring are recorded. Cantilevers are typically made by microlithography from silicon or silicon nitride. The dimensions of cantilevers are 100-200 $\mu$m long, 10-20 $\mu$m wide and 0.5 $\mu$m thick. The pointed tips are a few micrometers high. Tips may be 30 nm in radius but when the contact force is small the area of contact between the tip and surface is on the order of atomic dimensions. The forces measured are in the range of $10^{-13}$ to $10^{-6}$ N. SPM contrasts with stylus profilometry, which uses forces of $10^{-4}$ N over submicrometer areas. The deflection of the cantilever has been monitored by a variety of techniques but laser beam deflection is often used. The force constant of the cantilever can be selected for the range of forces under study. Once position and deflection data have been recorded by a computer, a surface map can be constructed. Low deflections are usually recorded as dark tones (shades) and high deflections as light tones. The range of forces reflecting discernible tones is selected to be appropriate to the sample.

SPM can be used on a wide variety of surfaces. The surfaces may either be electrical conductors or insulators. The apparatus can be operated using either dry samples or samples immersed in liquid media. As the exact nature of the sample is relatively unimportant SPM has been used on a wide variety of samples.

One of the most straightforward ways to use SPM is for the construction of topographic maps. The tip and surface experience attractive and repulsive intermolecular forces whose origins have been discussed above in regard to theories for surface free energy. In order to construct a topographic map, the SPM tip is brought into close proximity to the surface where it experiences a repulsive force.

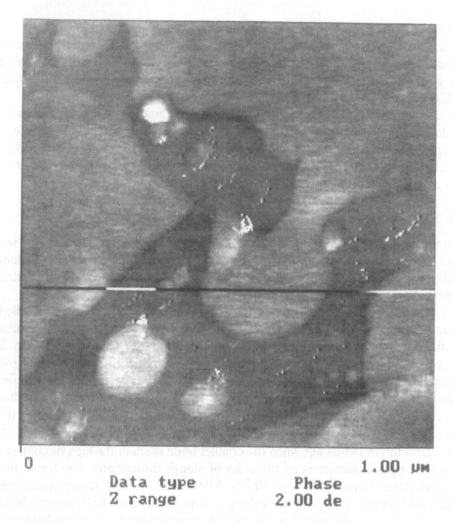

**Figure 6.** SPM phase image of a Pharmatose 200M particle [138]. Dark regions appear to be surface contaminants. Light colored specks are a second contaminant with different mechanical properties.

The tip can then be scanned in a raster pattern at constant force or constant height. Changes in height or photocurrent are collected for the calculation of topographic information.

Other information about surfaces can also be obtained by applying, for instance, oscillating (tapping) forces, lateral forces or varying the probe surface chemistry. These other useful techniques are discussed elsewhere [140]. The present author has together with other co-workers characterized pharmaceutical surfaces using phase, friction and adhesion imaging [141-143].

For those wishing to better understand surfaces, SPM provides a number of techniques that may be used to understand the nature of surfaces particularly with regard to surface chemical heterogeneity [144-149].

In phase imaging the cantilever is oscillated near its resonance frequency. The oscillation causes the probe to tap along the surface. In general, the oscillation of the tip will be out of phase with respect to the applied oscillating force. The shift in phase reflects the mechanical properties of the surface (elasticity, adhesion). A phase map of the surface can distinguish regions of different natures. In addition to phase imaging, both adhesion and friction maps may give insight into surface chemical heterogeneity (see Fig. 6).

SPM tips which have been chemically modified can be used to assess the surface chemical nature of surfaces by measuring, for instance, the relative adhesion of tips with different surface chemical modifications.

SPM is a powerful tool for investigation of the surface properties of materials. Drelich and co-workers have recently used SPM to measure surface free energies [150].

## 6.5. Distribution of adsorption energies from adsorption isotherms

Solid surfaces provide an assortment of adsorption sites. Even in systems where the surface would be regarded as being chemically pure both the presence of defect structures and the atomic architecture of the surface atoms provide for the existence of energetically diverse adsorption sites. It is, of course, possible to have surfaces that are surface chemical mosaics. In any event, the adsorption energy diversity can be represented by a function, $\chi(\varepsilon)$. $\chi(\varepsilon)$ is the probability of finding a site having adsorption energy, $\varepsilon$. It follows that

$$\int_{i}^{\infty} \chi(\varepsilon) \, d\varepsilon = 1 \qquad (88)$$

where $i = 0$ or $-\infty$.

The experimental adsorption isotherm, $\vartheta(T, p)$ (relative coverage) represents the average adsorption isotherm for all values of $\varepsilon$. The experimental isotherm may thus be expressed in terms of the adsorption isotherms of sites with fixed adsorption energy (also called the local isotherm), $\theta(\varepsilon, T, p)$. Thus

$$\vartheta(T, p) = \int_{i}^{\infty} \theta(\varepsilon, T, p) \, \chi(\varepsilon) \, d\varepsilon \qquad (89)$$

Eqn. 89 is known as the integral form of the adsorption isotherm [151].

Many analytical solutions for $\chi(\varepsilon)$ in Eqn. 89 have been offered in the literature [152]. Such analytical solutions require the functional form of $\chi(\varepsilon)$ to be guessed.

Methods to determine the shape of $\chi(\varepsilon)$ also have been explored in the literature. The various methods may be divided into analytical and numerical solutions.

Sips [153, 154], as well as Todes and Bondareva [155], suggested a solution involving the Stieltjes transform (also see the work of Misra [156]). The solution provides an analytical function describing $\chi(\varepsilon)$ but the method can only be applied in limited circumstances where the local isotherm can be represented by the Langmuir equation.

The condensation approximation also provides an analytical solution and appears to give good results [152]. In this it is assumed that the local isotherm, $\theta(p,\varepsilon)$ rises steeply about the condensation pressure, $p_c$, so that

$$\theta(p,\varepsilon)=\begin{cases}0, & p<p_c(T,\varepsilon)\\1, & p>p_c(T,\varepsilon)\end{cases}. \tag{90}$$

Furthermore, it may be assumed that $p_c$ is a monotonic function of $\varepsilon_c$, the condensation energy. Thus

$$\theta(p,\varepsilon)=\begin{cases}0, & \varepsilon<\varepsilon_c(T,p)\\1, & \varepsilon>\varepsilon_c(T,p)\end{cases}. \tag{91}$$

Combining Eqns. 89 and 91 the experimental isotherm can be expressed as follows

$$\vartheta(p)=\int_{\varepsilon_c(T,p)}^{\infty}\chi(\varepsilon)\,d\varepsilon \tag{92}$$

where $0\le\varepsilon\le\infty$. Differentiating Eqn. 92 with respect to $p$ yields the expression

$$\left(\frac{\partial\vartheta}{\partial p}\right)_T=-\chi(\varepsilon_c)\left(\frac{\partial\varepsilon_c}{\partial p}\right)_T \tag{93}$$

and furthermore recognizing the relation between $p_c$ and $\varepsilon_c$

$$\chi(\varepsilon)=-\left(\frac{\partial\vartheta}{\partial p}\right)_{p=p_c}\left(\frac{\partial p}{\partial\varepsilon}\right)_{p=p_c} \tag{94}$$

$\left(\dfrac{\partial\vartheta}{\partial p}\right)$ can be evaluated from the fit of an analytical expression to the experimental data (i.e., Langmuir, Freundlich, Dubinin–Radushkevich isotherms, etc.). The function $p=p_c(\varepsilon)$ must be known to evaluate $\left(\dfrac{\partial p}{\partial\varepsilon}\right)$. Harris [157, 158] and Cerofolini [159] have offered methods to do this. Both models state that if the local isotherm is of the Langmuir type then

$$p_c(\varepsilon)=K^L\,e^{-\varepsilon/kT}, \tag{95}$$

where $K^L$ is the Langmuir constant.

The condensation approximation is best applied at lower temperatures (those approaching absolute zero). The errors associated with the calculation have been discussed by Harris [157, 158] and Rudzinski and Everett [152].

Numerical solutions of Eqn. 89 are possible when accurate experimental adsorption isotherms covering a wide range of pressures (4-5 orders of magnitude) are available. As all experimental data are subject to error, the experimental data must be smoothed. Improper smoothing of the data leads to experimental artifacts in the calculated energy distribution [152, 160].

Data from GC measurements made at finite concentration give $\left( \dfrac{\partial \vartheta}{\partial p} \right)$ directly and can be used to calculate the energy distribution function [161, 162]. The development of chromatographic methods to study surfaces has been discussed in the literature [162-166]. IGC, in particular, is a valuable experimental tool [167-169].

In IGC experiments, the absolute retention volume, $V_N$ may be expressed using the following relation:

$$V_N = \frac{kT}{j} \left( \frac{\partial N_t}{\partial p} \right) \tag{96}$$

where $j$ is the James-Martin coefficient, $k$ is Boltzmann's constant and $N_t$ is the amount of vapor adsorbed at pressure, $p$. Eqn. 96 can be rewritten in terms of the amount of vapor needed for monolayer coverage, $N_m$. Thus,

$$V_N = \frac{N_m kT}{j} \left( \frac{\partial \vartheta}{\partial p} \right) \tag{97}$$

If the condensation approximation is applied, the adsorption energy distribution is a simple function of the adsorption isotherm. In this instance

$$\chi_{CA} = \frac{p}{kT} \left( \frac{\partial \vartheta}{\partial p} \right) = \frac{jp}{N_m (kT)^2} V_N \tag{98}$$

Eqn. 98 illustrates the utility of IGC measurements for the determination of surface energy distributions as experimental adsorption isotherms do not need to be differentiated.

A more accurate approximation of $\chi(\varepsilon)$ is given by the asymptotically correct condensation approximation (ACCA) [170-173]. This approximation results in the following expression.

$$\chi_{ACCA} = \frac{j\,p^2}{N_m (kT)^2} \left( \frac{\partial V_N}{\partial p} \right) \tag{99}$$

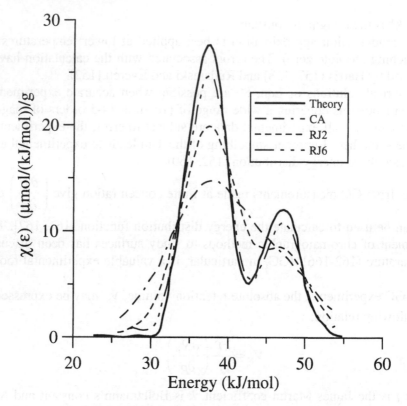

**Figure 7.** Figure showing an energy distribution function redrawn from work of Balard [179]. Solid line is theoretical distribution. Dashed lines represent distributions calculated using differential models. The plot illustrates the model dependence of the calculated energy distribution. CA is condensation approximation, RJ denotes RJ method. The plot also illustrates the use of the RJ to make increasingly better approximations to the theoretical result. RJ6 is a better approximation than RJ2.

The pressure dependence of $V_N$ necessary for the solution of Eqn. 99 can be expressed using the relation suggested by Gawdzik *et al.* [174]:

$$V_N = \exp\left[\sum_{m=0}^{s} A_m p^m\right] \qquad (100)$$

where $A_m$ is a coefficient of best fit to the experimental data. Combining Eqn. 99 and 100 results in the relation

$$\chi_{ACCA} = -\frac{y}{(kT)^2} \exp\left[\sum_{m=1}^{s} A_m y^{m-1}\right] \exp\left[\sum_{m=0}^{s} A_m y^m\right] \qquad (101)$$

where $y = Ke^{-\varepsilon/kT}$ .

Several other approximations for the solution of Eqn. 89 have been offered in the literature. Approximations have been offered by Adamson and Ling [175], House and Jaycock [176], Ross and Morrison [177] as well as by Rudzinski and

Jagiello [178]. The Rudzinski–Jagiello (RJ) method is particularly suitable with data obtained via IGC at room or higher temperatures. For a recent discussion on the merits and assumptions asscoiated with the RJ method, see the work of Balard [179]. The RJ method is particularly useful in that successively more accurate approximations can be made in order to achieve a convergent result (see Fig. 7).

The principal difficulties associated with obtaining a correct energy distribution [179] are related to:

1. Choosing experimental conditions so that isotherms over a wide pressure range can be obtained.

2. Choosing an adsorption model based on physical assumptions consistent with the experimental isotherm.

3. Choosing a suitable mathematical method that allows for the calculation of a stable (unique) solution.

For a more detailed discussion of the calculation of energy distribution the reader is directed to the recent review by Charmas and Leboda [180]. Changes in the surface energy distribution may assist the understanding of accompanying surface free energy changes.


## 7. SUMMARY

In this work the literature relevant to understanding the surface free energy and surface chemistry of solids is reviewed. An understanding of the surface properties of materials is necessary for the functioning and the manufacturing of a number of industrial products. In general products or processes that involve wetting, adhesion and capillary penetration require a knowledge and control of the surface chemistry of materials.

In this work the fundamental concepts relative to wetting, adhesion, capillary penetration and surface free energy are reviewed. The relations between the macroscopic phenomena and surface free energy are discussed.

Theoretical models useful for determining the surface free energy of materials have been discussed. The current discussions in the literature indicate that, at present, experimental data analyzed using different theoretical models may give results that seem to be contradictory. The various models used to describe Lewis acid–base interactions, for instance, use different reference states to assign numerical values to the acidity or basicity to a surface. At least part of the observed conflict results from differences in the definitions of acid and base between the models. The researcher must use care when comparing data calculated using different models. It will be preferable to characterize surfaces by more than one method. Experimental factors also contribute to conflict between results obtained by different methods. IGC measurements, for instance, tend to give $\gamma^{LW}$ values which are higher that those obtained using contact angles. It is believed that IGC measurements reflect only adsorption at high energy sites. The optimal model to describe the surface free energy has not yet been agreed upon.

The nature of the material to be studied may influence the type of measurements possible. Here several experimental approaches for the determination of surface free energy are discussed. The choice of experimental method may dictate or influence the theoretical model chosen to calculate surface free energy. The researcher will need to consider, in particular, the solubility of the material in the various probe liquids required to determine the surface free energy. The ability to measure contact angles on pharmaceuticals by wicking, by the Wilhelmy method, by heats of immersion and by traditional goniometry are limited by the solubility of the solid in the various probe liquids. Heats of immersion and wicking experiments, on the other hand, may be good choices for mineral pigments. IGC, despite the associated problems with the high values of $\gamma^{LW}$, appears to be a technique suitable for use with a wide variety of materials.

Because of the interrelation between surface chemistry and surface free energy some experimental methods which measure surface chemical properties and which can be used to better understand surface free energy are discussed. FMC is, for instance, an excellent complement to IGC as both techniques assess surface acid–base properties. SPM and surface energy distributions can be used, for instance, to monitor surface heterogeneity. SPM can also be used to make direct measurements of adhesion strength. ESCA is able to monitor the atomic and oxidation state composition of surfaces. The surface chemical changes associated with surface oxidation, reduction or contamination can be monitored by ESCA.

It is my hope that this work will serve as a resource to those researchers wishing to understand the relations between surface free energy and surface chemistry and the macroscopic properties of materials which contribute to ultimate product performance.

## REFERENCES

1. A. W. Adamson, *Physical Chemistry of Surfaces*, John Wiley, New York, NY (1990).
2. G. Beilby, *Aggreation and Flow of Solids*, MacMillan and Co., London (1921).
3. R. Eötvös, *Weid. Ann.* **27**, 456 (1886).
4. E. A. Guggenheim, *J. Chem. Phys.* **13**, 253 (1945).
5. I. Langmuir, in: *Colloid Symposium Monograph*, p. 48. The Chemical Catalog Company, New York, NY (1925).
6. *Miscelleanous Works of T. Young*, G. Peacock (Ed.), Vol. 1, p. 418. Murray, London (1855).
7. A. Dupre, *Theorie Mechanique de la Chaleur*, Gauthier-Villars, Paris (1869).
8. P. S. Laplace, *Mechanique Celeste*, Paris (1806).
9. E. W. Washburn, *Phys. Rev. Ser. 2* **17**, 273 (1921).
10. M. Mulqueen and P. D. T. Huibers, in: *Handbook of Applied Surface and Colloid Chemistry*, K. Holmberg ( Ed.), Vol. 2, p. 217, John Wiley, Chichester (2002).
11. D. Y. Kwok, Y. Lee and A. W. Neumann, *Langmuir* **14**, 1584 (1998).
12. H.-J. Butt and R. Raiteri, in: *Surface Characterization Methods, Surfactant Science Series* **87**, A. J. Milling (Ed.) p. 1, Marcel Dekker, New York, NY (1999).

13. W. Thompson, *Proc. R. Soc. (London)* **9**, 255 (1858).
14. W. Thompson, *Phil. Mag.* **17**, 61 (1858).
15. J. W. Gibbs, *Collected Works*, Vol. 1. Longmans Green, New York, NY (1928).
16. W. A. Zisman, in: *Contact Angle, Wettability and Adhesion*, Adv. Chem. Ser. No. 43, p. 1. American Chemical Society, Washington, DC (1964).
17. S. Banerjee and F. M. Etzler, *Langmuir* **11**, 4141 (1995).
18. A. Fredenslund, R. L. Jones and J. M. Prausnitz, *Am. Inst. Chem. Eng. J.* **21**, 1086 (1975).
19. F. M. Fowkes and M. A. Mostafa, *Ind. Eng. Chem. Prod. Res. Dev.* **17**, 3 (1978).
20. F. M. Fowkes, K. L. Jones, G. Li and T. B. Lloyd, *Energ. Fuels* **3**, 97 (1989).
21. F. M. Fowkes, *J. Adhesion Sci. Technol.* **1**, 7 (1987).
22. F. M. Fowkes, *J. Colloid Interf. Sci.* **28**, 493 (1968).
23. F. M. Fowkes, *J. Phys. Chem.* **66**, 382 (1962).
24. R. J. Good, in: *Contact Angle, Wettability and Adhesion*, K. L. Mittal (Ed.), p. 3. VSP, Utrecht (1993).
25. L. A. Girifalco and R. J. Good, *J. Phys. Chem.* **61**, 904 (1957).
26. C. J. van Oss, *Interfacial Forces in Aqueous Media*, Marcel Dekker, New York, NY (1994).
27. F. Chen and W. V. Chang, *Langmuir* **7**, 2401 (1991).
28. D. Berthelot, *Compt. Rend.* **126**, 1857 (1898).
29. W. V. Chang and X. Qin, in: *Acid-Base Interactions: Relevance to Adhesion Science and Technology*, K. L. Mittal (Ed.), Vol. 2, p. 3. VSP, Utrecht (2000).
30. J. Panzer, *J. Colloid Interf. Sci.* **44**, 142 (1973).
31. D. K. Owens and R. C. Wendt, *J. Appl. Polym. Sci.* **13**, 1741 (1969).
32. S. Wu, *J. Polym. Sci., Part C* **34**, 265 (1971).
33. S. Wu, *Polymer Interface and Adhesion*. Marcel Dekker, New York, NY (1982).
34. F. M. Fowkes, *J. Adhesion* **4**, 155 (1972).
35. C. Della-Volpe and S. Siboni, in: *Acid-Base Interactions: Relevance to Adhesion Science and Technology*, K. L. Mittal (Ed.), Vol. 2, p. 55. VSP, Utrecht (2000).
36. C. Della-Volpe, *J. Colloid Interf. Sci.* **195**, 121 (1997).
37. C. Della-Volpe, A. Deimichei and T. Ricco, *J. Adhesion Sci. Technol.* **12**, 1141 (1998).
38. C. J. van Oss, in: *Acid-Base Interactions: Relevance to Adhesion Science and Technology*, K. L. Mittal (Ed.), Vol. 2, p. 173. VSP, Utrecht (2000).
39. D. Y. Kwok, *Colloids Surf. A* **156**, 191 (1999).
40. R. J. Good, in: *Acid-Base Interactions: Relevance to Adhesion Science and Technology*, K. L. Mittal (Ed.), Vol. 2, p. 167. VSP, Utrecht (2000).
41. D. J. Gardner, S. Q. Shi and W. T. Tze, in: *Acid-Base Interactions: Relevance to Adhesion Science and Technology*, K. L. Mittal (Ed.), Vol. 2, p. 363. VSP, Utrecht (2000).
42. E. N. Dalal, *Langmuir* **3**, 1009 (1987).
43. F. M. Etzler, J. Simmons, N. Ladyzhynsky, V. Thomas and S. Maru, in: *Acid-Base Interactions: Relevance to Adhesion Science and Technology*, K. L. Mittal (Ed.), Vol. 2, p. 385. VSP, Utrecht (2000).
44. M. D. Vrbanac and J. C. Berg, in: *Acid-Base Interactions: Relevance to Adhesion Science and Technology*, K. L. Mittal and H. R. Anderson Jr. (Eds.), p. 67, VSP, Utrecht (1991).
45. R. S. Drago, G. C. Vogel and T. E. Needham, *J. Am. Chem. Soc.* **93**, 6014 (1971).
46. V. Gutmann, A. Steininger and E. Wychera, *Montash. Chem.* **97**, 460 (1966).
47. V. Gutmann, *The Donor-Acceptor Approach to Molecular Interaction*. Plenum, New York, NY (1978).
48. F. L. Riddle and F. M. Fowkes, *J. Am. Chem. Soc.* **112**, 3259 (1990).
49. R. S. Mulliken, *J. Chem. Phys.* **19**, 514 (1951).
50. J. C. Berg, in: *Wettability*, J. C. Berg (Ed.), p. 75. Marcel Dekker, New York, NY (1993).
51. L. H. Lee, *J. Adhesion Sci. Technol.* **5**, 71 (1991).
52. R. J. Pearson, *J. Am. Chem. Soc.* **85**, 3533 (1963).
53. U. Mayer, V. Gutmann and W. Gerger, *Montash. Chem.* **106**, 1235 (1975).

54. D. Y. Kwok and A. W. Neumann, in: *Acid-Base Interactions: Relevance to Adhesion Science and Technology*, K. L. Mittal (Ed.), Vol 2, p. 91. VSP, Utrecht (2000).
55. D. Li and A. W. Neumann, *J. Colloid Interf. Sci.* **148**, 190 (1992).
56. D. Li, M. Xe and A. W. Neumann, *Colloid Polym. Sci.* **271**, 573 (1993).
57. D. Y. Kwok and A. W. Neumann, *Colloids Surfaces A* **89**, 181 (1994).
58. C. A. Ward and A. W. Neumann, *J. Colloid Interf. Sci.* **49**, 286 (1974).
59. D. Li, J. Gaydos and A. W. Neumann, *Langmuir* **5**, 293 (1989).
60. D. Li and A. W. Neumann, *Adv. Colloid Interf. Sci.* **49**, 147 (1994).
61. D. Li, E. Moy and A. W. Neumann, *Langmuir* **6**, 885 (1990).
62. J. Gaydos and A. W. Neumann, *Langmuir* **9**, 3327 (1993).
63. D. Li and A. W. Neumann, *Langmuir* **9**, 3728 (1993).
64. D. Li and A. W. Neumann, *J. Colloid Interf. Sci.* **137**, 304 (1990).
65. R. N. Wenzel, *Ind. Eng. Chem.* **28**, 988 (1936).
66. A. B. D. Cassie, *Trans. Faraday Soc.* **40**, 546 (1944).
67. S. Baxter and A. B. D. Cassie, *J. Textile Inst.* **36**, 67 (1945).
68. A. B. D. Cassie, *Discuss. Faraday Soc.* **3**, 11 (1948).
69. K. Grundke, T. Bogumil, T. Gietzelt, H.-J. Jacobasch, D. Y. Kwok and A. W. Neumann, *Progr. Colloid Polym. Sci.* **101**, 58 (1996).
70. G. Antonow, *J. Chim. Phys.* **5**, 372 (1907).
71. D. Y. Kwok and A. W. Neumann, *J. Phys. Chem. B.* **104**, 741 (2000).
72. D. Y. Kwok, D. Li and A. W. Neumann, *Langmuir* **10**, 1323 (1994).
73. M. J. Morra, *J. Colloid Interf. Sci.* **182**, 312 (1996).
74. I. G. Salisbury and H. P. Huxford, *Phil. Mag. Lett.* **56**, 35 (1987).
75. D. R. Hamann, in: *Ordering at Surfaces and Interfaces*, A. Yoshimory, T. Shinjo and H. Watanabe (Eds.), Springer-Verlag, Heidelberg (1992).
76. A. M. Stoneham, *J. Amer. Ceram. Soc.* **64**, 54 (1981).
77. W. C. Mackrodt, *Phil. Trans. R. Soc. Lon. A* **341**, 107 (1992).
78. W. C. Mackrodt, *J. Chem. Soc, Faraday Trans. 2*, **85**, 541 (1989).
79. S. M. Foiles, in: *Equlibrium Structure and Properties of Surfaces and Interfaces*, A. Gonis and G. M. Stocks (Eds.). Plenum Press, New York, NY (1992).
80. P. J. Lawerence and S. C. Parker, in: *Computer Modeling of Fluids, Polymers and Solids*, C. R. A. Catlow (Ed.), p. 219. Kluwer Academic Publishers, New York, NY (1990).
81. G. Wulff, *Z. Kristallogr.* **34**, 449 (1901).
82. C. Herring, *Phys. Rev.* **82**, 87 (1951).
83. G. C. Benson and D. Patterson, *J. Chem. Phys.* **23**, 670 (1955).
84. M. Drechler and J. F. Nicholas, *J. Phys. Chem. Solids* **28**, 2609 (1967).
85. H. G. Muller, *Z. Phys.* **96**, 307 (1935).
86. R. S. Nelson, D. J. Mazey and R. S. Barnes, *Phil. Mag.* **11(109)**, 91 (1965).
87. A. W. Neumann and R. J. Good, in: *Surface and Colloid Science*, R. J. Good and R. R. Stromberg (Eds.), Vol 11, pp. 31-91, Plenum Press, New York, NY (1979).
88. W. C. Bigelow, D. L. Pickett and W. A. Zisman, *J. Colloid Sci.* **1**, 513 (1946).
89. P. Chen, D. Y. Kwok, R. M. Prokop, O. I. del Rio, S. S. Susnar and A. W. Neumann, in: *Drops and Bubbles in Interfacial Science*, D. M. R. Miller (Ed.), Elsevier Science, New York, NY (1998).
90. L. Wilhelmy, *Ann. Phys.* **119**, 177 (1863).
91. J. T. Davies and E. K. Rideal, *Interfacial Phenomena*. Academic Press, New York, NY (1961).
92. D. O. Jordan and J. E. Lane, *Austr. J. Chem.* **17**, 7 (1964).
93. A. W. Neumann and W. Tanner, *Tenside* **4(7)**, 220 (1967).
94. F. M. Etzler and J. J. Conners, in: *Surface Analysis of Paper*, T. E. Conners and S. Banerjee (Eds.), CRC Press, Boca Raton, FL (1995).
95. D. Y. Kwok, in: *Surface Characterization Methods*, M. J. Milling (Ed.), p. 37. Marcel Dekker, New York, NY (1999).
96. R. J. Roe, *J. Colloid Interf. Sci.* **50**, 70 (1975).

97. A. M. Schwartz and F. W. Minor, *J. Colloid Sci.* **14**, 572 (1959).
98. C. J. van Oss, R. F. Giese, Z. Li, K. Murphy, J. Norris, M. K. Chaudhury and R. J. Good, *J. Adhesion Sci. Technol.* **6**, 413 (1992).
99. R. F. Giese, P. M. Costanzo and C. J. van Oss, *Phys. Chem. Miner.* **17**, 611 (1991).
100. C. A. Ku, J. D. Henry, R. Siriwardane and L. Roberts, *J. Colloid Interf. Sci.* **106**, 377 (1985).
101. N. Yan, Y. Maham, J. H. Masliyah, M. R. Gray and A. E. Mather, *J. Colloid Interf. Sci.* **228**, 1 (2000).
102. J. M. Douillard, *J. Colloid Interf. Sci.* **182**, 308 (1996).
103. D. A. Spagnolo, Y. Maham and K. T. Chaung, *J. Phys. Chem.* **100**, 6626 (1996).
104. J. W. Whalen, *J. Phys. Chem.* **65**, 1676 (1961).
105. R. L. Venable, W. H. Wade and N. Hackerman, *J. Phys. Chem.* **69**, 317 (1965).
106. R. V. Siriwardane and J. P. Wightman, *J. Adhesion* **15**, 225 (1983).
107. T. Morimoto, Y. Suda and M. Nagao, *J. Phys. Chem.* **89**, 4881 (1985).
108. A. V. Kiselev, in: *Advances in Chromotography*, J. C. Giddings and R. A. Keller (Eds.). Marcel Dekker, New York, NY (1967).
109. D. R. Lloyd, T. C. Ward and H. P. Schreiber (Eds.), *Inverse Gas Chromatography: Characterization of Polymers and Other Materials*. American Chemical Society, Washington, DC (1989).
110. P. Mukopadhyay and H. P. Schreiber, *Colloids Surf. A* **100**, 47 (1995).
111. J. M. Braun and J. E. Guilett, *Adv. Polym. Sci.* **21**, 107 (1976).
112. J. E. G. Lipson and E. J. Guilett, in: *Development in Polymer Characterization*, J. V. Dawkins (Ed.), Vol III, Applied Science, London (1982).
113. B. M. Mandal, C. Bhattacharya and S. N. Bhattacharya, *J. Macromol. Sci. Chem. A* **26**, 175 (1989).
114. A. Voelkel, *Crit. Rev. Anal. Chem.* **22**, 411 (1991).
115. C. R. Hegedus and I. L. Kamel, *J. Coatings Technol.* **65(820)**, 23 (1993).
116. C. R. Hegedus and I. L. Kamel, *J. Coatings Technol.* **65(822)**, 31 (1993).
117. C. Kembell and E. K. Rideal, *Rroc. R. Soc. Ser. A* **187**, 53 (1946).
118. J. H. De Boer and S. Druyer, *Proc. Kon. Ned. Akad. Wet. B.* **55**, 451 (1952).
119. G. M. Dorris and D. G. Gray, *J. Colloid Interf. Sci.* **77**, 353 (1980).
120. M. Nardin and E. Papirer, *J. Colloid Interf. Sci.* **137**, 534 (1990).
121. J. Schultz, L. Lavielle and C. Martin, *J. Adhesion* **23**, 45 (1987).
122. H. Balard, E. Brendle and E. Papier, in: *Acid-Base Interactions: Relevance to Adhesion Science and Technology*, K. L. Mittal (Ed.), Vol. 2, p. 299. VSP, Utrecht (2000).
123. A. J. Groszek and C. E. Templer, *Fuel* **67**, 1658 (1988).
124. A. J. Groszek, *Carbon* **27**, 33 (1989).
125. C. E. Templer, in: *Proceedings, Particle Size Analysis*, Vol. 26, p. 301. The Society for Analytical Chemistry, London (1970).
126. G. I. Andrews, A. J. Groszek and N. Hairs, *ASLE Trans.* **15**, 184 (1972).
127. I. M. Veiga, A. C. Fernandes and B. S. Alimeda, *J. Mater. Sci. Lett.* **12**, 1206 (1993).
128. A. J. Groszek, S. Partyka and D. Cot, *Carbon* **29**, 821 (1991).
129. A. J. Groszek, in: *Physical Chemistry of Colloids and Interfaces in Oil Production*, H. Toulhoat and J. Lecourtier (Eds.), Editions Technip, Paris (1992).
130. A. J. Groszek, *Carbon* **25**, 717 (1987).
131. F. M. Fowkes, Y. C. Haung, B. Shah, M. J. Kulp and T. B. Lloyd, *Colloids Surf.* **29**, 243 (1988).
132. F. M. Fowkes, D. W. Dwight, D. A. Cole and T. C. Huang, *J. Non-Crystall Solid.* **120**, 47 (1990).
133. S. T. Joslin and F. M. Fowkes, *IEC Prod. Res. Dev.* **24**, 369 (1985).
134. G. W. Woodbury and L. A. Noll, *Colloid. Surf.* **28**, 233 (1987).
135. M. F. Finlayson and B. A. Shah, *J. Adhesion Sci. Technol.* **4**, 431 (1990).
136. R. S. Farinato, S. S. Kaminski and J. L. Courter, *J. Adhesion Sci. Technol.* **4**, 633 (1990).
137. T. L. Barr, *Modern ESCA: The Practice and Principles of X-ray Photoelectron Spectroscopy*. CRC Press, Boca Raton, FL (1994).

138. F. M. Etzler, R. Deanne, T. Burk, T. H. Ibrahim, G. Willing and R. D. Neuman, in: *Contact Angle, Wettability and Adhesion*, K. L. Mittal (Ed.), Vol. 2, pp. 13-43, VSP, Utrecht (2002).
139. J. Frommer, *Angew. Chem. Eng. Ed.* **31**, 1298 (1992).
140. E. zur-Muelen and H. Nieus, in: *Particle and Surface Characterization Methods*, R. H. Mueller, W. Mehnert and G. E. Hildebrand (Eds.), p. 99. Medpharm Scientific Publishers, Stuttgart (1997).
141. T. H. Ibrahim, T. R. Burk, F. M. Etzler and R. D. Neuman, *J. Adhesion Sci. Technol.* **14**, 1225 (2000).
142. G. A. Willing, T. R. Burk, F. M. Etzler and R. D. Neumann, *Colloid. Surf. A* **193**, 117 (2001).
143. G. A. Willing, T. H. Ibrahim, F. M. Etzler and R. D. Neuman, *J. Colloid Interf. Sci.* **226**, 185 (2000).
144. R. Nessler, *Scanning* **21**, 137 (1999).
145. S. N. Magnov and D. Reneker, *Annu. Rev. Mater. Sci.* **27**, 175 (1997).
146. S. N. Magnov and M.-H. Whangbo, *Surface Analysis with STM and AFM*. VCH, Weinheim (1996).
147. M. Radmacher, M. Fritz, J. D. Cleveland, D. A. Walters and P. K. Hansma, *Langmuir* **10**, 3809 (1994).
148. G. B. Binning, C. F. Quate and C. Gerber, *Phys. Rev. Lett.* **12**, 930 (1986).
149. P. Lemoine, R. W. Lamberton and A. A. Ogwu, *J. Appl. Phys.* **86**, 6564 (1999).
150. E. R. Beach, G. W. Tormoen and J. Drelich, *J. Adhesion. Sci. Technol.* **16**, 845 (2002).
151. S. Ross and J. P. Olivier, *On Physical Adsorption*. Wiley, New York, NY (1964).
152. W. Rudzinski and D. H. Everett, *Adsorption of Gases on Heterogeneous Surfaces*. Academic Press, New York, NY (1992).
153. R. Sips, *J. Chem. Phys.* **16**, 490 (1948).
154. R. Sips, *J. Chem. Phys.* **18**, 1024 (1950).
155. O. M. Todes and A. K. Bondareva, *Zhur. Priklad. Khim.* **21**, 693 (1947).
156. D. N. Misra, *J. Chem. Phys.* **52**, 5499 (1970).
157. L. B. Harris, *Surf. Sci.* **10**, 129 (1968).
158. L. B. Harris, *Surf. Sci.* **13**, 377 (1969).
159. G. F. Cerofolini, *J. Low Temp. Phys.* **6**, 473 (1972).
160. P. Brauer, M. Fassler and M. Jaroniec, *Thin Solid Films* **123**, 245 (1985).
161. E. Papirer, H. Balard and C. Vergelati, in: *Adsorption on Silica Surfaces*, E. Papirer (Ed.), Marcel Dekker, New York, NY (2000).
162. M. Heuchel, M. Jaroniec and R. K. Gilpin, *J. Chromatogr.* **628**, 59 (1993).
163. R. K. Gilpin, M. Jaroniec and M. B. Martin-Hopkins, *J. Chromatogr.* **513**, 1 (1990).
164. J. R. Conder and C. L. Young, *Physicochemical Measurement by Gas Chromatography*. Wiley, New York, NY (1979).
165. R. A. Pierotti, *The Interactions of Gases with Solids*, Wiley, New York, NY (1971).
166. M. Jaroniec, X. Lu and R. Madey, *J. Phys. Chem.* **94**, 5917 (1990).
167. D. R. Williams, *Eur. Chromatogr. Anal.* **9** (Feb. 1991).
168. H. P. Schreiber and D. R. Lloyd, in: *Inverse Gas Chromatography*, D. R. Lloyd, T. C. Ward and H. P. Schreiber (Eds.), p. 185. American Chemical Society, Washington, DC (1989).
169. D. R. Williams, in: *Controlled Interphases in Composite Materials*, H. Ishida (Ed.), Elsevier, New York, NY (1990).
170. G. F. Cerofolini, *Surf. Sci.* **52**, 195 (1975).
171. J. P. Hobson, *Can. J. Phys.* **43**, 1934 (1965).
172. J. P. Hobson, *Can. J. Phys.* **43**, 1941 (1965).
173. G. F. Cerofolini, *Surf. Sci.* **24**, 391 (1971).
174. J. Gawdzik, Z. Suprynowicz and M. Jaroniec, *J. Chromatogr.* **121**, 185 (1976).
175. A. W. Adamson and I. Ling, in: *Solid Surfaces and the Gas-Solid Interface; Adv. Chem Ser.* Vol. 33, p. 51, American Chemical Society, Washington, DC (1961).
176. W. A. House and M. J. Jaycock, *Colloid. Polym. Sci.* **256**, 52 (1978).
177. S. Ross and I. D. Morrison, *Surface Sci.* **52**, 103 (1975).
178. W. Rudzinski, J. Jagiello and Y. J. Grillet, *J. Colloid Interf. Sci.* **87**, 478 (1982).
179. H. Balard, *Langmuir* **13**, 1260 (1997).
180. B. Charmas and R. Leboda, *J. Chromatogr. A* **886**, 133 (2000).

# Part 3

# Wetting and Spreading:
# Fundamental and Applied Aspects

*Contact Angle, Wettability and Adhesion*, Vol. 3, pp. 267–291
Ed. K.L. Mittal
© VSP 2003

# Merging two concepts: Ultrahydrophobic polymer surfaces and switchable wettability

K. GRUNDKE,* M. NITSCHKE, S. MINKO, M. STAMM, C. FROECK, F. SIMON, S. UHLMANN, K. PÖSCHEL and M. MOTORNOV

*Institute of Polymer Research, Hohe Straße 6, D-01069 Dresden, Germany*

**Abstract**—We report a detailed study of the influence of roughness of plasma etched poly(tetrafluoroethylene) (PTFE) surfaces and a subsequent chemical modification on the macroscopic contact angle behavior using scanning electron microscopy (SEM), scanning force microscopy (SFM), X-ray photoelectron spectroscopy (XPS) and contact angle measurements. The aim of our investigations was to obtain a better understanding of the interplay between roughness and chemical composition on the wettability by quantifying the surface morphology, surface chemistry and contact angle hysteresis.

It was shown that plasma etching was a suitable technique to fabricate ultrahydrophobic PTFE surfaces with specifically designed surface roughness. Contact angle hysteresis was strongly dependent on the geometrical nature of the roughness features: the type of roughness rather than its absolute size was the determining factor to obtain ultrahydrophobicity. "Composite" (air trapped) surfaces, characterized by pin- or spire-like roughness features and open spaces between these roughness features, were necessary to obtain hysteresis-free contact angle patterns. The rms roughness values of these surfaces were in the range of half a micrometer up to more than one micrometer. By lowering the surface free energy of the fluoropolymer, ultrahydrophobicity was obtained at lower rms roughness values of about 300 nm.

By grafting mixed polymer brushes of carboxyl-terminated poly(styrene-co-2,3,4,5,6-pentafluorostyrene) (PSF-COOH) and carboxyl-terminated poly(2-vinyl pyridine) (PVP-COOH) onto the structured surfaces we could reversibly tune the surface properties of PTFE resulting in a sharp wettability transition from ultrahydrophobicity to hydrophilicity upon exposure to different liquids. This approach enables us to regulate wetting and adhesion in one and the same material over a wide range.

*Keywords*: Plasma etched PTFE surfaces; polymer brushes; ultrahydrophobicity; surface roughness; contact angles.

## 1. INTRODUCTION

To control wettability and adhesion of polymers numerous surface modification techniques are used, such as exposure to flames and low-pressure non-equilibrium

---

*To whom all correspondence should be addressed: Phone: ++49 351 4658 475, Fax: ++49 351 4658 284, E-mail: grundke@ipfdd.de

plasma and chemical modification (grafting) [1-3]. In many cases, the wettability is regulated by changes in the chemical composition of the surfaces. But, it has also been long recognized that surface roughness can be important for wettability. Wenzel was the first who discussed the influence of surface roughness on contact angle [4]. He introduced a roughness factor, $r_w$, into the Young equation (eq. 1) because he argued that in the case of a rough solid surface, the interfacial tensions $\gamma_{sv}$ and $\gamma_{sl}$ should not be referred to the geometric area, but to the actual surface area. Thus,

$$r_w = \frac{true\ surface\ area}{geometric\ surface\ area} \tag{1}$$

or

$$r_w(\gamma_{sv} - \gamma_{sl}) = \gamma_{lv}\cos\theta_w \tag{2}$$

For the contact angle on a rough surface he obtained

$$\cos\theta_w = r_w\cos\theta \tag{3}$$

where the Wenzel contact angle, $\theta_w$, is, therefore, the equilibrium contact angle on a rough solid surface having the intrinsic angle $\theta$. It corresponds to the absolute minimum in the free energy of the system. Based on this equation it can be predicted that roughness should have a major effect on the contact angle and, hence, on the wettability of surfaces. However, one has to take into account that the Wenzel equation does not describe contact angle hysteresis and hence the relation between roughness and the phenomenon of hysteresis.

At high level of roughness the surface becomes "composite" due to trapped air. This means that the liquid drop is in contact with a surface composed, in part, by the solid and, in part, by air. Such a porous surface is a special case of a heterogeneous surface. From considerations similar to those leading to eq. 3, Cassie and Baxter [5] derived the following equation assuming that the contact angle of the liquid on air would be 180°

$$\cos\theta_{cb} = a_1\cos\theta_1 - a_2 \tag{4}$$

where $a_1$ and $a_2$ are the fractional surface areas occupied by the material and the air respectively and $\theta_1$ is the corresponding intrinsic contact angle on the solid surface. $\theta_{cb}$ is the Cassie-Baxter contact angle, or, like $\theta_w$ for a rough surface, the equilibrium contact angle for the composite solid surface. It should be pointed out that $\theta_w$ and $\theta_{cb}$ normally will not be amenable to experimental determination [6]. $\theta_w$ and $\theta_{cb}$ are only a conceptual measure of wettability and cannot be realized in practice. For rough and heterogeneous surfaces, a number of stable contact angles can be measured. Two relatively reproducible angles are the largest ($\theta_a$) and the smallest ($\theta_r$) for the advancing and the receding contact line. The difference, $\theta_a$ - $\theta_r$, is called contact angle hysteresis. Johnson and Dettre predicted a transition from a nonporous (Wenzel regime) to a porous surface (Cassie-Baxter regime) with increasing surface roughness that could be verified by experimental results [7, 8]. According to these results hysteresis increases until this critical roughness is reached and then decreases.

There is a renewed interest in the effect of surface roughness on wettability. Recent papers from a variety of research fields have shown that surface topography is of great interest to create super water-repellent surfaces with self-cleaning ability [9-17]. In particular, rough polymer materials possessing a low surface free energy are very promising as super water-repellent synthetic materials with reduced adhesion properties. Morra *et al.* [18] reported that PTFE surfaces became ultrahydrophobic when they were treated by oxygen plasma. At longer teatment times of PTFE, they observed deeply etched surfaces without chemical modifications so that the wettability was controlled by roughness. However, the roughness features were only characterized qualitatively by SEM. They reported contact angles of $\theta_a/\theta_r = 170°/160°$ ($\theta_a$ is the advancing angle and $\theta_r$ the receding angle).

Though, a number of attempts have been made to correlate measured contact angle hysteresis with roughness, most of them have not succeeded in quantifying the effect. The reasons for this situation are, on the one hand, idealized models of roughness which, therefore, have limited practical applicability. On the other hand, there are experimental difficulties: (i) the difficulty in assigning the "correct" apparent contact angles; (ii) the inability to distinguish between variations in surface roughness and chemical heterogeneity and anisotropy of the surface; and (iii) poor roughness characterization.

We report here a detailed study of the influence of roughness of plasma etched PTFE surfaces and a subsequent chemical modification on the macroscopic contact angle behavior using SEM, SFM, XPS and contact angle measurements. Our aim was to obtain a better understanding of the interplay between roughness and chemical composition on the wettability by quantifying the surface morphology, surface chemistry and contact angle hysteresis. In the first part of our study we investigated the influence of roughness and chemical composition of plasma etched PTFE surfaces on the contact angle hysteresis of water droplets. By coating a thin layer of Teflon® AF 1600 on the plasma etched PTFE surfaces the surface chemistry could be kept constant. This enabled us to investigate only the influence of roughness on the contact angle hysteresis. In the second part of our study the strategy was to combine the concept of ultrahydrophobicity induced by roughness with the concept of switching surface properties caused by grafting mixed polymer brushes onto the plasma etched PTFE surfaces and subsequent exposure to different liquids [19]. By combining these two concepts we expected a sharp transition in wettability from ultrahydrophobicity to completely wettable surfaces upon exposure to different solvents.

## 2. EXPERIMENTAL

### 2.1. Materials

Poly(tetrafluoroethylene) (PTFE) foils, 0.5 mm thick, were purchased from PTFE Nünchritz GmbH, Nünchritz, Germany. The material as received was cut into pieces of size 2 x 2 cm$^2$ and subsequently cleaned in CHCl$_3$ for 10 min using an ultrasonic bath to remove any contaminants.

Commercially available Teflon® AF1600 (DuPont) was used to prepare thin films onto PTFE by dip coating PTFE substrates into a diluted solution of Teflon® AF1600 (solvent: Fluorinert FC-75). Teflon® AF1600 is an amorphous copolymer poly(tetrafluoroethylene-co-2,2-bis-trifluoro-methyl-4,5-difluoro-1,3-dioxole).

Carboxyl-terminated poly(styrene-co-2,3,4,5,6-pentafluorostyrene) (PSF-COOH) (0.75:0.25), $M_n$ = 16 000 g/mol, $M_w$ = 29 500 g/mol, was synthesized by radical copolymerization using 4,4´-azobis(4-cyanopentanoic acid) as the initiator to introduce end-carboxyl groups in the copolymer chains.

Carboxyl-terminated poly(2-vinyl pyridine) (PVP-COOH) $M_n$ = 49 000 g/mol, $M_w$ = 51 500 g/mol was purchased from Polymer Source, Inc. (synthesized by anionic polymerization).

Oxygen and ammonia for plasma etching with purities of 99.95% and 99.999%, respectively, were purchased from Messer Griesheim, Krefeld, Germany.

### 2.2. Plasma etching

For plasma etching of PTFE a cylindrical vacuum chamber made of stainless steel with a diameter of 250 mm and a height of 250 mm was used. The base pressure obtained with a turbomolecular pump was <10$^{-4}$ Pa. Oxygen was introduced into the chamber via a gas flow control system. The samples were introduced by a load-lock system and placed on an aluminum holder near the center of the chamber which was coupled capacitively to a 13.56 MHz radio frequency (RF) generator via an automatic matching network. The metallic wall of the whole chamber worked as a grounded electrode, i.e. the electrode configuration was higly asymmetric causing significant self-bias voltages and ion energies at the RF electrode [20]. The following parameters were used: oxygen flow 10 sccm, pressure 2 Pa, and effective RF power 200 W. The resulting self-bias voltage as displayed at the RF generator was approximately 1000 V. After plasma etching the samples were rinsed in an ultrasonic bath for 10 min in CHCl$_3$.

For ammonia plasma treatment of PTFE a cylindrical vacuum chamber made of stainless steel with a diameter of 350 mm and a height of 350 mm was used. The base pressure obtained with a turbomolecular pump was <10$^{-5}$ Pa. On the top of the chamber a 2.46 GHz electron cyclotron resonance (ECR) plasma source RR160 by Roth&Rau, Germany, with a diameter of 160 mm and a maximum power of 800 W was mounted. Ammonia was introduced into the active volume of the plasma source via a gas flow control system. The samples were introduced by a load-lock system and placed on a grounded aluminum holder near the center of the chamber. The distance between the sample and the excitation volume of the plasma source was about 200 mm. The following parameters were used: ammonia flow 15 sccm, pressure 7 x $10^{-1}$ Pa, and power 220 W. The treament time varied from 20 to 120 s.

## 2.3. Preparation of mixed brushes

Polymer chains of two carboxyl terminated incompatible polymers of different polarities: poly(styrene-co-2,3,4,5,6-pentafluorostyrene) (PSF-COOH) (0.75:0.25) and poly(2-vinylpyridine) (PVP-COOH) were attached to the PTFE substrate by end functional groups. Hydroxyl and amino functionalities, introduced covalently by ammonia plasma treatment onto the PTFE surface, were used to graft the mixed polymer brush. The brush was fabricated using a two-step "grafting" procedure [19]. In the first step, a thin film (10 nm) of carboxyl terminated PSF-COOH was spin-coated onto the surface of the PTFE foil and heated for 6 hours at 150°C to graft the polymer from the melt. Nongrafted polymer was removed by Soxhlet extraction and the second polymer PVP-COOH was grafted using the same procedure. Fourier transform infrared spectroscopic measurements in the attenuated total reflection mode (FTIR-ATR) evidenced the grafting of both polymers: characteristic bands of PSF-COOH (1601, 2923, and 3027 $cm^{-1}$) and PVP-COOH (1586 and 1590 $cm^{-1}$) were identified in the spectra. Control experiments showed that under these conditions about 3.5 $mg/m^2$ of each polymer was grafted onto the PTFE substrate which corresponds to 6 nm thickness of a film of a mixed polymer brush.

## 2.4. Surface characterization

The surface topography was investigated using scanning electron microscopy (SEM) and scanning force microscopy (SFM). SEM micrographs were obtained without metallization of the sample at a beam voltage of 1 kV with a low voltage scanning electron microscope.

SFM experiments were carried out under ambient conditions using Dimension 3100 microscope from Veeco/ Digital Instruments, Inc., Santa Barbara, CA, USA. The SFM was used in the tapping mode to reduce tip induced surface degradation and to avoid sample damage. Standard and ultra-sharp silicon tips were used with a resonance frequency of 300 kHz and 60 kHz, respectively. The surface root mean square (rms) roughness values $S_q$ were determined using a scan size area of 20 x 20 $\mu m^2$ and a commercial software. As per definition it represents standard deviation of the elevation, Z values at given coordinates $(x_i; y_j)$, within the given area, i.e.,

$$S_q = \sqrt{\frac{1}{MN} * \sum_{j=1}^{N}\sum_{i=1}^{M} |Z^2(x_i; y_j)|} \qquad (5)$$

where M is the number of scan lines in x direction and N is the number of scan lines in y direction. The rms roughness describes only structures vertical to the surface. However, the surface topography reflects both vertical and lateral structures. To include also the lateral distribution of the height features, the two-dimensional power spectral density function (PSD) was determined. The PSD represents a spatial frequency analysis of the surface topography. The PSD intensity describes the probability of a spatial frequency q. Thus, a peak in the PSD corresponds to the most prominent frequency which is related to a dominant in-plane length $\lambda$ using the relationship

$$\lambda = 2\pi/q \tag{6}$$

In addition, a roughness coefficient $r_s$ was calculated

$$r_S = \frac{\text{actual surface area}}{\text{geometric surface area}} \tag{7}$$

characterizing the ratio of the actual surface area and the geometric surface area. The $r_s$ parameter represents both vertical and lateral changes of the roughness features.

Images of the PTFE surfaces were obtained at three different locations and the average rms and $r_s$ roughness were determined for a particular surface-treated PTFE sample based on these images from three different locations. The surface roughness was also quantified using section analysis. By section analysis one can obtain vertical distance (depth), horizontal distance, angle between two or more points, and roughness along a section line.

Changes in the elemental surface composition of the treated samples were determined using X-ray photoelectron spectroscopy (XPS). An Axis Ultra spectrometer (Kratos Analytical, UK) equipped with a monochromatized Al $K_\alpha$ X-ray source of 300 W at 15 kV was used. The kinetic energy of photoelectrons was determined using a hemispherical analyzer with a constant pass energy of 160 eV for survey spectra and 20 eV for high-resolution spectra. All spectra were obtained at an electron take-off angle of 90°. Hence, the information depth of XPS in our case is no more than 5 nm. An effective charge compensation unit to compensate for charging effects was used during all measurements. Spectra were referenced to the C 1s peak of the $-CF_2-$ structure in PTFE at a binding energy BE = 292.48 eV. Quantitative elemental compositions were determined from peak areas using experimentally determined sensitivity factors and the spectrometer transmission function.

The wettability of the surfaces was characterized by contact angle measurements using sessile water droplets. Two techniques were employed: a goniometer technique and the axisymmetric drop shape analysis-profile (ADSA-P). Using a goniometer (DSA10 Krüss) advancing ($\theta_a$) and receding ($\theta_r$) contact angles from 6 individual drops placed on 6 fresh surface areas were measured by adding or withdrawing small volumes of water through a syringe. The needle was maintained in contact with the drop during the experiments. All readings were then averaged to give mean advancing and receding contact angles for each sample. The accuracy of this technique is of the order of ± 2°. ADSA-P technique determines liquid-fluid interfacial tensions and contact angles from the shapes of axisymmetric menisci, i.e., from sessile as well as pendant drops. The details of the methodology and experimental set-up can be found elsewhere [21]. Low-rate dynamic contact angle measurements were carried out by supplying liquid to the sessile drop from below the solid surface using a motorized syringe device. It is a good strategy to first deposit a drop of liquid on a given solid surface covering a small hole, which is needed to supply liquid from below. This experimental procedure is necessary since ADSA-P determines the contact angles based on a complete and undisturbed drop profile. Compared to the goniometer technique, the precision of the contact angle measurement is distinctly higher (± 0.5°). While the drop is growing at very slow motion of the three-phase contact line, a sequence of images are recorded by the computer (typically 1 image every 2-5 seconds). Since ADSA-P determines the contact angle and the three-phase contact radius simultaneously for each image, the advancing dynamic contact angle as a function of the three-phase contact radius (i.e., location on the surface) can be obtained. Furthermore, both the drop volume and the liquid surface tension are determined for each image, and can also be recorded. If the polymer surface is not very smooth or other complexities due to swelling, stick/slip, etc. occur, irregular and inconsistent contact angle or liquid surface tension values will be seen as a function of the three-phase contact radius. The details of the procedure and the experimental set-up for low-rate dynamic contact angle measurements are given elsewhere [21-24]. During the experiments the temperature and relative humidity were maintained, respectively, at (23 ± 0.5)°C and about 40%.

# 3. RESULTS

## 3.1. Surface morphology of differently treated PTFE foils

The morphology of an untreated PTFE surface and of $O_2$ plasma treated samples after different treatment times obtained by SEM is shown in Fig. 1. All scanning electron micrographs have the same magnification. The untreated PTFE foil has a rough and, to some extent, porous surface. With increasing treatment time in oxygen plasma, the original morphology of the PTFE surface is altered and a pin-like surface structure is observed. For longer treatment times, the spires become gradually coarser and taller while the general appearance of the features is preserved. After 10 minutes treatment time, a widely spaced spire-or cone-like structure of the PTFE surface is observed (Fig. 1g).

The etched PTFE surfaces were additionally treated in ammonia plasma in order to functionalize these surfaces with reactive groups for a subsequent grafting procedure. No change in surface morphology was observed. Then, the

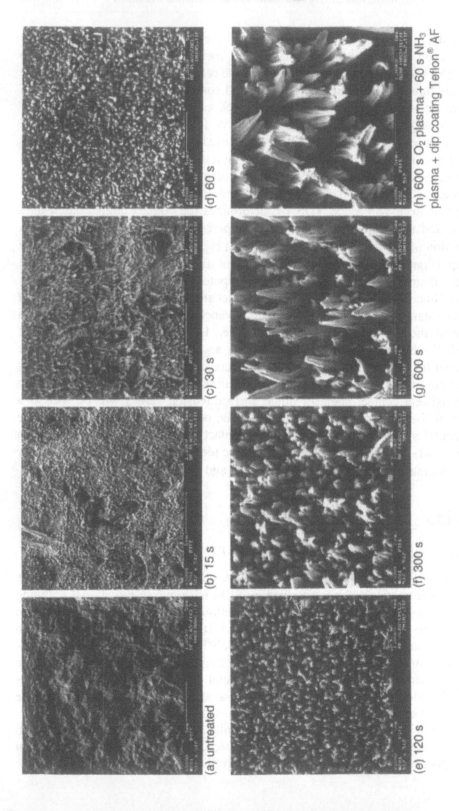

**Figure 1.** Scanning electron micrographs of untreated and oxygen plasma treated PTFE foils after different treatment times (a–g); (h) after 10 min oxygen plasma etching, plus subsequent treatment in ammonia plasma for 1 min, plus additional dip-coating in a Teflon® AF1600 solution.

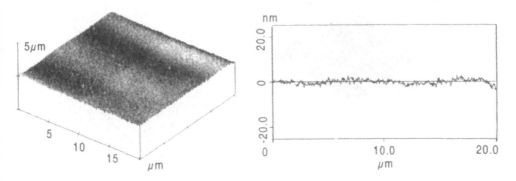

**Figure 2.** SFM topography image and corresponding cross section of a Teflon® AF1600 film prepared by spin-coating onto a silicon wafer. The rms roughness value calculated for the scan area of 20 x 20 $\mu m^2$ is 1.0 nm.

etched PTFE surfaces were also dip-coated with Teflon® AF1600 with the aim to investigate the influence of roughness on the contact angle hysteresis without changing the chemical composition of the surfaces. Fig. 1h shows that the roughness was obviously not altered by a thin Teflon® AF1600 film.

Since scanning electron micrographs do not provide any information concerning the absolute values of the roughness features, SFM measurements were carried out. Fig. 2 shows a typical SFM topography image and the corresponding cross section of a very smooth Teflon® AF1600 surface prepared by spin coating from solution onto a silicon wafer and subsequent annealing of the film for 1 hour at 165°C. In this case, the rms roughness value (20 x 20 $\mu m^2$) is 1.0 nm.

SFM topography images and the corresponding cross sections of the untreated and oxygen plasma treated PTFE surfaces are presented in Fig. 3. These topographical images confirm the results obtained qualitatively by SEM: the original morphology of the PTFE surface is replaced by surface structures possessing tips of varying sizes and spaces between them. These changes can already be observed after 30 sec treatment time.

The corresponding mean roughness values calculated for a scan size of 20 x 20 $\mu m^2$ and averaged at three different locations on each surface are shown in Fig. 4. It can be seen that the rms roughness values of the treated PTFE surfaces increased with increasing treatment time (Fig. 4a). These values describe only roughness features vertical to the surface. If the lateral distribution of the elevations is included in the roughness analysis by calculating the ratio between the actual surface area and the geometric surface area, $r_s$, the plot shown in Fig. 4b is obtained. From this plot it can be concluded that the actual surface area of the PTFE surface is already increased significantly after very short treatment times (15 sec). Up to a treatment time of 60 sec, the actual surface area does not change significantly. Then, at a treatment time of 120 sec, it increases again to distinctly higher values. From the fact that at longer treatment times (300 sec) the rms

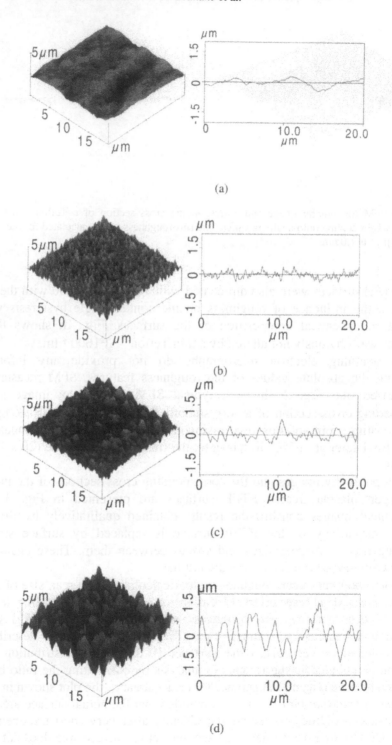

**Figure 3.** SFM topography images (20 x 20 $\mu m^2$) and corresponding cross sections of the untreated (a) and oxygen plasma treated PTFE surfaces after 30 sec (b), 60 sec (c) and 120 sec (d) treatment time.

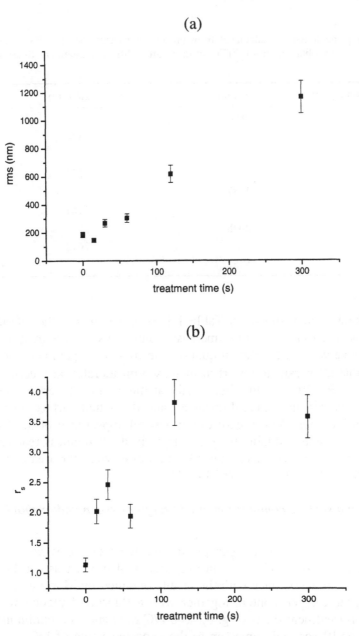

**Figure 4.** A plot of rms roughness (a) and $r_s$ roughness (b) values for PTFE surfaces with the treatment time in oxygen plasma. rms and $r_s$ values were calculated from a scan size of 20 x 20 $\mu m^2$.

roughness values are strongly increasing while the $r_s$ roughness is nearly constant we can conclude that the roughness features became coarser and widely spaced. This conclusion was also supported by the dominant in-plane length $\lambda$, calculated from the power spectral density (PSD) curves.

**Table 1.**
Dominant in-plane length $\lambda$, calculated from the power spectral density (PSD) curves of SFM topography images obtained from PTFE surfaces after different treatment times in an oxygen plasma

| Treatment time (sec) | $\lambda$ (nm) | $\Delta\lambda$ (nm) |
|---|---|---|
| 15 | 806 | |
| | | 324 |
| 30 | 1130 | |
| | | 357 |
| 60 | 1487 | |
| | | 1161 |
| 120 | 2648 | |
| | | 2613 |
| 300 | 5261 | |

From the $\lambda$ values shown in Table 1 it is apparent that the lateral distances between the roughness features increase continuously with treatment time. A comparison between the different quantitative roughness parameters (rms, $r_s$, and $\lambda$) shows that all parameters start to increase considerably at a certain treatment time (60 sec → 120 sec). In other words, at this point the vertical height of the roughness features, their lateral distances and the actual surface area all start to change considerably. With regard to the general appearence of the morphology, SEM results indicated qualitatively a change in the roughness features from the original morphology to a more pin-like structure when the treatment time was increased from 30 sec to 60 sec (cf. Fig. 1).

### 3.2. Elemental surface composition of PTFE after oxygen and ammonia plasma treatments

XPS investigations of the oxygen plasma treated PTFE surfaces revealed only minor changes in the F/C ratio and only traces of oxygen as can be seen from Figs. 5 and 6. In accordance with the results obtained by Morra et al. [18], some increase in the oxygen content together with a fluorine depletion was observed only at short treatment times (Fig. 6). The F/C ratio shows a minimum at a treatment time of 10 sec and comes up to the expected value of F/C = 2 for longer treatment times. It is known that oxygen plasma promotes etching via preferential attack of the carbon-carbon bonds [25]. The high resolution C1s spectra of the oxygen plasma treated samples (Fig. 7) show these structural changes. In addition to the characteristic $CF_2$ peak at 292.48 eV, increased amounts of $CF_3$ and CF groups are found at 294.3 eV and 290.6 eV, respectively, in the surface region of the etched PTFE samples compared to the untreated surface. Hence, it can be concluded that the oxygen plasma induces some chemical modification. But at

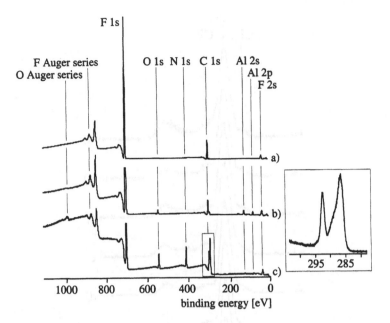

**Figure 5.** XPS survey spectra of untreated (a) and plasma treated PTFE surfaces after 600 sec oxygen plasma treatment (b) plus 60 sec ammonia plasma treatment (c). The inset shows the high-resolution C 1s spectrum of sample (c).

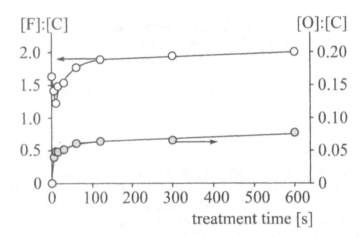

**Figure 6.** Elemental surface composition of the untreated and oxygen plasma treated PTFE samples in dependence of the treatment time; [F]:[C] (○), [O]:[C] (●).

longer treatment times, fast etching process predominates chemical surface modification resulting in morphology changes which was confirmed by SEM and SFM measurements. Due to the difference in susceptibility of crystalline and amorphous polymer regions to plasma, crystalline regions etch more slowly than amorphous regions [26-28].

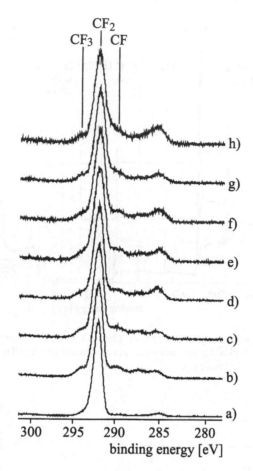

**Figure 7.** High-resolution C 1s spectra of untreated (a) and oxygen plasma treated PTFE; treatment time: 5 s (b), 15 s (c), 30 s (d), 60 s (e), 120 s (f), 300 s (g), and 600 s (h).

The survey spectrum together with the high-resolution C1s spectrum of a PTFE sample treated for 10 min in an oxygen plasma and subsequently by ammonia plasma for 60 sec is shown in Fig. 5c. From the O1s, N1s and C1s peaks of this spectrum, it can be concluded that both amino and hydroxyl groups were incorporated covalently into the PTFE surface after this treatment procedure. In this way, we obtained a rough, functionalized PTFE surface.

### 3.3. Wettability of differently treated PTFE surfaces

#### 3.3.1. Influence of oxygen plasma etching on contact angle hysteresis of PTFE surfaces

To characterize the wetting behavior of the PTFE surfaces, contact angle measurements with water were carried out using two sessile drop techniques, as described in Section 2.4.

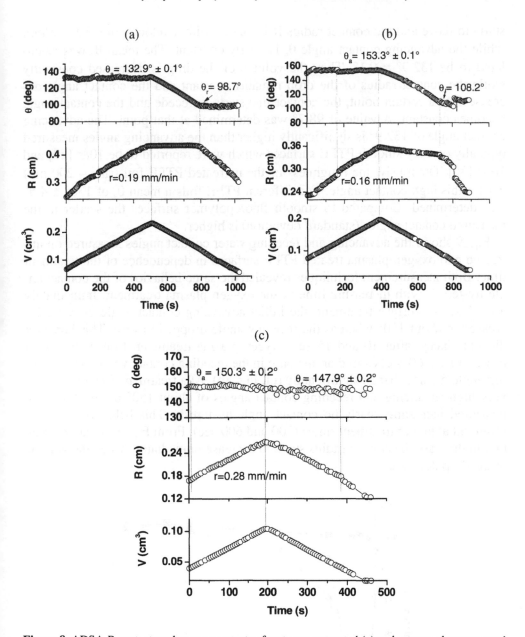

**Figure 8.** ADSA-P contact angle measurements of water on untreated (a) and oxygen plasma treated PTFE surfaces. Treatment time: (b) 60 sec; (c) 600 sec. θ is the contact angle, R is the contact radius of the sessile drop, r is the rate of motion of the three-phase contact line, and V is the drop volume. In c, the contact radius starts to decrease immediately when the volume of the drop starts to decrease indicating nearly no contact angle hysteresis.

Fig. 8a shows a typical ADSA-P contact angle pattern obtained for untreated PTFE foils. When the drop volume V is increased the three-phase contact line starts to move and the contact radius R increases with a velocity of 0.19 mm/min while the advancing contact angle $\theta_a$ is nearly constant. The mean $\theta_a$ was calculated to be $132.9° \pm 0.1°$. Then the volume of the drop is decreased constantly while the contact radius of the drop remains constant and the contact angle decreases. At a certain point, the contact line starts to recede and the contact angle becomes constant. A value of $98.7°$ was determined at that point. The advancing contact angle of $132.9°$ is significantly higher than the advancing angles measured typically on very smooth PTFE surfaces which were reported to be $108°$ [29] and $104°$ [30]. Obviously, the roughness of the untreated PTFE foil (rms $\approx 200$ nm) causes this high contact angle. For different PTFE foils a mean $\theta_a$ of $138° \pm 4.5°$ was determined. Compared to smooth fluoropolymer surfaces the scatter in the measured contact angles (standard deviation) is higher.

Fig. 9 shows the advancing and receding water contact angles measured on untreated and oxygen plasma treated PTFE surfaces in dependence of the treatment time. Both measurement techniques revealed the same behavior of the contact angle hysteresis with increasing time of the oxygen plasma treatment. Immediately after 5 sec of surface treatment, the initial advancing contact angle of water increased to about $150°$ whereas the receding angle dropped to zero. This behavior did not change after 10 and 15 sec oxygen plasma treatment. Longer treatment times (30 and 60 sec) caused an increase in the receding contact while the advancing angle did not change significantly (cf. Fig. 8b). At a treatment time of 120 sec, very high advancing and receding contact angles of about $150°$ up to $160°$ were measured indicating nearly no contact angle hysteresis. This behavior was also observed at longer treatment times (300 and 600 sec). From Fig. 8c it can be seen that in this case the contact radius starts to decrease immediately when the volume of the drop decreases.

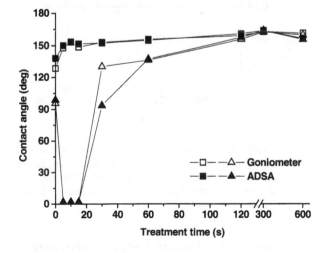

**Figure 9.** Advancing (■, □) and receding (▲, △) contact angles of water on oxygen plasma treated PTFE as a function of treatment time.

### 3.3.2. Influence of roughness on contact angle hysteresis of Teflon® AF1600 surfaces

To keep the surface chemistry constant the surfaces were dip-coated with a thin film of Teflon® AF1600. Fig. 10a shows the ADSA-P contact angle pattern of water measured on a very smooth Teflon®AF1600 surface (rms = 1 nm) in comparison to a Teflon®AF1600 surface dip-coated onto the untreated PTFE foil

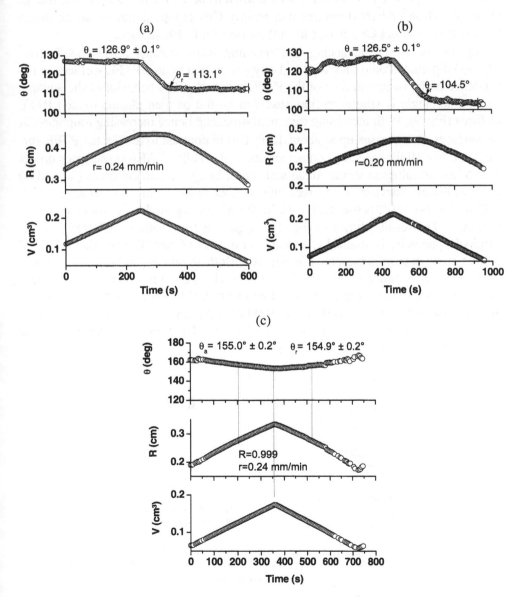

**Figure 10.** ADSA-P contact angle measurements of water on Teflon® AF1600: (a) film spin-coated onto a silicon wafer surface; film dip-coated (b) on an untreated PTFE foil, (c) on an oxygen plasma treated PTFE foil. treatment time 60 sec.

(Fig. 10b). For both surfaces, the same mean advancing contact angles $\theta_a$ were determined. But it is apparent that the contact angle hysteresis is lower on the smooth surface. This is caused by a higher receding angle. From the advancing contact angle of water ($\theta_a = 127°$) the surface free energy $\gamma_{sv}$ is calculated to be 8 mJ/m$^2$ for the Teflon®AF1600 surface using the equation-of-state approach for solid-liquid interfacial tensions [31]. This value is distinctly lower than the surface free energy of PTFE which is well known to be 20 mJ/m$^2$ [30]. The lower $\gamma_{sv}$ value of Teflon® AF1600 implies that mainly CF$_3$ groups dominate at the outermost surface and not CF$_2$ groups as in the case of a PTFE surface.

Fig. 11 shows the advancing and receding water contact angles on Teflon® AF1600 dip-coated onto untreated and oxygen plasma treated PTFE surfaces in dependence of the treatment time of the PTFE substrates. The behavior of the advancing contact angle is similar to that observed on the oxygen plasma treated PTFE surfaces (Fig. 9). With increasing treatment time, and hence increasing roughness of the surfaces, $\theta_a$ increases up to about 150°. But in contrast to the treated PTFE surfaces (cf. Fig. 9) the receding contact angles of the Teflon® AF1600 surfaces do not drop to lower values at shorter treatment times. Longer treatment times (30 and 60 sec) caused an increase in the receding contact angle from 105° to about 130° (ADSA-P) and 150°, respectively, while the advancing angle did not change significantly. The results shown in Fig. 11 suggest that on Teflon® AF1600 surfaces ultrahydrophobicity is attained at a treatment time of 60 sec. The detailed ADSA-P contact angle pattern is presented in Fig. 10c. Both advancing and receding contact angles were determined to be 155°. There is no change in this contact angle pattern with increasing treatment time. It should be mentioned that the plasma etched PTFE surfaces without a smooth Teflon® AF1600 film did not show ultrahydrophobicity at a treatment time of 60 sec (cf. Fig. 9 and Fig. 8b). They became ultrahydrophobic at a treatment time of 120 sec (cf. Fig. 9).

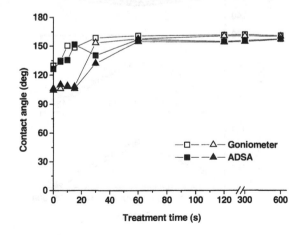

**Figure 11.** Advancing (■, □) and receding (▲, △) contact angles of water on oxygen plasma treated PTFE surfaces dip-coated with a thin film of Teflon® AF1600 as a function of treatment time.

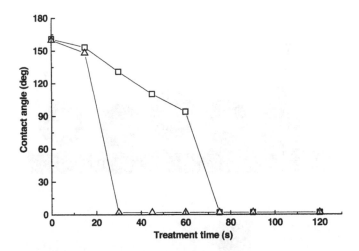

**Figure 12.** Advancing (□) and receding (△) water contact angles measured on surface treated PTFE foils as a function of the treatment time in ammonia plasma. The PTFE surfaces were etched in oxygen plasma for 600 sec and then treated with ammonia plasma.

### 3.3.3. Contact angle behavior of rough, NH₃ plasma treated PTFE surfaces

The advancing and receding water contact angles measured on oxygen plasma etched PTFE surfaces (600 sec) which were subsequently treated with ammonia plasma are plotted against treatment time in Fig. 12. It can be seen that both advancing and receding contact angles decrease with increasing treatment time in the ammonia plasma. At first the receding contact angles dropped to zero. At longer treatment times, the PTFE surfaces became completely wettable by water.

These results imply that we can change the surface properties of PTFE foils from ultrahydrophobicity to completely wettable surfaces in contact with water when the surfaces were etched in oxygen plasma and subsequently treated in ammonia plasma. The ultrahydrophobicity could be retained after dip-coating these surfaces from a Teflon® AF1600 solution.

### 3.3.4. Wettability of rough PTFE surfaces grafted with demixed polymer brushes

The wetting behavior of a PTFE substrate in contact with water is shown in Fig. 13a after different modification steps including a structuring and chemical modification of the surface. At first, the PTFE foil was etched in oxygen plasma, then treated in ammonia plasma, subsequently grafted by two carboxyl terminated incompatible polymers of different polarities (PSF-COOH, P2VP-COOH) and then exposed to toluene before the wetting experiment. The same sample was also exposed to acidic water (pH = 3). After drying the sample, the wettability in contact with water was again checked and can be seen in Fig. 13b. Figs. 13a and b show that the wettability of this PTFE substrate could be switched from ultrahydrophobicity to hydrophilicity upon exposure to different liquids. From Fig. 14 it can be seen that switching of wettability between $\theta_a=150°/\theta_r=150°$ and $\theta=0°$ is completely reversible by dipping the sample in toluene and water (at pH=3), respectively.

(a)

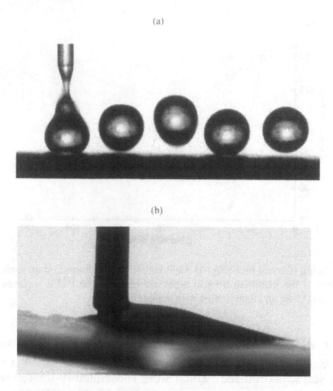

(b)

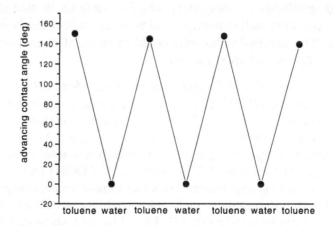

**Figure 13.** Photographs showing the different wetting behavior of a structured PTFE substrate grafted with two carboxyl terminated incompatible polymers of different polarities (PSF-COOH, P2VP-COOH) against water. The wettability of the PTFE substrate can be switched from ultrahydrophobicity (a) to hydrophilicity (b) by dipping the sample in toluene and acidic water (pH = 3), respectively.

**Figure 14.** Reversible switching between ultrahydrophobic and hydrophilic surface properties by dipping the PTFE sample in toluene and water (pH = 3), respectively. The hydrophobicity/hydrophilicity was measured by water contact angles. The PTFE sample was a structured substrate grafted with two carboxyl terminated incompatible polymers of different polarities (PSF-COOH, P2VP-COOH).

## 4. DISCUSSION

The wetting behavior of the fluoropolymer surfaces investigated in the present study confirms the experimental findings of other authors that contact angle hysteresis increases with roughness until a certain roughness is reached and then it decreases (cf. Fig. 9 and Fig. 11) [7, 18]. This experimentally observed behavior is consistent with the Johnson and Dettre theory according to which surfaces having an intrinsic contact angle greater than 90° show, at first, an increase in hysteresis with roughness and then a dramatic decrease of hysteresis at high roughness. This behavior should reflect a transition to a "composite" (air trapped) surface, characterized by such a huge roughness that a liquid with high intrinsic contact angle cannot completely wet the crevices. In this case, the surface beneath the liquid drop is constituted of either solid or air. Using a model of an idealized sinusoidal surface Johnson and Dettre calculated that on "composite" surfaces the height of the energy barriers between metastable states decreased dramatically and contact angle hysteresis was greatly lowered [8].

The contact angle data in the present study show this transition to a "composite" surface for both fluoropolymer surfaces (PTFE and Teflon® AF1600) when the oxygen plasma treatment time was increased from 15 sec to 30 sec. From the quantitative roughness analysis it is apparent that within the range where the contact angle hysteresis is increased (up to 15 sec treatment time) the actual surface area is strongly increased without any change in the vertical height of the roughness features (cf. Fig. 4). The significantly higher contact angle hysteresis of the PTFE surfaces which is due to very low receding angles is obviously caused by an additional chemical modification of the surfaces as could be quantified by XPS measurements (Fig. 6). The incorporation of oxygen into the surface region should result in a small amount of high-energy sites on the surface and, hence, in low receding contact angles caused by the "pinning" of the three-phase contact line to the surface. When we kept the surface chemistry constant by coating with a thin film of Teflon® AF1600, then the advancing angle increased with an increase of the $r_s$ parameter, but the receding angle did not change (cf. Fig. 11).

The transition to the wetting behavior typical for a "composite" surface was observed when the vertical height of the roughness features increased together with a further increase in the actual surface area of the rough surfaces. The absolute value of the roughness features in the vertical direction changed from a mean value of about 150 nm to about 250 nm. The dominant in-plane length as a measure of the lateral distances between the roughness features increased from about 800 nm to about 1100 nm. In other words, there was obviously no significant roughness change necessary to cause the contact angle hysteresis to decrease on both fluoropolymer surfaces. A further increase of the vertical roughness features to a mean value of about 300 nm with lateral distances of about 1500 nm caused ultrahydrophobicity for the fluoropolymer with the lower surface free energy. In this case, a hysteresis-free contact angle pattern was observed for the thin Teflon® AF1600 film (see Fig. 10c). Without the thin Teflon® AF1600 film, the wetting behavior of this rough PTFE surface was typical of a composite surface, but no ultrahydrophobicity was observed (Fig. 8b).

Even though we could not observe a significant change in the absolute roughness values to cause ultrahydrophobicity of the Teflon® AF1600 surface, SEM photographs indicate a change in the original morphology of the PTFE substrate to a pin-like structure at this point (Fig. 1, 60 sec treatment time). From these results we conclude that contact angle hysteresis is not only dependent on the absolute values of the roughness features but also on the geometric nature of the roughness. A certain type of roughness rather than its absolute size was the determining factor to obtain ultrahydrophobic Teflon® AF1600 surfaces. This finding is in agreement with an experimental study of Mason and coworkers [32-34]. They found that the contact angle hysteresis depended strongly on the geometrical nature of the roughness features which was in agreement with the theory developed by these authors. Our results are also consistent with the work reported by Chen *et al.* [15]. They give intuitive arguments that topology of the roughness is important and will control the continuity of the three-phase contact line and thus the hysteresis. In the case of separated pins that are close enough that water cannot intrude between them, a very discontinuous contact line will form. Thus, the contact line cannot make continuous contact with the surface and the drop would not remain pinned (in a metastable state) before advancing or receding and would move spontaneously with no contact angle hysteresis. This is exactly what we observed experimentally on the rough, pin-like Teflon® AF1600 surface.

The ultrahydrophobicity of PTFE surfaces was attained after 120 sec oxygen plasma treatment (cf. Fig. 8c and Fig. 9 ). These surfaces possess the same general appearence of a pin- or cone-like structure but with coarser roughness features of the same type (Fig. 1). At this point the rough PTFE surface is characterized by an rms roughness value of about 600 nm, the ratio of the actual surface area and the geometric area is about 4 and the dominant in-plane length is about 2600 nm. The results show that a further increase of the vertical heights and the lateral distances of the roughness features at longer treatment times (300 sec, 600 sec) did not change the ultrahydrophobic character of the Teflon® AF1600 and the PTFE surfaces, respectively. However, if these very rough PTFE surfaces were functionalized by amino and hydroxyl groups in ammonia plasma (cf. Fig. 5c) without changing their roughness (cf. Fig. 1h) wettability becomes a capillary phenomenon [35]. Due to the hydrophilic surface groups the ultrahydrophobicity of the PTFE surfaces disappeared after short treatment time in ammonia plasma (30 sec). Longer treatment times caused the typical contact angle behavior of a rough surface with a very small amount of polar groups at the PTFE surface: high advancing contact angles and very low receding angles result in high hysteresis similar to the oxygen plasma treated surfaces (cf. Fig. 9 and Fig. 12). If the amount of hydrophilic surface groups is further increased then water penetration promoted by capillary forces occurs causing also the advancing contact angle to decrease which results in complete wetting of the PTFE surface (Fig. 12).

This specifically designed structured PTFE surface (10 min oxygen plasma treatment plus ammonia plasma treatment) characterized by spire-like features and open spaces between the roughness features in the micrometer range was used for further chemical modifications. At first, it was shown that these structured PTFE surfaces could be used to switch the surface properties between hydrophilicity and ultrahydrophobicity when the surface chemistry was altered by an ammonia plasma modification and a subsequent dip-coating of this surface from a Teflon® AF1600 solution.

Another strategy to control the wettability of the structured PTFE substrates by chemical modifications is the "grafting to" approach of polymer brushes. In this case, a second level of structure was formed by nanoscopic self-assembled domains of demixed polymer brushes irreversibly grafted onto the roughness features. It was shown that the mixed brush formed domains of nanometer size and its (average) surface composition stemmed from an interplay between lateral and perpendicular segregations [36-38].

When PSF-COOH and PVP-COOH were grafted onto PTFE substrates that were plasma treated at short treatment times then the wettability could be reversibly switched from an advancing contact angle of 118° to 25° after exposure to toluene and water (pH=3), respectively. A much more pronounced switching effect of the wettability was observed for the rough PTFE substrate that consisted of spire-like structure. In this case, an advancing contact angle of 150° was measured after exposure to toluene. A drop of water rolls easily on this surface (Fig. 13), a fact that ultrahydrophobicity was preserved after the grafting procedure. After we immerse the same sample in an acidic water (pH = 3) bath for five minutes and dry the sample, however, a drop of water spreads on the surface because of capillary penetration effect. These observations demonstrate the usefulness of combining large scale roughness with the self-assembled structure of a binary polymer brush in the range of nanometers. The large scale surface structure amplifies the response and enables one to control wettability and adhesion over a wide range.

## 5. CONCLUSIONS

Plasma etching is a suitable technique to fabricate ultrahydrophobic polymer surfaces. To produce ultrahydrophobic PTFE surfaces with specifically designed surface roughness an interplay of the size scale of the roughness features, their morphology and the surface free energy is important. Contact angle hysteresis was strongly dependent on the geometrical nature of the roughness: the type of roughness rather than its absolute size was the determining factor to obtain ultrahydrophobicity. "Composite" (air trapped) surfaces, characterized by pin- or spire-like roughness features and open spaces between these roughness features were necessary. The rms roughness values of these surfaces were in the range of half a micrometer up to more than one micrometer. The transition to ultrahydrophobicity can be attained at lower rms roughness values by lowering the surface free energy.

By grafting demixed polymer brushes of PSF-COOH and PVP-COOH onto the structured surfaces we could reversibly tune the surface properties of PTFE resulting in a sharp transition of wettability from ultrahydrophobicity to hydrophilicity upon exposure to different liquids.

## REFERENCES

1. K. L. Mittal (Ed.), *Polymer Surface Modification: Relevance to Adhesion*, VSP, Utrecht (1996).
2. K. L. Mittal (Ed.), *Polymer Surface Modification: Relevance to Adhesion*, Vol. 2, VSP, Utrecht (2000).
3. K. Grundke, H.-J. Jacobasch, F. Simon and St. Schneider, *J. Adhesion Sci. Technol.* **9**, 327-350 (1995).
4. R. N. Wenzel, *Ind. Eng. Chem.* **28**, 988 (1936).
5. A. B. D. Cassie and S. Baxter, *Trans. Faraday Soc.* **3**, 16 (1944).
6. D. Li and A. W. Neumann, in: *Applied Surface Thermodynamics*, A. W. Neumann and J. K. Spelt (Eds.) pp. 109-168, Marcel Dekker, New York (1996).
7. R. H. Dettre and R. E. Johnson, Jr, in: *Contact Angle, Wettability and Adhesion*, Adv. Chem. Ser., 43, p. 136, Amer. Chem. Soc., Washington D.C. (1964).
8. R. E. Johnson, Jr and R. H. Dettre, in: *Contact Angle, Wettability, and Adhesion,* Adv. Chem. Ser. 43, p. 112, Amer. Chem. Soc., Washington D.C. (1964).
9. S. Shibuichi, T.Yamamoto,T. Onda and K. Tsujii, *J. Colloid Interface Sci.* **208**, 287-294 (1998).
10. K. Tsujii, T. Yamamoto, T. Onda and S. Shibuichi, *Angew. Chem.* **109**, 1042-1044 (1997).
11. A. Hozumi and O. Takai, *Thin Solid Films* **303**, 222-225 (1997).
12. S. Veeramasuneni, J. Drelich, J. D. Miller and G. Yamauchi, *Prog. Org. Coatings* **31**, 265-270 (1977).
13. J. P. Youngblood and T. J. McCarthy, *Macromolecules* **32**, 6800-6806 (1999).
14. S. R. Coulson, I. Woodward, J. P. S. Badyal, S. A. Brewer and C. Willis, *J. Phys. Chem.* **104**, 8836-8840 (2000).
15. W. Chen, A. Y. Fadeev, M. C. Hsieh, D. Öner, J. Youngblood and T. J. McCarthy, *Langmuir* **15**, 3395-3399 (1999) and references therein.
16. D. Öner and T. J. McCarthy, *Langmuir* **16**, 7777-7782 (2000).
17. M. Miwa, A. Nakajima, A. Fujishima, K. Hashimoto and T. Watanabe, *Langmuir* **16**, 5754-5760 (2000).
18. M. Morra, E. Occhiello and F. Garbassi, *Langmuir* **5**, 872-876 (1989).
19. S. Minko, S. Patil, V. Datsyuk, F. Simon, K.-J. Eichorn, M. Motornov, D. Usov, I. Tokarev and M. Stamm, *Langmuir* **18**, 289-296 (2002) and references therein.
20. K. Köhler, J. W. Coburn, D. E. Horne and E. Kay, *J. Appl. Phys.* **57**, 59 (1985).
21. D. Y. Kwok, T. Gietzelt, K. Grundke, H.-J. Jacobasch and A. W. Neumann, *Langmuir* **13**, 2880-2894 (1997) and references therein.
22. A. Augsburg, K. Grundke, K. Pöschel, H.-J. Jacobasch and A. W. Neumann, *Acta Polymerica* **49**, 417-426 (1998).
23. D. Y. Kwok, A. Li, C. N. C. Lam, R. Wu, S. Zschoche, K. Pöschel, T. Gietzelt, K. Grundke, H.-J. Jacobasch and A. W. Neumann, *Macromol. Chem. Phys.* **200**, 1121-1133 (1999).
24. M. Wulf, K. Grundke, D. Y. Kwok and A. W. Neumann, *J. Appl. Polym. Sci.* **77**, 2493-2504 (2000).
25. S. R. Kim, *J. Appl. Polym. Sci.* **77**, 1913-1920 (2000).
26. T. Yasuda, H. Yasuda, T. Okuno and M. Miyama, *Polym. Mater. Sci. Eng.* (Am. Chem. Soc.) **62**, 457 (1990).

27. D. W. Dwight and S. R. F. McCartney, *Org. Coat., Appl. Polym. Sci.* (Am. Chem. Soc.) **50**, 459 (1984).
28. D. W. Dwight in: *Characterization of Metal and Polymer Surfaces*, Vol. 2, L. H. Lee (Ed.), p. 313, Academic Press, New York (1977).
29. S. Wu, *Polymer Interface and Adhesion*, Marcel Dekker, New York (1982).
30. K. Grundke, T. Bogumil, T. Gietzelt, H.-J. Jacobasch, D. Y. Kwok and A. W. Neumann, *Progr. Colloid Polym. Sci.* **101**, 58-68 (1996).
31. D. Y. Kwok and A. W. Neumann, *Adv. Colloid Interface Sci.* **81**, 167 (1999).
32. J. F. Oliver, C. Huh and S. G. Mason, *Colloids Surfaces* **1**, 79-104 (1980).
33. C. Huh and S. G. Mason, *J. Colloid Interface Sci.* **60**, 11 (1977).
34. J. F. Oliver, C. Huh and S. G. Mason, *J. Colloid Interface Sci.* **59**, 1568 (1977).
35. K. Grundke in: *Handbook of Applied Surface and Colloid Chemistry*, K. Holmberg (Ed.), Chap. 7, pp. 129-134, John Wiley (2001).
36. S. Minko, D. Usov, E. Goreshnik and M. Stamm, *Macromol. Rapid Commun.* **22**, 206-211 (2001).
37. S. Minko, M. Müller, D. Usov, A. Scholl, C. Froeck and M. Stamm, *Phys. Rev. Lett.*, **88**, 035502-1 - 035502-4 (2002).
38. M. Müller, *Phys. Rev., E* **65**, 030802(R)-1 - 4 (2002).

*Contact Angle, Wettability and Adhesion*, Vol. 3, pp. 293–317
Ed. K.L. Mittal
© VSP 2003

# Self-propelled drop movement: Chemical influences on the use of kinetic or equilibrium approaches in reactive wetting

GARY C.H. MO,[1] JUN YANG,[1] SEOK-WON LEE,[2]
PAUL E. LAIBINIS[2] and DANIEL Y. KWOK[1, *]

[1] *Department of Mechanical Engineering, University of Alberta, Edmonton, Alberta T6G 2G8, Canada*
[2] *Department of Chemical Engineering, Massachusetts Institute of Technology, Cambridge, MA 02139, USA*

**Abstract**—We report studies of reactive wetting employing droplets of a non-polar liquid (decahydronaphthalene) on chemically patterned surfaces. The drops contain an *n*-alkylamine that adsorbs onto surfaces with exposed carboxylic acid groups and produces surfaces exposing methyl groups. The change in surface energy that occurs concurrent with the formation of an oriented monomolecular film of alkylamine during this process is sufficient to produce a self-propelled movement of decahydronaphthalene (DHN) drops on the surface. We employed patterning to direct the movement of the drops on the surface, thereby allowing determination of the relationships between the macroscopic fluidic behavior of the droplets and microscopic adsorption events. Specifically, we examined the effects of the unbalanced surface tension force and the influence of adsorbate concentration on drop movement. In the latter case, both kinetic and thermodynamic arguments are applied to describe the system. We compared the predictions by analyzing data from the present system and those reported by Domingues Dos Santos and Ondarcuhu that exhibited opposite trends in behavior. The present analysis provides insight into the influence of chemical reaction kinetics on adsorption-mediated drop movement (i.e., reactive wetting).

*Keywords*: Molecular self-assembly; self-assembled monolayers; wetting; adsorption kinetics; surface energy gradient; solid–liquid interface; surface tension.

## 1. INTRODUCTION

Solid–liquid interactions play an important role in the design and operation of numerous processes. Current interests in microfluidics, micro-droplet handling, micro-array generation, and integrated microsystems share common requirements of handling and manipulating fluidic behavior in cases dominated by surface effects [1–9]. The behavior of these systems is influenced by a variety of microscopic

---

*To whom all correspondence should be addressed: Phone: (1-780) 492-2791, Fax: (1-780) 492-2200, E-mail: daniel.y.kwok@ualberta.ca

factors and cannot be easily described. Capillary forces and other surface effects are sufficient to alter fluidic behavior due to the proportionately large areas of solid–liquid contact on these affected small volumes.

A number of recent theoretical and experimental studies [10–20] have shown that drop motion can be initiated and self-propelled by a surface energy gradient on a substrate. Such gradients can be generated either passively using surfaces with spatial variations in free energy [17, 18] or actively using surfactant-like agents that adsorb onto the contacted surface and induce localized dewetting events [10, 16, 19, 20]. The former approach has been effectively developed by Chaudhury and co-workers, using oxide surfaces that have been modified to expose a monotonically decreasing coverage of immobilized alkyl chains along the direction that propels drop movement [17, 18]. Examples of the latter include the self-propelled movement on glass surfaces of hydrocarbon drops containing fatty acids, or alkyltrichlorosilanes [10, 16]. In these experiments, the adsorbate modifies the glass surface and decreases the wettability of the substrate. The change in wettability effected by contact with the reactive drop results in its self-propelled movement on the glass surface. Domingues Dos Santos and Ondarcuhu [10] have studied the movement of octane and dodecane drops containing 1,1,2,2-tetrahydroperfluorodecyltrichlorosilane ($CF_3(CF_2)_7(CH_2)_2SiCl_3$) on glass and examined the effects of silane concentration and drop length on drop movement. These different approaches provide methods for moving liquids within microscale systems and avoid the need for mechanical parts, pumps, or external forces for driving liquid flow.

In this paper, we examine the effects of adsorbate chain length, concentration, and drop length on reactive wetting on chemically patterned surfaces. We employed decahydronaphthalene (DHN) droplets that contained various amounts of an $n$-alkylamine to reactively wet and move about on surfaces that expose a dense packing of carboxylic acid functionalities. The amine compounds adsorb onto this surface and produce a surface with a lower energy that exposes methyl groups, thereby causing a local surface energy gradient that is sufficient to induce a self-propelled movement of the contacting droplets on the surface. We employ patterning methods (micro-contact printing [21]) to confine the direction of drop movement on these surfaces, thereby allowing direct measurement of fluidic movement and velocity. This ability allowed examination of the relationships between macroscopic droplet behavior and microscopic adsorption events. Specifically, we examined the effects of the force due to the unbalanced surface tension and of drop composition (adsorbate concentration and drop size) on the drop velocity, and analyzed these results using both thermodynamic and kinetic considerations. The former effect was examined by varying the molecular structure of the adsorbate as a means to produce functionalized surfaces with different wetting properties.

## 2. MATERIALS AND METHODS

### 2.1. Materials

Silicon wafers were test grade and obtained from Silicon Sense (Nashua, NH, USA). Gold shot (99.99%) and chromium-coated filaments were obtained from American Precious Metals (East Rutherford, NJ, USA) and R.D. Mathis (Long Beach, CA, USA), respectively. Stamps for generating patterned surfaces by micro-contact printing [21] were made of poly(dimethylsiloxane) (PDMS) sold by Dow Corning (Midland, MI, USA) as SYLGARD Silicone Elastomer-184. Decahydronaphthalene (anhydrous, mixture of *cis* and *trans*) and *n*-alkylamines ($C_nNH_2$, $n$ = 6, 8, 10, 14, and 18) were obtained from Aldrich (Milwaukee, WI, USA) and used as received. Ethanol (190 proof) was obtained from Pharmco (Weston, MO, USA). Octadecanethiol ($CH_3(CH_2)_{17}SH$) was obtained from Aldrich and purified by recrystallization. 16-Mercaptohexadecanoic acid ($HS(CH_2)_{15}CO_2H$) was synthesized according to literature procedures [22].

### 2.2. Substrate preparation

Supported gold films were prepared by sequentially evaporating chromium (approx. 10 nm) and gold (approx. 100 nm) onto silicon wafers in a diffusion-pumped vacuum chamber at about $10^{-6}$ torr. The chamber was backfilled with air and the substrates were used within 48 h of preparation.

### 2.3. Formation of patterned surfaces

Figure 1 provides a schematic overview of the micro-contact printing process and its use to generate patterned surfaces for directing drop movement. In the first step, masters for fabrication of the elastomeric stamps used for micro-contact printing a self-assembled monolayer film pattern on the gold surface were prepared using conventional photolithography methods as described elsewhere [21]. The masters, consisting of a developed photoresist film with the desired features (negative 2 × 60 $mm^2$ rectangular tracks), were placed in a plastic or glass petri dish. A 10:1 ratio (w/w) mixture of SYLGARD silicone Elastomer-184 and its curing agent was poured over the master to a thickness of 5–10 mm. The mixture was allowed to cure either at room temperature overnight or at 65°C for 1–3 h. Sections of the polymer were cut with a razor blade and then peeled from the master, producing a negative copy of the master on the PDMS surface. The peeled sections were washed several times with ethanol and dried with a stream of $N_2$ before use. The generated PDMS stamp was "inked" by directly pouring a 5 mM ethanolic solution of octadecanethiol on the patterned PDMS stamp. After inking, the stamp was placed gently on a gold substrate and removed after 3–5 min to produce a pattern of octadecanethiol-coated areas and bare gold regions. The remaining unfunctionalized surface (2 × 60 $mm^2$ tracks) was derivatized by immersion of the surface into a 5 mM ethanolic solution of 16-mercaptohexadecanoic acid ($HS(CH_2)_{15}CO_2H$) for approx. 5 min. This step

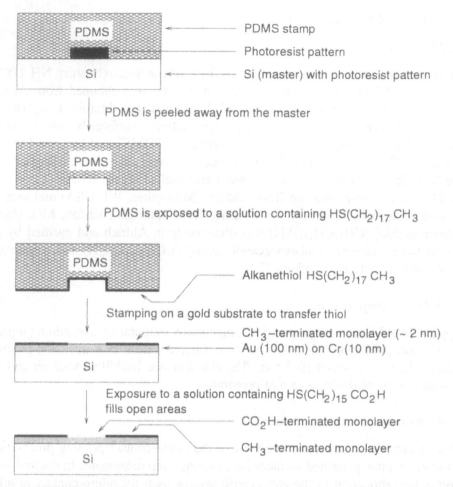

**Figure 1.** Schematic overview of the process for preparing the patterned elastomeric stamp and its use in the generation of patterned areas exposing $CH_3$ and $CO_2H$ surfaces. These different regions are illustrated as black and gray coated areas, respectively, on the Au/Cr/Si substrate.

completed formation of a patterned surface consisting of a series of $2 \times 60$ mm$^2$ tracks that exposed a $CO_2H$ surface, each surrounded by areas expressing a low energy $CH_3$-terminated surface. The samples were rinsed sequentially with ethanol and deionized water, and blown dry with $N_2$ before use.

### 2.4. Moving drop velocity measurements

Moving drop experiments were performed in the open laboratory under atmospheric conditions. Drops (approx. 1 $\mu$l) were deposited at one end of a $CO_2H$, exposing track using a microliter syringe and monitored. These drops spontaneously moved unassisted in a straight path to the opposite end of the track. Images of the moving drops (in top view) were taken using a CCD camera (60 frames/s) equipped with a microscopic lens, synchronized strobe illumination and SVHS recorder. These temporal images were analyzed to determine steady-state drop velocities.

## 3. RESULTS AND DISCUSSION

### 3.1. Background on experimental approach

Our generation of self-propelled drop movement relies on the conversion of a high energy surface to a low energy surface that induces a change in wettability sufficient to provide the energetic driving force for drop movement (see Fig. 2). For our base surface, the adsorption of mercapto-alkanoic acid ($HS(CH_2)_nCO_2H$) onto gold generates a densely packed supported monolayer film that exposes carboxylic acid groups at its surface. These high-energy surfaces are wetted by most liquids including water (for our investigations, the carboxylic acid surface provides a more controlled and reproducible high energy surface than glass whose surface charge and adsorption characteristics are influenced by environmental factors and cleaning procedures used). Previously, Lee and Laibinis reported [20] that these surfaces could be modified by contact with a non-polar solution containing an *n*-alkyl amine. In this process, the amino group undergoes an acid–base reaction with the surface carboxylic acid group and generates a non-covalently attached oriented monolayer of the alkyl amine at its surface. The resulting bilayer exposes the methyl groups ($CH_3$) of the alkyl amine at its surface, as evidenced by its wetting properties by water and various hydrocarbon liquids [20]. Exposure of the bilayer to a polar solvent removes the amine layer and regenerates the high energy carboxylic acid surface. The resulting surfaces are less wettable (i.e., exhibit higher values of $\theta_a$ and $\theta_r$) when alkyl amines of longer chain lengths or solutions of higher amine concentrations are employed [20].

The adsorption process to generate the amine adlayer can effect a self-propelled drop movement on the carboxylic acid surface when the applied droplet contains sufficient amine to effect adsorption and the liquid has a sufficiently high surface tension ($\gamma_{lv} > 30$ mJ/m$^2$) so as not to wet the generated methyl surface. Under these conditions (for example, a 1 mM drop of dodecylamine in benzene), a drop

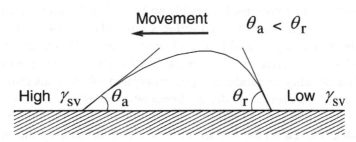

**Figure 2.** Schematic illustration of a moving droplet under the influence of a surface energy gradient. A requirement is that the advancing contact angle ($\theta_a$) on the forward-moving side of the drop be lower than the receding contact angle ($\theta_r$) on the backside.

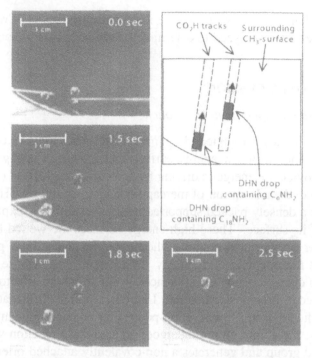

**Figure 3.** Series of images for two *n*-alkylamine-containing DHN drops on patterned $CO_2H/CH_3$ surfaces. Upper right: schematic illustration of the patterned surface comprising $2 \times 60$ mm$^2$ $CO_2H$-tracks surrounded by $CH_3$-terminated domains that restricted the drop movement. Upper left and counterclockwise: Images following the movement of $C_6NH_2$- and $C_{18}NH_2$-containing DHN drops deposited. Drops were deposited sequentially 1.5 s apart at the ends of different tracks, with the images showing their spontaneous movement and a faster velocity of the $C_{18}NH_2$-containing DHN drop (right track).

applied to an unpatterned carboxylic acid surface will spontaneously move about the surface in a meandering self-propelled and self-avoiding path as it converts its contacted areas to lower energy $CH_3$ surface regions.

To control the process, we employed micro-contact printing [21] to generate patterned surfaces that presented definable paths — $2 \times 60$ mm$^2$ rectangular tracks — for drop movement. The patterned surfaces exposed tracks of a $CO_2H$-terminated self-assembled monolayer (SAM) and a boundary of a $CH_3$-terminated SAM (see Fig. 3). This latter surface was not wetted by the droplet and exposed the same chemical functionality as produced upon adsorption of the *n*-alkylamine onto the $CO_2H$ surfaces. A droplet of decahydronaphthalene (DHN), a non-polar liquid with a surface tension of 41 mJ/m$^2$, containing an *n*-alkylamine placed at one end of a track, moved on the surface along the path defined by the micro-contact printing process (Fig. 3). The movement of the drops along linear paths eased determination of drop velocities and allowed examination of factors that influenced the fluid behavior.

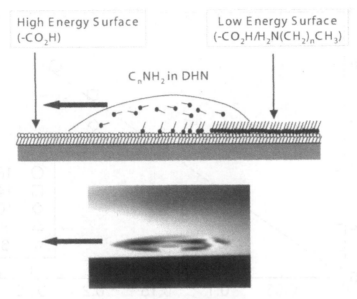

**Figure 4.** Schematic illustration of the surface energy change induced by adsorption of an *n*-alkylamine onto a $CO_2H$ surface. The lower image shows the side-view of a moving drop (including its reflection on the gold substrate) and its apparent asymmetric shape (i.e., $\theta_a < \theta_r$).

## 3.2. Effect of adsorbate chain length

Figure 3 shows a sequence of images for two DHN droplets, each applied to the end of one of two parallel $CO_2H$-exposing tracks in sequence, that contained a 1 mM solution of either $C_6NH_2$ or $C_{18}NH_2$. The images show that the movements of the drops are confined by the micro-contact printed $CH_3$ surfaces. These drops were deposited sequentially 1.5 s apart at the ends of different tracks, with images showing their spontaneous movement. An analysis of images from such experiments found that the DHN drop containing the $C_{18}NH_2$ moved faster (approx. 0.8 cm/s) than that containing the $C_6NH_2$ (approx. 0.5 cm/s). Studies of DHN drops containing *n*-alkylamines of intermediate chain lengths ($n = 8$, 10 and 14) exhibited velocities that showed a general increase with chain length. For all amines, these molecules adsorb onto the $CO_2H$ surface and form oriented monolayer films on the contacted areas with $CH_3$ groups exposed at the monolayer–air (liquid) interface. The effected change in surface functionality and corresponding change in surface energy provide the energetic driving force for drop movement. The longer-chained *n*-alkylamines form a thicker molecular layer on the $CO_2H$ surface to produce coatings that better mask the presence of the underlying polar functionalities and yield surfaces that are less wetted by DHN.

In these experiments, we expect the dominant force responsible for drop movement to be the unbalanced surface tension force $F_Y$ that results from the difference in wettability or surface energy between the front and the back sides of the drop (see

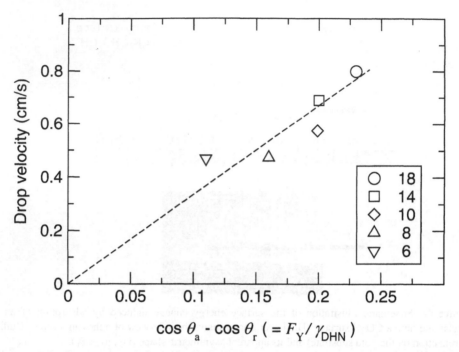

**Figure 5.** The velocity of various *n*-alkylamine ($C_nNH_2$)-containing DHN drops on $CO_2H$ surfaces versus the change in wetting induced by adsorption of the amine ($\cos\theta_a - \cos\theta_r$). The dashed line is a linear fit to equation (2).

Figs 2 and 4):

$$F_Y = \gamma_{lv}(\cos\theta_a - \cos\theta_r) \tag{1}$$

where $\gamma_{lv}$ is the surface tension of the liquid, and $\theta_a$ and $\theta_r$ are the advancing and receding contact angles of DHN on the native $CO_2H$ surface and on the amine-derivatized surface, respectively. In these experiments, DHN wets the $CO_2H$ surface (i.e., $\theta_a \simeq 0°$) and recedes from the amine-derivatized surface with a non-zero contact angle (i.e., $\theta_r > 0°$). Thus, equation (1) predicts that the unbalanced surface force imposed on drops applied to a bare $CO_2H$ surface depends directly on the receding contact angle on the resulting amine-derivatized surface.

Figure 5 plots the steady-state velocities of DHN drops containing different *n*-alkylamines as a function of the value of $\cos\theta_a - \cos\theta_r$. The data exhibit a linear relationship between the applied surface force and the measured steady-state velocity. Under these conditions, the forces on the drop must balance, and the viscous drag force must be equal to the unbalanced surface force [20] on the moving drop (equation (1)):

$$W\gamma_{lv}(\cos\theta_a - \cos\theta_r) = \frac{A\mu V}{h} \tag{2}$$

where $\mu$ is the viscosity of the liquid; $W$ is the length of contact lines at the front and rear of the drop and is roughly the width of track; $A$ is the contact area between the drop and substrate; $V$ is the drop velocity; and $h$ is a characteristic length that approximates the mean height of the drop as averaged over the drop/substrate contact area with respect to the local drag force. From Fig. 5, we obtain a value for the proportionality constant, $Wh\gamma_{lv}/A\mu$, of $\simeq 3.6$ cm/s and a value for $h$ of about 1.0 $\mu$m, suggesting that the viscous drag on the moving drop is localized primarily near the liquid–solid interface [20]. The assumed values for $\gamma_{lv}$ and $\mu$ are 41 mJ/m$^2$ and $1 \times 10^{-3}$ N·s/m$^2$, respectively. The suggestion from Fig. 5 and equation (2) is that changes in the wetting properties of the converted surface induce direct and correlated changes in the velocity of self-propelled drops on these surfaces. We include it here because equation (2) and the parameters determined are used in our subsequent analyses for the kinetic factors that influence drop movement.

### 3.3. Effect of adsorbate concentration on drop velocity

#### 3.3.1. Propulsion of DHN drops on $CO_2H$ surfaces by adsorption of $C_{18}NH_2$

A factor that will influence the effected change in surface energy upon adsorption of an $n$-alkylamine onto a $CO_2H$ surface is the resulting surface coverage of the adsorbate. Higher coverages of the alkylamine would be expected to generate lower energy surfaces and thus produce greater surface forces for (faster) drop movement. We observed that increases in the amount of an $n$-alkylamine present in the DHN drop effected faster drop movements. Figure 6a plots the effect of $C_{18}H_{37}NH_2$ concentration on the velocities of approx. 1-$\mu$l DHN drops applied to patterned $CO_2H$ surfaces. The data show a relatively linear increase for concentrations less than approx. 0.5 mM of this amine and an asymptotic value for concentrations greater than approx. 2 mM. We hypothesize that the differences in drop velocities result from variations in the surface coverage of the alkylamine, as effected either by equilibrium or kinetic factors on the adsorption process, both of which would be affected by adsorbate concentration. Below, we consider both regimes separately after development of expressions to relate drop velocities and adsorbate surface compositions.

For drop-movement experiments performed on $CO_2H$ surfaces, equation (2) can be rewritten as

$$V = \kappa(\cos\theta_a - \cos\theta_r) \simeq \kappa(1 - \cos\theta_r) \tag{3}$$

where $\kappa = Wh\gamma_{lv}/A\mu$ and $\theta_a = 0°$. In equation (3), $\theta_r$ will depend on the composition of the derivatized surface. To describe this relationship, we use the Cassie type of equation [23]

$$\cos\theta_r = \sum_i x_i \cos\theta_{i,r} = x\cos\theta_{100,r} + (1-x)\cos\theta_{0,r} \simeq 1 - (1 - \cos\theta_{100,r})x \tag{4}$$

where $x$ is the fractional composition of adsorbed amines, and $\theta_{100,r}$ and $\theta_{0,r}$ are the receding contact angles for complete adlayers (i.e., 100%) and bare $CO_2H$ surfaces (i.e., 0%), respectively. As DHN wets a pure $CO_2H$ surface, $\theta_{0,r} \simeq 0°$. Equation (4) provides a description of the incremental influence of adsorbed amine on the wetting properties of the modified surfaces. Substitution of equation (4) into equation (3) yields

$$V = \kappa(1 - \cos\theta_{100,r})x \tag{5}$$

for relating drop velocities on pure $CO_2H$ surfaces to the adsorbate coverage $x$ and the wetting properties of a complete amine adlayer.

### 3.3.2. Equilibrium analysis

The adsorption of the amine onto the $CO_2H$ surface proceeds by a non-covalent association between the amine adlayer and the underlying $CO_2H$ surface. The adsorption process is not permanent as the amine layer can be removed by rinsing the assembly with a polar solvent. Further, the amine adlayer can be replaced by exposure to a solution containing a different amine [20]. Because of these associative and dissociative factors, the coverage of adsorbed amine on the $CO_2H$ surface is likely to be affected by equilibrium influences. Within this framework, we relate the fractional surface coverage of the amine $x$ in equation (5) to its solution concentration in the DHN drop through out the Langmuir adsorption equation (equation (6)).

$$x = \frac{Kc}{1 + Kc} \tag{6}$$

where $K$ is the adsorption coefficient and $c$ is the concentration of amine in the contacting DHN drop. In our experiments, the amount of adsorbed amine is small compared to that in solution, and we consider $c$ to be unchanged through the adsorption process. We selected the Langmuir adsorption isotherm for describing the amine adsorption process as this approach assumes that the adsorption is limited to a monolayer coverage, there are specific sites for adsorption and the dominant interaction for adsorption is between the surface sites and the adsorbate. For the present system, the amines adsorb stoichiometrically onto the $CO_2H$ sites, they form a monolayer on the $CO_2H$ surface, and the acid–base interaction between the amine and $CO_2H$ groups is the primary energetic contributor that directs assembly. Further, the Langmuir isotherm provides one of the simplest descriptions of adsorption and avoids the introduction of additional parameters.

By combining equations (5) and (6), we obtain an equilibrium-based expression for describing the dependence of drop velocity on concentration as

$$V = \kappa(1 - \cos\theta_{100,r})\frac{Kc}{1 + Kc} \tag{7}$$

A useful limit is at high amine concentrations, where equation (7) reaches a maximum drop velocity $V_{c\to\infty}$ of

$$V_{c\to\infty} = \kappa(1 - \cos\theta_{100,r})\lim_{c\to\infty}\frac{Kc}{1 + Kc} = \kappa(1 - \cos\theta_{100,r}) \tag{8}$$

**Table 1.**
Summary of the determined adsorption constant $K$ and rate constant $k$ values from equations (7) and (13), respectively

| Reactive wetting system (Liquid/adsorbate/surface) | $K$ (l/mol) | $k$ (l/mols) |
|---|---|---|
| DHN/$C_{18}H_{37}NH_2$/$CO_2H$ | 2800 | 3500 |
| $n$-Octane/$CF_3(CF_2)_7(CH_2)_2SiCl_3$/glass[a] | 95 | 1100[a] |

[a]Data from Ref. [10].

Using the value of $\kappa = 3.6$ cm/s obtained from the data in Fig. 5 and an experimental value of $\theta_{100,r} \simeq 47°$ on a complete $C_{18}NH_2$ adlayer, equation (8) yields a maximum drop velocity of 1.15 cm/s. Using these values for equation (7), the data in Fig. 6a are well fitted by this equation using an adsorption coefficient $K$ of 2800 l/mol. The good agreement between the experimental data and equation (7) suggests that the use of a Langmuir adsorption isotherm and equilibrium considerations are reasonable for describing the underlying chemical process that induces drop movement for the system investigated here. We also consider the application of equation (7) to describe the data reported by Domingues Dos Santos and Ondarchuhu [10] for the self-propelled movement of alkane drops containing an alkyltrichlorosilane on patterned glass surfaces [10]. Figure 6b shows that equation (7) is moderately effective in fitting these data with a value for $K$ of 95 l/mol (Table 1). The greater value of $K$ for amine adsorption (2800 l/mol) compared to that for silane adsorption (95 l/mol) is surprising given that the silane forms a more robust, covalently attached molecular film. This result, which contrasts with chemical intuition, suggests that the use of equation (7) to describe the data in Fig. 6a may be inappropriate for one or both data sets.

### 3.3.3. Kinetic analysis
An alternative description of reactive wetting that is based on kinetic arguments has been provided by Domingues Dos Santos and Ondarcuhu [10]. Their derivation is based on equation (9) proposed by Raphaël [24] for the movement of a drop on an ideal surface

$$V = \frac{\gamma_{lv} \tan \theta^*}{6l\mu} (\cos \theta_a - \cos \theta_r) \qquad (9)$$

where $\gamma_{lv}$ and $\mu$ are the surface tension and viscosity of the liquid, respectively, $l$ is a logarithmic scaling factor ($\ln(x_{max}/x_{min})$, a ratio of macroscopic and molecular lengths) and $\theta^*$ is a dynamic contact angle given by

$$\cos \theta^* = \frac{\cos \theta_a + \cos \theta_r}{2} \qquad (10)$$

Brochard and de Gennes [14] extended this approach by incorporating a kinetic element in equation (9) that provided a dependence of $\theta_r$ on the coverage of the

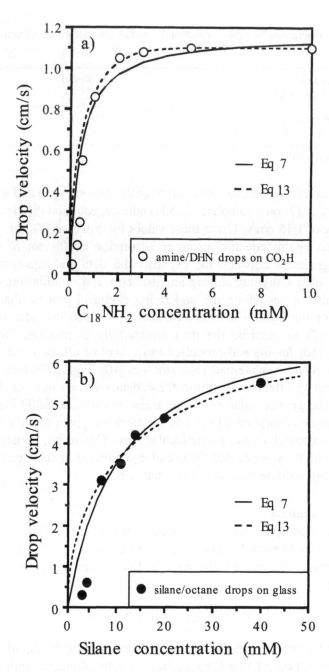

**Figure 6.** Effect of adsorbate concentration on the velocity of self-propelled moving drops. (a) Data for the movement of DHN drops induced by the adsorption of $C_{18}NH_2$ onto $CO_2H$ surfaces. (b) Data taken from Ref. [10] for the movement of *n*-octane drops induced by the adsorption of $CF_3(CF_2)_7(CH_2)_2SiCl_3$ onto a glass surface. The fits in both panels are to equations generated by considering equilibrium and kinetic influences (equations (7) and (13), respectively) (see text). Table 1 summarizes the relevant fitting parameters for the two data sets.

adsorbate. Specifically, they assumed that the surface coverage $\phi_s$ was given by

$$\gamma_{lv} \cos \theta_r = \gamma_{lv} \cos \theta_a - \gamma_1 \phi_s \tag{11}$$

where $\gamma_1$ is a constant that can be determined experimentally. Brochard and de Gennes assumed a first-order kinetic process for adsorption, thereby describing the surface coverage of the adsorbate as

$$\phi_s = 1 - \exp(-kct) = 1 - \exp\left(\frac{-kcL}{V}\right) \tag{12}$$

where $k, c, t$ and $L$ are rate constant for adsorption, the adsorbate concentration, the time for adsorption (for a moving drop, its length divided by its velocity) and the length of drop, respectively. Substitution of equations (11) and (12) into equation (9) yields

$$V = \frac{\gamma_{lv} \tan \theta^*}{6l\mu}\left[1 - \exp\left(\frac{-kcL}{V}\right)\right] \tag{13}$$

We applied equation (13) to the data in Fig. 6a and obtained a reasonable fit using values of $L = 3.5$ mm, $\gamma_{lv} \tan \theta^*/6l\mu = 1.11$ cm/s and a rate constant $k$ of 3500 l/mols. For the investigated system, the similar goodness of fits by the equilibrium (equation (7)) and kinetic (equation (13)) analyses do not allow determination of the more appropriate approach for describing the $DHN/C_{18}NH_2/CO_2H_{surf}$ system.

The derivation of equation (13) was provided by Domingues Dos Santos and Ondarcuhu [10] who suggested its ability to describe the concentration-dependent drop velocity data for experiments on glass using drops of octane. In their experiments, their drops contained 1,1,2,2-tetrahydro-perfluorodecyltrichlorosilane ($CF_3(CF_2)_7(CH_2)_2SiCl_3$), a compound that forms a covalently attached monolayer on glass surfaces. The derivatization proceeds with a change in wettability that is sufficient to induce self-supported drop movement. Figure 6b presents their experimental data [10] along with fits to equations (7) and (13); Table 1 summarizes the parameters for these fits and suggests a faster adsorption rate for the amine than for the trichlorosilane. As the latter requires chemical transformation both in solution and at the surface for reaction and the former relies on a simpler acid–base process, the relative ordering of rate constants is compatible with expectations based on chemical arguments. The data in Fig. 6b are well fitted by both equations, with little difference in the goodness of the two fits. As a result, the fits do not compel a preference for one approach over the other. However, as trichlorosilanes react irreversibly with glass surfaces [10], the use of equation (13) (kinetic model) to describe the data in Fig. 6b is likely more appropriate based on chemical arguments. In contrast, the adsorption of an $n$-alkylamine on a $CO_2H$ surface may include desorption events, thereby preventing selection of one equation as preferred over another based on chemical arguments or a comparison of the fits to the data in Fig. 6a to equations (7) and (13).

## 3.4. Drop length effect

In the above experiments, as the liquid surface tension defines the drop shape for a given track geometry, the drop volumes and, by association, their lengths were held fixed. We examined the ability to move drops of larger volumes on our patterned surfaces, holding the width of the $CO_2H$ tracks and the concentration of the alkylamine constant. Figure 7a shows that the surface reaction provides sufficient propulsion to effect the self-propelled movement of larger drops and that increases in DHN drop size resulted in decreases in drop velocity. We rationalize the observed trend in drop velocity by suggesting that an increase in drop length increases the contact area between the drop and the substrate, thereby causing a greater drag force on the drop. The result is a decrease in drop velocity to balance the forces of propulsion and resistance on the drop (equation (2)). Based on these arguments, the drop velocity would be expected to decrease as the drops are increased in length as observed in Fig. 7a; however, we note that this trend is opposite to that observed by Domingues Dos Santos and Ondarcuhu [10] in related experiments. Specifically, these authors observed that octane drops containing $CF_3(CF_2)_7(CH_2)_2SiCl_3$ on glass surface increased in their velocity as their length was increased (Fig. 7b) [10]. In both cases, the drops contain a reactive species that adsorbs onto the contacting high energy (wettable) surface and alters the wetting properties of the surface sufficiently to produce an unbalanced surface force that induces drop movement. Given the similarity in the two processes, we were interested in the source of their opposite behaviors (Figs 7a and 7b).

Based on the above discussion of the factors that affect drop movement, we assign the difference in Figs 7a and 7b to whether the moving drop is operating primarily in a kinetic regime or has reached its final state structure. For example, for drops that produce adlayers at their early stages of formation, longer drops provide more contact time with the surface and allow generation of adlayers with higher surface coverage and less wettable characteristics that can induce faster drop movement. At the later stages of adlayer formation, the additional contact time provided by an increase in drop length effects little additional driving force for drop movement while the increased drop length introduces a greater resistance to movement. Given these two regimes and their expected influence with the progress of the adsorption reaction, we have replotted the data from Figs 7a and 7b on a dimensionless reaction coordinate (Fig. 8) that takes into account the rates of adsorption ($kc$) and the contact time ($t$) between the drop and the surface ($t = L/V$). As the two systems employ liquids with different surface tensions and produce different values of $\theta_r$, we present the velocity data as normalized to their maximum values.

In Fig. 8, the replotted data from the two systems exhibit unifying elements. On this dimensionless axis, the data from Fig. 6a appear to be at late reaction times, the lower concentration data from Fig. 6b appear to be at early reaction times and the higher concentration data from Fig. 6b appear to transition across the early and late reaction times. This analysis provides a framework for connecting the two reaction regimes and demonstrates the transition in drop movement behavior. The transition

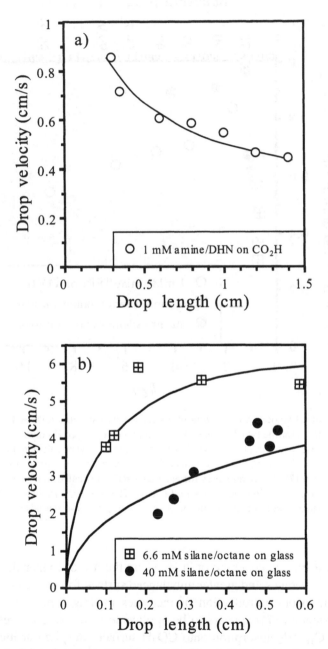

**Figure 7.** Effect of drop length on the velocity of self-propelled moving drops. (a) Data for the movement of DHN drops containing 1 mM $C_{18}NH_2$ onto $CO_2H$ surfaces. The line is provided as a guide to the eye. (b) Data taken from Ref. [10] for the movement of $n$-octane drops induced by the adsorption of $CF_3(CF_2)_7(CH_2)_2SiCl_3$ onto glass surfaces. The fits in the lower panel are to equation (13) where $k = 1100$ l/mols.

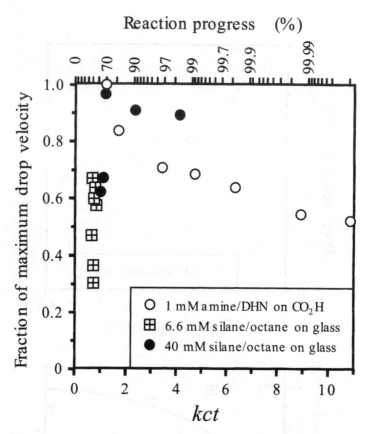

**Figure 8.** Normalized velocities of moving drops for the two systems relative to a dimensionless kinetic parameter ($kct$) for the two different adsorption processes: $C_{18}NH_2$ onto a $CO_2H$ surface and $CF_3(CF_2)_7(CH_2)_2SiCl_3$ onto a glass surface. Adsorption times $t$ were calculated using drop length and velocity data ($t = L/V$), and rate constants $k$ for the two processes were estimated from data in Fig. 6 using equation (7). Drop velocities for each data set were scaled to the maximum velocities predicted by equation (13) using the values in Table 1. The top axis provides a conversion of the $kct$ values to an extent of completion for the adlayer assuming a first-order adsorption process (% = $1 - \exp(-kct)$).

occurs around a surface coverage of 70% where the use of longer drops to achieve higher surface coverages and greater unbalanced surface forces is unable to compete with the simultaneous introduction of increases in drag force. The equilibrium description (equation (7)) appears to be applicable to the results presented earlier in Fig. 6a for $C_{18}NH_2$ adsorption onto $CO_2H$ surfaces despite our inability to note a better fit to the data using equation (7) or (13). For the data in Fig. 6b, the kinetic analysis appears to be more appropriate; however, the data point at highest concentration ($c = 0.04$ mol/l and $L = 0.35$ cm) may approach the limit for this analysis. Notably, the kinetic analysis by Domingues Dos Santos and Ondarcuhu of the higher concentration data in Fig. 7b may not be valid as evidenced by the decrease in drop velocity upon increasing $L$ above 0.2 cm and the correspondence of this decrease with the extent of reaction progress (Fig. 8). Figure 7b captures phenomenologically the decrease in drop velocity exhibited in Fig. 8 that could not be described by the fitted model (equation (13)). A complete description of the data in Fig. 8 would require suitable combination of the kinetic and thermodynamic regimes rather than approaching them separately, as done here. The evidence from this paper is that the two regimes are accessible experimentally and exhibit different phenomenological behaviors that are amenable to a single chemical explanation.

### 3.5. Alternative description

In this section, we study if a unified relationship for the effect of drop length on velocity exists. Both experiments revealed that the velocities of the droplets were constant, suggesting a force balance during motion. Starting with the simple notion that viscous-shear force $F_v$ and unbalanced surface force $F_Y$ are the only forces involved in the system, we write $F_v = F_Y$. An expression for the unbalanced surface force can be obtained using classical Young's equation [25] at both advancing and receding edges of the moving droplet. We, thus, write the unbalanced surface force as

$$F_Y = \gamma_{lv}(\cos \theta_a - \cos \theta_r)$$

Because of the nature of the reactive wetting mechanism considered, $\theta_a < \theta_r$. The expression for viscous-shear drag can be derived directly from Navier–Stokes momentum equations [26] assuming flow in two dimensions. Let us consider $x$, $z$ and $v$ as the horizontal axis, vertical axis and velocity profile inside the droplet, respectively, and the supporting substrate as the datum/zero for the $z$ axis (see Fig. 9a). We use $\zeta$ to represent the height profile of the droplet along its length and $\mu$ as the viscosity of the liquid. Employing boundary conditions $v|_{z=0} = 0$ and $dv/dz|_{z=\zeta} = 0$, we obtain an expression for $v(z)$:

$$v(z) = \frac{dP}{dx} \cdot \frac{z^2 - 2z\zeta}{2\mu} \tag{14}$$

where $P$ is the pressure inside the droplet. To obtain the value of $dP/dx$, we express the velocity profile $v(z)$ as a differential of fluid flux at any position $x_0$ with an average droplet velocity $V$ in the $x$ direction:

$$V\zeta(x_0) = \int_0^{\zeta(x_0)} v(z)dz \tag{15}$$

The pressure gradient can then be written as $dP/dx = -3\mu V/\zeta^2$. We then arrive at the expressions for the velocity profile $v(z)$ and shear force $F_v$ as

$$v(z) = \frac{3V \cdot (2z\zeta - z^2)}{2\zeta^2}$$

$$F_v = \int \mu \frac{dv}{dz}dx = 3\mu V \int \frac{dx}{\zeta(x)} \tag{16}$$

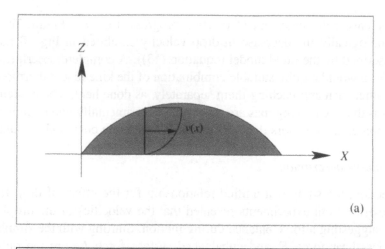

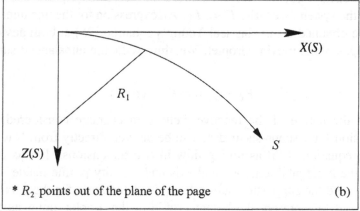

**Figure 9.** Schematics of the coordinate systems used.

Equation (16) is in a form of the known lubrication approximation. Its applicability to our problem depends on the validity of the boundary conditions. For the droplets in consideration, the dimension of the system is much larger than micrometers, hence slippage is not significant compared to the bulk. It is also reasonable to assume $dv/dz$ at the liquid–air interface to be zero, which ensures the time-invariance of the interface line. Equating $F_v$ and $F_Y$ yields

$$3\mu V \int_a^b \frac{dx}{\zeta(x)} = \gamma_{lv}(\cos\theta_a - \cos\theta_r) \tag{17}$$

The limits '$a$' and '$b$' of the above integration refer to the contact edges of the droplet at the substrate. As equation (17) assumes a two-dimensional flow, droplet length becomes the only important droplet-size-dependent variable that affects both $F_v$ and $F_Y$. We will treat them separately in the following sections.

### 3.5.1. Viscous force dependence on drop length

As the droplet increases in length, its contact area increases and height profile of the droplet varies. Since the droplet is in motion, the definition of viscosity results that velocity profile along its height at a given position determines the shear stress at that discrete position. The total viscous-shear force then depends on the cross-sectional shape of the droplet, hence on left-hand-side of equation (17). We first assume that drop length is less or sufficiently close to its capillary length $\lambda^{-1} = (\rho g / \gamma_{lv})^{1/2}$. Through pressure equilibrium, we can access the droplet shape using the Laplace equation of capillarity [25]:

$$\Delta P = \gamma_{lv} \left( \frac{1}{R_1} + \frac{1}{R_2} \right) \tag{18}$$

where $\Delta P$ is the pressure difference across the liquid–vapor interface; $R_1$ and $R_2$ are the two principal radii of curvature. We again use the $x$ and $z$ coordinates, but redefining the $z$-datum at the apex of the droplet (see Fig. 9b). Equation (18) can be reduced to a system of differential equations:

$$\frac{d\phi}{ds} = 2b + Cz - \frac{\sin \phi}{x}$$
$$\frac{dx}{ds} = \cos \phi$$
$$\frac{dz}{ds} = \sin \phi \tag{19}$$

where $C$ is the capillary constant, defined as $\rho g / \gamma_{lv}$; $b$ is the curvature at the apex of the droplet; $\phi$ is the tangent at any point on the parametric curve $s$. The Laplace equation of capillarity is suited to describe interfacial lines, and this formulation is especially suitable for smaller droplets where gravity is negligible. The droplet is assumed to be axisymmetric as the difference between the advancing and receding contact angles is small. We apply the boundary conditions $x(0) = z(0) = \phi(0) = 0$ and by definition $d\phi/ds|_{s=0} = b$. In a two-dimensional flow, the term $(\sin \phi / x) \to 0$ as $R_2 \to 0$ in equation (19). To further simplify the system, we assume that $\phi(s)$ has an absolute value smaller than 1 radian (57.3°) over the length of the sessile droplet and obtain the estimation for $\phi(s)$

$$\phi(s) = \frac{b}{\epsilon} \sinh(\epsilon s) \tag{20}$$

where $\epsilon^2 = \rho g / \gamma_{lv}$. Simultaneous integrations yield the location of the parametric curve:

$$x(s) = \int \cos \left( \frac{b}{\epsilon} \sinh(\epsilon s) \right) ds$$
$$z(s) = \int \sin \left( \frac{b}{\epsilon} \sinh(\epsilon s) \right) ds \tag{21}$$

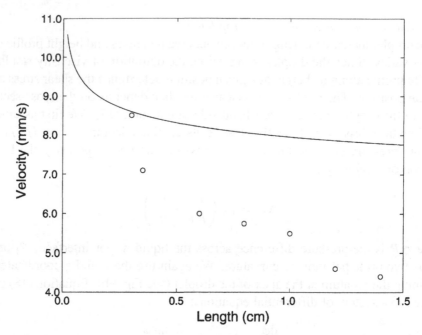

**Figure 10.** Numerical result without consideration of reaction mechanism; circles represent data from Lee *et al*. [27].

Equations (21) were used to numerically generate complete Laplacian curves, and the values of $\int_a^b \mathrm{d}x / \zeta(x)$ required in equation (16). We then compute the effect of length on viscous-shear drag and illustrate that the intuitive result — length increase leads to viscous drag increase — is correct.

*3.5.2. Adsorption reaction dependence on drop length*
The effect of droplet length on adsorption can either be positive or negative, depending on the adsorption system in question. As droplet length and the associated contact area increase, more sites become available for adsorption. Contact area increase corresponds to longer contact time between the adsorbate-carrying liquid and substrate, hence a positive effect on the percentage of adsorbate coverage $x_c$, surface energy gradient, and eventually on droplet velocity. However a limit must exist beyond which increase in contact area has no additional effect on the adsorption reactions and hence equilibrium. Furthermore, an excessive length can result in large viscous-shear drag, as discussed above. Thus we speculate the effect of adsorption reaction may eventually become negligible, and $(\cos \theta_a - \cos \theta_r)$ in equation (17) is approximately constant.

Brochard-Wyart and de Gennes [28] commented on Dos Santos and Ondarçuhu's experiments, and described the adsorption mechanism as a first-order kinetic process:

$$x_c = 1 - \exp\left(\frac{-kcL}{V}\right) \qquad (22)$$

Lee *et al*. [27] employed a Langmuir adsorption isotherm to describe the adsorbate coverage at equilibrium condition as

$$x_c = \frac{Kc}{1 + Kc} \tag{23}$$

In the above equations, $x_c$ is the percentage of adsorbate coverage, $k$ is the kinetic rate constant, $K$ is the adsorption constant, $c$ is the adsorbate concentration and $L$ is the drop length. Both of these descriptions worked well within the framework of their respective experiments. Yet neither description is able to predict the contrasting trends in velocity between the two systems. Here we propose a transition for the separate experimental systems where a suitable description of velocity shifts from kinetic to equilibrium or *vice versa*. Qualitatively speaking, the above two descriptions provide the ideal case studies of the length–velocity relationship. Since rate description equation (22) is likely to dominate when the viscous force is small, we predict that this description is more accurate at smaller lengths. Similarly, we expect the equilibrium description equation (23) to be true at larger lengths, where the viscous-shear force is much larger compared to the increase in unbalance surface force due to higher adsorbate coverage.

Here we briefly describe the numerical calculations employed. We have discussed about using the Laplacian curves in order to calculate the profile-dependent integral in equation (17). Romberg integrations were used to quickly obtain O(6) estimations to equations (21). Tabulated data from Hartland and Hartley [29] suggest that $b$, the apex curvature, can be used as a parameter to obtain a series of Laplacian curves. Contact angles for each profile were kept constant using a second-degree polynomial fit to the generated Laplacian curve. For the value of integral $\int dx/\zeta(x)$, there are possibilities of division-by-zero at each edge of the droplet where the height is, by definition, zero. We, therefore, employed truncated height profiles to estimate the integral values.

### 3.5.3. Combined description of the two adsorption experiments

We first test the hypothesis that reaction mechanics may be neglected at some length scales. For the $n$-alkylamine(DHN)/thiol/gold system, we found that numerical solutions using equation (17) do not fit the experimental data well when reaction effects are neglected ($\cos\theta_a - \cos\theta_r$ assumed constant), as evidenced in Fig. 10. We, thus, conclude that the reaction mechanics must be taken into account. It is interesting to note that our early attempts using qualitative polynomial approximations of droplet profiles, which had little physical relevance, was successful to obtain a fit to this same set of experimental data (also neglecting reaction effects). The necessity of the Laplacian curve calculation is emphasized here — one needs to perform the more complex, accurate Laplace solution for droplet cross section profile in order to avoid misleading results. For adsorption by either equilibrium or kinetic description in equation (17), and assuming a Cassie-type relationship in the contact angles at

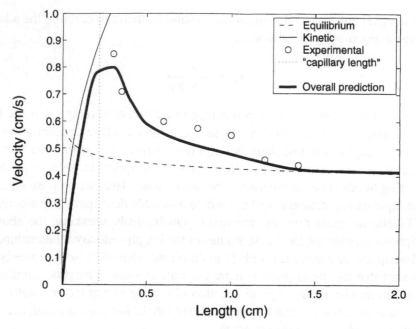

**Figure 11.** Overall length effect for the data from Lee *et al.* [27].

equilibrium yields

$$V_{\mathrm{eq}} = \frac{\gamma_{\mathrm{lv}}(1 - \cos\theta_{100,\mathrm{r}})}{3\mu \int_a^b \frac{\mathrm{d}x}{\zeta}} \cdot \frac{Kc}{1 + Kc} \tag{24}$$

$$V_{\mathrm{r}} = \frac{\gamma_{\mathrm{lv}}}{3\mu \int_a^b \frac{\mathrm{d}x}{\zeta}} \cdot \left[1 - \exp\left(\frac{-k_{\mathrm{r}}cL}{V_{\mathrm{r}}}\right)\right] \tag{25}$$

where $V_{\mathrm{eq}}$ and $V_{\mathrm{r}}$ represent, respectively, the velocities by equilibrium and kinetic descriptions; $\theta_{100,\mathrm{r}}$ is the experimental contact angle when the adsorbate surface coverage is 100%; $K$ and $k_{\mathrm{r}}$ are the adsorption constant and rate constant, respectively.

We plotted in Figs 11 and 12 qualitative combination of the two reaction regimes. In both figures, the kinetic (expressed by equation (25)), equilibrium (expressed by equation (24)) and the hypothesized overall length–velocity relationship are represented by solid, dash and bold lines, respectively. Specifically, Fig. 11 was obtained with an assumed adsorption constant $K$ of 550 l/mol and a rate constant $k_{\mathrm{r}}$ of 1000 l/mols. Similarly, Fig. 12 presents the case of the silane(n-alkane)/glass system, with an assumed adsorption constant $K$ of 700 l/mol and a rate constant $k_{\mathrm{r}}$ of 1500 l/mols. Stipulation of covalent, more robust silane adsorption [27] is supported by these parameters. The values of these constants, however, differ from those reported here and by Dos Santos and Ondarçuhu. To construct the overall relationship for the two sets of experiments, we no longer require a general goodness of fit over all data points. We wish to find the case in each adsorption system where

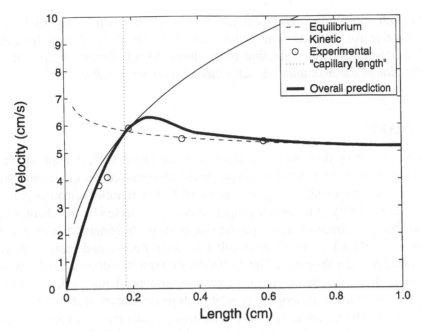

**Figure 12.** Overall length effect for data from Dos Santos and Ondarçuhu [10].

the curves generated by equations (24) and (25) bracket the experimental data, acting more-or-less like asymptotes. Equilibrium assumptions idealize the case of long droplet. Since the adsorption reaction is not restricted by the available sites, the reverse argument can be made about the kinetic assumptions. At a given length, the experimental velocity response may be attributed (unevenly) to both idealized cases. By using the two idealized cases as asymptotes, the predicted overall relationship is essentially a weighted average of the contribution from both regimes.

Given the adsorption system (includes carrier liquid, adsorbate concentration, etc.), the predicted overall relationship between droplet length and velocity suggests a maximum possible droplet velocity. This maximum velocity and the gradual transition from kinetic to equilibrium regime is reflected in the results reported by Dos Santos and Ondarçuhu, as shown in Fig. 12. From the reported data, we believe that the transition phase was overlooked by Dos Santos and Ondarçuhu [10], and was not observed by Lee *et al.* [27]. Clearly, our results illustrate the existence of such a phase for droplet lengths still smaller than those tested in experiments described here and those of Lee *et al.* A prediction of the overall relationship is shown in Fig. 11.

The capillary length $\lambda^{-1}$ for DHN is 2.16 mm; for octane (used by Dos Santos and Ondarçuhu) it is approx. 1.78 mm. These values are marked on the respective figures by dotted vertical lines. We note that in both Figs 11 and 12, the capillary length closely corresponds to the predicted maximum velocity for the given system. We believe that capillary lengths have an indicative value in these self-propulsion experiments. Phenomenologically, droplets with lengths greater than $\lambda^{-1}$ are better treated with equilibrium (equation (24)); adsorption kinetics (equation (25)) is more suitable for droplets with lengths smaller than $\lambda^{-1}$. Since the capillary length corresponds to an inflection from kinetic to equilibrium regime, it is also an indication of the maximum velocity that exists in its vicinity.

## 4. SUMMARY

Reactive wetting by droplets of decahydronaphthalene (DHN) containing an alkylamine can be used to direct the self-propelled drop movement on chemically patterned surfaces exposing contiguous areas of $CO_2H$ termini surrounded by $CH_3$ termini. The velocity of these self-propelled drops is affected by the chain length of the adsorbate, its concentration, and the size of drop. In general, the drop velocity increases with the adsorbate chain length and concentration, and achieves a limiting value for a particular drop size. The drop velocity exhibits a decrease with the length of the DHN drop. We found that the concentration-dependence of drop velocity for both data sets could be described by relationships developed based on purely thermodynamic or kinetic considerations. A difficulty is that these two analyses predict sufficiently similar behaviors that are not readily distinguished by experimental results. An examination of the results from the influence of drop length (or more appropriately drop-substrate contact time) provides a better ability to reveal kinetic and equilibrium regimes. Phenomenological differences between these regimes are evidenced by the observation of increasing or decreasing drop velocities for drops of increasing length. The competition between reaction kinetics and increased drag force yields maximum drop velocities for drop contact times that provide surface functionalization of $\approx 70–90\%$ (cf., Fig. 8). For longer contact times, incremental increases in resulting surface coverages and receding wetting characteristics are offset by increases in drag force that produce a net decrease in reactively driven drop velocity. We also presented a relationship for the effect of drop length on velocity. It was found that capillary length might be an indicator as a transition point from kinetic to equilibrium approaches.

*Acknowledgements*

This research was supported financially, in part, by the Alberta Ingenuity Establishment Fund, Natural Sciences and Engineering Research Council (NSERC) of Canada, Canada Research Chair Program (CRCP), and Canada Foundation for Innovation (CFI). G.M. and J.Y. acknowledge financial support from the Province of Alberta through the Alberta Ingenuity studentships.

## REFERENCES

1. J. Simon, S. Saffer and C. J. Kim, *J. Microelectromech. Syst.* **6**, 208 (1997).

2. C. A. Chen, J. H. Sahu, J. H. Chun and T. Ando, in: *Science and Technology of Rapid Solidification and Processing*, M. Otooni (Ed.), pp. 123–134. Kluwer Academic Publishers, Dordrecht (1995).
3. F. Gao and A. A. Sonin, *Proc. R. Soc. London A* **444**, 533 (1994).
4. M. Orme, C. Huang and T. Courter, in: *Melt Spinning, Strip Casting and Slab Casting*, E. Matthys and W. Truckner (Eds). The Mineral, Metals and Materials Society, Warrendale, PA (1996).
5. K. Schmaltz and C. H. Amon, in: *Melt Spinning, Strip Casting and Slab Casting*, E. Matthys and W. Truckner (Eds). The Mineral, Metals and Materials Society, Warrendale, PA (1996).
6. M. Vardelle, A. Vardelle, A. C. Leger, P. Fauchais and D. Gobin, *J. Thermal Spray Technol.* **4**, 50 (1994).
7. K. Ajito, *Appl. Spectrosc.* **52**, 339 (1998).
8. C.-Y. Kung, M. D. Barnes, N. Lermer, W. B. Whitten and J. M. Ramsey, *Appl. Opt.* **38**, 1481 (1990).
9. K. E. Miller and R. E. Synovec, *Talanta* **51**, 921 (2000).
10. F. Dominges Dos Santos and T. Ondarcuhu, *Phys. Rev. Lett.* **75**, 2972 (1995).
11. T. Ondarcuhu and M. Veyssié, *Nature* **352**, 418 (1991).
12. T. Ondarcuhu and M. Veyssié, *J. Phys. II* **11**, 75 (1991).
13. F. Brochard, *Langmuir* **5**, 432 (1989).
14. F. Brochard and P.-G. de Gennes, *C. R. Acad. Sci. Ser. B* **321**, 285 (1995).
15. M. E. R. Shanahan and P.-G. de Gennes, *C. R. Acad. Sci. Ser. B* **324**, 261 (1997).
16. C. D. Bain, G. D. Burnett-Hall and R. R. Montgomerie, *Nature* **372**, 414 (1994).
17. M. K. Chaudhury and G. M. Whitesides, *Science* **256**, 1539 (1992).
18. S. Daniel, M. K. Chaudhury and J. C. Chen, *Science* **291**, 633 (2001).
19. K. Ichimura, S.-K. Oh and M. Nakagawa, *Science* **288**, 1624 (2000).
20. S.-W. Lee and P. E. Laibinis, *J. Am. Chem. Soc.* **122**, 5395 (2000).
21. Y. Xia and G. M. Whitesides, *Angew. Chem. Int. Ed.* **37**, 550 (1998).
22. C. D. Bain, E. B. Troughton, Y.-T. Tao, J. Evall, G. M. Whitesides and R. G. Nuzzo, *J. Am. Chem. Soc.* **111**, 321 (1989).
23. A. B. D. Cassie, *Discuss. Faraday Soc.* **3**, 11 (1948).
24. E. Raphaël, *C. R. Acad. Sci. Ser. B* **306**, 751 (1988).
25. S. Lahooti, O. I. del Río, P. Cheng and A. W. Neumann, in: *Applied Surface Thermodynamics*, J. K. Spelt and A. W. Neumann (Eds), pp. 441–507. Marcel Dekker, New York, NY (1996).
26. F. M. White, *Fluid Mechanics*, 4th ed. WCB McGraw-Hill, New York, NY (1999).
27. S. W. Lee, D. Y. Kwok and P. E. Laibinis, *Phys. Rev. E* **65**, 051602 (2002).
28. F. Brochard-Wyart and P.-G. de Gennes, *C. R. Acad. Sci. Ser. B* **321**, 285 (1995).
29. S. Hartland and R. W. Hartley, *Axisymmetric Fluid-Liquid Interfaces*. Elsevier, Amsterdam (1976).

*Contact Angle, Wettability and Adhesion*, Vol. 3, pp. 319–338
Ed. K.L. Mittal
© VSP 2003

# Effects of surface defect, polycrystallinity and nanostructure of self-assembled monolayers on wetting and its interpretation

JUN YANG, JINGMIN HAN, KELVIN ISAACSON and DANIEL Y. KWOK *

*Nanoscale Technology and Engineering Laboratory, Department of Mechanical Engineering, University of Alberta, Edmonton, Alberta T6G 2G8, Canada*

**Abstract**—We report low-rate dynamic contact angle data of various liquids on self-assembled monolayers (SAMs) of octadecanethiol on annealed and non-annealed gold. It was found that interpretation of solid surface tensions using contact angle data on less well-prepared polycrystalline non-annealed gold surfaces can be misleading. Our findings were supported by reflectance infrared spectra and atomic force microscopy data that the surface nanostructure, defect and polycrystallinity can be important factors for a systematic study of wettability on SAMs in terms of surface energetics. We found the contact angle and adhesion patterns of various liquids on SAMs of octadecanethiol adsorbed onto annealed gold substrates to be consistent with recent experimental data for the relatively thick polymer-coated surfaces. The variation of surface structure in terms of surface energetics can be estimated only when a fundamental understanding of contact angles and surface tensions is known.

*Keywords*: Contact angles; solid surface tensions; self-assembled monolayers; adhesion; surface nanostructures; surface defect; surface crystallinity; low-energy solid surfaces.

## 1. INTRODUCTION

The determination of solid–vapor ($\gamma_{sv}$) and solid–liquid ($\gamma_{sl}$) interfacial tensions is of importance in a wide range of problems in pure and applied science. For example, the process of particle adhesion can be modeled by the net free-energy change $\Delta F^{adh}$ of the system during the adhesion process, which depends explicitly on the solid (particle) surface tensions: the net free-energy change per unit surface area for the adhesion process of a square particle is given by

$$\Delta F^{adh} = \gamma_{pv} - \gamma_{pl} - \gamma_{sl} \tag{1}$$

where $\gamma_{pv}$ and $\gamma_{pl}$ are, respectively, the particle–vapor and particle–liquid interfacial tensions. If $\Delta F^{adh} < 0$, the adhesion process is thermodynamically favorable. The

---

*To whom all correspondence should be addressed. Phone: (1-780) 492-2791, Fax: (1-780) 492-2200, E-mail: daniel.y.kwok@ualberta.ca

work of adhesion $W^{adh}$ is simply

$$W^{adh} = -\Delta F^{adh} \tag{2}$$

Because of the difficulties involved in measuring the surface tension involving a solid phase directly, indirect approaches are called for. Several independent approaches have been used to estimate solid surface tensions, including direct force measurements [1–9], contact angles [10–17], capillary penetration into columns of particle powder [18–21], sedimentation of particles [22–25], solidification front interaction with particles [26–33], film flotation [34–38], gradient theory [39–42], Lifshitz theory of van der Waals forces [42–45] and theory of molecular interactions [46–51]. Among these methods, contact angle measurements are believed to be the simplest.

Contact angle measurement is easily performed by establishing the tangent (angle) of a liquid drop with a solid surface at the base. The attractiveness of using contact angles $\theta$ to estimate the solid–vapor and solid–liquid interfacial tensions is due to the relative ease with which contact angles can be measured on suitably prepared solid surfaces. It is apparent in the literature that this seeming simplicity can be misleading.

The possibility of estimating solid-surface tensions from contact angles relies on a relation which was recognized by Young [52] in 1805. The contact angle of a liquid drop on an ideal solid surface is defined by the mechanical equilibrium of the drop under the action of three interfacial tensions: solid–vapor, $\gamma_{sv}$, solid–liquid, $\gamma_{sl}$, and liquid–vapor, $\gamma_{lv}$. This equilibrium relation is known as Young's equation:

$$\gamma_{lv} \cos \theta_Y = \gamma_{sv} - \gamma_{sl} \tag{3}$$

where $\theta_Y$ is the Young contact angle, i.e., a contact angle which can be inserted into Young's equation. The experimentally accessible contact angles may or may not be equal to $\theta_Y$, as there exist many metastable contact angles which are not equal to $\theta_Y$.

Several contact angle approaches [10–17], of current interest, were largely inspired by the idea of using Young's equation for the determination of surface energetics. While these approaches are, logically and conceptually, mutually exclusive, they share, nevertheless, the following basic assumptions:

(1) All approaches rely on the validity and applicability of Young's equation for surface energetics from experimental contact angles.

(2) Pure liquids are always used; surfactant solutions or mixtures of liquids should not be used, since they would introduce complications due to preferential adsorption.

(3) The values of $\gamma_{lv}$, $\gamma_{sv}$ (and $\gamma_{sl}$) are assumed to be constant during the experiment, i.e. there should be no physical/chemical reaction between the solid and the liquid.

(4) The surface tensions of the test liquids should be higher than the anticipated solid surface tension.

(5) The values of $\gamma_{sv}$ in going from liquid to liquid are also assumed to be constant, i.e., independent of the liquids used.

Recent contact angle studies [53–55] have shown the complexities of contact angle phenomena, which prevent use of contact angle measurements for surface energetic calculations, when low-rate dynamic contact angle measurements are made using an automated drop shape analysis. It was found that there were three apparent contact angle complexities which violated the basic assumptions made in the interpretation of contact angles for solid surface tensions:

(1) slip/stick of the three-phase contact line;

(2) contact angle increases/decreases as the drop front advances; and

(3) liquid surface tension changes as the drop front advances.

These contact angles should not be used for the determination of solid surface tensions. With respect to the first point, slip/stick of the three-phase contact line indicates that Young's equation is not applicable. Increase/decrease in the contact angle and change in the liquid surface tension as the drop front advances violate the expectation of no physical/chemical reaction. Therefore, when the experimental contact angles and liquid surface tensions are not constant, they should be disregarded. After eliminating the meaningless (non-constant) data, experimental contact angles on a large number of polymer surfaces yield smooth curves of $\gamma_{lv} \cos \theta$ *versus* $\gamma_{lv}$, $\cos \theta$ *versus* $\gamma_{lv}$ and $W_{sl}$ *versus* $\gamma_{lv}$ for one and the same solid surface [53–64]. Changing the solid surface shifts the curve in a very regular manner, suggesting the following relations

$$\gamma_{lv} \cos \theta = f_1(\gamma_{lv}, \gamma_{sv}), \tag{4}$$

$$\cos \theta = f_2(\gamma_{lv}, \gamma_{sv}) \quad \text{and} \tag{5}$$

$$W_{sl} = f_3(\gamma_{lv}, \gamma_{sv}) \tag{6}$$

In Fig. 1 we reproduce the adhesion and contact angle patterns for six surfaces: fluorocarbon FC722 [53], hexatriacontane [65–68], cholesteryl acetate [65–67], poly(*n*-butyl methacrylate) [57], poly(methyl methacrylate/*n*-butyl methacrylate) [69] and poly(methyl methacrylate) [58]. This and other results suggest the existence of universal adhesion and contact angle patterns for low-energy solid surfaces. However, the question arises as to whether the low-rate dynamic measurements giving constancy of $\theta$ and $\gamma_{lv}$ will always guarantee that the above contact angle assumptions are fulfilled and the patterns in Fig. 1 are unique.

Since essentially all measurements were performed on relatively thick polymer-coated surfaces, it is legitimate to question if the selection of surfaces was biased. In the present study, we purposely selected thin monomolecular films as our substrates, prepared by self-assembled monolayers (SAMs) of octadecanethiol adsorbed onto gold. The choice of octadecanethiol SAM is due to its popularity and ease of generating different wettabilities [70–75]. To our knowledge, a systematic study of the low-rate dynamic contact angle measurements on SAMs in terms of solid surface tensions has not been performed. A review of literature contact angle data

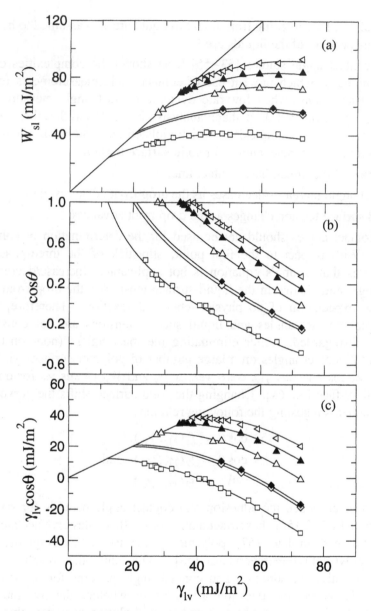

**Figure 1.** (a) The solid–liquid work of adhesion $W_{sl}$, (b) cosine of the contact angle $\cos\theta$ and (c) liquid vapor surface tension times cosine of the contact angle $\gamma_{lv} \cos\theta$ *versus* the liquid–vapor surface tension $\gamma_{lv}$ for a fluorocarbon FC722 ($\square$), hexatriacontane ($\diamond$), cholesteryl acetate ($\blacklozenge$), poly($n$-butyl methacrylate) ($\triangle$), poly(methyl methacrylate/$n$-butyl methacrylate) ($\blacktriangle$) and poly(methyl methacrylate) ($\triangleleft$) surfaces.

of octadecanethiol $CH_3(CH_2)_{17}SH$ SAM on Au suggests that the dependence of $W_{sl}$, $\gamma_{lv} \cos\theta$ and $\cos\theta$ on $\gamma_{lv}$ is not unique and is different from those shown in Fig. 1. The question arises as to whether the patterns in Fig. 1 do not exist for these monomolecular films or are there other reasons which complicate our interpretation.

The aim of this work is to investigate if the relationships and patterns shown in Fig. 1 exist only for the thick polymer surfaces and not for the relatively thin monolayers by means of low-rate dynamic contact angle measurements. As a second objective, we have also studied how surface nanostructure and defect of octadecanethiol SAM adsorbed onto gold affect wetting in terms of solid surface tension interpretation.

## 2. EXPERIMENTAL

### 2.1. Materials

Silicon wafers of test grade were obtained from Wafer World (West Palm Beach, FL, USA) in circular discs of about 10-cm diameter and were cut into rectangular shapes of about 2.5 cm × 5 cm. Gold shot (99.999%) and titianium shot (99.995%) were obtained from Kurt J. Lesker (Clairton, PA, USA). Ethanol (100%) was obtained from the Chemistry Dept. at the University of Alberta. Octadecanethiol $(CH_3(CH_2)_{17}SH)$ was obtained from Aldrich and used as received. Six liquids were chosen for contact angle measurements. Selection was based on the following criteria: (1) they should include a wide range of intermolecular forces; (2) they should be non-toxic; and (3) the liquid surface tension should be higher than the anticipated solid surface tension [10, 14, 21, 76]. They are listed in Table 1.

### 2.2. Preparation of SAMs

Supported gold films were prepared by sequentially evaporating titanium (approx. 10 nm) and gold (approx. 100 nm) onto small rectangular silicon wafers in a diffusion-pumped vacuum chamber at about $10^{-6}$ torr. The chamber was backfilled with air and the substrates were used within 48 h of preparation. The evaporated surfaces were rinsed with ethanol before SAMs formation. SAMs were prepared by immersing into 1 mM of $CH_3(CH_2)_{17}SH$ in ethanol overnight. The resulting surfaces were rinsed with ethanol and blown dry by nitrogen before use. Evaporated gold substrates were also flame annealed for about 30 s using a Bunsen burner under

**Table 1.**
Experimental advancing and receding contact angles on SAMs of octadecanethiol $CH_3(CH_2)_{17}SH$ adsorbed onto Au

| Liquid | $\gamma_{lv}$ (mJ/m$^2$) | $\theta_a$ (deg) | $\theta_r$ (deg) |
|---|---|---|---|
| water | 72.70 ± 0.09 | 119.1 ± 0.8 | 100.2 ± 0.7 |
| formamide | 59.08 ± 0.01 | 88.7 ± 0.8 | 63.0 ± 1.4 |
| ethylene glycol | 47.55 ± 0.02 | 81.5 ± 0.6 | 66.4 ± 1.1 |
| bromonaphthalene | 44.31 ± 0.05 | 67.2 ± 0.8 | 44.1 ± 0.8 |
| decanol | 28.99 ± 0.01 | 50.7 ± 0.5 | 38.2 ± 1.1 |
| hexadecane | 27.62 ± 0.01 | 45.4 ± 0.4 | < 20.0 |

Error bars are the 95% confidence limits.

ambient laboratory condition. After approx. 1 min, the annealed substrate was then immersed into 1 mM of $CH_3(CH_2)_{17}SH$ in ethanol overnight.

## 2.3. Characterization of SAMs

SAMs were first characterized by a Sopra GES5 Variable Angle Spectroscopic Ellipsometer. The ellipsometry measurements were performed using a rotating polarizer in the tracking analyzer mode. A broad band of light (300 to 850 nm) from a 75 W Xe-arc lamp is linearly polarized and directed onto the film surface at an incident angle of 75° from the surface normal. The $\tan \Psi$ and $\cos \Delta$ for each bare gold substrate were measured as references immediately after evaporation. After immersion into 1 mM of octadecanethiol/ethanol solution overnight, a new set of $\tan \Psi$ and $\cos \Delta$ for each substrate were measured again using an ambient-film-substrate model for regression with known refractive index ($n$ and $k$) for octadecanethiol adsorbed onto gold. The refractive index for octadecanethiol adsorbed onto gold as a function of wavelength was independently obtained from a Sopra GXR Grazing X-ray Reflectometer, rather than assuming an index of refraction (e.g., $n = 1.46$) at a given wavelength (e.g., $\lambda = 6328$ Å) as typically performed in the literature. Such spectroscopic measurements are expected to provide more accurate results in ellipsometer thickness since the optical constants for a range of wavelengths were used simultaneously. The thickness was calculated according to the following equation:

$$(\tan \Psi)e^{(i\Delta)} = f(n_i k_i T_i),$$

where $n$ and $k$ are the optical constants of the film and $T$ is its thickness; subscript $i$ represents different wavelength. The averages of three measurements made at each location on the sample were used to calculate the thickness of each sample.

Reflectance Infrared spectra of SAMs of octadecanethiol onto Au were obtained using a ThermoNicolet Nexus 670 spectrometer equipped with a VeeMax grazing angle accessory. A p-polarized light was incident at 70° from the surface normal and the reflected light was detected by means of an MCT-A detector cooled with liquid nitrogen. The spectra resolution was 2 $cm^{-1}$. Spectra were referenced to the corresponding bare (anneal and non-annealed) Au substrates and 1024 scans were obtained for good signal-to-noise ratios. An infrared gain of 2 was selected for all reflectance infrared (IR) measurements to ensure that the input IR signals were constant. Samples were rinsed with ethanol and blown dry by $N_2$ prior to characterization.

The Atomic Force Microscopy (AFM) measurements were performed using a Digital Instruments Nanoscope IIIa atomic force microscope (Digital Instruments, Santa Barbara, CA, USA). Standard silicon nitride cantilevered probes were used with a force/spring constant in the range between 0.06–0.58 N/m. The AFM images of annealed and non-annealed Au surfaces were captured by using contact mode under ambient laboratory conditions. The surfaces were cut into 1 cm × 1 cm samples to fit onto a 1.5 cm × 1.5 cm sample stage.

## 2.4. Contact angle measurements

Contact angle measurements were chosen as the last step for the characterization of SAMs. A Linux version of the Axisymmetric drop shape analysis-profile (ADSA-P) was used for sessile drop contact angle measurements. ADSA-P is a technique to determine liquid–fluid interfacial tensions and contact angles from the shapes of axisymmetric menisci, i.e., from sessile as well as pendant drops [77, 78]. Assuming that the experimental drop is Laplacian and axisymmetric, ADSA-P finds a theoretical profile that best matches the drop profile extracted from an image of a real drop, from which the surface tension, contact angle, drop volume, surface area and three-phase contact radius can be computed. The strategy employed is to fit the shape of an experimental drop to a theoretical drop profile according to the Laplace equation of capillarity, using surface/interfacial tension as an adjustable parameter. The best fit identifies the correct surface/interfacial tension from which the contact angle can be determined by a numerical integration of the Laplace equation. Details of the methodology and experimental set-up can be found elsewhere [53, 54, 77–79].

Sessile drop experiments were performed by ADSA-P to determine the advancing and receding contact angles. The temperature and relative humidity were maintained, respectively, at $23.0 \pm 0.5°C$ and at about 40%, by means of an independent central air-conditioning unit in the laboratory. It has been found that, since ADSA-P assumes an axisymmetric drop shape, the values of liquid surface tensions measured from sessile drops are very sensitive to even a very small degree of surface imperfection, such as roughness and heterogeneity, while contact angles are less sensitive. Therefore, the liquid surface tensions used in this study were independently measured by applying ADSA-P to a pendant drop, since the axisymmetry of the drop is enforced by using a circular capillary. Results of the liquid surface tension from previous studies [53, 57, 58] are reproduced in Table 1.

In this study, at least 5 and up to 15 dynamic contact angle measurements at velocities of the three-phase contact line in the range from 0.1 to 1.0 mm/min were performed for each liquid. The choice of this velocity range was based on previous studies [53, 54, 80, 81] which showed that low-rate dynamic contact angles at these velocities were essentially identical to the static contact angles, for these relatively smooth surfaces. Liquids were supplied from below the surface through a hole of approx. 1 mm in diameter on the substrate by means of a motorized-syringe system. Details of this setup can be found elsewhere [53, 54].

## 3. RESULTS AND DISCUSSION

### 3.1. Formation of SAMs on polycrystalline gold

Figure 2 displays typical low-rate dynamic contact angles of water on SAM of octadecanethiol $CH_3(CH_2)_{17}SH$ adsorbed onto Au. As can be seen, increasing the drop volume $V$ linearly from 0.11 cm$^3$ to 0.12 cm$^3$ increases the apparent contact

*J. Yang* et al.

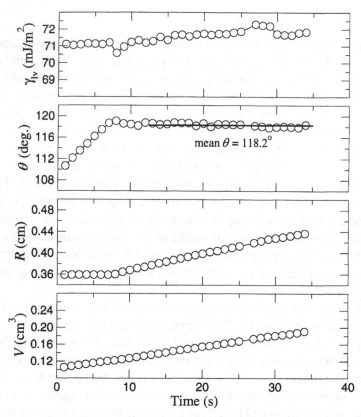

**Figure 2.** Low-rate dynamic contact angles of water on SAMs of octadecanethiol $CH_3(CH_{17})_2SH$ adsorbed onto Au.

angle $\theta$ from ca 110 to 118° at essentially constant three-phase contact radius $R$. This is due to the fact that even carefully placing an initial water drop from above on a solid surface can result in a contact angle somewhere between advancing and receding. Further increase in the drop volume causes the three-phase contact line to advance, with $\theta$ essentially constant as $R$ increases. Increasing the drop volume in this manner ensures the measured $\theta$ to be an advancing contact angle. Since the contact angles are essentially constant after $R = 0.37$ cm and according to Ref. [54], these contact angles can be used for the determination of solid surface tensions. The averaged contact angle of 118° in Fig. 2 is consistent with the literature values [82, 83]. The experimental results for the other five liquids also yield essentially constant contact angles; these results together with the receding angles are summarized in Table 1. The solid–liquid work of adhesion $W_{sl}$, $\cos\theta$ and $\gamma_{lv}\cos\theta$ *versus* $\gamma_{lv}$ for these liquids are plotted in Fig. 3. However, there are significant scatters in these data even though the procedures of low-rate dynamic contact angles were used to distinguish meaningful angles from meaningless ones. Comparison of this figure with Fig. 1 suggests that the two patterns are different. The question arises as

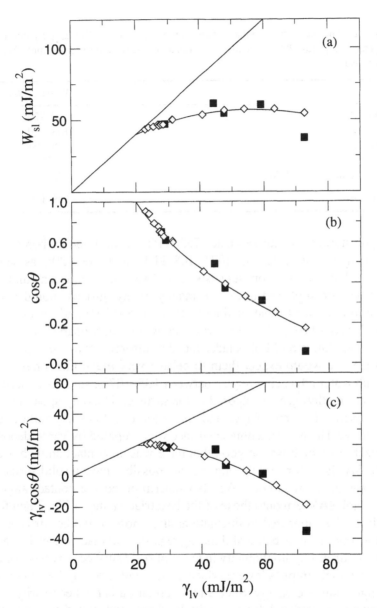

**Figure 3.** (a) The solid–liquid work of adhesion $W_{sl}$, (b) cosine of the contact angle $\cos\theta$ and (c) liquid–vapor surface tension times cosine of the contact angle $\gamma_{lv}\cos\theta$ *versus* the liquid–vapor surface tension $\gamma_{lv}$ for hexatriacontane (◇) and SAMs of octadecanethiol $CH_3(CH_2)_{17}SH$ adsorbed onto Au (■).

to whether or not the patterns shown in Fig. 1 are unique only for thick polymer surfaces and not for other surfaces such as monolayers.

From the point of view of surface energetics, we note that a water contact angle of 119.1° (Table 1) for a surface exposing purely methyl groups ($CH_3(CH_2)_{17}S/Au$) appears to be too high. For example, the experimental advancing contact angles of

**Table 2.**
Comparison of expected water contact angles on fluorocarbon, PTFE, polystyrene (PS) and poly(methyl methacrylate) (PMMA) with that measured on SAMs of octadecanethiol $CH_3(CH_2)_{17}SH$ adsorbed onto Au

| Surface | Water contact angle (deg) |
|---|---|
| Fluorocarbon | 118–120 |
| PTFE | 108–110 |
| Hexatriacontane | 105–107 |
| Polystyrene, PS | 90–92 |
| Poly(methyl methacrylate), PMMA | 73–75 |
| $CH_3(CH_2)_{17}S/Au$ | 118–120 |

water on polystyrene, hexatriacontane, Teflon PTFE and fluorocarbon are expected to be, respectively, 90–92°, 105–107°, 108–110°, and 118–120°, as summarized in Table 2. This comparison suggests that SAMs of octadecanethiol adsorbed onto Au which are supposed to expose mainly methyl groups should have a solid surface tension similar to that of fluorocarbon (118–120°) and lower than that of Teflon PTFE (108–110°). These interpretations are questionable. From a surface energetic viewpoint, if SAM of octadecanethiol adsorbed onto Au exposes only the methyl group, we would expect them to behave very much like a hexatriacontane surface (with water advancing angles between 105–107°), since both surfaces have predominately methyl groups exposed to the surface. The only apparent difference is that the former is a monolayer and the latter is a thick crystalline and well-ordered surface. The hexatriacontane surface was prepared by vapor deposition the surface quality of which was so good that there was no contact angle hysteresis for water [67]. SAMs with methyl groups cannot possibly have a solid surface tension as low as that of fluorocarbon. We also superimposed the contact angle data of octadecanethiol SAM/Au onto those of the hexatriacontane in Fig. 3 and found that the monolayer data appeared to fluctuate around those of the hexatriacontane. We wish to point out that conventional thinking regarding contact angles is that they are indicators for surface hydrophobicity in terms of solid surface tensions; for example, higher contact angle implies lower surface energy and similarly lower contact angle suggests higher surface energy. Such an interpretation is not necessarily conclusive for the systems considered here as will be illustrated and discussed later. We speculate that the discrepancy may come from additional and unexpected effect(s) of changing solid–liquid interfacial tensions on the contact angles through the variation of surface structures, even though the solid–vapor surface tension might have been constant. If $\gamma_{sv}$ and $\gamma_{lv}$ are constant, changes in $\theta$ suggest that $\gamma_{sl}$ is changing. In the next section, we will quantify such effects by looking into the surface structures of octadecanethiol SAMs adsorbed onto Au.

It is commonly known in the literature [84, 85] that thermally evaporated gold yields smoother and better polycrystalline structures than that by sputtering. It has also been found that SAMs on thermally annealed gold have larger terraces with

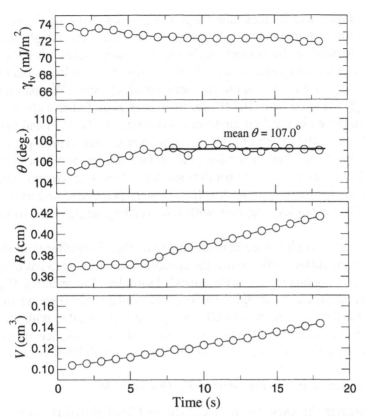

**Figure 4.** Low-rate dynamic contact angles of water on SAMs of octadecanethiol $CH_3(CH_2)_{17}SH$ adsorbed onto annealed gold.

less surface defects [86, 87]. Thus, SAMs on annealed gold are typically used in Atomic Force Microscopy (AFM) study to obtain atomic resolution pictures. As a matter of fact, most contact angle studies of SAMs on gold were prepared either by sputtering or thermal evaporation due to the relatively simple procedures [82, 88–90]. To our knowledge, no systematic contact angle study on SAMs has yet been performed and has looked into the details of how surface structures affect wetting in terms of surface energetic interpretation using thermally annealed gold. Thus, in the next section we will investigate low-rate dynamic contact angles on octadecanethiol monolayers formed on thermally annealed gold, in an attempt to isolate any (possible) effects of surface structures and defects on solid surface tension interpretation.

### 3.2. Formation of SAMs on annealed gold

Figure 4 shows the low-rate dynamic contact angle results of water on SAMs of octadecanethiol adsorbed onto a thermally-annealed-gold substrate. Similar to the experimental results in Fig. 2, as $V$ increases at the beginning, $\theta$ increases at constant $R$. This is due to the fact that such contact angles are not truly advancing angles and it takes time for the drop front to advance. As $V$ increases further, the three-phase contact radius $R$ moves and the contact angle remains rather constant. Averaging the contact angle yields a mean value of 107.0°. We see that this water contact angle value (107.0°) is significantly lower than that shown in Fig. 2 (118.2°) on the evaporated (non-annealed) gold. We also note that the contact angle obtained here (107.0°) for the $CH_3(CH_2)_{17}S$/thermally-annealed-gold is similar to those obtained on the hexatriacontane (cf., Table 2) and paraffin. This result agrees with the expectation that a monolayer surface exposing predominately methyl groups should have a similar solid surface tension and hence contact angle as those of hexatriacontane and paraffin surfaces. Low-rate dynamic contact angle measurements for the remaining five liquids were performed and found to be also constant. These angles, together with the receding angles, are summarized in Table 3.

It is of interest to plot these contact angles in Fig. 5 together with those on the hexatriacontane surface. We see that the apparent scatter in Fig. 3 has disappeared in Fig. 5 using the contact angles on the annealed samples. The resulting $W_{sl}$, $\cos \theta$ and $\gamma_{lv} \cos \theta$ versus $\gamma_{lv}$ curves are quite smooth and similar to those shown in Fig. 1. We conclude that the experimental results in Fig. 5 are compatible with the functional relationships in equations (4)–(6). The origin of the difference in experimental patterns between the non-annealed and annealed gold will be discussed below.

### 3.3. Characterizations by ellipsometry, FT-IR and AFM

The ellipsometer thickness for octadecanethiol $CH_3(CH_2)_{17}SH$ adsorbed onto annealed gold was consistently 21 Å; whereas that formed on the non-annealed Au varied between 20–21 Å. While this difference is not statistically significant, we noticed that the experimental $\tan \Psi$ and $\cos \Delta$ for the non-annealed-Au samples did not always match those of the theoretical curves. Nevertheless, our thicknesses are consistent with those reported in the literature [83].

The reflectance spectra for SAMs derived from octadecanethiol on Au and annealed Au are shown in Fig. 6. In the both spectra of Fig. 6, the asymmetric

**Table 3.**

Experimental advancing and receding contact angles on SAMs of octadecanethiol $CH_3(CH_2)_{17}SH$ adsorbed onto evaporated (non-annealed) and annealed gold

| Liquid | $\gamma_{lv}$ | Non-annealed | | Annealed | |
|---|---|---|---|---|---|
| | | $\theta_a$ (deg) | $\theta_r$ (deg) | $\theta_a$ (deg) | $\theta_r$ (deg) |
| water | 72.7 | $119.1 \pm 0.8$ | $100.2 \pm 0.7$ | $106.9 \pm 0.5$ | $92.3 \pm 0.9$ |
| formamide | 59.1 | $88.7 \pm 0.8$ | $63.0 \pm 1.4$ | $92.4 \pm 1.5$ | $69.2 \pm 1.9$ |
| ethylene glycol | 47.6 | $81.5 \pm 0.6$ | $66.4 \pm 1.1$ | $81.6 \pm 2.4$ | $68.2 \pm 1.6$ |
| bromonaphthalene | 44.3 | $67.2 \pm 0.8$ | $44.1 \pm 0.8$ | $76.1 \pm 0.9$ | $64.3 \pm 1.3$ |
| decanol | 28.9 | $50.7 \pm 0.5$ | $38.2 \pm 1.1$ | $53.2 \pm 0.9$ | $45.1 \pm 1.3$ |
| hexadecane | 27.6 | $45.4 \pm 0.4$ | $< 20.0$ | $45.7 \pm 0.8$ | $35.4 \pm 2.2$ |

Error bars are the 95% confidence limits.

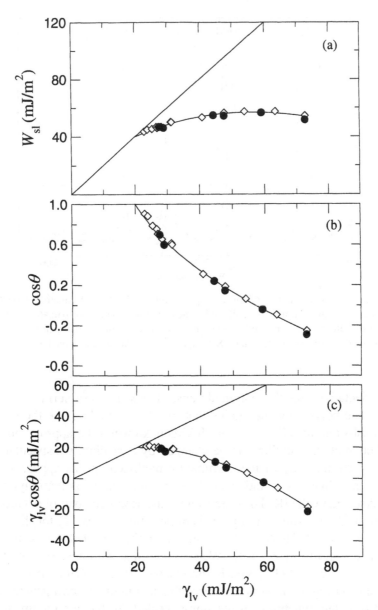

**Figure 5.** (a) The solid–liquid work of adhesion $W_{sl}$, (b) cosine of the contact angle $\cos\theta$ and (c) liquid–vapor surface tension times cosine of the contact angle $\gamma_{lv}\cos\theta$ *versus* the liquid–vapor surface tension $\gamma_{lv}$ for hexatriacontane ($\Diamond$) and SAMs of octadecanethiol $CH_3(CH_2)_{17}SH$ adsorbed onto thermally annealed Au ($\bullet$).

methylene peaks appeared at about 2918 cm$^{-1}$. This indicates a primarily *trans-* zigzag-extended hydrocarbon chain with few *gauche* conformers. Both spectra demonstrate that SAMs of octadecanethiol adsorbed onto Au and annealed Au are highly crystalline. However, the intensities of the methylene peaks are larger on Au and smaller on the annealed Au. The difference in the peak intensity could reflect

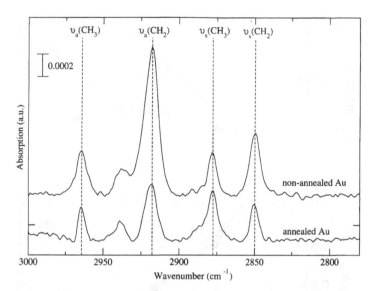

**Figure 6.** Grazing incidence polarized infrared spectra for SAMs of octadecanethiol $CH_3(CH_2)_{17}SH$ adsorbed onto evaporated (non-annealed) and annealed gold. The approximate positions of the methylene modes are 2918 (asym) and 2850 (sym) $cm^{-1}$, and those for the methyl modes are 2964 (asym), 2935 (sym, Fermi resonance) and 2879 (sym) $cm^{-1}$. The spectra have been offset vertically for clarity.

different orientations for the polymethylene chains on these surfaces or different amounts of the polymethylene chains that the IR detected. Since the tilt of chain on the Au substrate for alkanethiolate SAMs is known to be approx. 30° and this structural orientation is unlikely to be changed by annealing, we speculate that the difference in the intensity of the asymmetric methylene peaks appearing at about 2918 $cm^{-1}$ and 2850 $cm^{-1}$ indicates different amounts of polymethylene chains that were detected by the IR. The spectrum for the adsorbed layer of octadecanethiol on non-annealed-Au exhibits a higher dichroic ratio (approx. 2) for the methylene absorption modes ($\nu_a(CH_2)/\nu_s(CH_2)$); and that on the annealed-Au exhibits a much lower dichroic ratio (approx. 1.3). We also note that the intensities of the methyl modes at 2964 (asym), 2935 (sym, Fermi resonance) and 2879 (sym) $cm^{-1}$ for the two substrates in Fig. 6 are similar. These features in the spectra provide evidence that SAM of octadecanethiol on the non-annealed Au has a structure that is not the same as that on the annealed Au. Independent AFM images shown in Fig. 7 suggest that the annealed Au has larger terraces (as much as 200 nm); while the non-annealed Au has much smaller gold steps. From the interpretation of the above IR and AFM results, we constructed a model in Fig. 8 that illustrates a possible arrangement of octadecanethiol adsorbed onto non-annealed and annealed Au. From the schematic, it is expected that the reflectance IR would detect more methylenes per unit projected area on the non-annealed Au than that for the annealed Au. This is due to the polycrystalline nature of non-annealed Au that causes variation of the methyl and methylene groups exposed to water. The schematic

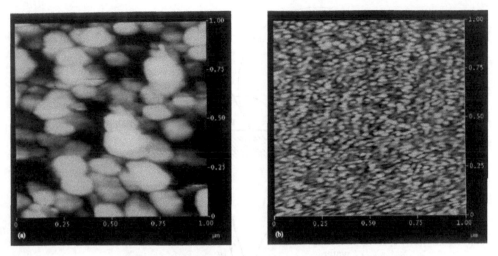

**Figure 7.** AFM images of (a) annealed Au (b) non-annealed Au for a scan size of 1 μm.

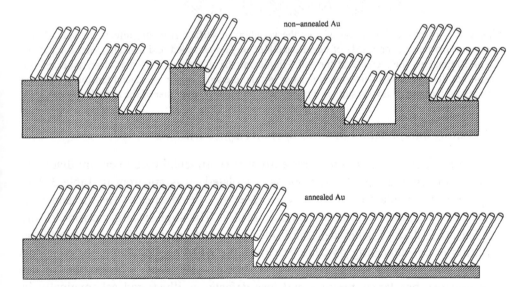

**Figure 8.** Schematic illustration of SAM assembly on two different Au substrates. The top part demonstrates SAM assembly of octadecanethiol adsorbed onto non-annealed Au with smaller gold steps. The bottom part illustrates SAM assembly of octadecanethiol adsorbed onto annealed Au with larger terraces.

also supports the IR results that the intensities for the methyl absorption peaks on both substrates should be similar, as the amount of methyl groups exposed to water per unit projected area would be more or less similar.

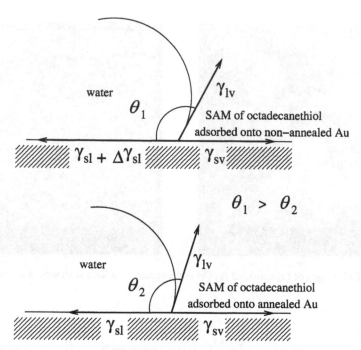

**Figure 9.** Schematic illustration of how the variation of surface structure affects the solid–liquid interfacial tension $\gamma_{sl}$ for SAMs of octadecanethiol on non-annealed and annealed Au. $\Delta\gamma_{sl}$ in the top part represents the increase in the solid–liquid interfacial tension from that of an annealed Au. Given that the liquid–vapor interfacial tension $\gamma_{lv}$ and the solid–vapor interfacial tension $\gamma_{sv}$ remain unchanged, increase in $\Delta\gamma_{sl}$ results in a higher contact angle $\theta_1$ than that of the annealed Au, $\theta_2$.

### 3.4. Wetting interpretation in terms of solid surface tensions

Contact angle interpretation requires extreme experimental care to ensure that none of the commonly accepted assumptions is violated. It is apparent in Table 3 that, in general, the contact angle hysteresis $H = \theta_a - \theta_r$ decreases for the annealed gold substrate, suggesting better surface quality. The slightly larger errors for the contact angle data of octadecanethiol adsorbed onto annealed Au were due to the variation of our annealing procedures. From the AFM results above, we conclude that the non-annealed surface consists of smaller gold steps, whereas that of the annealed one has larger terraces and less defects, as illustrated schematically in Fig. 8. In the case of the non-annealed samples, water could "see" a deeper surface layer (the methylenes) under the methyl groups due to the relatively less packed monolayers as a result of smaller gold steps or defects. This allows formation of additional intermolecular interactions between water and the methylenes in the solid–liquid interface. In this case, the surface no longer consists of predominately methyl groups, but a mixture of methyl and methylene groups in the solid–liquid interfacial region. These molecular interactions with methylenes would result in a higher, additional, interfacial tension between the solid–liquid interface, $\Delta\gamma_{sl}$, and is illustrated schematically in Fig. 9. This additional effect is believed to have less influence on the solid–vapor interface, due to presence of the relatively less dense water vapor on the solid–vapor interface. Assuming that $\gamma_{sv}$ is roughly constant for both annealed and non-annealed surfaces, increase in $\gamma_{sl}$ for the latter would cause the contact angle to increase from that of a surface where water can only "see" most of the outer methyl groups and less methylene groups. Thus, it appears that the change in the contact angle is a result of additional and unexpected formation of intermolecular interactions at the solid–liquid interface for less packed monolayers. The magnitude would depend, of course, on the intermolecular interactions strength, polarity of the liquid and structures of the monolayers. In this case, even though we apparently fix our substrate, the intermolecular interactions would not be the same for a given pair of liquid and solid, depending on surface defect, structure, polycrystallinity and arrangement of monolayers. In Table 3, we also note that the advancing contact angles for the annealed and non-annealed surfaces are nearly identical and did not change very much, for both ethylene glycol and hexadecane. However, the molecular interactions between hexadecane/$CH_3(CH_2)_{17}SH$ and ethylene glycol/$CH_3(CH_2)_{17}SH$ on two different Au surfaces are not clearly known. We conclude that, whatever these interactions are, they appear to be insensitive to the structural changes and surface defects of the monolayers. If the intermolecular interactions change drastically as we change the substrate surface preparation for a given liquid, this violates our expectation of constant solid properties along the curves shown in Figs 1, 3 and 5. Interpretation of such angles on the non-annealed surfaces would be difficult [91–93]. Since the annealed surfaces contain less defects, additional variation of the solid–liquid interfacial interactions $\Delta\gamma_{sl}$ would be minimal. We conclude that the surface structures of the underlying gold on which the monolayers of octadecanethiol are formed can affect directly the interpretation of contact angles. Only when a fundamental framework such as that shown in Figs 1 or 5 is established, the variation of the interfacial interactions from liquid to liquid as an additional effect can be studied more systematically. With the above stipulation, the variation of $\gamma_{sl}$ as a result of surface structural change can be estimated if we assume $\gamma_{sv}$ for the non-annealed and annealed Au to be the same. Therefore, a $\Delta\gamma_{sl}$ of 12.9 mJ/m$^2$ can be estimated by taking the difference in the water advancing angles, $\Delta\gamma_{sl} = (72.7$ $(\cos 107 - \cos 118))$, using Young's equation. This value represents an increase in $\gamma_{sl}$ due to the variation of surface structures of octadecanethiol SAM adsorbed onto Au when interacting with water. However, conventional thinking would have interpreted the water contact angle of 118° for octadecanethiol SAM on Au to have a much lower solid surface tension similar to that of fluorocarbon ($\gamma_{sv} \approx 12$ mJ/m$^2$), rather than a $\gamma_{sv}$ of 19–20 mJ/m$^2$ for methyl-terminated surfaces.

## 4. CONCLUSIONS

We conclude that surface defect, structure and crystallinity of Self-Assembled Monolayers (SAMs) of octadecanethiol adsorbed onto gold can affect the interpretation of contact angles in terms of solid surface tensions. The contact angle and adhesion patterns of various liquids for SAMs of octadecanethiol adsorbed onto annealed gold are consistent with recent experimental findings on the relatively thick polymer-coated surfaces. We also found that interpretation of solid surface tensions using contact angle data on less well-prepared polycrystalline gold surfaces can be misleading. The variation of surface structure in terms of surface energetics can be estimated only when a fundamental understanding of contact angles and surface tensions is known.

## Acknowledgements

This research was supported financially, in part, by the Natural Sciences and Engineering Research Council (NSERC) of Canada, Canada Research Chair Program (CRCP) and Canada Foundation for Innovation (CFI). J.Y. acknowledges support from the Province of Alberta through an Alberta Ingenuity studentship. D.Y.K. thanks Professor P. E. Laibinis, Dr. R. Michalitsch and Dr. N. Kim for helpful discussions and Professor J. Malisyah for letting us to use his AFM.

## REFERENCES

1. B. V. Derjaguin, V. M. Muller and Yu. P. Toporov, *J. Colloid Interface Sci.* **73**, 293 (1980).
2. K. L. Johnson, K. Kendall and A. D. Roberts, *Proc. R. Soc. (London)* **A324**, 301 (1971).
3. V. M. Muller, V. S. Yushchenko and B. V. Derjaguin, *J. Colloid Interface Sci.* **92**, 92 (1983).
4. A. Fogden and L. R. White, *J. Colloid Interface Sci.* **138**, 414 (1990).
5. R. M. Pashley, P. M. McGuiggan, R. G. Horn and B. W. Ninham, *J. Colloid Interface Sci.* **126**, 569 (1988).
6. H. K. Christenson, *J. Phys. Chem.* **90**, 4 (1986).
7. P. M. Claesson, C. E. Blom, P. C. Horn and B. W. Ninham, *J. Colloid Interface Sci.* **114**, 234 (1986).
8. P. M. Pashley, P. M. McGuiggan and R. M. Pashley, *Colloids Surfaces* **27**, 277 (1987).
9. R. M. Pashley, P. M. McGuiggan, B. W. Ninham and D. F. Evans, *science* **229**, 1088 (1985).
10. W. A. Zisman, in: *Contact Angle, Wettability and Adhesion, Advances in Chemistry Series, No. 43*. American Chemical Society, Washington, DC (1964).
11. F. M. Fowkes, *Ind. Eng. Chem.* **56(12)**, 40 (1964).
12. O. Driedger, A. W. Neumann and P. J. Sell, *Kolloid - Z. Z. Polym.* **201**, 52 (1965).
13. A. W. Neumann, R. J. Good, C. J. Hope and M. Sejpal, *J. Colloid Interface Sci.* **49**, 291 (1974).
14. J. K. Spelt and D. Li, in: *Applied Surface Thermodynamics*, J. K. Spelt and A. W. Neumann (Eds), pp. 239–292. Marcel Dekker, New York, NY (1996).
15. D. K. Owens and R. C. Wendt, *J. Appl. Polym. Sci.* **13**, 1741 (1969).
16. C. J. van Oss, M. K. Chaudhury and R. J. Good, *Chem. Rev.* **88**, 927 (1988).
17. R. J. Good and C. J. van Oss, in: *Modern Approaches to Wettability: Theory and Applications*, M. Schrader and G. Loeb (Eds), pp. 1–27. Plenum Press, New York, NY (1992).
18. H. G. Bruil, *Colloid Polym. Sci* **252**, 32 (1974).
19. G. D. Cheever, *J. Coating Technol.* **55**, 53 (1983).
20. H. W. Kilau, *Colloids Surfaces* **26**, 217 (1983).
21. K. Grundke, T. Bogumil, T. Gietzelt, H.-J. Jacobasch, D. Y. Kwok and A. W. Neumann, *Progr. Colloid Polym. Sci.* **101**, 58 (1996).

22. E. I. Vargha-Butler, T. K. Zubovits, D. R. Absolom and A. W. Neumann, *J. Dispersion Sci. Technol.* **6(3)**, 357 (1985).
23. E. I. Vargha-Butler, E. Moy and A. W. Neumann, *Colloids Surfaces* **24**, 315 (1987).
24. E. I. Vargha-Butler, T. K. Zubovits, D. R. Absolom and A. W. Neumann, *Chem. Eng. Commun.* **33**, 255 (1985).
25. D. Li and A. W. Neumann, in: *Applied Surface Thermodynamics*, J. K. Spelt and A. W. Neumann (Eds), pp. 509–556. Marcel Dekker, New York, NY (1996).
26. S. N. Omenyi and A. W. Neumann, *J. Appl. Phys.* **47**, 3956 (1976).
27. D. Li and A. W. Neumann, in: *Applied Surface Thermodynamics*, J. K. Spelt and A. W. Neumann (Eds), pp. 557–628. Marcel Dekker, New York, NY (1996).
28. A. E. Corte, *J. Geophys. Res.* **67**, 1085 (1962).
29. P. Hoekstra and R. D. Miller, *J. Colloid Interface Sci.* **25**, 166 (1967).
30. J. Cissé and G. F. Bolling, *J. Crystal Growth* **10**, 67 (1971).
31. J. Cissé and G. F. Bolling, *J. Crystal Growth* **11**, 25 (1971).
32. A. M. Zubko, V. G. Lobonov and V. V. Nikonova, *Sov. Phys. Crystallogr.* **18**, 239 (1973).
33. K. H. Chen and W. R. Wilcox, *J. Crystal Growth* **40**, 214 (1977).
34. D. W. Fuerstenau and M. C. Williams, *Colloids Surfaces* **22**, 87 (1987).
35. D. W. Fuerstenau and M. C. Williams, *Part. Character.* **4**, 7 (1987).
36. D. W. Fuerstenau and M. C. Williams, *Int. J. Miner. Process.* **20**, 153 (1987).
37. D. W. Fuerstenau, M. C. Williams, K. S. Narayanan, J. L. Diao and R. Urbina, *Energy Fuels* **2**, 237 (1988).
38. D. W. Fuerstenau, J. Diao and J. Hanson, *Energy Fuels* **4**, 34 (1990).
39. S. J. Hemingway, J. R. Henderson and J. R. Rowlinson, *Faraday Symp. Chem. Soc.* **16**, 33 (1981).
40. R. Guermeur, F. Biquard and C. Jacolin, *J. Chem. Phys.* **82**, 2040 (1985).
41. B. S. Carey, L. E. Scriven and H. T. Davis, *AIChE J.* **26**, 705 (1980).
42. E. Moy and A. W. Neumann, in: *Applied Surface Thermodynamics*, J. K. Spelt and A. W. Neumann (Eds), pp. 333–378. Marcel Dekker, New York, NY (1996).
43. H. C. Hamaker, *Physica* **4**, 1058 (1937).
44. J. N. Israelachvili, *Proc. R. Soc. London A.* **331**, 39 (1972).
45. A. E. van Giessen, D. J. Bukman and B. Widom, *J. Colloid Interface Sci.* **192**, 257 (1997).
46. T. M. Reed, *J. Phys. Chem.* **55**, 425 (1955).
47. B. E. F. Fender and G. D. Halsey, Jr., *J. Chem. Phys.* **36**, 1881 (1962).
48. D. E. Sullivan, *J. Chem. Phys.* **74**, 2604 (1981).
49. D. V. Matyushov and R. Schmid, *J. Chem. Phys.* **104**, 8627 (1996).
50. D. Y. Kwok and A. W. Neumann, *J. Phys. Chem. B* **104**, 741 (2000).
51. J. Zhang and D. Y. Kwok, *J. Phys. Chem. B* **106**, 12594 (2002).
52. T. Young, *Philos. Trans. R. Soc. London* **95**, 65 (1805).
53. D. Y. Kwok, R. Lin, M. Mui and A. W. Neumann, *Colloids Surfaces A* **116**, 63 (1996).
54. D. Y. Kwok, T. Gietzelt, K. Grundke, H.-J. Jacobasch and A. W. Neumann, *Langmuir* **13**, 2880 (1997).
55. D. Y. Kwok, C. N. C. Lam, A. Li, A. Leung, R. Wu, E. Mok and A. W. Neumann, *Colloids Surfaces A* **142**, 219 (1998).
56. D. Y. Kwok, C. N. C. Lam, A. Li, A. Leung and A. W. Neumann, *Langmuir* **14**, 2221 (1998).
57. D. Y. Kwok, A. Leung, A. Li, C. N. C. Lam, R. Wu and A. W. Neumann, *Colloid Polym. Sci.* **276**, 459 (1998).
58. D. Y. Kwok, A. Leung, C. N. C. Lam, A. Li, R. Wu and A. W. Neumann, *J. Colloid Interface Sci.* **206**, 44 (1998).
59. D. Y. Kwok, C. N. C. Lam, A. Li, K. Zhu, R. Wu and A. W. Neumann, *Polym. Eng. Sci.* **38**, 1675 (1998).
60. D. Y. Kwok, A. Li, C. N. C. Lam, R. Wu, S. Zschoche, K. Pöschel, T. Gietzelt, K. Grundke, H.-J. Jacobasch and A. W. Neumann, *Macro. Chem. Phys.* **200(5)**, 1121 (1999).

61. D. Y. Kwok, C. N. C. Lam and A. W. Neumann, *Colloid J.* **62(3)**, 324 (2000).
62. D. Y. Kwok, R. Wu, A. Li and A. W. Neumann, *J. Adhesion Sci. Technol.* **14(5)**, 719 (2000).
63. D. Y. Kwok, A. Li and A. W. Neumann, *J. Polym. Sci. B: Polym. Phys.* **16**, 2039 (1999).
64. O. I. del Río, D. Y. Kwok, R. Wu, J. M. Alvarez and A. W. Neumann, *Colloids Surfaces A* **143**, 197 (1998).
65. G. H. E. Hellwig and A. W. Neumann, in: *Proc. 5th Int. Congr. on Surface Activity, Section B*, p. 727, Barcelona (1968).
66. G. H. E. Hellwig and A. W. Neumann, *Kolloid-Z. Z. Polym.* **40**, 229 (1969).
67. A. W. Neumann, *Adv. Colloid Interface Sci.* **4**, 105 (1974).
68. D. Y. Kwok and A. W. Neumann, *Adv. Colloid Interface Sci.* **81**, 167 (1999).
69. D. Y. Kwok, C. N. C. Lam, A. Li and A. W. Neumann, *J. Adhesion* **68**, 229 (1998).
70. A. Ulman, in: *An Introduction to Ultrathin Organic Films from Langmuir-Blodgett to Self-Assembly*. Academic Press, Boston, MA (1991).
71. R. Bhatia and B. J. Garrison, *Langmuir* **13**, 765 (1997).
72. F. P. Zamborini and R. M. Crooks, *Langmuir* **14**, 3279 (1998).
73. M. H. Schoenfisch and J. E. Pemberton, *J. Am. Chem. Soc.* **120**, 4502 (1998).
74. A. T. Lusk and G. K. Jennings, *Langmuir* **17**, 7830 (2001).
75. A. N. Parikh and D. L. Allara, *J. Chem. Phys.* **96**, 927 (1992).
76. D. Y. Kwok, H. Ng and A. W. Neumann, *J. Colloid Interface Sci.* **225**, 323 (2000).
77. Y. Rotenberg, L. Boruvka and A. W. Neumann, *J. Colloid Interface Sci.* **93**, 169 (1983).
78. P. Cheng, D. Li, L. Boruvka, Y. Rotenberg and A. W. Neumann, *Colloids Surfaces* **43**, 151 (1990).
79. S. Lahooti, O. I. del Río, P. Cheng and A. W. Neumann, in: *Applied Surface Thermodynamics*, J. K. Spelt and A. W. Neumann (Eds), pp. 441–507. Marcel Dekker, New York, NY (1996).
80. D. Y. Kwok, D. Li and A. W. Neumann, in: *Applied Surface Thermodynamics*, J. K. Spelt and A. W. Neumann (Eds), pp. 413–440. Marcel Dekker, New York, NY (1996).
81. D. Y. Kwok, C. J. Budziak and A. W. Neumann, *J. Colloid Interface Sci.* **173**, 143 (1995).
82. P. E. Laibinis, R. G. Nuzzo and G. M. Whitesides, *J. Phys. Chem.* **96**, 5097 (1992).
83. P. E. Laibinis, G. M. Whitesides, D. L. Allara, Y. T. Tao, A. N. Parikh and R. G. Nuzzo, *J. Am. Chem. Soc.* **113**, 7152 (1991).
84. R. G. Nuzzo, F. A. Fusco and D. L. Allara, *J. Am. Chem. Soc.* **109**, 2358 (1987).
85. W. Guo and G. K. Jennings, *Langmuir* **18**, 3123 (2002).
86. W. Haiss, D. Lackey, J. K. Sass and K. H. Bescoke, *J. Chem. Phys.* **95**, 2193 (1991).
87. E. Delamarche, B. Michel, H. Kang and C. Gerber, *Langmuir* **10**, 4103 (1994).
88. Y. S. Shon and T. R. Lee, *Langmuir* **15**, 1136 (1999).
89. D. J. Vanderah, C. W. Meuse, V. Silin and A. L. Plant, *Langmuir* **14**, 6916 (1998).
90. R. D. Weinstein, D. Yan and G. K. Jennings, *Ind. Eng. Chem. Res.* **40**, 2046 (2001).
91. J. Drelich and J. D. Miller, *J. Colloid Interface Sci.* **167**, 217 (1994).
92. A. Amirfazli, D. Y. Kwok, J. Gaydos and A. W. Neumann, *J. Colloid Interface Sci.* **205**, 1 (1998).
93. J. Drelich, J. L. Wilbur, J. D. Miller and G. M. Whitesides, *Langmuir* **12**, 1913 (1996).

*Contact Angle, Wettability and Adhesion*, Vol. 3, pp. 339–360
Ed. K.L. Mittal
© VSP 2003

# Wetting of a substrate by nanocrystallites

## ELI RUCKENSTEIN[*]

*Department of Chemical Engineering, University at Buffalo, The State University of New York, Buffalo, NY 14260, USA*

**Abstract**—Transmission electron microscopy and electron diffraction experiments were carried out to investigate the behavior of nanocrystallites of iron and platinum on an alumina substrate during heating in oxygen and hydrogen at various temperatures. In oxygen, both the small and the large iron crystallites emitted patches of multilayer thin films. These patches coalesced to generate a contiguous film connecting a large number of crystallites. During subsequent heating in hydrogen, the films were reduced and fractured, generating patches surrounding the existing particles, or independent patches that contracted to generate particles in locations that were initially devoid of particles. During the experiments involving platinum on alumina, the crystallites appeared to be smaller during the heating in oxygen and almost regained their initial sizes during the subsequent heating in hydrogen. They actually extended as undetectable films during the heating in oxygen and contracted to their initial sizes during the subsequent heating in hydrogen. Explanations of the observed phenomena are offered here.

*Keywords*: Supported metals; nanocrystallites; thin films; wetting.

## 1. INTRODUCTION

Supported metal catalysts, which are employed extensively in the chemical industry, consist of a porous support, such as alumina, silica, etc. and small metal crystallites (e.g., Pt, Au, Fe, Ag, etc.) in the nano-size range distributed over the internal surface of the support. Such systems are metastable from a thermodynamic viewpoint because their free energies can decrease as a result of sintering of the crystallites. Although metastable, they can be practically useful if the sintering can be minimized and if a sintered catalyst can be redispersed. Understanding of the mechanisms of sintering and redispersion is, therefore, important.

A number of mechanisms of sintering have been proposed: (i) migration and coalescence of crystallites [1]; (ii) emission of single atoms by the small crystallites and their capture by large ones by ripening [2-7]; and (iii) a combination of the two [5]. The ripening can be global (Ostwald) [2-5, 7] or direct [6]. In global ripening, the small crystallites lose atoms to a surface phase of single atoms dis-

---

[*]Phone: 716-645-2911, Fax: 716-645-3822, E-mail: feaeliru@acsu.buffalo.edu

persed over the substrate, and the large crystallites capture them from this phase. In this case, the surface phase of single atoms is supersaturated with respect to the large crystallites but is undersaturated with respect to the small ones. In direct ripening, which is a result of local fluctuations, the atoms released by a small particle move directly to a neighboring large crystallite, although the surface phase of single atoms is, on average, undersaturated with respect to all the particles of the system. These sintering mechanisms have been reviewed in detail [3, 8-11].

Transmission electron micrographs [12] have revealed that during heating in oxygen, Pd crystallites extended on an alumina substrate, changed their shape and exhibited tearing and fragmentation. They also showed that for various metals supported on alumina, films with thicknesses in the nanometer range coexisted with crystallites during heating in oxygen [13-15]. These films connected several crystallites and/or formed a contiguous phase on the substrate. They have sometimes been detected by electron microscopy, but frequently they were too thin to detect. During subsequent heating in hydrogen, the interconnecting films either fractured and merged with the particles or even triggered the coalescence of neighboring particles. The undetectable films that covered the substrate in regions containing only few particles could also fracture, thus generating patches, the contraction of which led to particles in regions initially lacking them.

The general tendency for the decrease in the free energy of a system is responsible for these phenomena, and the surface and interfacial free energies are expected to make a major contribution to the free energies of such dispersions of small crystallites on substrates. Consequently, wetting and spreading, which provide the means by which crystallites can decrease the free energy of the system, are expected to play a major role in catalyst aging by sintering and in redisperion of supported metal catalysts.

To demonstrate the role of wetting and spreading in the sintering and redispersion of supported metal catalysts, some experimental results characterizing iron on alumina and platinum on alumina are presented in the following sections. Then qualitative explanations of the spreading and the coexistence of multi-atomic surface films with crystallites in an oxygen atmosphere and their fracture into crystallites in a hydrogen atmosphere are presented. Four specimens, denoted A, B, C, and D, are reported in detail; specimens A and B represent iron on alumina, and specimens C and D platinum on alumina.

## 2. EXPERIMENTAL RESULTS

### 2.1. Iron on alumina

Iron was selected for an investigation of the role of interactions between gas, crystallite, and substrate in wetting and spreading, because iron is the most reactive of the group VIII metals. Therefore, it was expected that its behavior would reveal changes occurring when the chemical atmosphere was changed. In an oxidizing atmosphere at 700°C [13], thick films were detected around the crystallites, and the possible existence on the substrate of multi-atomic thin films (either of iron oxide or iron aluminate) was inferred from the electron micrographs and the electron diffraction patterns. The expectation that at higher temperatures the latter films would become thick enough to be observed in the micrographs over a larger area prompted us to perform high temperature experiments (>800°C) [15]. A summary of the results of these experiments is presented to emphasize the coexistence of thin films with crystallites in an oxygen atmosphere and the fracture of these films in a hydrogen atmosphere.

The experimental procedure employed was as follows. Electron-transparent and nonporous films of amorphous $Al_2O_3$ were prepared by anodically building up the oxide on a clean, thin, high-purity aluminum foil (99.999%, Alfa Products, Inc.) followed by stripping off of the oxide film by dissolving the unoxidized aluminum in a dilute mercuric chloride aqueous solution [13]. The oxide films were washed in distilled water and placed on gold electron microscope grids. They were subsequently heated in air at 800°C for 72 h to transform the alumina into γ or η form and also to ensure that no further changes would occur during subsequent treatment.

Iron films of various thicknesses were deposited onto the alumina substrates by evaporating the corresponding amounts of 99.998% pure iron wire (Alfa Products, Inc.) from a tungsten basket in an Edwards vacuum evaporator under a vacuum (pressure $< 2 \times 10^{-6}$ Torr). The substrate was kept at room temperature during the deposition. The two specimens, A and B, had an initial metal film thickness of 7.5 Å, but different heating histories. They were subjected to alternate heating in hydrogen and oxygen. The ultra-high purity hydrogen (supplied by Linde Division, Union Carbide Corporation, 99.999% pure, < 1 ppm $O_2$ and < 2 ppm moisture) was additionally purified (by passage through a Deoxo unit (Engelhard Industries), followed by a silica gel column, a bed of particles of 15 wt.% MnO on $SiO_2$, and finally beds of 13X and 5A molecular sieve particles immersed in liquid nitrogen), to further reduce the amount of $O_2$ and moisture, because iron is readily oxidized even in the unpurified ultra-high purity hydrogen.

### 2.1.1. *Evidence of a thick surface film. Specimen A*

Specimen A was first subjected to alternate oxidation and reduction at temperatures between 500 and 700°C, after which it was heated successively for 1, 2, and 5 h in hydrogen additionally purified as described above. Fig. 1 is a micrograph of the sample following this treatment. The subsequent heat treatment in oxygen generated interconnected films surrounding the crystallites on the substrate (Fig. 2). Further heating in oxygen increased the thickness of the films around the particles (Fig. 3). Away from the particles, the film was not clearly visible, but the substrate grain boundaries were considerably less sharp than those observed after heating in hydrogen. Therefore, it seems that a film is present over a large part of the support, with a thickness that is larger closer to the particles and much smaller

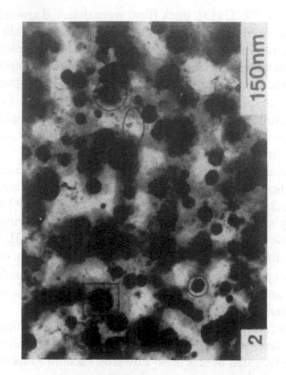

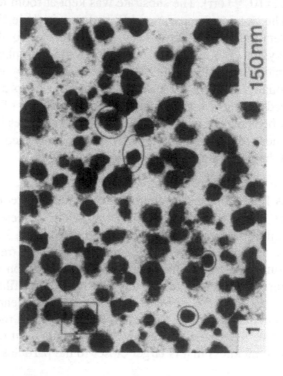

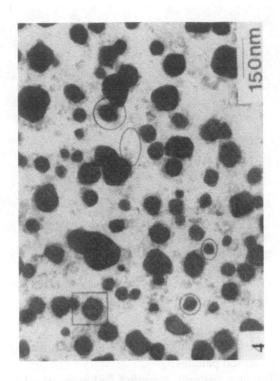

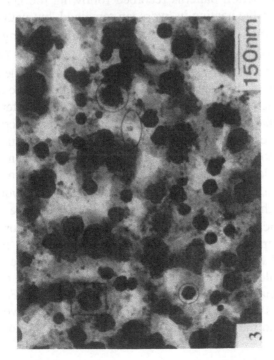

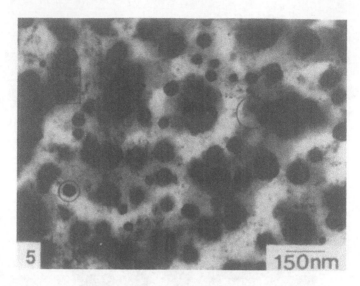

**Figures 1-5.** TEM micrographs for specimen A after heated successively in hydrogen (Fig. 1); oxygen (Fig. 2); oxygen (Fig. 3); hydrogen (Fig. 4); oxygen (Fig. 5).

further away from them. The film may not be continuous throughout, but rather may consist of interconnected patches.

In the electron diffraction patterns recorded following the heating in oxygen, the rings detected after the heating in hydrogen remained, but with d-values larger and approaching those of $\gamma - Fe_2O_3$. A number of additional rings appeared, the most intense corresponding to the major d-values of $AlFeO_3$. These additional rings were less bright and were spotty, indicating that only some orientations of crystallites or patches of thick films (more likely the latter) were diffracting. Because the film and the additional spotty rings appeared only after the heating in oxygen, it is likely that the film was composed of $AlFeO_3$. The crystallites themselves were probably covered by such a film. Indeed, some crystallites in contact with each other did not coalesce on prolonged heating in oxygen, indicating that perhaps a protective layer of $AlFeO_3$ was present on the crystallites, hindering their coalescence. The subsequent heating of this specimen in hydrogen at 500°C for 5 h did not affect the composition but somewhat decreased the thickness of the film around the particles. The heating in hydrogen at 800°C for 1 h, however, caused almost a complete withdrawal of the films around the particles, which most likely coalesced with the latter. The particles that were close to each other or already in contact coalesced (probably because of the elimination of the protective layer), leading to a small decrease in the number of crystallites (Fig. 4). The additional diffraction rings observed under the oxidizing conditions were almost completely eliminated, and the d-values of the other rings reverted to those observed before oxidation.

Thick films connecting neighboring particles reappeared on heating for only 2 h in oxygen at 800°C (Fig. 5), with the film appearing to cover the particles. On increasing the temperature to 900° and heating in oxygen for 1 h, the smaller particles (including those as large as 300 nm) either decreased in size or vanished. On continued heating for 4 more hours, a number of additional particles (up to 50 nm in diameter) disappeared or decreased in size, most likely because of their spreading and merging with the surface film, which appeared to become thicker.

One may be tempted to attribute the disappearance of a large number of crystallites to ripening. However, there was only little change in the size of the larger particles and, in fact, a marginal decrease in size. The rapid and total disappearance of large particles (those up to about 50 nm) suggests the possibility of loss of material by evaporation. However, this is not likely, because the melting temperatures of iron and its oxides are much higher than the temperatures of the experiments and, furthermore, the material was mostly recovered during the subsequent heating in hydrogen.

Only 1 h of heating in hydrogen at 900°C caused the withdrawal of the films, which merged with the particles, as shown by the observation that all the particles increased significantly in size. Only a few pairs of neighboring crystallites coalesced. The drastic decrease in the amount of material present as crystallites, observed following heating in oxygen, was mostly reversed during the reduction in hydrogen. The net effect appears to be a drastic sintering induced via the contact between the crystallites through the thick surface film generated in the oxidizing atmosphere, and also by the coalescence of the particles, because of the contraction of the interlinking surface film in the reducing atmosphere. There was, therefore, a large decrease in the number of crystallites, but not in the total number of atoms or molecules. The higher-temperature reduction yielded $FeAl_2O_4$ and possibly also some $\alpha$-Fe.

### 2.1.2. Specimen B

Another $Fe/Al_2O_3$ specimen, with a 7.5 Å initial Fe film thickness, was heat treated in hydrogen first at 500°C and later at 800°C. Its state after these treatments is shown in Fig. 6. The specimen was subsequently heated in oxygen for 1 h at 500°C to determine whether the surface film detected at higher temperatures was also formed at lower temperatures. A film was indeed generated around and probably even on the particles, as shown by Fig. 7. A number of small particles disappeared and a few relatively large particles appeared. Some of these particles resulted from the growth of small but detectable particles, and others appeared in places where there were no detectable particles before. $FeAl_2O_4$ was clearly identified in the electron diffraction patterns. The subsequent heating in hydrogen for 8 h at 500°C led to a complete withdrawal of the film. A few particles disappeared, some coalesced with neighboring crystallites, and a number of smaller particles appeared on the substrate. $FeAl_2O_4$ was still identified as the major compound in the electron diffraction patterns. When the specimen was heated again in hydrogen for 4 h at the higher temperature of 800°C (to ensure a more complete reduction), a few of the small crystallites grew considerably, while other (small as well as large) crystallites vanished. Subsequently, the specimen was heated in oxygen at successively higher temperatures of 600, 700, and 800°C, and prior to each oxidation step it was reduced in hydrogen at 800°C. There was virtually no change in the particle size or shape after any of these stages of heating, and only narrow film rings of lower contrast were observed around the particles after oxidation at 700 and 800°C. It is emphasized that the first heating in oxygen at 500°C, after a high-temperature reduction at 800°C, led to the formation of a detectable film; the subsequent heat treatments in oxygen at 600, 700, and 800°C did not, however, generate such detectable films. Either undetectable films spread out on the substrate at the higher temperatures, and/or the wetting characteristics of the substrate were modified during the previous heat treatments.

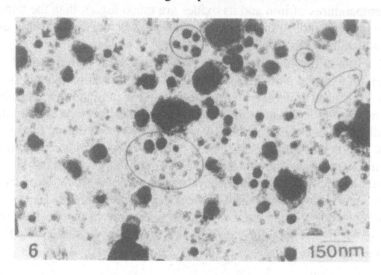

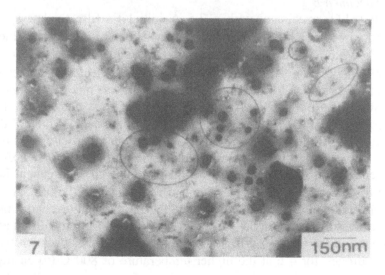

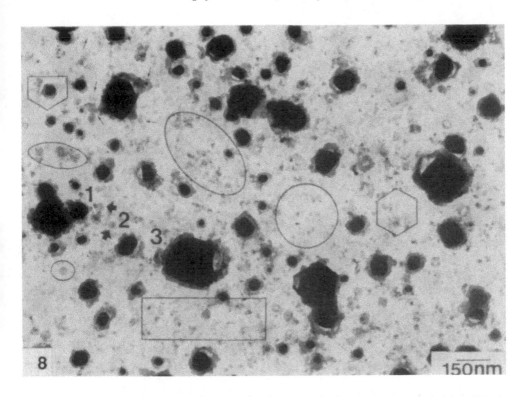

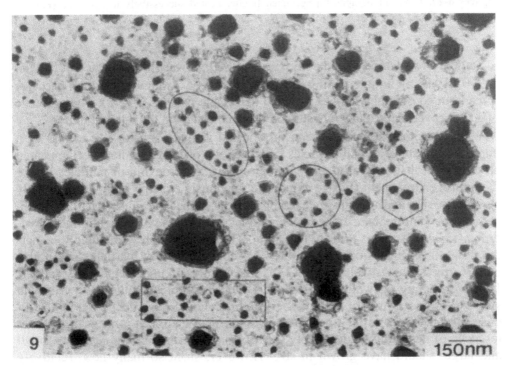

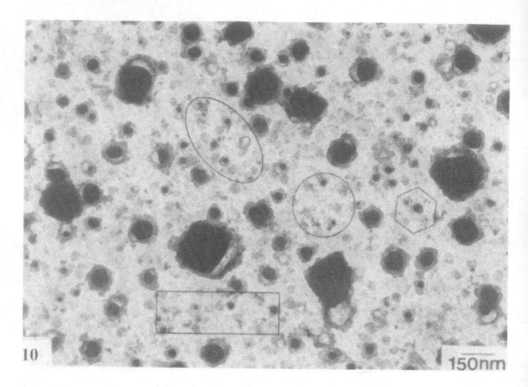

**Figures 6-10.** TEM micrographs for specimen B after heated successively in hydrogen (Fig. 6); oxygen (Fig. 7); oxygen (Fig. 8); hydrogen (Fig. 9); oxygen (Fig. 10).

Sample B, which was heated in oxygen at 800°C, was further heated in oxygen at 900°C without an intervening hydrogen treatment. A number of small particles spread out considerably, leaving behind detectable patches (Fig. 8). The presence of a film covering the substrate surface was inferred from the micrograph.

When specimen B was subsequently heated in hydrogen for 2 h at 800°C, a number of relatively small particles (about 15 nm or larger in diameter) appeared all over the substrate, particularly in regions that were devoid of particles before the hydrogen treatment (Fig. 9). Since the previously existing particles grew only slightly in size and the substrate grain boundaries became much sharper, it appears that contiguous patches of multilayer films coexisted with the three-dimensional crystallites prior to reduction. The films probably fractured to generate new, three-dimensional crystallites. Of course, the new particles could also have formed via Ostwald ripening or coalescence of undetectable crystallites. However, during Ostwald ripening, the larger crystallites should have grown in size and at least a few of the small crystallites should have decreased in size or even disappeared. On the other hand, if the growth had been a result of the coalescence of migrating, undetectably small crystallites, then all the particles, small and large, should have grown in size. However, since so many new, relatively large particles formed everywhere, while the previously existing three-dimensional crystallites (both small as well as large) changed only slightly, it is most likely that the new particles formed via the fracture of a surface film and/or by the contraction of undetectable surface patches.

The fact that the particles most likely originated from a surface film was further verified, indirectly, by heating specimen B again in oxygen at 800°C and subsequently at 900°C. At 800°C, the particles extended marginally, and narrow film rings appeared around them, but the number of particles remained virtually the same. However, after heating in oxygen at 900°C, most of the small particles spread considerably, losing their particulate appearance. Only thick extended patches could be seen where particles had been present before (Fig. 10). The extended patches contracted, tending to form particles again, on subsequent heating in hydrogen at 800°C. The extension to patches of film and recontraction to particles were observed repeatedly during the subsequent cycles of treatment in oxygen and hydrogen at 900°C.

The heating history appears to affect the specimen behavior. Following the heating in oxygen at 900°C, either detectable thick patches of interconnected films (specimen A) or thin, undetectable films (specimen B) were formed on the substrate. In the former case, the films contracted and merged with the neighboring crystallites on subsequent heating in hydrogen, whereas in the latter new particles were formed on the substrate.

## 2.2. Platinum on alumina

In what follows the results obtained with model $Pt/Al_2O_3$ catalysts heated alternatively in hydrogen and oxygen at temperatures in the range 300-750°C are presented. They reveal that there is an alternation in the size, namely, an apparent decrease in the size of the particles on heating in oxygen and a recovery of almost the original size on subsequent heating in hydrogen. This size alternation continued over a number of cycles and was observed at temperatures between 500 and 700°C. The results can be explained in terms of the redispersion of the crystallites via the formation of disconnected individual films around the particles that result from the strong interactions between crystallites and support.

The method of preparation of model $Pt/Al_2O_3$ catalysts and the treatment procedure were similar to those for $Fe/Al_2O_3$ and are presented in detail in Ref. [16]. The term hydrogen means the as-received ultra-high-purity hydrogen ($\leq$ 1 ppm $O_2$ and $\leq$ 3 ppm moisture) from Linde Division, Union Carbide Corporation, and "purified hydrogen" refers to the above-mentioned hydrogen purified as stated in a previous section. The experiments were carried out with a number of specimens. The results obtained with two of them are presented below in detail.

### 2.2.1. Specimen C

A sample with an initial 2-nm thick film was heated at 500°C for 8 h in purified hydrogen; this treatment generated small particles of about 1.6 nm average diameter.

To more easily identify the processes that occurred, the particle size was increased further by heating the sample in hydrogen at 600°C for 4 h followed by heating at 700°C for 5 h, at 800°C for 2 h, and finally again at 600°C for 2 h. This treatment led to relatively large particles, about 7 nm average size. On subsequent heating of the sample in oxygen at 300°C, there was a decrease in size of all the particles. Additional heating in oxygen at progressively higher temperatures in steps of 50°C up to 600°C (with a 2-h duration at each temperature) did not result in significant changes in either the size distribution or the shapes of the particles. However, when the sample was subsequently heated in hydrogen at 500°C, all the particles, small and large, increased significantly in size. Only a few particles disappeared, indicating that neither ripening nor migration and coalescence involving detectable particles was responsible for the growth. Further heating in oxygen at 500°C resulted again in a decrease in the sizes of all the particles. When the sample was subsequently heated in hydrogen at 500°C, the particles again increased in size and regained almost the size they had prior to the heating in oxygen. Alternate heating in hydrogen and oxygen at 500°C was carried out over six cycles, and the above alternation in size was observed in each cycle (Figs. 11-13). A decrease in size on heating in oxygen and an increase in size on heating in hydrogen were also observed when the same specimen was subsequently heated at 700°C.

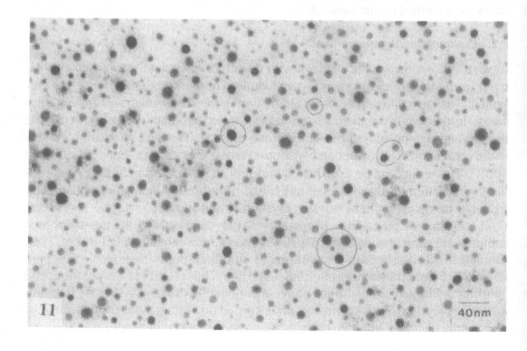

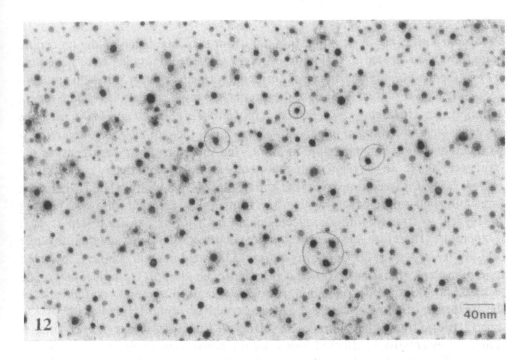

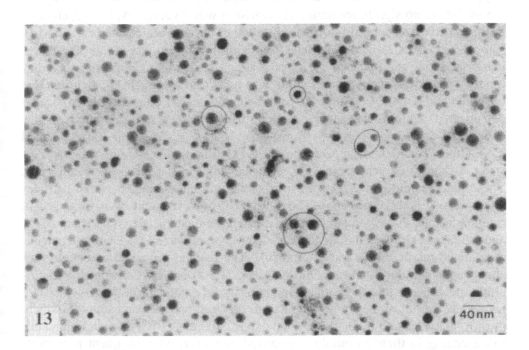

**Figures 11-13.** TEM micrographs for specimen C after heated successively in hydrogen (Fig. 11); oxygen (Fig. 12); hydrogen (Fig. 13).

## 2.2.2. Specimen D

A sample with 2 nm thick initial film and from the same batch as the preceding sample (C) was heated in a slightly different sequence. Particles were generated by heating the Pt/alumina sample in purified hydrogen at 500°C for 12 h and subsequently at 600°C for 5 h. Before the alternate heating in hydrogen and oxygen, sample C described above was heated in oxygen between 300 and 600°C in steps of 50°C, without hydrogen treatment in between. If the particles had already been oxidized at the lower temperature of 300°C, the subsequent heat treatments of the oxidized sample in oxygen at higher temperatures would have caused little change, as was observed up to 600°C with sample C. To verify whether there is a threshold temperature beyond which film formation is detected following heating in oxygen, we started with a reduced sample prior to the oxidation step at each temperature. Therefore, the present sample was also heated in oxygen in the range 300-600°C in steps of 50°C, but in contrast to the treatment of sample C, between oxidation steps, the sample was heated in hydrogen at 500°C to reduce the oxide to metal prior to the next oxidation. No detectable decrease in the particle size or the formation of films around the particles could be found following the 300 to 450°C oxygen treatments. Particles as small as 0.8 nm in diameter or even smaller remained essentially unaffected following the above treatments in hydrogen and oxygen. These results indicate that even though an oxide might have formed at or below 500°C, no significant molecular emission was involved. At 500°C, changes in the particle sizes and film formation started to become noticeable (Figs. 14-17). The decrease in size of the particles following heating in oxygen at 500°C was marginal during the first two cycles of heating, but it was more pronounced during the subsequent cycles; the extent of size reduction was again smaller than that observed for sample C. Some large particles decreased considerably in size or disappeared, while nearby smaller particles had only decreased marginally in size (Regions 1-5 in Figs. 14 and 15). Subsequent heating in oxygen at 700°C resulted in an almost complete disappearance of the existing particles on the substrate. Instead, a number of very large, thin, irregular hexagonal patches were observed. These patches could have resulted from the strong interactions between particles and substrate (see Discussion). They survived, however, and even increased in size, on heating in oxygen at 800°C. The patches are characterized by sharp edges and are facetted in irregular, hexagonal shapes, as shown in Fig. 17. Some patches overlapped (particles marked (a) and (b)). The patches remained essentially unchanged on subsequent heating in hydrogen at 500 or 700°C. From the micrographs, it appears that the patches represent large extended particles, which are formed as a result of the strong interactions between particles and support (which, in turn, are enhanced by the high-temperature, 700 or 800°C, oxygen treatment). Little change in their morphology occurred, however, on subsequent heat treatment in hydrogen or oxygen. It is noteworthy that a thin, hexagonal, pill-box morphology of platinum particles was observed on heating in hydrogen at tem-

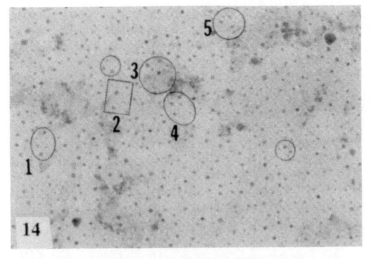

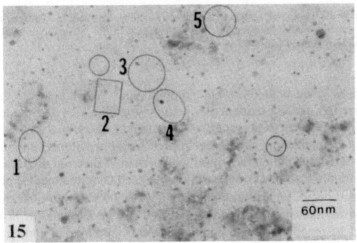

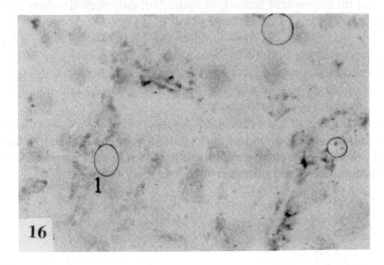

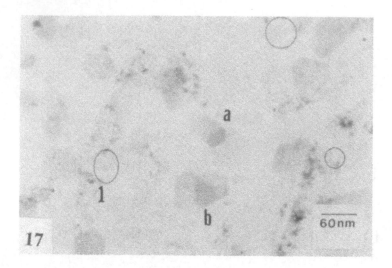

**Figures 14-17.** TEM micrographs for specimen D after heated successively in hydrogen at 500°C for 2 h (Fig. 14); oxygen at 600°C for 5 h (Fig. 15); oxygen at 700°C for 5 h and hydrogen at 500°C for 2 h (Fig. 16); oxygen at 800°C for 3 h (Fig. 17).

peratures $\geq$ 500°C when reducible oxide supports such as $TiO_2$ were employed [17]. Apart from the indirect evidence, via film formation (see Discussion), for strong interactions between the crystallites and support, there is also some direct evidence for such interactions from electron diffraction patterns, although the compounds formed could not be identified with certainty. Electron diffraction patterns were recorded for all the samples following each heat treatment. Following heating in oxygen, especially at temperatures above 500°C, additional rings appeared in the diffraction patterns, a few of which can be assigned to $Pt_3O_4$ and $Al_2Pt$. The rings assigned to the compound formed from platinum and aluminum did not disappear on subsequent heating in hydrogen; only their d-values changed (and only marginally). This result indicates that once such a compound is formed it is not easily decomposed. This compound formation may involve the diffusion of platinum species into the bulk of the alumina substrate and may be difficult to reverse. However, at the interface between the particle and the substrate, the compound could be relatively easily reduced (even if only partly) and could lead to the observed contraction of the particles.

Although the identities of the compounds formed could not be completely ascertained, the appearance of the additional rings in the electron diffraction patterns (some of which can be attributed to some known Pt-Al compounds) nonetheless suggests compound formation between $PtO_x$ and $Al_2O_3$.

## 3. DISCUSSION

A few mechanisms can be considered for the redispersion of sintered supported metal catalysts in an oxidizing atmosphere. The sintered crystallites can emit atoms or molecules to the substrate which can then coexist with the crystallites as a two-dimensional dispersed phase, thereby increasing the degree of dispersion [11]. Such a molecular emission mechanism is expected to be favored by high temperatures and by an oxidizing atmosphere rather than by a reducing one, since in oxidizing atmospheres the possibility of a decrease in the free energy of the system via the strong interactions between the iron oxide or the platinum oxide molecules and the alumina substrate creates a driving force for the emission of molecules from the particles to the substrate. On subsequent heating in hydrogen, the dispersed oxide molecules and the small crystallites were reduced (to metal), and the interactions with the support became weaker. Two possibilities come to mind. (i) Either new small particles were formed (not necessarily in the same locations where the particles might have disintegrated by emission of atoms or molecules to the substrate), and/or (ii) the reduced atoms were captured by the existing three-dimensional crystallites. Since new particles were not generated in our experiments either following heating in oxygen or on subsequent heating in hydrogen, one can infer that as a result of heating in oxygen, dispersed single molecules were formed on the substrate and that they were captured by the existing crystallites on subsequent heating in hydrogen. However, the extent of capture of atoms by various crystallites will be different because of the differences in their radii of curvature. Since the rate of emission is larger for crystallites with large curvatures, the large crystallites are expected to undergo a greater growth than the smaller ones. Consequently, a size distribution different from that generated during heating in oxygen was expected to develop during reduction. However, as noted above, all the detectable particles increased in size and attained almost the size they had prior to the oxygen treatment. Therefore, it appears that a mechanism based on emission and capture of single atoms from a two-dimensional phase of single atoms located on the substrate surface is unlikely to be responsible for the observed behavior.

Another possible and in the present case more likely mechanism for redispersion in an oxygen atmosphere is by the spreading of a film. It is well known that the surface free energy $\left(\gamma_{cg}\right)$ of a metal is larger than the surface free energy of the corresponding oxide. Furthermore, since, in general, a metal oxide interacts more strongly with the substrate (which is another oxide) than does a metal, the interfacial free energy $\gamma_{cs}$ between the crystallite and substrate is also smaller in an oxidizing than in a reducing atmosphere. Indeed, the interfacial free energy $\gamma_{cs}$ is related to the net interaction energy $U_{cs}$ between crystallite and substrate via the expression

$$\gamma_{cs} = \gamma_{cg} + \gamma_{sg} - (U_{int} - U_{str}) = \gamma_{cg} + \gamma_{sg} - U_{cs}, \qquad (1)$$

where $U_{int}$ is the interaction energy per unit contact area between substrate and supported material; $U_{str}$ is the strain energy per unit contact area due to the mismatch of their lattices; $\gamma$ is the interfacial free energy, and the subscripts c, s, and g refer to the crystallite, substrate, and gas, respectively. The net interaction energy is larger when the support and the supported material are similar in structure (two oxides, for instance, rather than an oxide and a metal), or when there are covalent interactions between the two. Since $\gamma_{cg}$ and $\gamma_{cs}$ are both smaller in an oxygen atmosphere, the contact angle $\theta$ is also smaller (as one can see from Young's equation, given below), and the particle is expected to be more extended in an $O_2$ atmosphere than in a $H_2$ atmosphere:

$$cos\ \theta = \frac{\gamma_{sg} - \gamma_{cs}}{\gamma_{cg}}. \qquad (2)$$

The increase in volume through oxidation also contributes to a greater extension in an oxygen atmosphere. If $\gamma_{sg} - \gamma_{cs} > \gamma_{cg}$, the particle will spread out completely on the substrate because cos $\theta$ cannot be larger than unity and no equilibrium wetting angle can be obtained.

Consequently, if the interactions between the iron oxide or platinum oxide and the alumina are strong enough, the small particles (which will be more completely oxidized than the larger ones) could spread out completely [because of the large driving force $\gamma_{sg} - \gamma_{cs} - \gamma_{cg}$ for spreading when $U_{cs}$ is very large ($\gamma_{cs}$ is very small)], whereas the larger particles would only extend. Therefore, the small particles would no longer be detected in the micrographs following heating in oxygen, and the larger particles would appear smaller if the films extending from the leading edge of the particles were so thin as not to be detectable by electron microscopy, or they would appear larger if the films extending from the leading edges of the particles were sufficiently thick. If the above interactions are not very strong (i.e., if the driving force for spreading is not very large), or if the rate of spreading is low, the particles would only be extended and could, therefore, still be detected in the micrographs (of course, they would appear smaller and have an undetectable film around them or appear larger because of a detectable film around them). These films would contract and merge with the respective crystallites on subsequent heating in hydrogen and, depending upon the reconstruction that occurs, the particles could regain the sizes they had prior to the oxygen treatment. This scenario appears to explain the observed results better than the ripening mechanism.

In the case of Fe/alumina, iron was easily oxidized, the oxidized compounds that formed reacted with the substrate and, as a result of spreading, detectable films were formed around the particles. Platinum is much less readily oxidized than iron, and the films formed by spreading were undetectable. The experimental results might suggest that iron and platinum behave differently; in reality, the difference is a matter of degree. Being more reactive, iron leads to detectable films, whereas the much less reactive platinum gives only undetectable films.

In the above considerations, macroscopic concepts have been extended to systems involving nanoscale sizes and thicknesses to explain some of the experimental observations. Of course, the use of macroscopic concepts for nanoscale particles can have at most a qualitative validity. There are, however, some experimental results that cannot be explained on the basis of macroscopic concepts because the nanoscale size induces new phenomena. Why, for instance, during heating in hydrogen that follows heating in oxygen, numerous particles were formed in regions which were free of particles before the heating in oxygen and why pill-box morphologies were generated under some conditions?

To explain the first observation, let us start from a thin film of catalyst uniformly distributed over a substrate. In a thick film, the atoms (molecules) located at its free surface are not affected by the presence of the substrate, whereas in a thin film the range of the interaction forces is greater than the thickness of the film. As a result, the free energy of a system containing a thin film also depends on the thickness of the film. For illustration purposes, London dispersion interactions will be assumed to act among the atoms (molecules). Of course, the interaction forces involved are more complex; nevertheless, for the qualitative considerations that follow, the simplicity of the expressions obtained makes the physics more transparent. The following expression can be derived for the interaction potential $\phi$ that acts on a unit volume located at the planar free surface of a film [11]

$$\phi = \phi_0 + A / 6\pi h^3, \tag{3}$$

where A is the Hamaker constant, h the thickness of the film, and $\phi_0$ the value of $\phi$ in the limit $h \to \infty$. For the continuum assumption involved in the derivation of Eq. (3) to be valid, the film should have at least 5 atomic (molecular) layers.

The Hamaker constant can be written in the form [11]

$$A = A_{11} - A_{12}, \tag{4}$$

where $A_{11}$ is the Hamaker constant for the interactions among the molecules of

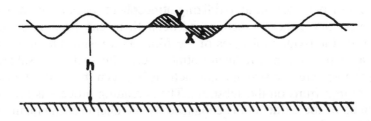

**Figure 18.** Schematic of the perturbation of the free surface of a thin film.

the film and $A_{12}$ is the Hamaker constant for the interactions between the molecules of the film and those of the substrate. Because the potential $\phi_0$ involves only attractive interactions, it has a negative value. The second term in the expression for $\phi$ can be positive or negative, depending on the strengths $A_{11}$ and $A_{12}$ of the two types of interactions. $A$ is positive when the interactions between the molecules of the film are stronger, and negative when the interactions between the molecules of the film and those of the substrate are stronger. In the former case, the film is not expected to wet the substrate, whereas in the latter the film is expected to wet the substrate.

Let us slightly perturb the free surface of the film to a wave shape (Fig. 18). Such perturbations are present in any system and are of thermal or mechanical origin. In the system considered here there are additional chemical causes, for instance, the nonuniform surface reduction. These perturbations can be either amplified in time or they can decay. In the former case the film ruptures, whereas in the latter the film is stable. The following thermodynamic considerations provide some insight regarding these two types of behavior. First, one should note that by increasing the area of the free interface the perturbation increases the free energy of the system associated with the macroscopic surface free energy. However, in addition, the perturbation displaces some molecules to greater distances from the substrate-film interface. Even though the above expression for $\phi$ is valid only for molecules located at a planar free interface, it nonetheless suggests that when $A > 0$ an increase in h decreases the potential energy of the displaced molecules by making their potential energy more negative. Consequently, the molecules displaced from the cusp (X in Fig. 18) to the crest (Y) contribute to the decrease of the total free energy due to the London interactions when $A > 0$, and to its increase when $A < 0$. In the former case, the perturbation will grow and the film will rupture when the free energy decrease exceeds the free energy increase caused by the increased surface area of the free interface. In the latter case, the perturbation will decay and the planar film will remain stable. In a reducing atmosphere, the film is reduced (at least to some extent) to metal, for which the interactions among atoms (and hence $A_{11}$) are expected to be much stronger than the interactions between the atoms of the film and those of the substrate-oxide (reflected in $A_{12}$). As a result, the film can become unstable and rupture, generating either patches that surround existing particles (with which they finally coalesce), or independent patches which contract to form particles located in places previously devoid of particles. In an oxidizing atmosphere, because both the film and substrate are oxides, the interactions between them can become comparable to those between the oxide molecules of the film. As a result, the films can remain stable. If after heating in a reducing atmosphere, the system is subjected to an oxidizing atmosphere, the stronger interactions between the oxides will lead to the spreading of thin films on the substrate. The spreading process is affected by the larger volume of the oxide and is enhanced by the chemical interactions between oxides with the formation of aluminates. The latter interactions can stimulate a complete spreading of the film over the substrate. However, the chemical interactions with the substrate bind the spreading films to the substrate, and as a result structures are generated involving crystallites and thin films. These structures can be metastable thermodynamically, but they exhibit some kinetic stability because the compounds formed retard the advancement of the films.

Consequently, the stronger interactions between the catalyst atoms (molecules) and those of the substrate in an oxygen atmosphere are responsible for the emission of multilayer films by the crystallites. The weaker interactions between the atoms and substrate in a hydrogen atmosphere lead to the rupture of the films formed previously in the oxygen atmosphere.

The kinetics of spreading is highly complex, because it involves an increase in volume during oxidation, different reactivities of the metal at the leading edge and on the surface of a particle, and the gradient in the surface tension that is thus generated (the Marangoni effect). If the surface tension of the oxidized metal is smaller than that of the substrate, the molecules of the former will be displaced in the direction of the substrate. When, however, the surface tension is lower, molecules of the substrate will be displaced towards the particles and can engulf them.

Regarding the second observation, the possibility of a planar structure morphology was anticipated theoretically by Ruckenstein and Lee [18], who noted that Young's equation implies the existence of a macroscopic wetting angle defined on a length scale that is large compared with atomic dimensions. However, because of short-range interactions near the leading edge, the internal angle between the horizontal and the line connecting the center of two successive molecules, at the solid-gas interface, located in the same vertical plane, can vary from a value $\theta_0$ at the leading edge to the macroscopic wetting angle $\theta$ at a large distance from the leading edge. If the short range interactions near the leading edge are repulsive, $\theta_0 > \theta$. If, in such cases $\cos\theta_0 > 1$, then no angle can be generated and the atoms will spread on the substrate over the entire available area. If $\cos\theta_0 < 1$, but $\cos\theta > 1$, then total spreading will occur at some distance from the leading edge, while the atoms near the leading edge will generate a wetting angle $\theta_0$ and the particles will acquire planar structures with a rapid variation of the angle near the leading edge.

## 4. CONCLUSION

Transmission electron microscopy and electron diffraction experiments revealed that iron and platinum nanocrystallites supported on alumina emitted thin films during heating in oxygen which fractured during the subsequent reduction in hydrogen. In case of iron, which is much more reactive, the films emitted were thicker and coalesced to generate a contiguous film connecting a large number of crystallites. During subsequent reduction, the film ruptured generating patches surrounding the crystallites or independent patches that contracted to particles in locations initially devoid of particles. The platinum crystallites also emitted films, but they were very thin and undetectable by electron microscopy. The platinum crystallites almost regained their initial sizes during the subsequent reduction in hydrogen.

## REFERENCES

1. E. Ruckenstein and B. Pulvermacher, J. Catalysis, **29**, 224 (1973).
2. B. K. Chakraverty, J. Phys. Chem. Solids, **28**, 2401 (1967).
3. P. Wynblatt, R. A. Dalla Betta and N. A. Gjostein, in *The Theoretical Basis of Heterogeneous Catalysis*, E. Drauglis and R. J. Jaffe (Eds.), p. 510, Plenum Press, New York (1975).
4. P. C. Flynn and S. E. Wanke, J. Catalysis, **33**, 233 (1974).
5. E. Ruckenstein and D. B. Dadyburjor, J. Catalysis, **48**, 73 (1977).
6. E. Ruckenstein and D. B. Dadyburjor, Thin Solid Films, **55**, 89 (1978).
7. H. H. Lee, J. Catalysis, **62**, 129 (1980).
8. S. E. Wanke and P. C. Flynn, Catalysis Rev. Sci. Eng., **12**, 93 (1975).
9. E. Ruckenstein and D. B. Dadyburjor, Chem. Eng. Rev., **1**, 251 (1983).
10. H. H. Lee and E. Ruckenstein, Catalysis Rev. Sci. Eng., **25**, 475 (1983).
11. E. Ruckenstein, in *Metal-Support Interactions in Catalysis, Sintering and Redispersion*, S. A. Stevenson, J. A. Dumesic, R. T. K. Baker and E. Ruckenstein (Eds.), p. 140, Van Nostrand Reinhold, New York (1987).
12. J. J. Chen and E. Ruckenstein, J. Phys. Chem., **85**, 1606 (1981).
13. I. Sushumna and E. Ruckenstein, J. Catalysis, **94**, 239 (1985).
14. I. Sushumna and E. Ruckenstein, J. Catalysis, **108**, 77 (1987).
15. E. Ruckenstein and I. Sushumna, J. Catalysis, **97**, 1 (1986).
16. I. Sushumna and E. Ruckenstein, J. Catalysis, **109**, 433 (1988).
17. S. A. Stevenson, J. A. Dumesic, R. T. K. Baker and E. Ruckenstein (Eds.), *Metal-Support Interactions in Catalysis, Sintering and Redispersion*, Van Nostrand Reinhold, New York (1987).
18. E. Ruckenstein and P. S. Lee, Surface Sci., **52**, 298 (1975); E. Ruckenstein, J. Colloid Interface Sci., **86**, 573 (1982).

*Contact Angle, Wettability and Adhesion*, Vol. 3, pp. 361–372
Ed. K.L. Mittal

# Light-induced reversible wetting of structured surfaces

NICOLA RICHARDS,[1] JOHN RALSTON[*, 1] and GEOFFREY REYNOLDS[2]

[1] *Ian Wark Research Institute, University of South Australia, Mawson Lakes Boulevard, Mawson Lakes, South Australia 5095 Australia*
[2] *School of Pharmaceutical, Molecular and Bio-medical Sciences, University of South Australia, Mawson Lakes, SA 5095, Australia*

**Abstract**—Derivatised 2,4-diketopyrimidine groups have been anchored onto planar gold surfaces as self-assembled monolayers (SAMs) for chemical and photochemical studies. Three derivatives: 5-methyluracil (thymine), 5-trifluoromethyluracil, and 5-nitrouracil were chosen for the specific moiety at position C-5, which can exert different inductive effects on the adjacent aromatic ring. The measured acidity constant decreases from 5-methyluracil to 5-nitrouracil as the inductive properties of the group at C-5 position increase. When grafted to the surface the acidity constant increases, albeit still obeying the expected inductive trend. The functionality also exerts an influence on the surface wetting properties. Surface wettabilities indicate that the orientation of the pyrimidine groups at the surface occurs so that the C-5 position has a direct effect upon the surface chemistry. Upon dimerisation by irradiation with ultraviolet light the surface contact angle increases significantly. The contact angle changes reflect that the dimer has a different surface charge than the monomer and a substantially different conformation leading to light induced changes in contact angle of up to 22°.

*Keywords*: Uracils; pyrimidines; wetting; photo-responsive; self-assembled monolayers; reversible wetting.

## 1. INTRODUCTION

Surfaces which respond to external stimuli such as light allow various interfacial properties, e.g., wetting, liquid crystal alignment and dispersability to be altered in a controlled fashion [1-3]. In our earlier studies, we have shown that wettability can be manipulated reversibly by light when a substrate surface is modified by a photo-responsive monolayer [1].

In 1960 Beukers and co-workers [4, 5] found that when frozen aqueous solutions of 5-methyluracil (thymine) were irradiated with UV light, a stable photoproduct was formed which lacked the characteristic UV absorption maximum at 264 nm. The photoproduct had a molecular weight twice that of thymine, suggesting that it was in fact two molecules joined by a cyclobutane type ring located be-

*To whom all correspondence should be addressed. Phone: +61 8 8302 3066,
Fax: +61 8 8302 3683, E-mail: john.ralston@unisa.edu.au

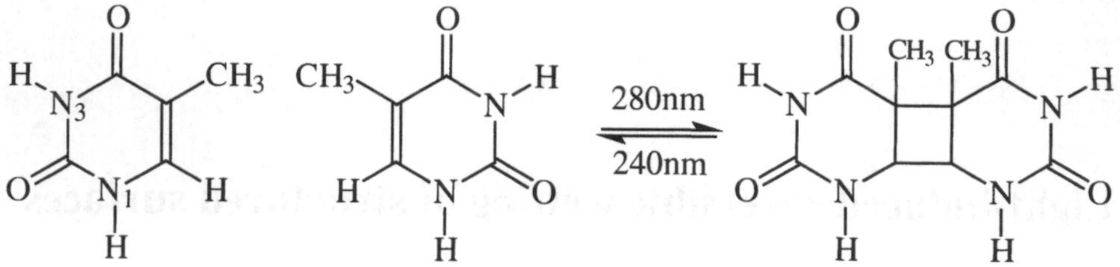

**Figure 1.** Thymine molecules dimerisation equilibrium.

tween positions C-5 and C-6. Thymine is thus referred to as a photochromic molecule and adjacent thymines in DNA form these dimers when exposed to UV radiation. The formation of specific photoproducts in DNA is the fundamental adverse effect of UV irradiation and leads to mutagenesis, carcinogenesis and cell necrosis in individual cells and complex organisms [6]. Today it is known that thymine will dimerise reversibly depending upon the wavelength of light to which it is exposed. Molecules possessing this capacity will dimerise in the range of 200-300 nm; however, below 250 nm, the shorter wavelengths of higher energy are able to cleave the photoproduct resulting in an eventual equilibrium (Figure 1). In order to control the position of this equilibrium the most efficient dimerisation is performed at 280 nm and the cleavage at 240 nm.

The novel aspect of a system where wetting properties are modified by light is that neither bulk phase needs to be altered in order to see a significant change in contact angle. Thin coatings of photochromic molecules can be used as photoresists, for data storage or for molecular recognition [7]. If a surface tension gradient can be established then the flow of liquids can be induced, providing a driving force for the operation of micro- and nano-scale chemical processes [8-10]. Even very small changes in contact angle are important. For example, Ichimura and co-workers [11, 12] utilised monolayer systems where a modest, five degree photo-induced change in contact angle led to an effective control of large-scale liquid crystal alignment in sandwich displays.

In this investigation, functionalised 2,4-diketopyrimidine groups have been anchored onto gold surfaces as self-assembled monolayers (SAMs) for chemical and photochemical studies. We report the results of a light-induced wettability investigation of 5-methyluracil (thymine), 5-trifluoromethyluracil, and 5-nitrouracil at various pHs. For the specific head groups chosen, the moiety present at position C-5 can exert strong or weak inductive effects on the adjacent aromatic ring, influencing the related acidity constant (see below).

## 2. MATERIALS AND METHODS

### 2.1. Materials

All reagents were of analytical grade and were obtained from Sigma-Aldrich, except 5-trifluoromethyluracil, provided by Pfaltz and Bauer, USA. In this study 2,4-diketopyrimidine head groups were selected based on the inductive properties of the moiety at position C-5. Water (ultra-high quality (UHQ)) used in the experiments was produced by reverse osmosis with two stages of ion exchange and two stages of activated carbon prior to final filtration (surface tension 72.8 mN/m at 20°C, resistivity $\kappa < 10^{-6}$ µS/m).

### 2.2. Methods

#### 2.2.1. Thiol preparation

2.2.1.1. Preparation of 5-methyluracil thiol

N-1-alkylester-5-methyluracil was prepared by adding NaH to 5-methyluracil in dimethylformamide (DMF) (DMF was dried over anhydrous magnesium sulphate for one hour and filtered prior to distillation). The white mixture was stirred at 50°C for one hour under an inert atmosphere, then ethylbromoacetate was added drop-wise and allowed to react for a further two hours at 50°C under nitrogen. Solvent was removed under vacuum to yield a pale yellow solid that was added to chloroform and stirred at room temperature for 30 minutes. After this time, insoluble organic salts were removed by vacuum filtration and the chloroform evaporated to yield a pale yellow solid which was recrystallised from water until the UV spectrum did not alter upon further recrystallisation. The NMR spectrum was recorded using a 200 MHz Varian Gemini Fourier Transform spectrometer over 16 scans. Relevant spectral features referred to DMSO-d6 were: $\delta$ 1.2(t, 3H, $CH_3$), 1.75(s, 3H, $CH_3$), 4.18(q, 2H, $CH_2$), 4.42(s, 2H, $CH_2$), 7.50(s, 1H, $C_6$-H), 11.40(s, 1H, N-3H).

N-1-alkylacid-5-methyluracil was prepared by hydrolysis of N-1-alkylester-5-methyluracil in NaOH (0.5 M). The pale white solution was refluxed for one hour under nitrogen. Upon cooling, HCl (conc) was added drop-wise, the pale white precipitate was removed by vacuum filtration and solid white rod-like crystals were obtained from UHQ water. Relevant NMR spectral features referred to DMSO-d6 were: $\delta$ 4.4(s, 2H, $CH_2$), 7.5(s, 1H, $C_6$-H), 11.4(s, 1H, N-3H).

N-1-alkylacid-5-methyluracil was dried in an oven and then added to DMF. After ten minutes sonication the mixture was stirred overnight under a nitrogen atmosphere to complete the dissolution. Carbonyldiimidazole was dissolved in DMF with sonication and then added to the mixture, which was subsequently heated at 40°C for 5/10 minutes to encourage formation of the activated ester. After 5/10 minutes, aminoethanethiol hydrochloride dissolved in DMF, was added and the mixture stirred overnight at ambient temperature under nitrogen. Solvent was removed under vacuum to yield a thin brown oil. The solid was separated in water/chloroform and the water layer was removed. Relevant NMR spectral features referred to DMSO-d6 were: $\delta$ 1.1(t, 1H, SH), 1.75(s, 3H, CH$_3$), 2.9(t, 2H, CH$_2$), 3.1(t, 2H, CH$_2$), 4.1(s, 2H, CH$_2$), 7.4(s, 1H, C$_6$-H), 7.98(s, 1H, N-H), N-3H deprotonated.

### 2.2.1.2. Preparation of 5-trifluoromethyluracil thiol

N-1-alkylester-5-trifluoromethyluracil was prepared by adding NaH to 5-trifluoromethyluracil in DMF (DMF was dried over anhydrous magnesium sulphate for one hour and filtered prior to distillation). The mixture was stirred at 90-100°C (the range is critical to avoid formation of di-substituted product) for one hour under an inert atmosphere, then ethylbromoacetate was added drop-wise and allowed to react for a further two hours at 50°C under nitrogen. Solvent was removed under vacuum to yield a pale yellow solid. The solid was separated in water/chloroform and the chloroform layer was removed, the ester precipitated and was recovered as white crystals. The NMR spectrum was recorded using a 200 MHz Varian Gemini Fourier Transform spectrometer over 16 scans. Relevant NMR spectral features referred to DMSO-d6 were: $\delta$ 1.2(t, 3H, CH$_3$), 4.2(q, 2H, CH$_2$), 4.6(s, 2H, CH$_2$), 8.42(s, 1H, C$_6$-H), 12.00(s, 1H, N-3H).

N-1-alkylacid-5-trifluoromethyluracil was prepared by hydrolysis of N-1-alkylester-5-trifluoromethyluracil in NaOH (0.5 M). The pale white solution was refluxed for two hours under nitrogen. Upon cooling, HCl (conc) was added drop-wise and the liquid was removed under vacuum to yield a pale yellow solid. Relevant NMR spectral features referred to DMSO-d6 were: $\delta$: 4.5(s, 2H, CH$_2$), 8.42(s, 1H, C$_6$-H), 12.00(s, 1H, N-3H).

N-1-alkylacid-5-trifluoromethyluracil was dried in an oven and then added to DMF. After ten minutes sonication the mixture was stirred overnight under nitrogen to complete the dissolution. Carbonyldiimidazole was dissolved in DMF with sonication and then added to the mixture, which was heated at 40°C for 5/10 minutes to encourage formation of the activated ester. After 5/10 minutes, amino-ethanethiol hydrochloride dissolved in DMF, was added and the mixture stirred overnight at ambient temperature under nitrogen. Solvent was removed under vacuum to yield a thin brown oil. The solid was separated in water/chloroform and the water layer was removed. Relevant NMR spectral features referred to DMSO-d6 were: $\delta$ 1.2(s, 1H, SH), 2.8(t, 2H, CH$_2$), 3.1(t, 2H, CH$_2$), 4.5(s, 1H, CH$_2$), 8.0(bs, 1H, N-H), 8.65(s, 1H, C$_6$-H), N-3H deprotonated.

### 2.2.1.3. Preparation of 5-nitrouracil thiol

N-1-alkylester-5-nitrouracil was prepared by adding NaH to 5-nitrouracil in DMF (DMF was dried over anhydrous magnesium sulphate for one hour and filtered prior to distillation). The pale yellow mixture was stirred at 50°C for one hour under an inert atmosphere, then ethylbromoacetate was added drop-wise and allowed to react for a further two hours at 50°C under nitrogen. Solvent was removed under vacuum to yield a pale yellow solid. The solid was insoluble in chloroform, therefore it could not be separated from inorganic salts in chloroform. The solid was recrystallised from water as pale yellow iridescent flakes. The

NMR spectrum was recorded using a 200 MHz Varian Gemini Fourier Transform spectrometer over 16 scans. Relevant NMR spectral features referred to DMSO-d6 were: δ 1.2(t, 3H, CH$_3$), 4.2(q, 2H, CH$_2$), 4.75(s, 2H, CH$_2$), 9.2(s, 1H, C$_6$-H), 12.2(s, 1H, N-3H).

N-1-alkylacid-5-nitrouracil was prepared by hydrolysis of N-1-alkylester-5-nitrouracil in NaOH (0.5 M). The pale yellow solution could not be heated because the reagent decomposed; therefore, the solution was stirred at room temperatures for five hours. HCl (conc) was added drop-wise, and the liquid was removed under vacuum to yield a pale yellow/orange solid. Relevant NMR spectral features referred to DMSO-d6 were: δ: 4.1(s, 2H, CH$_2$), 9.1(s, 1H, C$_6$-H), 12.2(s, 1H, N-3H).

N-1-alkylacid-5-nitrouracil, was dried in an oven and then added to DMF. After ten minutes sonication the mixture was stirred overnight under nitrogen to complete dissolution. Carbonyldiimidazole was dissolved in DMF with sonication and then added to the mixture, which was heated at 40°C for 5/10 minutes to encourage formation of the activated ester. After 5/10 minutes, aminoethanethiol hydrochloride dissolved in DMF was added and the mixture stirred overnight at ambient temperature under nitrogen. Solvent was removed under vacuum to yield a thin brown oil. The solid was separated in water/chloroform and the water layer was removed. Relevant NMR spectral features referred to DMSO-d6 were: δ 1.2(s, 1H, SH), 2.8(t, 2H, CH$_2$), 3.1(t, 2H, CH$_2$), 4.5(s, 1H, CH$_2$), 8.0(bs, 1H, N-H), 8.65(s, 1H, C$_6$-H), N-3H deprotonated.

### 2.2.2. Bulk acidity constants

Bulk acidity constants were determined spectroscopically. Ultraviolet (UV) spectra were recorded with a Varian Cary 1E UV/Vis. Spectrophotometer. Samples were placed in clean 10 mm quartz cuvettes with a tightly fitting cap (Starna Pty Ltd, Thornleigh, NSW, Australia). Aqueous 5 x 10$^{-5}$ M solutions of the pyrimidines were prepared in a background electrolyte of 0.1 M NaCl. pH values between 1.5 and 5.5 were achieved by addition of HCl(aq.) to a freshly prepared 500 ml aliquot of the solution. For values between pH 5.5 and 13, NaOH(aq.) was added to a separate aliquot of solution; both samples were continually stirred with a magnetic stirrer. Small aliquots of solution were removed for immediate sampling and then returned to the stock for further adjustment. The pH was adjusted and recorded at 25°C using a Hanna Instruments pH meter, equipped with a Hanna Instruments HI 1131B-pH electrode calibrated with Merck colour-coded buffers at pH values of 4.00 ± 0.05 and 7.00 ± 0.05 or pH 7.00 ± 0.05 and 10.00 ± 0.05. For each sample spectrum recorded the background was subtracted over the entire wavelength range.

### 2.2.3. Preparation of substrates

Prior to gold deposition, silicon wafers were rinsed in ethanol and water followed by drying in a stream of nitrogen. Each plate was then plasma cleaned for 2 minutes in a Harrick Scientific PDC-32G plasma cleaner (60 W). The wafers were

coated with a chromium (99.999%) adhesion layer prepared by resistive evaporation from a tungsten boat. Immediately after, gold was deposited upon this sample
surface from a molybdenum boat. The sample was kept at room temperature in an
evaporation chamber maintained at $1 \times 10^{-6}$ Torr. The deposition rate and layer
thickness were monitored using a quartz crystal oscillator. Typically the deposition rate of both the chromium and gold was 0.3 nm/s and the thicknesses of the
layers were approximately 10 nm and 100 nm, respectively.

### 2.2.4. Formation of self-assembled monolayers

Freshly prepared gold slides were immersed in a thiol solution ($1 \times 10^{-3}$ M in ethanol) and self-assembled monolayers were allowed to form for at least 24 hours,
after which point there was no further change in water contact angle. The slides
were removed and thoroughly rinsed successively in ethanol and water and then
dried in a stream of nitrogen.

### 2.2.5. Contact angle measurement and surface titration

Advancing water contact angles were measured using the sessile drop technique.
Images were recorded using a CCD camera with 5x magnification. Small drops of
approximately 20 µl in volume were placed directly onto the surface using a
micropipette and were illuminated using a diffuse light source. For surface titra-
tions, individual samples were fully immersed in the same pH buffer solution that
was used to measure the contact angle. After 45 seconds of spreading, the images
of the drops were saved and digitised and then edge detection software was used
to determine the sessile advancing angle. Recorded angles are the average of 5
readings and were taken from different areas on the surface. The readings were
taken under ambient conditions, 22°C ± 1, 50-60% humidity.

### 2.2.6. Irradiation

The samples were irradiated using a MM3 diffraction grating type illuminator
equipped with a 300 W Xenon lamp. Photo-dimerisation was initiated by irradiation of the sample at 280 nm (energy flux 5.5 $Jcm^{-2}sec^{-1}$). Photo-cleavage was
achieved by irradiation of the sample at 240 nm (energy flux 0.4 $Jcm^{-2}sec^{-1}$). The
bandwidth of the irradiated light was 20 nm.

## 3. RESULTS

### 3.1. Acidity constant (pKa)

The measured acidity constants decrease from 5-methyluracil to 5-nitrouracil as
the inductive property of the group at C-5 position increases. In solutions the pKa
is 9.9 for 5-methyluracil and decreases to 5.0 upon substitution with a nitro group
at the C-5 position. Upon N-1-alkylation, for all species, the pKa increases
slightly due to a shift in the electronic distribution in the pyrimidine ring. When
grafted to the surface the acidity constant increases further, still following the expected inductive trend (Table 1).

**Table 1.**
Acidity constant data for 5-methyluracil, 5-trifluoromethyluracil and 5-nitrouracil i) in solution, ii) N-1-alkyl derivatives in solution, and iii) N-1-alkyl thiols grafted onto the surface as a self-assembled monolayer

|  | pKa ± 0.2 | | |
|---|---|---|---|
|  | Solution | $N_1$-alkyl derivative in solution | $N_1$-alkyl derivative on surface |
| 5-methyluracil | 9.9 | 10.5 | 11.0 |
| 5-trifluoromethyluracil | 7.3 | 8.4 | 9.1 |
| 5-nitrouracil | 5.0 | 6.3 | 8.2 |

## 3.2. Contact angles

The contact angles at pH 5.8 are similar for 5-methyl and 5-trifluoromethyluracil, decreasing by 10° in the presence of a nitro group at the C-5 position (Table 2). Upon dimerisation all the contact angles increase to approximately 75°. Surface irradiation results in an approximate 10° change in contact angle for 5-methyl and 5-trifluoromethyl uracil and a 21° change in contact angle for 5-nitrouracil.

At pH 11 all contact angles are reduced compared to those measured on the same surfaces at pH 5.8. Upon dimerisation all the contact angles increase to approximately 68° (Table 3), hence, surface irradiation causes a 10° increase in contact angle for 5-methyluracil. In the case of 5-trifluoro and 5-nitro uracils, increases of around 16° and 22° occur.

**Table 2.**
Monomeric and dimeric surface water contact angles, in degrees, for 5-methyluracil, 5-trifluoromethyluracil and 5-nitrouracil at pH 5.8

|  | pH 5.8 | | |
|---|---|---|---|
|  | Monomer $\theta \pm 2$ | Dimer $\theta \pm 2$ | $\Delta\theta$ |
| 5-methyluracil | 67 | 76 | 9 |
| 5-trifluoromethyluracil | 67 | 75 | 8 |
| 5-nitrouracil | 56 | 77 | 21 |

**Table 3.**
Monomeric and dimeric surface water contact angles, in degrees, for 5-methyluracil, 5-trifluoromethyluracil and 5-nitrouracil at pH 11.0

|  | pH 11.0 | | |
|---|---|---|---|
|  | Monomer $\theta \pm 2$ | Dimer $\theta \pm 2$ | $\Delta\theta$ |
| 5-methyluracil | 59 | 69 | 10 |
| 5-trifluoromethyluracil | 50 | 66 | 16 |
| 5-nitrouracil | 45 | 67 | 22 |

## 4. DISCUSSION

### 4.1. Influence of molecular structure on pKa

Altering functionality at position C-5 of the pyrimidine derivative head group, in solution, has an effect on the acidity constant. As the inductive capacity of the R group at position C-5 is increased, the lability of the dissociable hydrogen at positions N-1 and N-3 increases, hence the pKa decreases from the 9.9 recorded for 5-methyluracil (R=CH$_3$), to 7.3 when R=CF$_3$ and 5.0 when R=NO$_2$ (Table 1). In the case of uracil products the first dissociation leaves the remaining hydrogen free to migrate between the annular nitrogen sites [13]. If dissociation is forced from position N-3, e.g. when the pyrimidine is functionalised at position N-1, then the first pKa of the compound generally increases since there is less available delocalisation for the resulting negative charge.

When the molecules are grafted to the surface after alkylation at N-1, the pKa alters as the local environment is altered. The reason for the increase in surface pKa is most likely due to the formation of a surface hydrogen bond network, combined with attractive electrostatic forces resulting from neighbouring negative charges which increase the energy required to remove the proton. Introduction of a group such as a nitro group which is able to assume a negative charge through resonance increases the attractive electrostatic force and thus further increases the ionisation energy.

### 4.2. Influence of molecular structure and pKa on contact angle

A change in surface wettability is achieved through manipulation of the molecular structure of the self-assembled monolayer. For example, the functionality at position C-5 can be altered from R=CH$_3$ to include more hydrophobic or hydrophilic species such as R=CF$_3$ or R=NO$_2$, respectively. In addition to the direct effects of molecular structure, the ionisation of the species at the surface can affect the wetting of the substrate, with different contact angles on ionised and unionised surfaces.

The effect of surface charge on wetting has been investigated to only a modest extent. In 1989 Fokkink and Ralston [14] determined that an electrical double layer would form spontaneously at a charged solid-liquid interface, altering the interfacial tension and thence the contact angle. The system then minimizes the free energy by increasing the area of the solid/liquid interface, i.e. on a charged surface a droplet of constant volume spreads, with the change in contact angle described by equation (1).

$$\cos\theta = \cos\theta(\text{pzc}) - \Delta F_{DL} / \gamma_{LV} \qquad (1)$$

Where pzc  = Point of zero charge

$\Delta F_{DL}$ = Free energy of the double layer

$\gamma_{LV}$ = Interfacial tension of the liquid/vapour interface.

At pH 5.8 the thymine monomer is protonated and has a water contact angle of 67°. At pH 11 the contact angle on the monomer decreases to 59° because the surface is partially charged. As the pH increases beyond the pKa of the N-1-alkylated monomer (pKa 11.0), the surface fully deprotonates and the contact angle decreases by a further 5°. A SAM with a lower pKa will ionise at a lower pH. Thus the type of substituent present at position C-5 was altered to increase the acidity of the pyrimidine head group. 5-trifluoromethyluracil monomer at pH 5.8 gave a water contact angle of 67°, the same as on 5-methyluracil. With a pKa of 9.1, however, the surface was fully deprotonated at pH 11 and the angle decreased below that of 5-methyluracil, to 50°.

In water at pH 5.8 the contact angle for 5-nitrouracil is 56°, some 10° lower than that observed for 5-methyluracil. Since the structure is similar to 5-methyluracil, this suggests that the nitro group has a direct effect on the contact angle through orientation of the ring at the surface. This reduced contact angle value indicates that the orientation of the pyrimidine groups at the surface occurs so that the C-5 position has a direct effect upon the surface chemistry.

### 4.3. Influence of UV irradiation on pKa and on contact angle

When thymine is irradiated with UV light at 280 nm, dimers form at the surface which then photo-split to yield the monomer at 240 nm. Aqueous solutions of thymine yield four isomeric dimers, [15-17] illustrated in Figure 2.

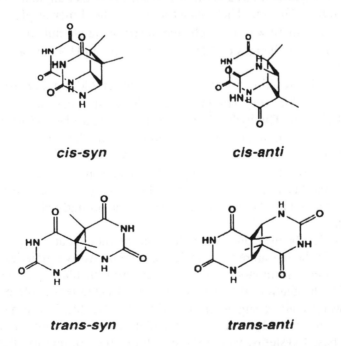

**Figure 2.** Four isomeric dimers cis-syn, trans-syn, trans-anti, cis-anti, resulting from 280 nm irradiation of thymines in solution [15].

In order to maximise the efficiency of the photo-dimerisation it is desirable to orient the bases so that the *cis-syn* isomer is the major isomer formed. Since the driving force for cleavage is steric repulsion of the bulky groups at C-5 the *cis-syn* isomer has the highest reactivity for photo-splitting [18]. In DNA, the conformation of thymine is controlled by connection to the DNA chain and by complementary base pair hydrogen bonding [18]. This fixed orientation ensures that photo-dimerisation in DNA yields only the *cis-syn* isomer. If thymine is grafted onto a smooth gold surface as a thin robust film or self-assembled monolayer (SAM), the fixed orientation should ensure that *cis-syn* isomer dominates and inert photo-dimer formation is minimal.

Although the photochromic properties of members of the pyrimidine family have been known for some time, Abbott *et al.* [1] were the first to realize that the pKa of the two distinct photo-stable species differed. Thus at a given pH the monomer may be charged whilst the dimer remains neutral. Hence irradiation with UV light can influence the surface charge.

Dimer formation on 5-methyluracil increases the contact angle of the surface to 76° (Table 2). The pKa of the dimer is expected to be higher than that of the monomer due to less available charge delocalisation upon loss of aromaticity; therefore at pH 11 there should be a lower proportion of surface deprotonation. Indeed the dimer contact angle at pH 11 is 69°. On a 5-methyluracil surface, the largest reversible contact angle increase upon irradiation is 10° at pH 11.

5-trifluoromethyluracil has a similar increase in contact angle to 5-methyluracil after irradiation at 280 nm. The contact angle on the dimer at pH 11 was measured as 66°, hence there was a 16° change in contact angle but still at an undesirably high pH. Smaller changes of 13° were observed at pH 9.5 a value less than 11 but greater than 9.1.

Irradiation of hydrophilic 5-nitrouracil surfaces at 280 nm increases the contact angle to 77°, a value very similar to the dimer surfaces of 5-methyl and 5-trifluoromethyluracil. The hydrophilicity of the group at the C-5 position is obviously lost, suggesting that the molecules reorganise on the surface to shield the C-5 position from the contacting liquid. This radiation-induced conformational change on the surface leads to a large 21° change in water contact angle. The change in wettability is independent of surface charge; in fact the contact angle changes associated with the deprotonated surface at around pH 9-11 are of a similar magnitude.

Contact angle changes due to photo-induced molecular reorganisation have been reported in the literature for a variety of photosensitive moieties [19-22]. For example, azo-benzene monolayers have been used on planar and colloidal substrates [21-24]. Irradiation of these monolayers effected reversible photo-induced changes in water contact angles of around 9-11° [21, 22]. These monolayer will only photo-switch provided that the azo moiety is not sterically hindered and, in alkane-thiol based systems, where there is little free volume to allow switching between the cis and trans conformers, the monolayer will not respond [23]. Other groups used to effect much larger changes in contact angle include spiropyrans [20] and photo-responsive dyes such as malachite green [19]. Contact angle changes due to apparent conformational changes have been reported to be as large as 40°, however the reversibility is either poor or non-existent and the values reported are suspect since the spin-cast films often degraded in contact with water. In contrast to these reports, 5-methyluracil dimerisation has been shown to be reversible over numerous irradiation cycles, without degradation of the film. Inaki and co-workers [7, 25, 26] have shown that in thin, spin-cast films the reversibility extends well beyond 10 irradiation cycles with little or no loss of efficiency provided that the thymines are optimally positioned for dimerisation equilibrium.

## 5. CONCLUSION

Self-assembled monolayers incorporating functionalised 2,4-diketopyrimidines have been successfully prepared on planar substrates. 5-Methyluracil, 5-trifluoro-methyluracil and 5-nitrouracil were anchored as alkylated short chain thiols through the annular N-1 position onto cleaned gold wafers. The surface acidity constants increased upon anchoring the molecules, most likely due to electrostatic forces in the fixed orientation environment and the formation of a hydrogen bond network between the closely packed head groups.

Surface wettabilities indicate that the orientation of the pyrimidine groups at the surface occurs so that the C-5 position has a direct effect upon the surface chemistry. Upon dimerisation, the contact angles increase significantly, reflecting a change in surface charge. Large changes in contact angle were observed for 5-nitrouracil, which were independent of surface charge and were apparently due to conformational changes alone.

*Acknowledgements*

Financial support from the Australian Research Council through the Special Research Centres Scheme is gratefully acknowledged, as are valuable discussions with Professor Y. Inaki and colleagues.

## REFERENCES

1. S. Abbott, J. Ralston, G. Reynolds and R. Hayes, *Langmuir* **15**, 8923-8928 (1999).
2. K. Ichimura, Y. Suzuki, T. Seki, A. Hosoki and K. Aoki, *Langmuir* **4**, 1214 (1988).
3. M. Ueda, H.-B. Kim and K. Ichimura, *J. Mater. Chem.* **4**, 883 (1994).
4. R. Beukers, J. Ijlistra and W. Berends, *Rec. Trav. Chim* **79**, 101 (1960).
5. R. Beukers and W. Berends, *Biochim. Biophys Acta* **41**, 550 (1960).
6. B. P. Ruzsicska and D. Lemaire, in: *CRC Handbook of Photochemistry and Photobiology* 1289-1317 (1995).
7. Y. Inaki, *Polymer News* **17**, 367-371 (1992).
8. M. K. Chaudhury and G. M. Whitesides, *Science* **256**, 1539-1541 (1992).

9. C. D. Bain, G. D. Burnett-Hall and R. R. Montgomerie, *Nature* **372**, 414-415 (1994).
10. B. S. Gallardo, V. K. Gupta, F. D. Eagerton, L. I. Jong, V. S. Craig, R. R. Shah, and N. L. Abbott, *Science* **283**, 57-60 (1999).
11. K. Aoki, T. Seki, M. Sakuagi and K. Ichimura, *Makromol. Chem.* **193**, 2163 (1992).
12. Y. Kawanishi, T. Tamaki, M. Sakuragi, T. Seki, Y. Suzuki and K. Ichimura, *Langmuir* **8**, 2601 (1992).
13. K. Nakanishi, N. Suzuki and F. Yamazaki, *Bull. Chem. Soc. Jpn* **34**, 53-57 (1961).
14. L. G. Fokkink and J. Ralston, *Colloids Surfaces* **36**, 69-76 (1989).
15. N. Tohnai, Y. Inkai, M. Miyata, N. Yasui, E. Mochizuki and Y. Kai, *Bull. Chem. Soc. Jpn.* **72**, 1143-1151 (1999).
16. D. L. Wulff and G. Fraenkel, *Biochim. Biophys. Acta* **51**, 332-339 (1961).
17. M. A. Herbert, J. C. LeBlanc, D. Weinblum and H. E. Johns, *Photochem. Photobiol.* **9**, 33-43 (1969).
18. Y. Inaki, E. Mochizuki, H. Donoue, M. Miyata, N. Yasui and Y. Kai, *J. Photopolym. Sci. Technol.* **12**, 725-734 (1999).
19. N. Negishi, K. Tsunemitsu and I. Shinohara, *Polym. J.* **13**, 411-412 (1981).
20. S. Hayashida, H. Sato and S. Sugawara, *Polym. J.* **18**, 227-235 (1986).
21. C. L. Feng, Y. J. Zhang, J. Jin, Y. L. Song, L. Y. Xie, G. R. Qu, L. Jiang and D. B. Zhu, *Langmuir* **17**, 4593-4597 (2001).
22. L. M. Siewierski, W. J. Brittain, S. Petrash and M. D. Foster, *Langmuir* **12**, 5838-5844 (1996).
23. S. D. Evans, S. R. Johnson, H. Ringsdorf, L. M. Williams and H. Wolf, *Langmuir* **14**, 6436-6440 (1998).
24. Z. Sekkat, J. Wood, Y. Geerts and W. Knoll, *Langmuir* **11**, 2856-2859 (1995).
25. Y. Inaki and Y. Wang, *J. Photopolym. Sci. Technol.* **4**, 259-266 (1991).
26. Y. Inaki, Y. Wang, M. Kubo and K. Takemoto, *Chemistry of Functional Dyes. Proceedings of the International Symposium on Chemistry of Functional Dyes 2*, Kobe, Japan 365 (1992).

*Contact Angle, Wettability and Adhesion*, Vol. 3, pp. 373–383
Ed. K.L. Mittal
© VSP 2003

# Wetting in porous media: Some theoretical and practical aspects

## ABRAHAM MARMUR[*]

*Department of Chemical Engineering, Technion - Israel Institute of Technology, 32000 Haifa, Israel*

**Abstract**—Wetting in porous media is discussed from two different perspectives: (a) the penetration criterion and kinetics of penetration in systems of limited size, and (b) wettability characterization of porous media. The special thermodynamic and kinetic effects in systems consisting of small liquid reservoirs or thin porous media are presented and explained. The various possibilities of characterizing porous media by the kinetics of penetration are discussed, emphasizing the advantages and disadvantages of each method.

*Keywords*: Capillary; porous medium; wetting; inkjet; paper.

## 1. INTRODUCTION

Wetting of solids by liquids is a ubiquitous phenomenon for the very simple reason that most biological, daily-life, and industrial systems include solid surfaces and liquids. A very large class of systems involves porous bodies that are or need to be wet by liquids. Examples include water transport in soil and rocks, printing on paper, coating of wooden surfaces, and many others.

The basic mechanism of liquid penetration into porous media can be demonstrated by studying penetration into a simple cylindrical capillary. However, the quantitative details are extremely complex because of the very intricate structure of real porous media. The problem has two aspects: one is the quantitative understanding of how liquids penetrate into a porous medium of a known structure; the other is the characterization of porous media by measurements of liquid penetration into them.

The literature on this topic is vast. The present paper is intended to be only a short review of two aspects of the author's work in this field: (a) capillary penetration and displacement in systems of limited size, and (b) characterization of porous media by the kinetics of penetration into them. The models discussed below were motivated by practical, industrial needs. They demonstrate how simple

---

[*]Fax: 972-4-829-3088, E-mail: marmur@tx.technion.ac.il

models may assist in understanding the essence of practical problems and help in formulating guidelines for solving them. This review is by no means intended to be exhaustive neither in contents nor in referencing. A more complete list of references can be found in the papers mentioned below and in review papers that exist in the literature [e.g. 1-3].

## 2. SYSTEMS OF LIMITED SIZE

For many years the study of wetting of porous media was concerned only with infinite porous media and infinite liquid reservoirs. However, the development of inkjet printing and need for understanding print quality triggered the realization that systems of limited size (such as tiny ink drops and very thin, porous printing layers) are of interest [2]. As a first-order approximation, systems consisting of a porous medium and a liquid have three characteristic dimensions: the typical size of the porous medium (such as thickness), $D$, the typical radius of curvature of the liquid reservoir, $R$, and the typical dimension of a pore, $r$. These three dimensions can be grouped into two independent, dimensionless ratios, $R/r$ and $D/r$, which are not necessarily infinite. Their actual, finite values determine the behavior of the system: $R/r$ is responsible for the criterion for penetration and its rate, while $D/r$ is responsible for the "re-exposure effect" to be explained below.

The above ratios are based on characteristic dimensions that, by definition, ignore distributions, e.g. pore size distribution. Therefore, the models to be reviewed below that are based on these ratios describe the essential phenomena, but do not necessarily yield accurate quantitative results. More realistic and complex models need to be developed for describing real porous media. However, as will be shown below, the finite size of the liquid reservoir may also have a major effect on the kinetics of penetration into real systems with a pore size distribution (the "re-distribution effect"). Thus the study of capillary penetration in systems of finite size is essential.

### 2.1. Penetration and displacement criteria

The driving force for capillary penetration is the pressure difference between the liquid reservoir pressure at the capillary entrance, $P_r$ (see Fig. 1), and the pressure in the liquid side of the meniscus, $P_m$ (see Fig. 1). The criterion for penetration is then based on the magnitude of this pressure difference: penetration proceeds as long as

$$P_r - P_m > P_h \tag{1}$$

where $P_h$ is the hydrostatic pressure difference between the liquid reservoir and the meniscus. The meniscus pressure is given by [4, 5]

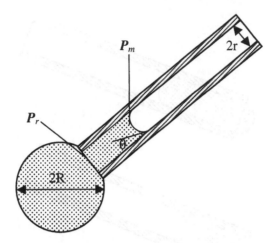

**Figure 1.** Penetration of a drop (limited-size liquid reservoir) into a capillary.

$$P_m = P_{at} - \frac{2\sigma\cos\theta}{r} \qquad (2)$$

where $P_{at}$ is the atmospheric pressure, $\sigma$ is the surface tension of the liquid, $\theta$ is the contact angle that the liquid forms with the solid internal surface of the capillary, and $r$ is the internal capillary radius (see Fig. 1). The liquid reservoir pressure depends on the radius of curvature of the liquid reservoir [6], and is given by

$$P_r = P_{at} + \frac{2\sigma}{R} \qquad (3)$$

where $R$ is the mean radius of the liquid reservoir (see Fig. 1).

The generalized criterion for penetration can now be rewritten by substituting Eqs. (2) and (3) into Eq. (1) to obtain

$$\frac{2\sigma}{R} + \frac{2\sigma\cos\theta}{r} > P_h \qquad (4)$$

It should be noted that $P_h$ and $R$ are usually not constant during penetration (the contact angle may also vary, however for simplicity it is assumed constant, because its dynamic behavior is not related to the main point discussed here). The generalized criterion for *initial* penetration is derived by equating $P_h$ to zero, since then the meniscus is just about to develop from the liquid reservoir. Thus

$$\frac{2\sigma}{R_i} + \frac{2\sigma\cos\theta}{r} > 0 \qquad (5)$$

where $R_i$ is the initial radius of curvature of the reservoir. In terms of the contact angle, the *initial* penetration criterion can be presented as

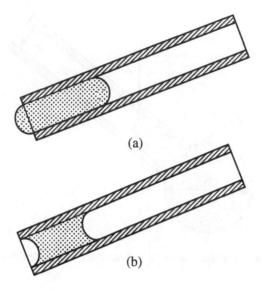

**Figure 2.** The final equilibrium shape of a drop in a capillary: (a) $\theta > 90°$ (b) $\theta < 90°$.

$$\cos\theta > -\frac{r}{R_i} \tag{6}$$

For an infinite reservoir ($r/R_i \rightarrow 0$), this criterion implies that $\theta$ must be smaller than 90°, as has been known for a very long time. However, for a small reservoir, initial penetration is possible also for obtuse contact angles [6]! This seemingly surprising result is a natural outcome of the extra pressure that exists inside the small reservoir due to its finite radius.

For applications such as printing or writing on paper, penetration has to be carefully optimized by adjusting the contact angle. If penetration is excessive, "feathering" occurs; if penetration is limited, the liquid stays outside the porous layer and may be smeared. For such applications, the knowledge that penetration may occur at contact angles higher than 90° is extremely important. However, there is a price to pay for this "extra" penetration: for contact angles higher than 90° the final equilibrium state of the liquid reservoir involves a residual part that is outside of the capillary (see Fig. 2a). This is in contrast to the case of contact angles lower than 90°, for which penetration is always complete (Fig. 2b). None-theless, numerical studies [6] showed that the fraction of liquid penetrating into the porous medium may be close to one, even for contact angles much higher than 90°. Of course, this fraction approaches zero as the contact angle increases and approaches 180°.

Similar considerations can be applied to the process of displacement of a liquid from a capillary into the ambient vapor atmosphere by another liquid. When the liquid inside the capillary is being displaced, the radius of curvature at the exit

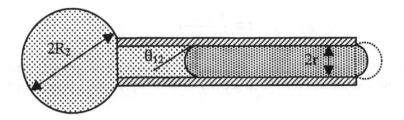

**Figure 3.** Displacement of a liquid from a capillary by another liquid.

first decreases, until the radius of the bulging drop equals that of the capillary; then the drop radius increases (see Fig. 3). During the radius decreasing stage, the pressure inside the bulging drop increases, thus the opposition to displacement of the liquid from the capillary increases. Consequently, complete displacement is possible only when the maximum pressure in the displaced drop is overcome. The general, approximate condition for complete spontaneous displacement of liquid 1 by liquid 2 is then [7]

$$r/R_2 > \approx \frac{1-(1+\sigma^*-2\sqrt{\sigma^*})\cos\theta_{12}}{\sigma^*} \qquad (7)$$

where $R_2$ is the radius of curvature of the displacing liquid, $\sigma^* \equiv \sigma_2/\sigma_1$, and $\theta_{12}$ is the contact angle that the displacing liquid makes with the capillary in the presence of the displaced liquid (see Fig. 3).

Interestingly, when the reservoir of the displacing liquid is infinite ($r/R_2 \rightarrow 0$), complete spontaneous displacement is possible, in principle, only when $\sigma^* > 4$. However, such a ratio of surface tensions hardly exists in nature. Thus, complete spontaneous penetration may be possible only with the additional displacing pressure that originates from the finite curvature of the displacing liquid.

## 2.2. Rate of penetration and displacement

The basis for studying the rate of penetration into capillaries and porous media is the Lucas-Washburn (LW) equation [4, 5]. For horizontal penetration into an empty capillary, or when the effect of gravity may be neglected (e.g. initial stages of vertical penetration), the LW equation reads

$$X^2 = 2\tau \qquad (8)$$

In this equation $X$ is the dimensionless penetration distance, defined as

$$X \equiv \frac{l}{r} \qquad (9)$$

where $l$ is the penetration distance into the capillary, and $\tau$ is the dimensionless time

$$\tau \equiv \frac{\sigma \cos \theta}{4\mu r} t \qquad (10)$$

where $t$ is the time, and $\mu$ is the viscosity of the liquid. For upward vertical penetration into a capillary, for which gravity is meaningful, the LW equation is written as

$$-\frac{X}{G} - \frac{\ln(1 - GX)}{G^2} = \tau \qquad (11)$$

where

$$G \equiv \frac{\rho g r^2}{2\sigma \cos \theta} \qquad (12)$$

In this equation $\rho$ is the density of the liquid, and $g$ is the gravitational acceleration. The LW equation has been experimentally verified for cases of negligible inertia and relatively constant contact angles, and extensively used [2]. However, in its above forms, it does not account for systems of limited size.

When a finite drop penetrates into a capillary, the process is self-accelerated. This is so, since the radius of the drop decreases during penetration, causing an increase in the pressure inside the drop, which, in turn, accelerates penetration. Thus, the finite size of the drop has an important effect on the kinetics, in addition to its effect on the penetration criterion.

When displacement from a capillary is considered, the increase in pressure of the displaced drop has the opposite effect: the kinetics of displacement is slowed down as long as the radius of the bulging drop is decreasing. Once the point of minimum radius is passed, or when the displaced liquid is connected to an infinite reservoir, penetration proceeds at a rate, which is determined by the curvature of the displacing reservoir, the viscosity ratio, the surface tensions of the two liquids, and the contact angle inside the capillary [8].

A different capillary penetration problem that is very important for the printing industry is the radial penetration of a drop placed on paper [9]. This process consists of two sub-processes that, to some extent, occur simultaneously: (a) penetration of the liquid in the drop into the paper, and (b) radial growth of the wet stain. The latter was found to obey the following simple power law [10]

$$A = kt^n \qquad (13)$$

where $A$ is the area of the wet stain, and $k$ and $n$ are empirical constants. In all existing data, $n$ seems to have the same value of 0.30-0.35.

This low value of $n$ could not be accounted for by existing theory. The reason for this could be one of two: (a) the radial geometry, which is not considered in the original LW theory, or (b) a different penetration mechanism than was assumed by Lucas and Washburn. To check the effect of the radial geometry, the radial capillary model was developed based on the Lucas-Washburn assumptions [11]. In this model, an axisymmetric drop was assumed to penetrate into the space between two infinite parallel plates, through a small hole in one of them, as shown in Fig. 4a. Using this model, the kinetics of radial penetration turned out to be given by $A \approx kt^{0.9}$ [9]. Thus, the slow kinetics characterized by $n \approx 0.3$ could not be explained simply on the basis of the radial geometry.

To check the possibility that a different mechanism of penetration is involved, a simple experiment was done [9], as shown in Fig. 4b. First, instead of studying radial penetration from a drop of a limited size, an infinite liquid reservoir was used, in order to test the theory of the radial capillary. This was achieved by using a vertical short glass capillary to drive the liquid from a Petri dish ("infinite" reservoir) into the paper. The kinetics of wet area growth fitted very well to that predicted by the radial capillary model: $A \approx kt^{0.9}$. Then, the glass capillary was disconnected. At this point there was a finite amount of liquid in the paper, which continued to spread according to Eq. (13), with $n$ being around 0.35. Thus, it was demonstrated that this value of $n$ was due to the finite amount of the liquid in the wet stain. The simple explanation of the mechanism of continued penetration in this case is capillary suction of the liquid by the small pores from the large pores, whereas the latter become unsaturated, i.e. filled again with air [9]. This mechanism was termed *"the re-distribution effect."*

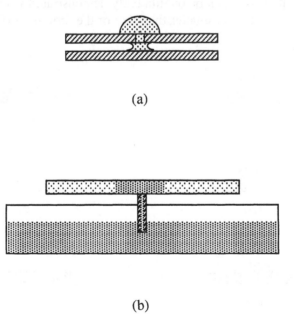

(a)

(b)

**Figure 4.** (a) The radial capillary model: an axisymmetric drop penetrates into the space between two infinite parallel plates, through a small hole in one of them; (b) Penetration of a liquid from a Petri dish ("infinite" reservoir) into paper through a short capillary.

## 2.3. The re-exposure effect

This effect is a surprising outcome of the limited thickness of some porous media, such as paper. The re-exposure effect can be qualitatively explained by referring to a simple, cylindrical capillary. In general, a liquid penetrates into a capillary because the formation of a solid-liquid interface is more favorable than that of a solid-vapor interface (disregarding for the moment the effect of a limited liquid reservoir, this statement is equivalent to saying that the contact angle should be acute). Imagine now that the wall of the capillary becomes perforated. If the liquid penetrates, it is re-exposed to the vapor phase through the holes in the wall, and the formation of liquid-vapor menisci at these holes requires surface energy. Therefore, wall perforation makes penetration less favorable.

The edges of a porous medium may be regarded as a "perforated capillary." If the porous medium is thick, this edge effect is negligible. However, if the porous medium is thin, the gain in liquid-vapor interfacial energy may be detrimental to liquid penetration. This effect was theoretically demonstrated for two different systems: a drop penetrating into a thin porous medium (see Fig. 5) [12], and capillary rise in such a medium [13]. For the former case, it was demonstrated that penetration of the drop might be partial even if the contact angle was less than 90°. Under proper conditions, the drop may penetrate only to the extent that its base radius equals the radius of the wet stain inside the porous medium (see Fig. 5d). The explanation can be qualitatively demonstrated with the aid of Fig. 5. If the wet stain has a radius smaller than that of the drop base (Fig. 5c), the liquid

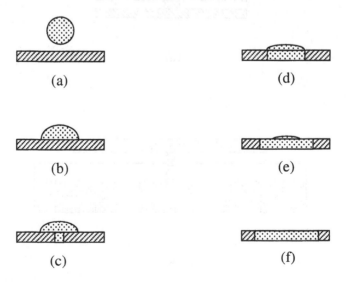

**Figure 5.** Stages in drop penetration into a thin porous medium: (a) drop before landing (b) initial landing of the drop (c) initial penetration into the porous medium (d) equal radii of drop base and wet stain (e) continued penetration (f) full penetration.

in the drop is exposed to the vapor inside the porous medium. If the wet stain has a radius larger than that of the drop base (Fig. 5e), the liquid in the wet stain is exposed to the outside vapor. Thus, it turns out that the system minimizes the liquid exposure to vapor by keeping the base and stain radii the same [12]. For capillary rise in thin porous media it was theoretically demonstrated that the height of rise might be as low as about 1% of what it should be for thick porous media, for contact angles that were relatively high, but less than 90° [13].

## 3. CHARACTERIZATION OF POROUS MEDIA BY THE KINETICS OF PENETRATION

Characterization of porous media with respect to capillary penetration is a major, important problem. Ideally, the results of a characterization process are expected to yield information on the pore size distribution and on the contact angle. However, in reality it is very difficult to achieve these goals, as discussed below. The most popular static method for measuring pore size distribution is mercury porosimetry. However, the contact angle cannot be measured by this method. Also, for many applications, the important parameter is the rate of penetration. Thus, it makes sense to characterize porous media by measuring the rate of penetration of a given liquid into the studied porous medium, using a standardized geometry. The simplest approach to the interpretation of the results of such a measurement is identifying a cylindrical capillary that is equivalent in terms of penetration rate to the real porous medium. In other words, the porous medium is characterized by an equivalent radius, $r_e$, and an equivalent contact angle, $\theta_e$ of a cylindrical capillary.

A frequently used, and practically attractive method, is the measurement of unidirectional penetration rates into a sample of the porous medium. If the sample is sufficiently short, it does not matter whether the penetration is horizontal or vertical, and Eq. (8) can be used to interpret the results. However, in this equation, $r_e$ and $\cos\theta_e$ appear as an inseparable product, so $r_e$ and $\theta_e$ cannot be independently determined. To correct for this problem, the two-liquid method has been recommended. In this method, the rate of penetration is measured twice. First, it is measured with a liquid that supposedly forms a zero contact angle with the solid matter of which the porous medium consists. From this measurement the equivalent radius $r_{e1}$ is calculated from Eq. (8), assuming $\cos\theta_{e1} = 1$. In the second experiment, the penetration rate of the actual liquid of interest is measured. In this experiment, the equivalent radius, $r_{e2}$, is assumed to be equal to $r_{e1}$, therefore $\theta_{e2}$ can be determined from Eq. (8). One disadvantage of this method is that it cannot be directly demonstrated that indeed $\theta_{e1} = 0$. In addition, it is not at all clear that always $r_{e2} = r_{e1}$. Theoretical calculations are currently being done to test this assumption.

Another possible solution to this problem of independently assessing $r_e$ and $\theta_e$ is the measurement of vertical penetration rates over a penetration length for which gravity is meaningful [14]. The main advantage is that the equivalent radius and contact angle are independently determined from Eq. (11), in which $r_e$ and $\theta_e$ appear in different combinations in the definitions of $G$ and $\tau$. The disadvantage of the method is that the time required for the experiments is on the order of hours rather than minutes as in the short-sample penetration method. In addition, it is not yet clear how to relate the equivalent contact angle to the true, intrinsic value of the contact angle. For example, the $\theta_e$ measured by the vertical penetration method of an oil (Brookfield viscosity standard oil) into Whatman filter paper no. 1 was about 77°, whereas this oil definitely makes a low contact angle with the fibers themselves [14]. To demonstrate that this was not an experimental error, the product $r_e\theta_e$ was measured by both vertical penetration and horizontal penetration and came out to be practically the same. Attempts to explain this seeming discrepancy are currently in progress.

## 4. SUMMARY

Wetting in porous media of limited size is a fascinating topic with many important practical applications. The following points shortly summarize the above discussion:

1. The consideration of a high-curvature liquid reservoir (small $R/r$ ratio) led to the generalization of the criterion for penetration into a capillary in terms of the radius of curvature rather than in terms of the contact angle. The generalized criterion allows for penetration at contact angles higher than 90°, though penetration may not be complete.

2. The kinetics of penetration is accelerated by the existence of a high-curvature liquid reservoir.

3. When a finite amount of the liquid is present, e.g. a liquid drop that penetrates into paper, the penetration may proceed by "re-distribution." This mechanism implies that liquid is moving from relatively large pores into smaller pores, due to the higher capillary pressure difference in the latter.

4. When the thickness of the porous medium is small (small $D/r$ ratio), penetration may be incomplete even when the contact angle is less than 90°, due to the "re-exposure" of the liquid inside the pores to the outside fluid.

Characterization of porous media in terms of wettability and pore size distribution is of utmost importance. Some approaches to characterization of porous media by the kinetics of penetration have been discussed. This short discussion demonstrates that the problem is yet far from being solved, and that the role of theoretical modeling is essential both in interpreting experimental data and in avoiding misinterpretations.

# REFERENCES

1. N.R. Morrow and G. Mason, *Current Opinion Colloid Interface Sci.* **6**, 321 (2001).
2. A. Marmur, in: *Modern Approaches to Wettability*, M. E. Schrader and G. Loeb (Eds.), Ch. 12, pp. 327-358, Plenum Press, New York (1992).
3. J. van Brakel, *Powder Technol.* **11**, 205 (1975).
4. R. Lucas, *Kolloid Z.* **23**, 15 (1918).
5. E. W. Washburn, *Phys. Rev.* **17**, 273 (1921).
6. A. Marmur, *J. Colloid Interface Sci.* **122**, 209 (1988).
7. A. Marmur, *J. Colloid Interface Sci.* **130**, 288 (1989).
8. A. Marmur, *Chem. Eng. Sci.* **44**, 1511 (1989).
9. D. Danino and A. Marmur, *J. Colloid Interface Sci.* **166**, 245 (1994).
10. E. Kissa, *J. Colloid Interface Sci.* **83**, 265 (1981).
11. A. Marmur, *J. Colloid Interface Sci.* **124**, 301 (1988).
12. A. Marmur, *J. Colloid Interface Sci.* **123**, 161 (1988).
13. A. Marmur, *J. Phys. Chem.* **93**, 4873 (1989).
14. A. Marmur and R. D. Cohen, *J. Colloid Interface Sci.* **189**, 299 (1997).

*Contact Angle, Wettability and Adhesion*, Vol. 3, pp. 385–406
Ed. K.L. Mittal
© VSP 2003

# Dynamic aspects of wetting in granular matter

DIMITRIOS GEROMICHALOS,* MIKA KOHONEN, FRIEDER MUGELE and
STEPHAN HERMINGHAUS

*Department of Applied Physics Laboratory, Ulm University, D-89069 Ulm, Germany*

**Abstract**—We have studied the impact of wetting on the dynamics of granular matter. In a first type of experiments, we focussed on the movement of liquid imbibition fronts in a fixed granular matrix. Roughness exponent morphology in both two-dimensional and three-dimensional systems calculated from correlation functions of the interface position compares favorably with recent theoretical models involving non-local effects. In a second type of experiments, we studied the effect of liquid on the mixing behavior of a granular system consisting of a mixture of glass beads of different sizes. We found that the extent of segregation upon horizontal shaking depends strongly on the liquid content.

*Keywords*: Granular matter; wetting; imbibition; non-equilibrium thermodynamics.

## 1. INTRODUCTION

Even though much progress has been made in the last years in describing the dynamics of dry granular materials [1, 2], the dynamic properties of "wet" systems composed of a granular material and a liquid remain largely unknown. This is, on the one hand, due to the complexity of these systems which makes theoretical approaches diffcult. On the other hand, only a few experiments have yet been conducted on these systems. In this paper we describe some of the experiments we have performed on the dynamic aspects of wet granular materials. They comprise not only the mechanical properties of the wet granulate, but also the dynamics of the liquid/gas interface within the material. A thorough understanding of the latter is indispensable for modelling the properties of the whole system, since the deformation of a wet granular matter entails the motion of liquid interfaces in its interior.

In order to analyze the dynamic properties of the fluid phase, in a first series of experiments we examined the flow of liquid through model granular materials. The imbibition of a liquid in a randomly distorted medium is interesting in itself,

---

*To whom all correspondence should be addressed. Phone: +49 (0)731 50 22930,
Fax: +49 (0)731 50 22958, E-mail: Dimitrios.Geromichalos@physik.uni-ulm.de

since it occurs, for example, in processes such as oil recovery, irrigation in agriculture, or wetting of raw material powders in industrial chemistry. For practical reasons, these experiments were mainly conducted in two-dimensional model systems [3]. The results were compared with theory and strongly support non-local theoretical models of imbibition [4, 5], showing a crossover behavior from a non-local regime at small length scales to a local regime at larger length scales. We were also able, for the first time, to tune the crossover length by changing the system parameters in accordance with non-local models [6]. Experiments for the three-dimensional case are the first of their kind and show a similar transition.

In a second series of experiments we studied the dynamic aspects of a granular material for the case in which a small amount of liquid had already spread uniformly throughout the sample. The mechanical properties of wet granular materials play a major role in industrial applications and geology. In our experiments we concentrated on the effect of liquid on segregation in mixtures of differently sized glass beads upon horizontal shaking. We observed a strong dependence of the segregation on the liquid content. Surprisingly, this dependence was not monotonic, as one might have expected.

## 2. IMBIBITION INTO POROUS MATERIALS

### 2.1. Theoretical and experimental background

Because of the tremendous practical importance of imbibition into porous materials, many experiments have been carried out with the aim of obtaining a thorough physical understanding of the mechanisms involved. However, satisfactory agreement with theory has not yet been achieved, partly because of inherent complications in many of the experiments performed so far. Although the three-dimensional case is of extremely great relevance, experiments have focused on the experimentally more accessible two-dimensional case, since the moving fronts in this case can be simply recorded by a camera. One can distinguish between two different types of experimental systems. One consists of paper as the random distorted medium [7–9]. The problems that occur in the case of paper are swelling, poorly defined roughness, and evaporation [10]. Other experiments using Hele-Shaw cells with randomly distributed glass beads [11, 12] provide a well-defined geometry, but exhibit strong pinning effects. Also the intrinsic length scales could be of the order of the diameter of the beads. Hence statistical features of the invading front may escape observation.

The first approach in describing the dynamics of liquid with a height $H$ rising into a capillary with radius $R$ was made by Washburn in 1921 [13]. Starting from simple force equilibrium arguments he obtained the equation

$$8\pi\eta H \frac{dH}{dt} = 2\pi R\gamma \cos\theta - \pi R^2 \rho g H$$

where $\eta$ is the viscosity, $\gamma$ the surface tension, $\theta$ the contact angle and $\rho$ the liquid density. The main feature of this equation is that the $H \propto \sqrt{R\gamma/2\eta}\sqrt{t}$ for short times $t$, and that the front asymptotically approaches a certain equilibrium height $H_{eq} = \dfrac{2\gamma}{R\rho g}$. For two parallel plates separated by a distance $d$ the Washburn equation takes the form

$$\frac{dH}{dt} = \frac{1}{12} \frac{d^2}{\eta H}\left(\frac{2\gamma}{d} - \rho g H\right)$$

with $H_{eq} = \dfrac{2\gamma}{d\rho g}$ and $H(t) \approx \dfrac{d\gamma}{3\eta}\sqrt{t}$ and for short times.

In porous media the situation is more complicated, but it can be shown [14] that at least for the average motion, Darcy's law, which may be viewed as a more general form of Washburn's law, is valid:

$$\langle \mathbf{v} \rangle = -\frac{k}{\eta}(\nabla P - \rho \mathbf{g})$$

where $\mathbf{v}$ is the velocity, $k$ the permeability, $P$ the driving pressure and $\mathbf{g}$ the gravitational acceleration vector. The behavior $H \propto \sqrt{t}$ for short times and the reaching of a saturation height was confirmed by many experiments both in two [15] and in three dimensions [16].

To gain a deeper understanding of the imbibition process one has to take into account explicitly how the roughness of the granular material influences the flow and eventually causes a roughening of the liquid front.

Assuming that the form of the front is at least self-affine, one obtains the Family-Vicsek scaling [17] for the width $w$ of the front:

$$w(L,t) \approx L^\chi f\left(t/L^z\right) \qquad z = \chi/\beta$$

where $L$ is the system size, $\chi$ the roughness exponent, $\beta$ the growth exponent, $z$ the dynamical exponent, and $f(u)$ a function with $f(u) \propto u^\beta$ for $u \ll 1$ and $f(u) = const$ for $u \gg 1$.

Since the local saturation width of the front is reached much faster than global width, one calculates in practice the exponents by determining the slope of the double logarithmic plot of the following spatial and temporal correlation functions:

$$C(l) = C(l,0) \propto l^\chi \qquad C(\tau) := C(0,\tau) \propto t^\beta$$

$$C(l,\tau) = \sqrt{\left\langle (\tilde{h}(l+x, \tau+t) - \tilde{h}(x,t))^2 \right\rangle}$$

$$\tilde{h}(x,t) = h(x,t) - \overline{h}(t)$$

$$\overline{h}(t) = \frac{1}{L} \int_0^L h(x,t)\mathrm{d}x$$

where $h(x, t)$ is the local height. $\tilde{h}(x,t)$ is the deviation of the front height at the time $t$ and the position $x$ from the mean value $\overline{h}(t)$ at the time $t$. To obtain the correlation function $C(l, \tau)$ one averages the squared differences of the height deviations. Averaging is performed over all $x$ and $t$, and over many samples.

The roughening processes can be described either with the percolation theory [3] (which yields $\chi \approx 0.6$ for the directed percolation depinning (DPD) model in two dimensions) or, in an analytical form by continuum models. A continuum model, which provides a general description of the roughening of surfaces, is expressed by the equation of Kardar, Parisi and Zhang [18] (KPZ) as:

$$\frac{\partial h(x,t)}{\partial t} = \nu \nabla^2 h + \frac{\lambda}{2}(\nabla h)^2 + \eta(x,t)$$

where $\nu$ is an effective surface tension, $\lambda$ characterizes the lateral growth, and $\eta$ is a non-correlated, Gauss-distributed random variable with $\langle \eta(x,t) \rangle = 0$ and $\langle \eta(x,t)\eta(x',t') \rangle = 2\mathcal{D}\delta^{d^*}(x-x')\delta(t-t')$ where $d^*$ is the dimension of the system and $\mathcal{D}$ a normalization factor. The values for $\chi$ obtained from the KPZ equation using the renormalization group theory lie, in general, around 0.6 for the two-dimensional case and 0.3 for the three-dimensional case [3]. The value for $\beta$ in the two-dimensional case is 0.3. One can further discern the case of the so-called "quenched noise" ($\eta(x, h)$; QKPZ). For a two-dimensional system the value for the roughness exponent lies between 0.6 and 0.7 [3].

It should be noted that neither the percolation theory nor the KPZ equation reproduces Washburn's law [5]. Furthermore, the experimentally determined roughness exponents are scarcely reproducible and lie between 0.6 for paper wetting experiments [3] and 0.8 for Hele-Shaw cells experiments [3]. Importantly, the percolation theory as well as the KPZ equation do not take into account mass conservation within the fluid, an important property which is typical for fluids and from which non-locality can result [19]. In recent years, there has been considerable improvement in the theoretical description of advancing liquid front phenomena [4, 5, 20, 21], with particular focus on the non-local effects on imbibition fronts. The regime observed at smaller scales is particularly interesting, since significant deviations from the QKPZ behavior are observed in experiments, and it seems worthwhile to explore its possible origin in non-local effects in the liquid front dynamics. As we will see later, our experiments show a crossover in the spatial correlation functions from a larger roughness exponent at short length scales to QKPZ-like behavior on larger length scales. We will denote the short scale roughness exponent by $\chi_1$ and the large scale roughness exponent by $\chi_2$. Such a behavior is indeed predicted by two recent non-local models [4, 5].

One type of model incorporates the non-local effects due to lateral pressure gradients in the liquid close to the front line. These are accounted for by Darcy's law, $\mathbf{v} = -\kappa\nabla p$ (gravity neglected). An equation of motion of the fluid front can be derived ([4, 22]) by invoking fluid incompressibility, $\nabla\mathbf{v} = 0$, and using the Grinstein-Ma expression for the random field Ising model,

$$\frac{\partial h}{\partial t} = F + \gamma\frac{\partial^2 h}{\partial x^2} + \eta(x, h)$$

which describes the front evolution in a rough material. Here $F$ is the driving force, $h$ the height of the front, $x$ the horizontal position, $\gamma$ the surface tension, and $\eta$ a term describing the roughness (see [4] and references therein). The equation of motion, as obtained by Ganesan and Brenner [4], is intrinsically non-local, non-linear and rather complicated, but the scaling properties of its solutions can be derived assuming small amplitudes $h(x, t) - H(t)$. It was found (for forced fluid invasion, FFI) that $\chi_1 = 0.75$. On the other hand, for spontaneous imbibition the authors give a roughness exponent of 0.633 which is the expected value for strong pinning by directed percolation [3]. (It could be that the experiments reported here fall in the FFI universality class, since the pinning was probably too weak.)

The other non-local model, the phase field model [5, 23], describes the problem of fluid invasion with a Ginzburg-Landau like scheme

$$\mathcal{F}(\phi) = \int((\nabla\phi)^2 / 2 + V(\phi))dx$$

$$V(\phi) = -\phi(x,t)^2 / 2 + \phi(x,t)^4 - \alpha(x)\phi(x,t)$$

with the phase field $\phi(x, t)$ which takes the value $-1$ in the dry and $+1$ in the wetted state. This model is analogous to other phase field models which are used to describe systems consisting of different phases (e.g. solid and liquid phases). The roughness of the material is modelled by $\alpha$, a stochastic and non-correlated variable. In order to obtain a solution for the distribution of the liquid and gas phases, $\phi(x, t)$, one has to minimalize $\mathcal{F}(\phi)$. Since the term $(\nabla\phi)^2/2$ has to vanish, therefore, it can be associated with a driving force. The potential $V(\phi)$ is chosen such that it has two minima which correspond to the dry and wet phases. The equation of motion for the phase field variable is

$$\frac{\partial\phi}{\partial t} = -\nabla^2(\nabla^2\phi + \phi - \phi^3 + \alpha)$$

This model yields the Washburn law for the case with no roughness. It is also elaborate enough to produce information about the crossover length $\xi$, at which the transition from the non-local short scale to the local large scale behavior occurs. Furthermore, it predicts that the crossover length $\xi$ should be time dependent

[5] for spontaneous imbibition, with $\xi = \sqrt{\dfrac{\gamma}{6\eta v}}d$ and $v \propto 1/\sqrt{t}$, i.e. $\xi \propto t^{1/4}$.
From Washburn's law we know that

$$\bar{V}(t) = \sqrt{\frac{d\gamma}{3\eta}}\frac{1}{\sqrt{t}}$$

for short times. Hence, we obtain

$$\xi \propto \sqrt{hd} ,$$

i.e. the crossover length is predicted to increase with the separation of the plates as $d^{1/2}$.

## 2.2. Experimental setup

We have studied spontaneous imbibition of a liquid front into a gap of prescribed width $d$ between two roughened glass plates. This geometry is similar to previous experiments using Hele-Shaw cells, but in our experiments the roughness is on a much smaller length scale. Furthermore, the roughness is close to a non-correlated Gaussian distribution. The plates were made from optical quality glass, obtained with a planarity of about one micrometer over a distance of ten centimeters. The thickness of the plates was 8 mm, thick enough to avoid any flexure of the glass surfaces by capillary forces. One surface of each plate was roughened by lapping with coarse sapphire powder (grain diameter $\approx$ 15 µm), thereby preserving the large scale planarity. The specifications for the roughness were: $R_a \approx 0.55$ and $R_z \approx 3.5$ µm [36]. The maximum height variation in the topography of the rough surfaces was $\approx$ 5 µm, as can be seen in Figure 1a, which was obtained by scanning force microscopy (SFM). The Fourier transform of the topography (Figure 1b) shows that the roughness is uniform and isotropic. To see this on a larger scale, we also conducted an optical Fourier transform of the rough glass surface using an expanded laser beam, and focussing the transmitted light onto a CCD camera chip. The intensity distribution, which is shown in Figure 1c, represents the square of the 2D Fourier transform of the rough glass surface [24]. Again, there is no indication of a characteristic length scale or preferential orientation of the roughness.

Before the experiments, the glass plates were ultrasonically cleaned in acetone, ethanol and then water, and blown dry with nitrogen gas. In order to remove possible residual organic contamination, they were immersed for several hours in concentrated sulfuric acid with an oxidant added, thoroughly washed with deionized water, and blown dry with nitrogen again.

The plates were mounted with the rough sides facing each other, using stainless steel ribbons with thicknesses ranging from 10 to 50 µm as spacers, in order to maintain a well defined plate separation. Spontaneous imbibition started after

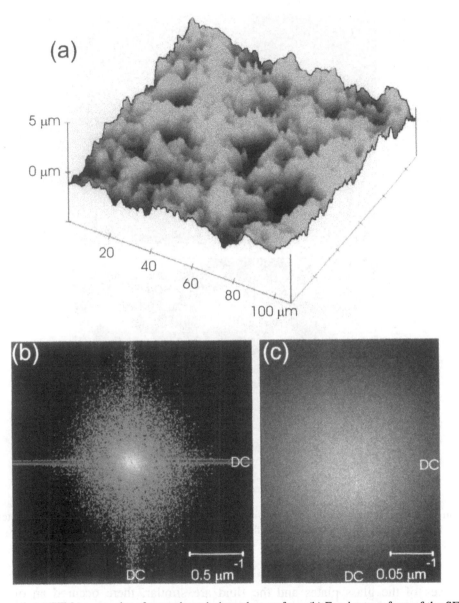

**Figure 1.** (a) SFM topography of a roughened glass plate surface. (b) Fourier transform of the SFM picture. The cross lines are artifacts arising from the transform procedure. *DC* marks the point of origin of the Fourier space. (c) Optical Fourier transform, showing the roughness on a larger length scale. In (b) and (c) the dimension is given in Fourier space units.

placing the lower edge of the vertically mounted pair of plates into a liquid reservoir. The liquid entered the gap between the plates spontaneously due to capillary forces. The plates were illuminated with a white luminescent screen and observed in transmission with a high resolution CCD camera. Since the optical refraction

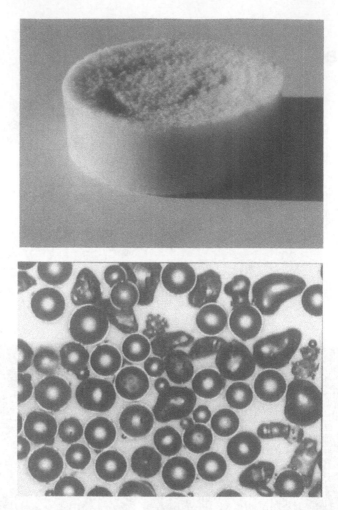

**Figure 2.** Upper picture: Wax-sand block with the frozen 3D imbibition front at the top (diameter: 8 cm); Lower picture: Microscopic view of the granular material used in the 3D imbibition experiments (typical grain diameter: 10 μm).

indexes of the glass plates and the fluid are similar, there occured an optical immersion of the roughness by the fluid which led to a sharp contrast at the invading front line. In order to improve the statistics, five runs were taken at each plate separation. The invading front line was imaged using a high resolution CCD camera, digitized and subsequently analysed numerically. We used water and 1-undecanol $(CH_3(CH_2)_{10}OH)$ as wetting liquids. The latter yields particularly smooth fronts which can easily be analysed by conventional procedures. Furthermore, it rises at a much slower rate than water, which eases data acquisition.

In the case of water, air entrainment at the moving front, leading to the inclusion of air bubbles in the rising liquid column, was frequently observed. The bubbles were observed to become "smeared" after some time, since the glass at their position became wetted by a precursor film. After that they moved with the liquid. This phenomenon is interesting in itself, but will not be discussed further here.

We were also able to produce and record spontaneous imbibition fronts for the three-dimensional case. Here we used a sample of glass beads with a diameter of approximately 100 μm as a granular material (Figure 2). The beads were poured into a glass pipe (diameter 8 cm) which was sealed with a porous tissue at the bottom, such that beads were held in the pipe. The pipe with the beads was placed in an oven, heated to 100°C and then dipped into a liquid wax reservoir at the same temperature. After some seconds the pipe was removed from the reservoir and cooled down. Subsequently, it was possible to push out the sand-wax block and blow away the glass beads which were not glued by the wax with nitrogen (Figure 2). The frozen imbibition front was then recorded with an optical profilometer. We should note that although wax contracts upon freezing, the wax-sand system displays only a negligible volume reduction since the sand resists the contraction of the wax.

## 2.3. Results

Figure 3a shows a typical rising front of undecanol. In general for larger plate separations $d$, the fronts became smoother because of the increased influence of the front line tension which is directly proportional to its "thickness", i.e. the plate separation. The height of the front was small as compared to the predicted value of $H_{eq}$ for all plate separations investigated. Thus the average height of the front should rise according to Washburn's law, i.e., $H = A\sqrt{t}$ [13]. This was, in fact, observed with satisfactory precision, as demonstrated in Figure 3b. The prefactors $A$ found experimentally were about one half of the corresponding theoretical values for smooth plates and perfect wetting, $A_{th} = \sqrt{\dfrac{d\gamma}{3\eta}}$ , according to Washburn's law. The difference is very probably due to the change of the hydrodynamic boundary conditions at the glass surfaces arising from the roughness. Small scale local variations in the surface energy of the roughened glass may also be important.

The roughness exponent $\chi$ [17] was obtained from the correlation function $C(l)$, as explained above. Figure 4a shows a typical result. It is clear that a transition in the roughness exponent occurs at length scale $\xi$. For length scales less than $\xi$, we obtain $\chi_1 \approx 0.8$. For length scales larger than $\xi$ we observed a crossing over of this exponent to the quenched noise Kardar-Parisi-Zhang (QKPZ) and DPD exponent, $\chi_2 \approx 0.6$ [3, 18]. The latter regime is indicative of local dynamics, i.e. the dynamics is influenced by lateral interactions only over distances which are small compared to the examined length scale. However, the exact results were scattered, this

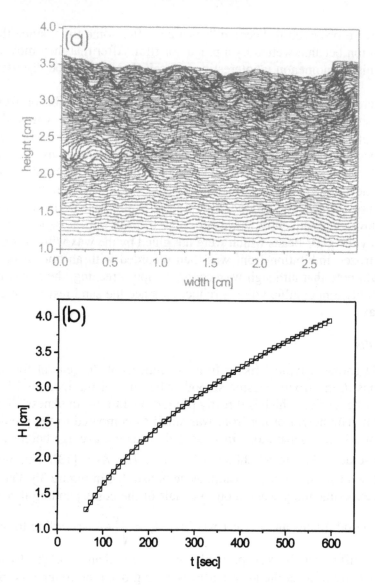

**Figure 3.** A rising undecanol front at different times, with a glass plate separation of 10 μm: (a) A series of snapshots of the front, each separated in time by 10 sec, decreasing from the top. (b) The mean height *H* of the front as a function of time. Washburn's law is superimposed as a solid line, and is in good agreement with the data.

crossing over of the exponents was reproduced in practically all runs conducted. Earlier experiments conducted with rougher glass plates ($R_a \approx 40$ μm) and unde-canol gave similar results. Similar crossover effects have been reported before, but were interpreted in terms of pinning effects [7]. It seems worthwhile to re-examine these results in view of the more advanced theories which are available now.

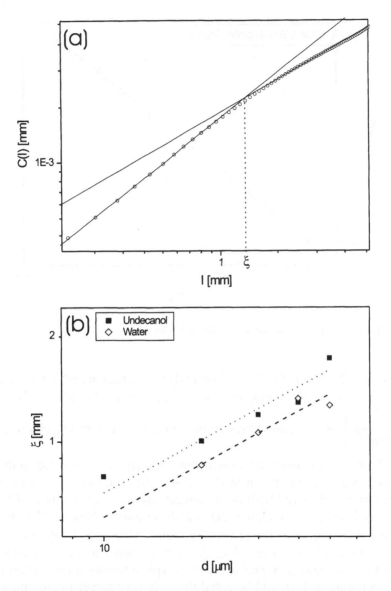

**Figure 4.** (a) Typical example of a spatial correlation function with two regimes (used liquid: water; plate separation: 40 μm). (b) Crossover length $\xi$ for different plate separations. The curves indicate the theoretical prediction, $\xi \propto \sqrt{d}$ .

As mentioned above, we were for the first time able to observe a dependence of the crossover length $\xi$ on the plate separation $d$. It is clear from Figure 4b (double logarithmic plot) that $\xi$ increases with increasing plate separation. The solid curve represents the above formula, $\xi \propto \sqrt{d}$ , as obtained from the phase field model, in agreement with the data within experimental scatter. Furthermore, we have reason

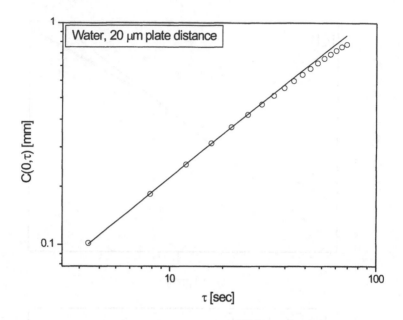

**Figure 5.** Typical example of a temporal correlation function.

to believe that $\xi = 0$ for $d = 0$ based on earlier experiments which we conducted with ethanol ($d = 0$; roughness: 40 μm). From the phase field model $\left( \xi = \sqrt{\dfrac{\gamma}{6\pi\eta\upsilon}}\,d \right)$, we can estimate $\xi$ from the measured front velocity to be of the order of 2 mm ($\upsilon$ was obtained experimentally). This is very close to the values found in our analysis, as shown in Figure 4b. Thus we have reason for the presumption that we observed both the non-local and the local regimes of liquid invasion into a disordered medium in 2D, and the crossover between the two [37].

We have also calculated the growth exponent $\beta$ from the temporal correlation function $C(\tau)$ (see above). The values differed for undecanol and water and were about 0.5 for undecanol and 0.6 for water. A typical temporal correlation function is shown in Figure 5. It should be noted that it is very surprising to obtain such a correlation function for the case of spontaneous imbibition. In theory, one obtains such a power law behavior if the motion is time-invariant, whereas in our case the velocity of the front decreased with time. Nevertheless, our result comes close to those obtained with forced fluid flow in Hele-Shaw cells [7].

We also examined the statistical behavior of the height deviation from the mean value $\delta h = h - \overline{h}$. We found that the obtained histograms could be fitted well with Gaussian curves for undecanol. For water the smallest (negative) values of $\delta$ occurred more often than the biggest (positive) ones, which indicates a stronger pinning (Figure 6). This would indicate that the undecanol system conforms more closely to theoretical models.

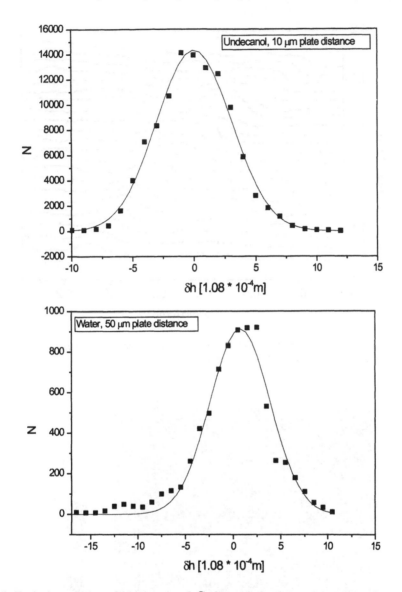

**Figure 6.** Typical examples of histograms of $\delta h$ for undecanol (upper graph) and water (lower graph). $N$ is the number of the observed values of $\delta h$. The curves represent Gaussian fits.

The equation of motion derived in ref. [4] is strongly nonlinear and its solutions at large amplitudes have not yet been examined. Thus, it is of interest to note here that we have frequently observed kinks in the rising liquid fronts which move laterally, as one might expect from a solitonic solution to the equation of motion. Figure 7 shows such an object (the small bump) moving with a velocity comparable to the velocity of the front. It may thus be of interest to search for solitonic solutions to the equation of motion of the front.

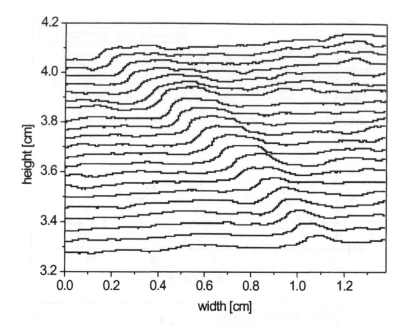

**Figure 7.** Rising undecanol front, with a plate separation of 10 μm. The curves represent a series of snapshots each separated in time by 4 sec, decreasing from top.

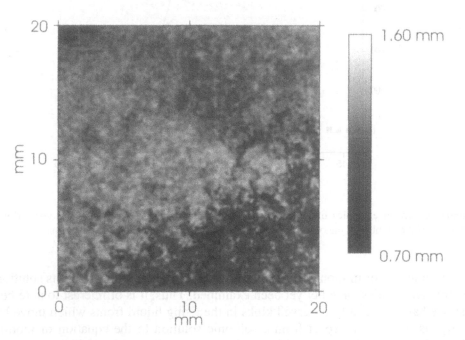

**Figure 8.** Topography of a frozen 3D imbibition front. (arbitrary absolute value, the total mean height $H$ was approximately 1 cm).

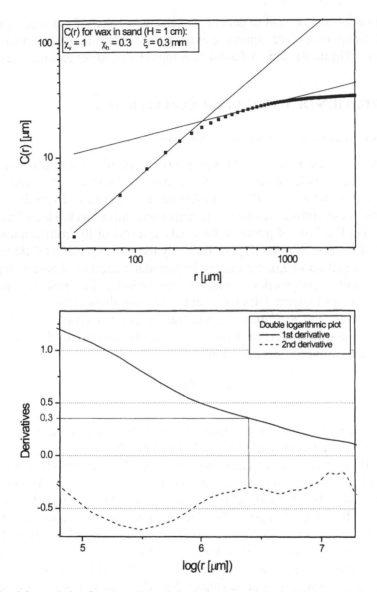

**Figure 9.** Spatial correlation function $C(r)$ and the derivatives of the double logarithmic plot (1st derivative: $d \log(C(r))/d \log(r)$; 2nd derivative: $d^2 \log(C(r))/d(\log(r))^2$).

In the 3D experiments mentioned above we recorded the surface topography with an optical profilometer (Figure 8). We determined the roughness exponent of the frozen imbibition fronts in a way analogous to the 2D case. Here we had instead of $C(l)$ a radial correlation function $C(r)$, i.e. we averaged over all azimuthal angles. For small distances we obtained a value of $\chi_1 \approx 1$ which seemed to cross over to the value predicted by the percolation theory with the Eden model, namely 0.3 [25] (Figure 9). Although the double logarithmic plot does not really display a straight line for this second exponent, the absolute value of the second derivative of this plot showed a clear minimum at the point where the first derivative had the value of 0.3 (Figure 9). This behavior was reproduced in several measurements.

## 3. SEGREGATION OF WET GRANULAR MATERIALS

### 3.1. Theoretical and experimental background

In general, there are many similarities between granular materials and fluids. In both cases the particle number and the particle volume is conserved. The flow patterns in granular media often resemble those observed in hydrodynamic flow. But despite these similarities there are important differences which have made a description of the flow of granular materials in terms of hydrodynamics impossible as yet (for a review see ref. [2]). Probably the most important difference is the nature of dissipation of kinetic energy in granular materials. There is friction between the grains, and particle collisions are inelastic. For instance, arches can form in a granular material which prevent flow through an orifice.

An interesting feature of granular materials is their behavior when a vibration is applied, such that they start moving. An important quantity to describe the shaking is the dimensionless "acceleration"

$$\Gamma = A \frac{\omega^2}{g}$$

where $A$ is the amplitude, $\omega$ the frequency of the shaking and $g$ the acceleration of gravity. Often a behavior called "fluidisation" can be observed where the vibration takes the role of an effective temperature and the material starts to "flow". A considerable amount of work has been done in this area [1, 26]. Another interesting effect caused by shaking, which occurs in mixtures of particles of different sizes, is the "Brasil nut" effect [27]. Here the particles segregate such that, typically, the bigger ones rise to the top and the smaller ones sink to the bottom. Several effects may contribute to producing segregation. For example, in the case of large differences in the diameters of the smaller and the larger particles, segregation may occur simply due to a sieving effect. Also, it is easy for small particles to fill the space beneath a large particle when it moves up because of the shaking. On the other hand, several smaller particles have to move away simultaneously before a big particle can sink down again [27]. Convective currents also play a role in the segregation [28]. It should also be noted that horizontal or vertical shaking can lead to completely different granular behaviors [1].

As is known from everyday experience, the mechanical properties of a granular material are very sensitive to the presence of even small amounts of liquid. This is due to the cohesion of the granular material which arises from the presence of capillary bridges between the grains [29]. It has been shown that the surface tension of the liquid is not the only relevant parameter, but that the viscosity also plays an important role. Another effect of capillary forces is the change of the packing of granular material, which leads, for small liquid amounts, to a bigger porosity, i.e. a looser packing, after some shaking [30]. To date most studies in this area have been concerned with the stability of wet sand piles (see e.g. [31, 32]).

Very recently there has been interest in the effect of liquid on the segregation behavior of granular mixtures [33, 34]. In these initial studies experiments were preformed by pouring wet mixtures of granular materials through an orifice and determining the extent of segregation in the resulting heap. Here, wet mixtures of particles with two different sizes flowed through an orifice and segregated. It was observed that with increasing amount of liquid added the tendency to segregate decreased. In our first experiments concerning the effect of liquid on the behavior of a granular material, we have also focussed on the segregation behavior. Importantly, we studied mixing in a sample which had been shaken horizontally.

## 3.2. Experimental setup

We used mixtures of 150 $\mu$l of glass beads with radius $r_1 = 2.5$ mm and 150 ml of smaller glass beads with radius $r_2$ which ranged from 0.05 to 0.5 mm. In order to prevent crystallization, we chose glass beads which were not exactly monodisperse. (The width of the size distribution was on the order of 10%.) The mixtures, to which known amounts of liquid were added, were placed in sealable jars (diameter 8 cm), put on a shaker and shaken for half an hour after the liquid had distributed uniformly. The shaking motion consisted of small horizontal circles with a diameter of approximately 0.5 cm, and a frequency of approximately 20 Hz, so that $\Gamma = 3.6$. The frequency was chosen such that the particles at the bottom of the jar moved very slowly. After shaking the particles for half an hour no further changes in the segregation pattern were observed and photographs of the jars were taken. It was very easy to see where the bigger and smaller particles were and hence to determine the extent of segregation. To obtain better statistics, several runs were made. As fluid we used water but we also used ethanol for control measurements with similar results.

## 3.3. Results

In qualitative agreement with the results of previous studies, we also observed segregation in the dry mixtures, with the large particles tending to collect at the top of the sample. However, the extent of the observed segregation was found to depend strongly on the amount of added liquid. As mentioned above, the extent of segregation in the mixtures could be easily judged from the photographs (see Figure 10). We define the segregation transition zone (mixture zone) as the zone in which both the smaller and the larger particles were present. Although there was a small concentration gradient of the smaller particles from the walls of the jar to its center, and although we observed mainly the particles at the walls, the results

**Figure 10.** Horizontally shaken mixture of beads (diameters: 1 mm and 5 mm) without water (right) and after adding 15 drops of water (left).

given below are qualitatively valid for the whole volume, as we confirmed by taking samples at the center.

In experiments with the smallest glass beads we observed that the extent of mixing increased with the amount of added liquid. This is what one would intuitively expect, due to the increased cohesion between the particles. Experiments with larger particles, however, revealed that the phenomena were in fact more complex, showing a decrease of propensity to mix when only a small amount of liquid (less than ca. 0.05% of the total volume; see also Fig. 11) was added. Figure 11 shows an example of this effect (here $r_2 = 1$ mm).

For spheres of intermediate sizes both effects could be seen. In order to obtain a better picture about these phenomena we performed several experiments with different amounts of liquid and plotted the height (normalized by the total height of the sample) of the mixing zone versus the liquid amount. Figure 11 shows such a plot for the case where $r_2 = 0.475$ mm. Here both effects mentioned above are visible. First the segregation increases, possibly due to the decreased mobility of the smaller particles in the presence of liquid bridges between the beads. Upon continued addition of the liquid the extent of segregation is reduced. Here the small particles begin to form larger aggregates and provide more resistance to the motion of the larger particles. At last a point of perfect mixture is reached, which remains the same even if 1000 drops are added. (When we conducted the same experiment under water we observed segregation again.)

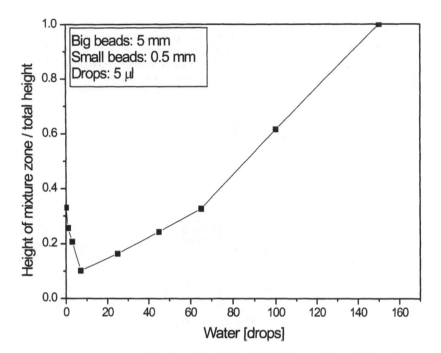

**Figure 11.** Normalized height of the mixture zone of horizontally shaken glass beads (diameters: 5 mm and 0.5 mm) as a function of the amount of water added.

We should mention that the general movement of the whole sample was a kind of circular convection around the axis of the jar. In the dry case all of the individual particles remained quite free in this convective movement and easily changed their relative position. Upon addition of the liquid the lower region of the sample, composed of small particles or a mixture of small and large particles, had more of a "viscoplastic" character. This lower "phase" coexisted with an upper "phase" of free particles for intermediate amounts of added liquid. The approach to complete mixing was characterized by a rapid increase in the height of the lower "phase".

Based on numerous measurements we were able to construct the phase diagram shown in Figure 12. This diagram summarizes the observed segregation behavior as a function of the radius of the small spheres and the amount of liquid added. The grey area denotes the combination of the parameters for which complete mixing occurs. The shape of this area is similar to that observed in a previous study [33], though the experimental setup was very different. The dotted horizontal line is an upper limit for the radius of the smaller spheres for which complete mixing can occur. No matter how much liquid we added for the case $r_2 = 0.78$ mm there always remained an area of freely moving large particles at the top of the sample. The steep straight line in the plot denotes the parameter combination at which the segregation reaches its maximum after adding more and more liquid.

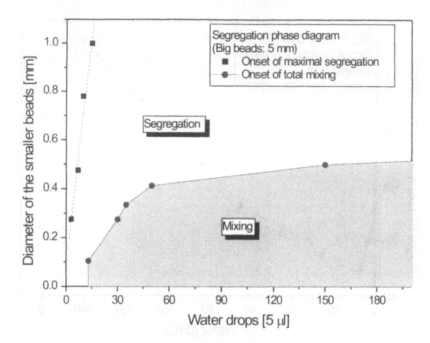

**Figure 12.** Segregation phase diagram for mixtures of glass beads.

## 4. CONCLUSION

In conclusion, our results may be interpreted as being consistent with a crossover from non-local to QKPZ or DPD behavior in two-dimensional liquid imbibition between rough parallel plates. We determined the roughness exponent $\chi$ of the fronts and found that it crossed over from a value of 0.8 at small scales to a value of 0.6 at larger scales. We were also able to vary the crossover length by varying the plate separation. This observation is in accordance with recent theoretical treatments proposed for liquid imbibition in disordered media. It should be noted that very recently a similar crossover behavior was found for the case of forced fluid flow [35]. We also found strong indications for a similar behavior in the three-dimensional case.

Furthermore, we examined the effect of small amounts of liquid on segregation in mixtures of different sized particles after horizontal shaking. We found that the segregation did not necessarily depend monotonically on the fluid content and deduced a phase diagram for the segregation behavior.

Both results are of importance for understanding the complex behavior of wet granular matter.

*Acknowledgements*

The authors appreciate inspiring discussions with Tapio Ala-Nissila and Sami Majaniemi from Helsinki University. Further we would like to thank Matthias

Mahlich, Werkstoffe der Elektrotechnik, Ulm University for providing the profilometer. Financial support from the German Science Foundation within the Priority Program 'Wetting and Structure Formation at Interfaces' is gratefully acknowledged. M.K. acknowledges financial support from the Alexander von Humboldt Foundation.

## REFERENCES

1. G. H. Ristow, *Pattern Formation in Granular Materials*, Springer, Berlin (2000).
2. L. P. Kadanoff, Rev. Mod. Phys. **71**, 435 (1999).
3. A.-L. Barabási and H. E. Stanley, *Fractal Concepts in Surface Growth*, Cambridge University Press, Cambridge (1995).
4. V. Ganesan and H. Brenner, Phys. Rev. Lett. **81**, 578 (1998).
5. M. Dubé, M. Rost, K. R. Elder, M. Alava, S. Majaniemi and T. Ala-Nissila, Phys. Rev. Lett. **83**, 1628 (1999).
6. D. Geromichalos, F. Mugele and S. Herminghaus, Phys. Rev. Lett. **89**, 104503 (2002).
7. V. K. Horváth, F. Family and T. Vicsek, Phys. Rev. Lett. **67**, 3207 (1991).
8. S. V. Buldyrev, A.-L. Barabási, F. Caserta, S. Havlin, H. E. Stanley and T. Vicsek, Phys. Rev. A **45**, R8313 (1992).
9. V. K. Horváth and H. E. Stanley, Phys. Rev. E **52**, 5166 (1995).
10. M. Dubé, M. Rost and M. Alava, Eur. Phys. J. B **15**, 691 (2000).
11. V. K. Horváth, F. Family and T. Vicsek, J. Phys. A **24**, L25 (1991).
12. A. Paterson, M. Fermigier, P. Jenffer and L. Limat, Phys. Rev. E **51**, 1291 (1995).
13. E. W. Washburn, Phys. Rev. **17**, 273 (1921).
14. M. Sahimi, Rev. Mod. Phys. **65**, 1393 (1993).
15. E. Schäffer and P.-Z. Wong, Phys. Rev. E **61**, 5257 (2000).
16. M. Lago and M. Araujo, Physica A **289**, 1 (2001).
17. F. Family and T. Vicsek, J. Phys. A **18**, L75 (1985).
18. M. Kardar, G. Parisi and Y.-C. Zhang, Phys. Rev. Lett. **56**, 889 (1986).
19. S. He, G. L. M. K. S. Kahanda and P.-Z. Wong, Phys. Rev. Lett. **69**, 3731 (1992).
20. C.-H. Lam and V. K. Horváth, Phys. Rev. Lett. **85**, 1238 (2000).
21. A. Hernández-Machado, J. Soriano, A. M. Lacasta, M. A. Rodriguez, L. Ramirez-Piscina and J. Ortin, Europhys. Lett. **55**, 194 (2001).
22. J. Koplik and H. Levine, Phys. Rev. B **32**, 280 (1985).
23. M. Dubé, M. Rost, K. R. Elder, M. Alava, S. Majaniemi and T. Ala-Nissila, Eur. Phys. J. B **15**, 701 (2000).
24. E. Hecht, *Optics*, 3rd ed., Springer, Berlin (1998).
25. D. E. Wolf and J. Kertész, Europhys. Lett. **4**, 651 (1987).
26. G. Metcalfe, S. G. K. Tennakoon, L. Kondic, D. G. Schaeffer and R. P. Behringer, Phys. Rev. E **65**, 031302 (2002).
27. A. Rosato, K. J. Strandburg, F. Prinz and R. H. Swendsen, Phys. Rev. Lett. **58**, 1038 (1987).
28. J. B. Knight, Phys. Rev. Lett. **70**, 3728 (1993).
29. T. Mikami, H. Kamiya and M. Horio, Chem. Eng. Sci. **53**, 1927 (1998).
30. C. L. Feng and A. B. Yu, J. Colloid Interface Sci. **231**, 136 (2000).
31. N. Fraysse, H. Thomé and L. Petit, Eur. Phys. J. B **11**, 615 (1999).
32. T. G. Mason, A. J. Levine, D. Ertaş and T. C. Halsey, Phys. Rev. E **60**, R5044 (1999).
33. A. Samadani and A. Kudrolli, Phys. Rev. Lett. **85**, 5102 (2000).
34. A. Samadani and A. Kudrolli, Phys. Rev. E **64**, 051301 (2001).
35. J. Soriano, J. Ortin and A. Hernández-Machado, Phys. Rev. E 66, 031603 (2002).

36. The roughness specifications are defined by DIN norms as follows: Mean roughness: $R_a = \dfrac{1}{X} \int |y(x)| dx$ ; mean peak-to-valley roughness: $R_z = \dfrac{1}{n} \sum_{i=1}^{n} R_{zi}$ ; single peak-to-valley roughness: $R_{zi} = y(\text{peak}) - y(\text{next valley})$. The local height of the surface is denoted here by $y(x)$. $X$ stands for the size of the sample and $n$ for the number of peaks.

37. The fact that $\xi$ is found here to be of the order of the capillary length has no deeper physical significance, since $\xi$ varies with plate distance, but the capillary length does not.

*Contact Angle, Wettability and Adhesion*, Vol. 3, pp. 407–426
Ed. K.L. Mittal
© VSP 2003

# The effect of rock surface characteristics on reservoir wettability

CHANDRA S. VIJAPURAPU and DANDINA N. RAO*

*The Craft and Hawkins Department of Petroleum Engineering, Louisiana State University,
3516 CEBA Bldg., Baton Rouge, LA 70803*

**Abstract**—Significant strides have been made in recent years in gaining a better understanding of the role of fluid compositions on reservoir wettability. However, our knowledge of the effects that the solid surface characteristics have on establishing and altering wettability is quite limited. This study examines the effect of rock mineralogy and surface roughness on wettability in rock-brine-hydrocarbon systems. The wettability is characterized by using two different techniques. The Wilhelmy plate technique has been used to obtain dynamic (advancing and receding) contact angles averaged over the surface area of the solid substrate used. These results are compared with point-values of dynamic contact angles measured using the dual-drop dual-crystal (DDDC) technique for both smooth and rough solid surfaces of different mineralogy and roughness. The surfaces have been characterized using an Optical Profilometer and a Scanning Electron Microscope. Fluids from the Yates reservoir in West Texas have been used. While the Wilhelmy Plate Technique displayed insensitivity, the DDDC Technique showed significant effects of mineralogy and roughness on dynamic contact angles and hysteresis.

*Keywords*: Contact angles; dual-drop dual-crystal technique; solid-liquid-liquid systems; surface roughness; wettability.

## 1. INTRODUCTION

One of the common criticisms of contact angle measurements in reservoir engineering is that they use smooth crystal surfaces in place of reservoir rocks, which are not only rough but also are mineralogically heterogeneous. The main purpose of studying the effects of surface roughness on contact angles has been to investigate the possibility of relating contact angles measured on smooth surfaces to those likely to be operative on rough surfaces within porous media. Although the importance of surface roughness to wetting behavior is recognized, difficulties have been reported in obtaining a definitive account of roughness effects. This study is an attempt to address these concerns.

---

*To whom all correspondence should be addressed. Phone: (225) 578-6037,
Fax: (225) 578-6039, E-mail: dnrao@lsu.edu

Wenzel [1] studied the effect of roughness on contact angle and proposed a theory, which is used to derive the relationship between the equilibrium angle observed on a smooth surface, $\theta_E$, and the advancing angle on a rough surface $\theta_A$:

$$\text{Cos } \theta_A = r \text{ Cos } \theta_E. \tag{1}$$

where r is the roughness factor, defined as the ratio of the surface area of the rough surface to that if the solid were microscopically smooth.

If the surface is rough enough with a large number of asperities, the angle measured with the horizontal will actually be the apparent contact angle because the asperity on the surface might not be horizontal as illustrated by Adamson [2]. The true contact angle can be much larger than what we measure with the horizontal.

Cassie and Baxter [3] in their study on waterproofing fabrics used the example of bird feathers to show the difference between true and apparent contact angles. They reported an apparent contact angle of 150° as opposed to a true contact angle of 100°.

Oliver *et al.* [4] and Mason [5] in their theoretical study of the influence of surface roughness on spreading and wettability showed by calculating the equilibrium shape of the liquid drop resting on a rough surface the relation between the true equilibrium contact angle at the three-phase contact line and the apparent contact angle observed microscopically at the geometrical contour plane of the solid. In their attempt to clarify the Wenzel's relation between the apparent and true contact angles they considered surfaces having random roughness to derive a statistical relation between the angles which introduced a factor for surface texture in addition to surface roughness.

Morrow [6] showed the importance of surface roughness on the apparent contact angle and contact angle hysteresis. He studied the effect of roughness on contact angle through capillary rise in Teflon tubes. He reported an increase in contact angle hysteresis between smooth tubes and rough ones. Furthermore, he reported that severe roughness did not change the value of previously obtained contact angles on rough surfaces.

Anderson [7] in his wettability review noted the following regarding the effect of surface roughness on wettability. Contact angle on a smooth surface remains fixed, whereas on rough surfaces, as in reservoir rocks, where there are sharp edges, the angle depends on the geometry of the rock surface. Surface roughness diminishes the apparent contact angle when the contact angle measured on a flat plate is less than 90°, and increases the apparent contact angle when the true contact angle is greater than 90°.

Mason [5] examined the equilibrium spreading by studying two extremes of surface roughness. On examining spiral grooves he observed a stick-jump in contact line movement, which agreed well with the theory of concentric grooves. On radial grooves the contact line movement was reversible and advancing contact angle agreed well with Wenzel's equation. On other forms, the contact line jump movement was less evident as channeling of fluids increased, with the result that spreading varied between limiting cases. On further study using a Scanning Electron Microscope, it was suggested that hysteresis might have been caused due to other roughness factors such as sharp edges, which inhibit contact line advance and recession.

Robin *et al.* [8] describe an experimental study of how chemical heterogeneity and surface roughness can affect wetting phenomenon. They used various heterogeneous substrates such as horizontal stripes, vertical stripes, checkerboard patterns, and low surface energy Teflon solids. Contact angle was measured using the Wilhelmy Plate technique. Their study led to the conclusion that uniform distribution of heterogeneities leads to larger advancing and receding angles than random distribution. They developed a model for the effects of both heterogeneity and roughness, and the results from their modeling were found to be consistent with experimental measurements of Cassie and Baxter [3]. They concluded that both heterogeneity and roughness had significant effects on contact angle hysteresis, and that in reservoir rocks the contact angle within pore space would exhibit more hysteresis than that observed on a smooth surface.

Yang *et al.* [9] in their study on mechanisms for contact angle hysteresis and advancing contact angles used an atomic force microscope to characterize the mica surfaces that had been equilibrated in brine and followed by aging in crude oil at an elevated temperature of 80°C. They used two crude oils, brine and mica surfaces. They recorded a linear correlation between the advancing contact angle and the surface mean roughness. The advancing angle increased from 20° for an average roughness of 0.08 nm to 144° for roughness of 28 nm, thereby showing that surface roughness has a significant effect on the advancing contact angle. They concluded that in addition to surface topology, there were other properties, like chemical property of the oil, which were important in determining the advancing contact angle.

Most of the above reported studies, except that by Yang *et al.* [9], have considered solid-liquid-air systems. Hence in spite of some experimental studies there is very little known about the impact of surface roughness on contact angles involving two immiscible liquids such as in reservoirs containing brine and crude oil. In this paper we attempt to report some recent findings on how surface roughness can affect the dynamic contact angles in rock-oil-brine systems.

## 2. MATERIALS AND METHODS

### 2.1. Materials

Experiments were conducted using stock-tank reservoir fluids obtained from Yates field, West Texas, USA (brine and crude oil), and the rock samples were obtained from Wards Scientific, Rochester, NY, USA. The solid surfaces used were glass, quartz, Berea sandstone, dolomite and calcite. Quartz, dolomite and calcite were cut and treated using hydrofluoric acid, hydrochloric acid, graded wheel plate, and sandblasted, to obtain two levels of roughness. Glass and Berea were taken as the two extreme cases of roughness for the silica-based surfaces.

The solid samples were cleaned by soaking them in a mixture of methanol (87% by vol) and chloroform (13% by vol) for two hours and then boiling them in deionized water for 30 minutes.

## 2.2. Dual-drop dual-crystal apparatus

The Dual-Drop Dual-Crystal cell [10-12] and the associated apparatus for carrying out the contact angle measurements at ambient conditions is shown in Figure 1. The apparatus is set up in such a way so that the upper crystal moves in the vertical direction and the lower one moves in the horizontal direction. The lower crystal can be rotated around its horizontal axis so that both of its surfaces could be used for experimentation. There is a provision for letting the crude oil into the cell and this is through a syringe, which is connected to a 1/16" tubing which can be inserted in the cell, and its height adjusted as required by the particular experiment.

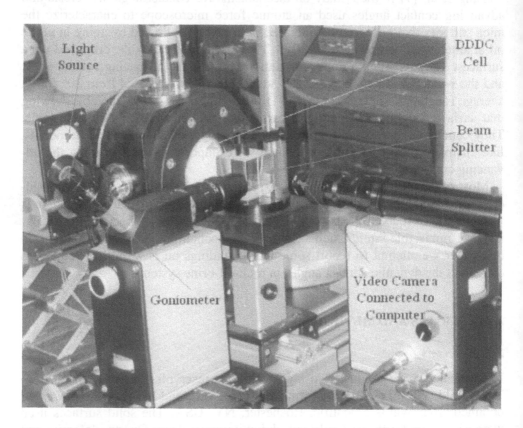

**Figure 1.** Assembled Dual-Drop Dual-Crystal Cell with Goniometer, Video Camera and Beam Splitter.

## 2.3. Preparation and cleaning of the apparatus

The cell is first cleaned with deionized water and is then filled with brine. After an experiment is completed, the inlet valve is opened to let some brine in so that it drains the oil floating at the top. This is done to avoid the floating oil from coming down into the cell and contacting the Teflon interior.

## 2.4. Experimental procedure

Previously cleaned rock substrates were placed in both the upper and lower holders, which were then assembled carefully into the thoroughly cleaned cell. The reservoir brine was taken in a large container, which was kept at a sufficient height to allow flow by gravity. The cell was gradually filled up and some fluid was even allowed to drain from the top to ensure that there were no air bubbles in the cell.

The crude oil was now let into the cell drop by drop. A few drops of oil were initially allowed to rise through the brine phase and rest at the top of the cell interior in order to attain oil-brine equilibrium before starting the experiment. Then two separate oil drops were placed on the two crystal surfaces. This was done by first sliding the lower crystal sideways, raising the oil-dropper tip and placing the drop on the upper crystal, lowering the tip, sliding the lower crystal back, and placing a drop on the lower crystal. The sizes of the drops were chosen so as to cover as much of the crystal width as possible without losing the drops. The cell was then set-aside with all the valves closed to age for a predetermined time for the oil-brine-crystal interactions to reach equilibrium.

The lower crystal holder was then rotated slowly so that the lower surface and the oil drop on it were now facing upwards. When the lower crystal is rotated there are three possible ways the drop can behave: (1) the drop stays attached to the surface due to adhesion, or (2) part of the oil drop floats to the top of the cell due to buoyancy while leaving behind some of the oil on the surface, or (3) all of the oil drop detaches cleanly from the lower crystal without leaving any oil on the surface. In the first two cases the upper crystal is brought down so that the two oil drops merge, and in the third instance the oil drop on the upper surface is made to contact the same area as was previously occupied by the oil drop that floated away.

Once the two drops are merged, or the upper drop is brought in contact with the lower crystal, the angles and the distances of the drop corners from the edge of the lower solid surface are measured to provide the initial values at zero time. Then the lower crystal is shifted sideways in small steps thus enabling water to advance onto a previously oil occupied surface. The contact angle measured here satisfies the definition of the water-advancing contact angle. This lateral shift in the lower crystal position creates a water receding angle on the other side of the oil drop as the oil advances. The advancing and receding contact angles are measured until they show negligible change with time. Then a second shift of the lower crystal is made and the two angles are measured until they become stable. This procedure is repeated until at least two consecutive shifts yield similar water-advancing and receding contact angles.

## 2.5. Wilhelmy plate apparatus

The Wilhelmy Plate Apparatus used in this study has been described elsewhere [12]. It consists of a control unit connected to a computer. The position of the platform holding the beaker containing the oil-water pair is controlled by a stepper motor which, in turn, is controlled to move the platform up or down by the computer. A microbalance is connected to the control unit to sense the change in weight as the solid substrate, which is suspended from the microbalance, travels through the liquid phases as the platform rises or falls. The computer shows the output as a plot of adhesion tension (which is the product of interfacial tension and cosine of the contact angle) vs. distance. The stepper motor is set to move the platform one-inch up in the advancing mode and one-inch down in the receding mode. The apparatus is protected from air currents by a specially designed cabinet, which encloses it.

## 2.6. Interfacial tension (IFT) measurements

Interfacial tension was measured using the DDDC cell to capture the drop profile image and using the Drop Shape Analysis Software. IFT values were checked against the du Noüy ring tensiometer measurements.

## 2.7. Roughness measurement and characterization

The selected solid substrates were carefully cut approximately to a size of 25 mm by 15 mm using a cutting blade and then were thinned by polishing to a thickness of about 2-3 mm. These cut rock samples were then cleaned by soaking them in a mixture of methanol (87% by vol) and chloroform (13% by vol) for two hours and boiling in deionized water for 30 minutes.

For characterization of roughness, these treated rock samples were coated with a gold and palladium mixture to enable them to reflect light when used on the Optical Profilometer for measuring their roughness. They were coated using the S150 Sputter coater, Edwards/Kiese, Germany. However, the coated samples were not used in any of the contact angle measurements.

The roughness of the various samples was characterized using an Optical Profilometer. This apparatus measures surface profiles by reflecting light from the coated samples. The Stylus Probe Profilometer measures the roughness by scanning a mechanical stylus across the sample. The Profilometer can be used to measure etch depths, deposited film thickness, and surface roughness.

The Optical Profilometer gives the value of roughness at a particular point on the rock surface, so a number of readings were taken on both sides of each of the surfaces and an average value of the roughness was used in the plots.

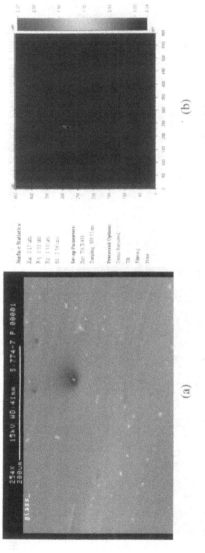

**Figure 2.** Surface Images of Glass ($R_a = 0.17$ μm) using (a) Scanning Electron Micrograph, (b) Optical Profilometry image.

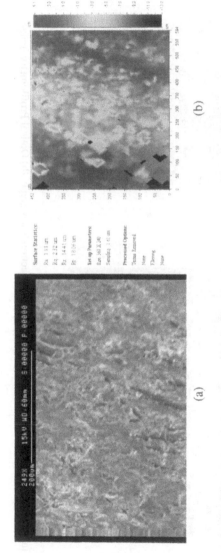

**Figure 3.** Surface Images of Quartz A ($R_a = 1.81$ μm) using (a) Scanning Electron Micrograph, (b) Optical Profilometry image.

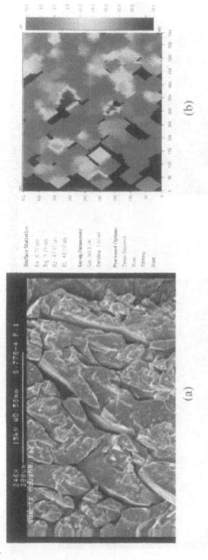

**Figure 4.** Surface Images of Quartz B ($R_a$ = 6.81 μm) using (a) Scanning Electron Micrograph, (b) Optical Profilometry image.

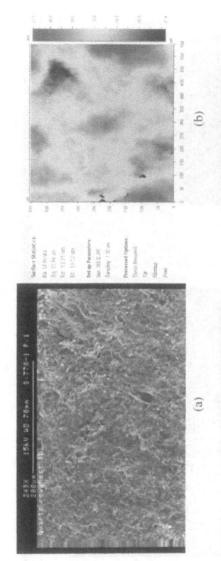

**Figure 5.** Surface Images of Quartz C ($R_a$ = 13.9 μm) using (a) Scanning Electron Micrograph, (b) Optical Profilometry image.

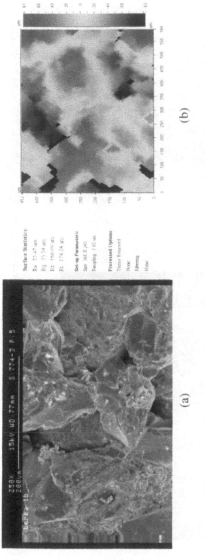

(a)        (b)

**Figure 6.** Surface Images of Berea ($R_a$ =30.23 µm) using (a) Scanning Electron Micrograph, (b) Optical Profilometry image.

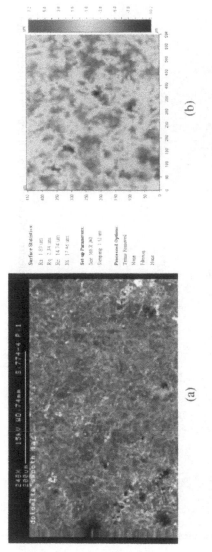

(a)        (b)

**Figure 7.** Surface Images of Dolomite A ($R_a$ =1.95 µm) using (a) Scanning Electron Micrograph, (b) Optical Profilometry image.

(a)                                                                    (b)

**Figure 8.** Surface Images of Dolomite B ($R_a$ =15.6 μm) using (a) Scanning Electron Micrograph, (b) Optical Profilometry image.

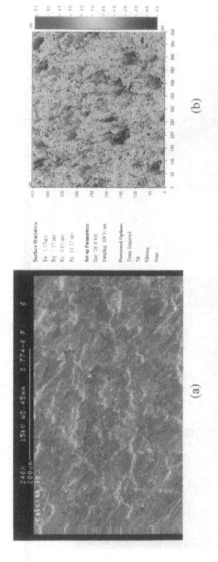

(a)                                                                    (b)

**Figure 9.** Surface Images of Calcite A ($R_a$ =1.17 μm) using (a) Scanning Electron Micrograph, (b) Optical Profilometry image.

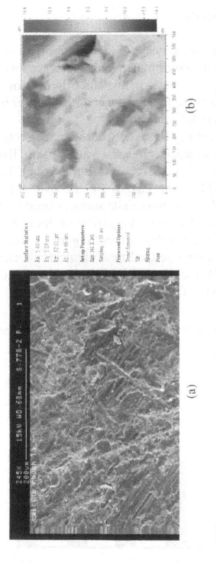

**Figure 10.** Surface Images of Calcite B ($R_a$ = 5.46 μm) using (a) Scanning Electron Micrograph, (b) Optical Profilometry image.

## 2.8. Surface topography

The surface topography of the samples used in this study was measured using a
Scanning Electron Microscope at various magnifications, viz., 250, 500, 1000 and
1500. Selected images are included in the paper.

## 3. RESULTS AND DISCUSSION

Rock mineralogy and surface roughness are the two main parameters we have examined in this study. Three different mineralogical surfaces were chosen, namely
silica, dolomite and calcite. For each of the three mineralogy groups, surfaces
with different roughnesses were obtained as mentioned earlier by etching with HF
or HCl, by polishing or by sandblasting.

Figures 2-6 show the Scanning Electron Micrographs as well as Optical Profilometer images along with average roughness values of the five silica-based surfaces (glass, quartz A, quartz B, quartz C, and Berea rock). Figures 7 and 8 are
SEM micrographs and Profilometer images for the two dolomite surfaces and
Figures 9 and 10 correspond to the two calcite surfaces used in the experiments.

The SEM images taken at the same magnification indicate the topographical
differences between the various substrates used. Figure 11 summarizes the roughness characteristics averaged over several measurements on each substrate. In all
the experiments involving Wilhelmy and DDDC measurements, brine and crude
oil from the Yates field in West Texas were used.

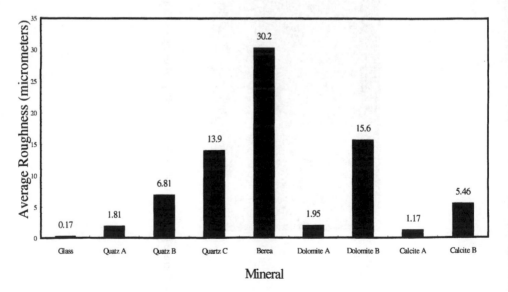

**Figure 11.** Roughness of Various Surfaces used in Experiments (Averaged over several separate
measurements).

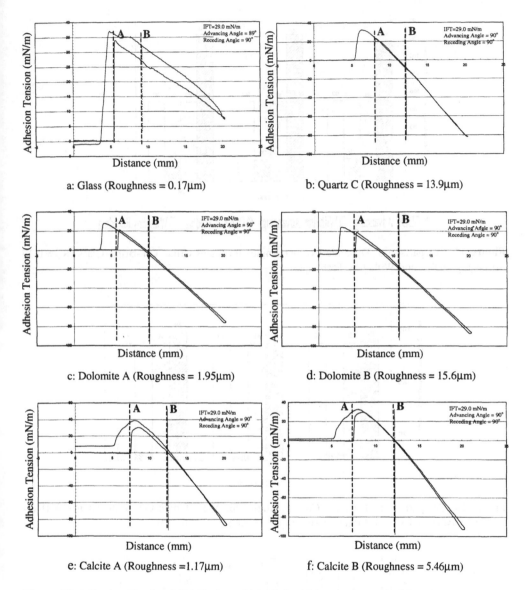

**Figure 12.** Adhesion Tension ($\gamma$Cos$\theta$) plots obtained for 24-hour aging period from Wilhelmy Tests with Yates Crude oil and Yates Brine (A: Air-Oil interface; B: Oil-Water interface).

Figure 12 shows the adhesion tension ($\gamma$Cos$\theta$, where $\gamma$ is the liquid-liquid interfacial tension, and $\theta$ is either the advancing or the receding contact angle) plots obtained for 24-hour aging period from Wilhelmy Tests with Yates Crude oil and Yates Brine from which it can be noticed that the Wilhelmy Technique was insensitive to the surface roughness of the samples and it consistently recorded an angle of 90° for both the advancing and receding modes. Figures 13-17 show the apparent dynamic (water-advancing and receding) contact angles measured on sil-

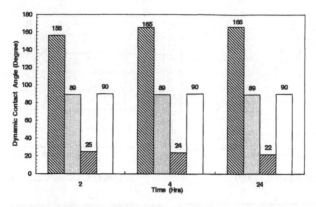

**Figure 13.** Comparison of Dynamic Contact Angles for Yates Crude Oil–Brine–Glass ($R_a = 0.17$ μm) using DDDC and Wilhelmy Techniques.

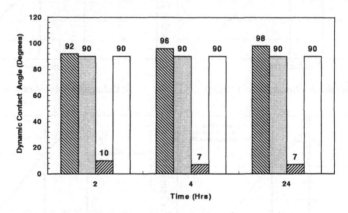

**Figure 14.** Comparison of Dynamic Contact Angles for Yates Crude Oil–Brine–Quartz A ($R_a = 1.81$ μm) using DDDC and Wilhelmy Techniques (Legend as in Fig. 13).

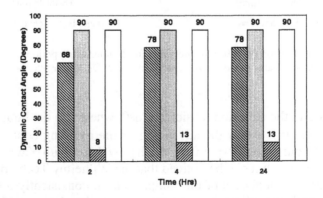

**Figure 15.** Comparison of Dynamic Contact Angles for Yates Crude Oil–Brine–Quartz B ($R_a = 6.81$ μm) using DDDC and Wilhelmy Techniques (Legend as in Fig. 13).

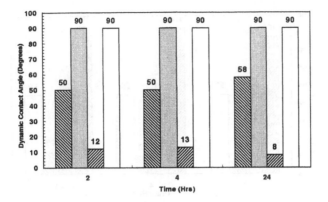

**Figure 16.** Comparison of Dynamic Contact Angles for Yates Crude Oil–Brine–Quartz C ($R_a$ = 13.9 μm) using DDDC and Wilhelmy Techniques (Legend as in Fig. 13).

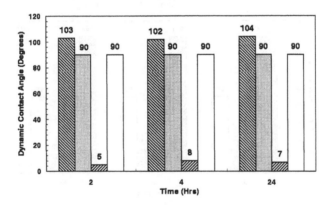

**Figure 17.** Comparison of Dynamic Contact Angles for Yates Crude Oil–Brine–Berea ($R_a$ = 30.23 μm) using DDDC and Wilhelmy Techniques (Legend as in Fig. 13).

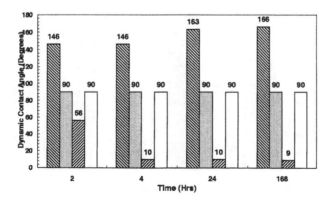

**Figure 18.** Comparison of Dynamic Contact Angles for Yates Crude Oil–Brine–Dolomite A ($R_a$ = 1.95 μm) using DDDC and Wilhelmy Techniques (Legend as in Fig. 13).

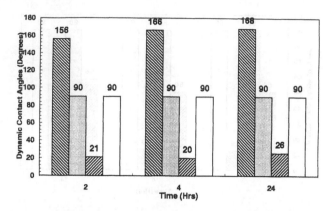

**Figure 19.** Comparison of Dynamic Contact Angles for Yates Crude Oil–Brine–Dolomite B ($R_a$ = 15.6 μm) using DDDC and Wilhelmy Techniques (Legend as in Fig. 13).

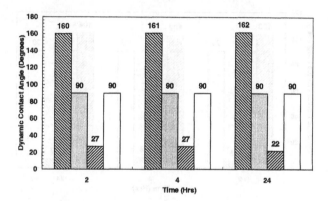

**Figure 20.** Comparison of Dynamic Contact Angles for Yates Crude Oil–Brine–Calcite A ($R_a$ = 1.17 μm) using DDDC and Wilhelmy Techniques (Legend as in Fig. 13).

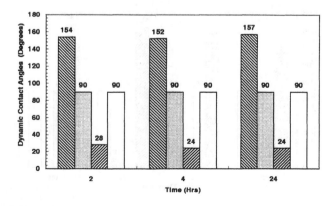

**Figure 21.** Comparison of Dynamic Contact Angles for Yates Crude Oil–Brine–Calcite B ($R_a$ = 5.46 μm) using DDDC and Wilhelmy Techniques (Legend as in Fig. 13).

ica-based surfaces using both the DDDC and Wilhelmy technique, while similar plots for dolomite surfaces are given in Figures 18-19, and for calcite surfaces in Figures 20-21. A striking feature is the 90° angles consistently indicated by the Wilhelmy measurements in all of the plots in Figures 13-21 indicating the insensitivity of the Wilhelmy technique to mineralogy, surface roughness and the mode of the movement of the three-phase contact-line (whether advancing or receding). Several tests were run to verify the proper functioning of the Wilhelmy apparatus by checking with strongly water-wet and oil-wet solid-fluid systems. The apparatus yielded excellent match with the past results indicating its ability to function properly. Next, several tests were run with thinner solid substrates to enable the microbalance to be more sensitive to the oil-water interface as it traversed across the solid substrate. These tests also resulted in 90° angles confirming the relative insensitivity of the Wilhelmy technique in such contact angle measurements. This ruled out the initial objective of our study to compare the dynamic angles measured using the DDDC technique with those from the well-established Wilhelmy apparatus. The reason for the observed insensitivity appears to be the rate at which the solid substrate is made to traverse across the oil-brine interface, especially when the surface is rough. Before the microbalance has had a chance to sense the interfacial force at a particular contact line on the surface, the surface could have moved to a new position, with different surface topography leading to the insensitivity of the instrument observed in out tests. This aspect needs to be further explored by conducting Wilhelmy tests at relatively slow rates of immersion.

In contrast to Wilhelmy tests, the DDDC tests yielded dynamic contact angle results that showed marked influence of mineralogy, surface roughness and the mode of contact-line movement. The DDDC test results are shown for each of the substrates in Figures 13 to 21, and a summary of the data is shown in Figure 22 for silica-based surfaces and Figure 23 for dolomite and calcite surfaces.

As shown in Figure 22, the advancing angle on silica-based surfaces continuously decreased (from 166° on smooth glass to 58° on rough quartz C) with increasing surface roughness. Berea rock, on the other hand, although being a silica-based rock, did not fall on this decreasing trend of Figure 22. This may be related to mineralogical composition of Berea - being somewhat different from pure silica. Since the receding angles stayed quite low in the range of 7°-26° with increasing roughness, the overall effect of roughness on hysteresis (which is the difference between advancing and receding contact angles) is one of a decreasing trend on silica surfaces. Decreasing trends in hysteresis with roughness were also observed, as shown in Figure 23, for dolomite and calcite surfaces, although the severity of the effect was much less than on silica-based surfaces.

Based on the studies involving solid-liquid-air systems our expectation was that the hysteresis would increase with increasing surface roughness. However, our experimental results presented in this paper dispelled this expectation and indicated the need to be cautious while extrapolating results and observations from solid-liquid-vapor systems to the more complex solid-liquid-liquid systems.

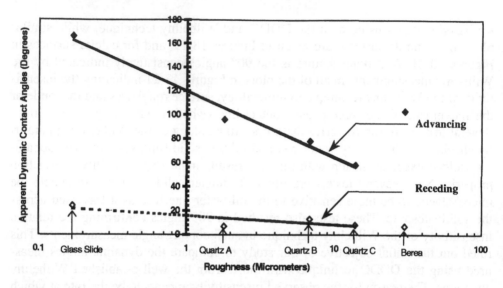

**Figure 22.** Effect of Surface Roughness on Dynamic Contact Angles Determined by DDDC Technique on Silica-Based Surfaces using Yates Brine and Crude Oil.

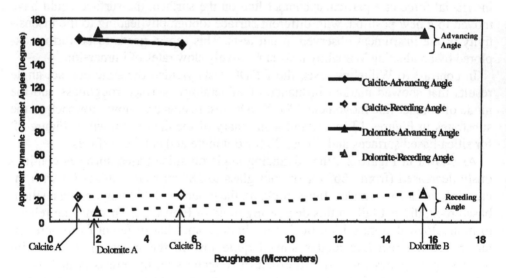

**Figure 23.** Effect of Surface Roughness on Dynamic Contact Angles Determined by DDDC Technique on Dolomite and Calcite Surfaces using Yates Brine and Crude Oil.

## 4. CONCLUSIONS

1. The Wilhelmy apparatus displayed insensitivity when used to infer dynamic contact angles and their variations with solid surface characteristics. It consistently indicated 90° angles irrespective of the surface type, roughness and whether the oil-water interface was advancing or receding.

**Table 1.**
Composition of Yates brine of pH 7.39

| Parameter | Concentration units (mg/l) |
|---|---|
| Total Dissolved Solids | 9200 |
| Calcium | 425 |
| Magnesium | 224 |
| Potassium | 50.5 |
| Sodium | 1540 |
| Hardness as CaCO₃ | 1500 |
| Hardness as Carbonate | 810 |
| Hardness as Non-carbonate | 730 |
| Bicarbonate | 800 |
| Alkalinity | 810 |
| Sulfate | 660 |
| Chloride | 3700 |

2. The DDDC technique indicated significant effects of mineralogy, surface roughness and mode of measurement (advancing or receding) on dynamic contact angles. The weeklong ambient condition DDDC tests of this study indicate an aging time of 24 hours to be sufficient for attaining solid-liquid-liquid equilibrium.

3. All three mineral surfaces, quartz, dolomite and calcite, showed strong oil-wet tendencies ($155° < \theta_a < 166°$) with Yates reservoir crude oil and brine.

4. The water-advancing contact angles measured using the DDDC technique on silica-based surfaces showed a sharp decline with increasing surface roughness indicating a shift from oil-wet nature on smooth surface to either intermediate-wet or weakly water-wet on rougher surfaces. Measurements on actual Berea rock samples, though not falling on the same trend as the other silica surfaces, also indicated intermediate-wettability. Dolomite and calcite surfaces, though displaying a slight decline in advancing angle with roughness, retained strongly oil-wet nature within the range of roughness examined. The receding angles on all surfaces remained low (7°-22°) indicating their insensitivity to surface type and roughness. Thus the overall effect of increasing surface roughness was to decrease the contact angle hysteresis. Further work in this area is aimed at investigating these effects at reservoir conditions of elevated pressures and temperatures using live reservoir crude oils containing dissolved gaseous components.

5. The results of this experimental study have not only dispelled our expectations from solid-liquid-vapor systems that contact angle hysteresis would increase with increasing surface roughness, but also indicated the need to be

cautious in extrapolating the results and observations from solid-liquid-vapor systems to the more complex solid-liquid-liquid systems.

*Acknowledgements*

The financial support from the Louisiana Board of Regents support fund (LEQSF 2000-03-RD-B-06) and Marathon Oil Company is gratefully acknowledged. Sincere thanks are due to British Petroleum for donating the computerized Wilhelmy apparatus to LSU. The authors would like to thank Rick Young, Geology Department (LSU) for his help in preparing the solid samples; Margaret Cindy Henk, Biological Sciences (LSU) in coating the samples and with the use of the SEM; Dan Lawrence, Department of Petroleum Engineering (LSU) for the technical help; Varshni Singh and Pankaj Gupta, Mechanical Engineering (LSU) for their help in using the optical profilometer; Paul Rodriguez Jr., Chemical Engineering (LSU) for his help in sandblasting the solid surfaces used in the experiments.

**REFERENCES**

1. R.N. Wenzel, *J. Phys. Colloid Chem.*, **53**, 1466 (1936).
2. A.W. Adamson, *Physical Chemistry of Surfaces*, Fifth Edition, John Wiley, New York (1990).
3. A.D.B. Cassie and S. Baxter, *Trans. Faraday Soc.*, **40**, 456 (1944).
4. J.F. Oliver, C. Huh and S.G. Mason, *J. Colloid Interface Sci.*, **59**, 568 (1977).
5. S.G. Mason, in *Wetting, Spreading and Adhesion*, J.F. Padday (Ed.), pp. 321-326, Academic Press, New York (1978).
6. N.R. Morrow, *J. Can. Petroleum Technol.*, **14**, 42 (1975).
7. W.G. Anderson, *J. Petroleum Technol.*, **38**, 1246 (1986).
8. M. Robin, A. Paterson, L. Cuiec and C. Yang, Paper No.37291, *Proc. SPE International Symposium On Oilfield Chemistry*, Houston, TX (1997).
9. S.Y. Yang, G.J. Hirasaki, S. Basu and R. Vaidya, *Proc. 5th International Symposium on Evaluation of Reservoir Wettability and its Effect on Oil Recovery*, Trondheim, Norway (1998).
10. D.N. Rao and M. Girard, *J. Canadian Petroleum Technol.*, **35**, 31 (1996).
11. D.N. Rao and R.S. Karyampudi, *J. Adhesion Sci. Technol.*, **16**, 579-596 (2002).
12. Z. Muhammad and D.N. Rao, Paper No. SPE 65409, *Proc. 2001 SPE International Symposium on Oilfield Chemistry*, Houston, TX (2001).

*Contact Angle, Wettability and Adhesion*, Vol. 3, pp. 427–439
Ed. K.L. Mittal
© VSP 2003

# Viscous dissipation and rheological behavior near the solid/liquid/vapor triple line. Application to the spreading of silicone oils

PIERRE WOEHL[*] and ALAIN CARRÉ

*Corning S.A., Fontainebleau Research Center, 7 bis, avenue de Valvins, 77210 Avon, France*

**Abstract**—A basic problem in liquid spreading is the hydrodynamic description of the viscous braking force near the moving contact line. With a sharp wedge profile, the viscous energy dissipation becomes infinite, which would mean that a drop is not able to spread on a solid surface. The problem is often currently solved by introducing a "cut-off" length, for example by considering that the drop profile is curved near the solid/liquid/vapor triple line by van der Waals long range forces.

We propose a new approach to solve the problem of divergence at the solid/liquid/vapor triple line. This approach is based on the hypothesis that the rheological behavior of a Newtonian liquid may be modified near the triple line due to extremely high shear rate. Above a critical value of the shear rate, near the triple line and near the solid surface, the liquid becomes shear-thinning, so that the apparent viscosity of the liquid decreases as the shear rate increases. As a result, there is no divergence in the calculation of the viscous energy dissipation and of the braking force.

This new description of the viscous braking phenomenon in liquid spreading is well supported by spreading experiments on two silicone oils on glass substrates of different wettability. The silicone oils are Newtonian below a critical value of the shear rate. Above this critical value, the viscosity of the liquid decreases according to a power-law of the shear rate.

As a basic result, the spreading kinetics is only described from the rheological properties of the liquid, without the need of introducing other parameters such as a "cut-off" or slip length.

*Keywords*: Spreading; Newtonian behavior; shear-thinning behavior; wetting; kinetics of spreading; contact angle; braking force; triple line.

## 1. INTRODUCTION

The spreading of liquids on solid surfaces is of considerable interest. The dynamic aspects of spreading are particularly relevant in several practical applications in industry, such as coating, adhesive bonding, composite manufacturing, painting and printing. When inks or paints are applied onto dry substrates, they must wet their substrates before solidifying. Similarly, good spreading is required to ensure

---

[*]To whom all correspondence should be addressed. Phone: 33 1 64 69 74 83,
Fax: 33 1 64 69 74 55, E-mail: woehlp@corning.com

interfacial contact between two phases when applying a polymeric adhesive or in the manufacturing of composite materials.

The kinetics of spreading of liquids on rigid solid surfaces can be described with two basic models. The molecular-kinetic theory proposed by Blake and Haynes [1] considers a molecular "hopping" mechanism, derived from the Eyring rate process theory [2]. The spreading of Newtonian liquids has also been described in the literature with the hydrodynamic theory [3-5]. When a liquid spreads spontaneously on a flat and rigid solid, a dynamic equilibrium is set up in which the non-equilibrated Young force, and the corresponding excess capillary energy, causes triple line motion. Simultaneously, viscous dissipation in the liquid, chiefly near the wetting front, reduces the spreading speed. This theory has been successfully verified in experiments corresponding to a noninertial, purely viscous regime. In these conditions, a hydrodynamic contribution to dissipation is considered dominant [6].

In this paper we develop the basic assumptions of the kinetic theory of liquid spreading in the case where the liquid has a Newtonian or linear behavior and the divergence problem raised by this approach. Then, we show that in the case where the liquid is shear-thinning, meaning that the liquid becomes more fluid under shear stress, the divergence of the energy dissipated at the triple line does not exist. This leads us to propose a new hypothesis related to the behavior of Newtonian fluids under high shear rates [7]. In this paper, this hypothesis is verified by new experimental results and molecular modeling of the behavior of liquids under high shear rates [8-10].

## 2. THEORETICAL

### 2.1. The hydrodynamic theory for Newtonian and shear-thinning liquids

The equilibrium of a small liquid drop resting on a solid surface is described by the Young equation. In the case where the drop does not completely spread on the solid, a contact angle $\theta_e$ is obtained at equilibrium. This angle satisfies the Young equation [11]:

$$\gamma_{SV} = \gamma_{SL} + \gamma_{LV} \cos \theta_e, \tag{1}$$

where $\gamma_{SV}$ is the solid surface tension, $\gamma_{SL}$ the interfacial tension between the solid and the liquid, and $\gamma_{LV}$ the liquid surface tension.

When a drop spreads on a solid surface, it is under the action of a driving force. As the dynamic contact angle $\theta_d$ is higher than the equilibrium value $\theta_e$, a force results from the unbalanced Young equation. A simple force analysis leads to the expression for the driving force for spreading, which depends on the liquid surface tension and on dynamic and equilibrium contact angles. As spreading occurs, we can define an excess capillary energy, $E_c$, per unit time and length of the solid/liquid/vapor triple line, given by the product of the spreading force and the spreading speed, U, as:

$$E_c(J/m.s) = \gamma_{LV}(\cos\theta_e - \cos\theta_d).U. \tag{2}$$

Let us consider now a liquid drop spreading on a solid surface. Some time is necessary to reach the equilibrium value of the wetting angle, and this time depends on the liquid viscosity. This means that a braking force exists due to viscosity. This force can be calculated from the hydrodynamic theory of liquid spreading [3]. One way to obtain the viscous braking force consists of calculating the energy dissipated by the liquid flow during spreading. Most of the viscous dissipation is located at the drop periphery in the liquid wedge around the drop. The liquid wedge is represented in Figure 1. The calculation is based on the following boundary conditions for the liquid flow:

•   no slip at the solid surface
•   no shearing at the free liquid surface.

The calculation of the viscous dissipation requires knowing the flow profile inside the moving liquid wedge. The calculation assumes that the wetting angles are small and that the lubrication approximation is valid. In these conditions, the vertical flow speed is negligible and the liquid flow is comparable to a Poiseuille flow resulting only from the horizontal pressure gradient. The expression for the

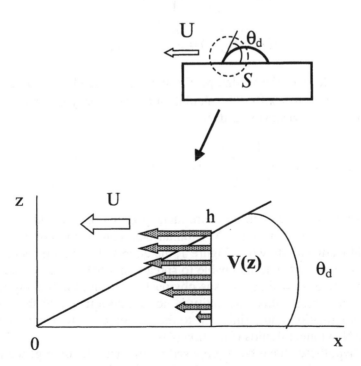

**Figure 1.** Flow profile inside a moving liquid wedge.

shear stress, $\tau(z)$, for a Newtonian liquid as a function of the liquid viscosity, $\eta$, and of the shear rate, $\dot{\gamma}(z)$, or speed gradient, $\dfrac{dv(z)}{dz}$, is given by:

$$\tau(z) = \eta \frac{dv(z)}{dz} = \eta \dot{\gamma}(z).$$ (3)

The pressure gradient, $\dfrac{dP}{dx}$, is equal to shear stress gradient according to the linear form of the Stokes equation (purely viscous flow) as:

$$\frac{dP}{dx} = \frac{d\tau(z)}{dz}.$$ (4)

The liquid flow, Uh, passing through a height h, is evaluated as the sum of $v(z)dz$. The speed profile in the liquid wedge, the shear rate and the shear stress can be obtained as:

$$v(z) = \frac{3U}{2h^2}(2hz - z^2),$$ (5)

$$\dot{\gamma}(z) = \frac{dv(z)}{dz} = \frac{3U}{h^2}(h - z),$$ (6)

$$\tau(z) = \eta\,\dot{\gamma}(z) = \frac{3\eta U}{h^2}(h - z).$$ (7)

Near the triple line, the wedge profile is approximated by a straight interface [3]. Assuming that $h \approx x.\theta_d$, the calculation of the energy dissipation, $E_d$, in the liquid wedge due to viscosity leads to:

$$E_d (J / m.s) = \int_0^r \left[ \int_0^{h = x.\theta_d} \tau(z).\dot{\gamma}(z)dz \right] dx = \frac{3\eta U^2}{\theta_d} \int_0^r \frac{dx}{x}.$$ (8)

This calculation raises one problem, namely Equation (8) leads to a divergent, infinite result. If we consider a Newtonian behavior in the entire wedge and a sharp liquid wedge as shown in Figure 1, the energy dissipation becomes infinite, implying that a drop would be unable to spread on a solid surface due to the infinite braking force, $F_b = E_d/U$. This is obviously not true; the braking force has to be finite. Furthermore, this singularity is not a consequence of the viscous flow regime approximation and is inherent to the no-slip condition imposed at the triple line for Newtonian liquids (Equation (4)).

Several hypotheses have been proposed to alleviate the divergence problem of the viscous dissipation in a spreading liquid wedge. One way consists of introducing a cut-off length, $x_m$, below which the liquid wedge profile is curved by the long range van der Waals forces [3]. Under this condition the energy dissipation due to viscosity is finite and is given by:

$$E_d(J/m.s) = \frac{3\eta U^2 \ln(r/x_m)}{\theta_d}. \qquad (9)$$

Another hypothesis to remove the singularity is to allow fluid slip near the moving contact line. In that case, the boundary conditions at the front include a slip length [4, 12, 13].

Assuming that the substrate is never perfectly dry, and that evaporation and condensation at the front always produces a thin precursor film [14-17] allows also alleviating the singularity.

In the case where the liquid is non-Newtonian but shear-thinning, the divergence problem does not exist, essentially because the apparent viscosity of a shear-thinning liquid decreases when the shear rate increases, to vanish under infinite shear rate.

The shear-thinning behavior can be described by a power-law between the shear stress and the shear rate as:

$$\tau(z) = a\,\dot{\gamma}(z)^n, \qquad (10)$$

where the exponent n is lower than 1 and the coefficient a is constant. The apparent viscosity, $\eta_a$, can be defined as:

$$\tau(z) = a\,\dot{\gamma}(z)^n = \eta_a\,\dot{\gamma}(z) \Rightarrow \eta_a = \frac{a}{\dot{\gamma}(z)^{1-n}}, \qquad (11)$$

showing that the apparent viscosity decreases as the shear rate increases. With the boundary conditions of no-slip at the solid surface and no shearing at the free liquid surface, integration of the Stokes equation gives the speed profile, the shear rate and the shear stress in the liquid wedge [18, 19] as:

$$v(z) = U\frac{(2+\frac{1}{n})}{(1+\frac{1}{n})}\left[1-(1-\frac{z}{h})^{1+\frac{1}{n}}\right], \qquad (12)$$

$$\dot{\gamma}(z) = \frac{dv(z)}{dz} = (2+\frac{1}{n})(1-\frac{z}{h})^{\frac{1}{n}}\frac{U}{h}, \qquad (13)$$

$$\tau(z) = a.\dot{\gamma}(z)^n = a(2+\frac{1}{n})^n(1-\frac{z}{h})\frac{U^n}{h^n}. \qquad (14)$$

The calculation of the energy dissipation can be made on the entire liquid wedge, the result converges and leads to the expression for $E_d$ for a shear-thinning fluid:

$$E_d (J/m.s) = \int_0^r \left[ \int_0^{h = x.\theta_d} \tau(z) . \dot{\gamma}(z) dz \right] dx = a.r^{1-n} \frac{(2 + \frac{1}{n}) U^{n+1}}{(1-n)\theta_d^n}. \qquad (15)$$

## 2.2. The Newtonian/shear-thinning transition

In this paper, we propose to examine the divergence problem differently, in particular from the evolution of the rheological properties of liquid under high shear rates. In Figure 2 is plotted the viscosity of six silicone oils (viscosity from 1,000 to 100,000 cSt) which are considered as good models of Newtonian liquids [20]. But above a critical value of the shear rate, the viscosity of these oils decreases when the shear rate increases. Therefore, all these silicone oils have the same behavior: they are Newtonian but they become shear-thinning above a critical shear rate. Then, the log-log representation in Figure 3 is a linear function corresponding to Equation (11). Therefore, silicone oils are Newtonian below a critical shear rate and become shear-thinning above it, satisfying exactly a power-law dependence with the shear rate.

This observation leads us to propose a new hypothesis that alleviates the divergence problem, i.e., the rheological behavior of the liquid is modified near the contact line due to the high shear rate. Above a critical value of the shear rate the liquid is no longer Newtonian but shear-thinning. This hypothesis is also supported by the results predicted by molecular simulation [8-10].

The consequences of the rheological transition in the moving liquid wedge can be simply evaluated. Below a critical shear rate, the liquid is Newtonian and has a constant viscosity and above the critical shear rate the liquid is shear-thinning.

The distance from the drop edge where the transition occurs can be defined from the expression for the shear rate for z = 0. Knowing the value of the critical shear rate, $\dot{\gamma}_c$, allows us to calculate, from Equation (6) for z = 0, the distance $x_{ST}$ where the transition between the two behaviors occurs:

$$\dot{\gamma}(0) = \frac{3U}{h} = \frac{3U}{x.\theta_d} \Rightarrow x_{ST} = \frac{3U}{\dot{\gamma}_c \theta_d}. \qquad (16)$$

The different regions and the Newtonian/shear-thinning transition in a moving liquid wedge are shown in Figure 4. In the example of Figure 4, the spreading speed is 1 cm/min, the dynamic contact angle is 1 rad, and the critical shear rate is 1000 s$^{-1}$ (PDMS of viscosity 10,000 cSt). As a result, the distance $x_{ST}$ is 0.5 μm. Obviously, the minimum value of $x_{ST}$ given by equation (16) cannot be lower than the molecular size to have a physical meaning.

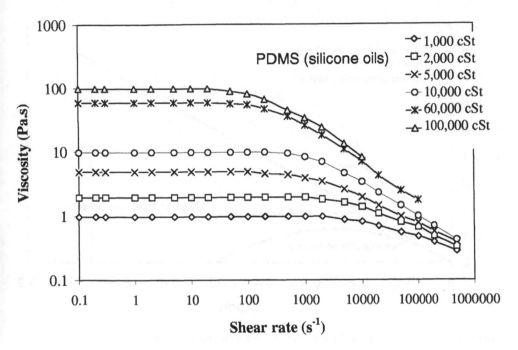

**Figure 2.** Rheological behavior of several silicone (PDMS) oils of viscosity ranging from 1,000 to 100, 000 cSt as a function of the shear rate.

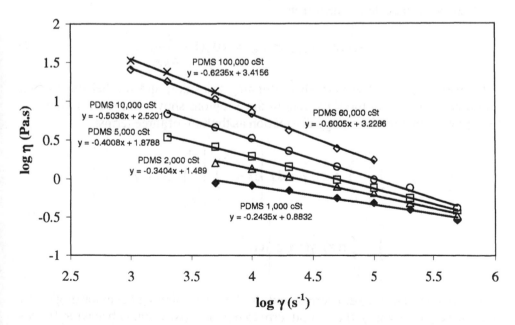

**Figure 3.** Verification of the power-law dependence in the shear-thinning region for the silicone oils of different viscosities.

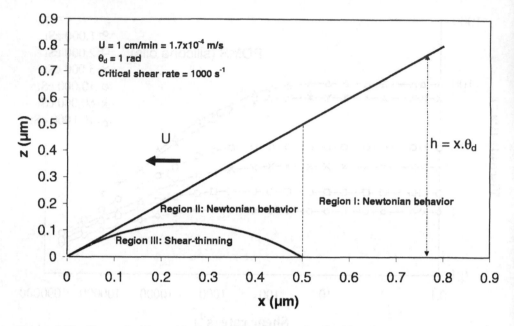

**Figure 4.** The Newtonian/shear-thinning transition in a moving liquid wedge.

The Newtonian/shear-thinning transition above the solid surface ($z > 0$ and $0 < x < x_{ST}$) can also be determined from:

$$\dot{\gamma}_c = \frac{3U}{h^2}(h - z_{ST}) \Rightarrow z_{ST} = x\theta_d(1 - \frac{x}{x_{ST}}). \tag{17}$$

Considering the Newtonian/shear-thinning transition and the behavior of the liquid as a function of the shear rate in the 3 regions shown in Figure 4 allows us to express explicitly the energy dissipation as the sum of three terms:

$$E_d(J / m.s) = \int_0^{x_{ST}}\left[\int_0^{z_{ST}}\tau(z).\dot{\gamma}(z)dz\right]dx + \int_0^{x_{ST}}\left[\int_{z_{ST}}^{h}\tau(z).\dot{\gamma}(z)dz\right]dx +$$

$$+ \int_{x_{ST}}^{r}\left[\int_0^{h}\tau(z).\dot{\gamma}(z)dz\right]dx. \tag{18}$$

The first term of the sum corresponds to the shear-thinning behavior (region III, $0 < x < x_{ST}$, $0 < z < z_{ST}$), the second term to region II (Newtonian behavior, $0 < x < x_{ST}$, $z_{ST} < z < h$) and the third term to region I (Newtonian behavior, $x_{ST} < x$, $0 < z < h$). Integration of Equation (18) with the respective expressions for the shear stress, $\tau(z)$ (Equation (7) or (14)), the shear rate, $\dot{\gamma}(z)$ (Equation (6) or (13)), $x_{ST}$ (Equation (16)) and $z_{ST}$ (Equation (17)) leads to the following converging result

$$E_d \quad (J/m.s) = \frac{3^{1-n}(2n+1)^{n+1} a \dot{\gamma}^{n-1}}{n^n(1-n)(3n+1-n^2)} \cdot \frac{U^2}{\theta_d} + \frac{\eta U^2}{\theta_d} + \frac{3\eta \ell U^2}{\theta_d}, \qquad (19)$$

with:

$$\ell = \ln(\frac{r\theta_d \dot{\gamma}_c}{3U}). \qquad (20)$$

Equations (19) and (20) describe the energy dissipation only from the rheological properties ($\eta$, $\dot{\gamma}_c$, a, n) and from the dynamic spreading parameters (U, $\theta_d$, r). It is important to point out that there is no need to hypothesize the existence of a molecular cut-off length introduced in the classical hydrodynamic theory [3].

The Newtonian/shear-thinning transition ($x_{ST}$ and $z_{ST}$) can also be localized from the shear rate expression for a shear-thinning behavior (Equation (13)). This leads to a slightly different estimation of $E_d$ (see reference [7]). The difference results from the assumption of a sharp transition between Newtonian and shear-thinning behavior leading to a discontinuity in the velocity profile. However, the region I being the main contributor to the energy dissipation, the way of defining the transition zone, from the Newtonian side or from the shear-thinning side, is of little practical consequence.

## 3. EXPERIMENTAL

The substrates were clean glass slides and glass slides treated with a fluorinated silane, the perfluorodecyltrichlorosilane (FDS). The liquids were two poly (dimethylsiloxane) (PDMS) oils of viscosities 10,000 and 100,000 cSt. Liquid drops of 0.2 µl were placed on the substrates and the contact angles were measured with a Ramé-Hart goniometer equipped with a video camera and a printer.

The measurements consisted of measuring both the static and dynamic contact angles, and the drop radius, r, as a function of time after drop deposition, t. The spreading speed, U, defined as dr/dt, was also determined as a function of time after deposition.

Figure 5 presents the rheological behavior of the silicone oil having a viscosity of 10,000 cSt (9.74 Pa.s). Above a critical shear rate, $\dot{\gamma}_c$, of 1,000 s$^{-1}$, the fluid becomes shear-thinning. Therefore, in spreading experiments, we will consider that the liquid is Newtonian for shear rates lower than 1000 s$^{-1}$ and is shear-thinning above it.

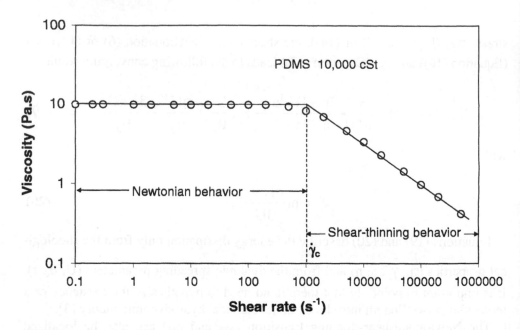

**Figure 5.** Modeling of the rheological behavior of silicone of viscosity 10,000 cSt. The vertical dotted line represents the modeled transition between the Newtonian and shear-thinning behaviors.

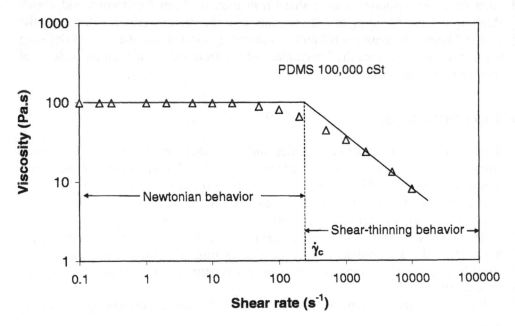

**Figure 6.** Modeling of the rheological behavior of silicone of viscosity 100,000 cSt. The vertical dotted line represents the modeled transition between the Newtonian and shear-thinning behaviors.

Figure 6 shows the rheological behavior of the silicone oil having a higher viscosity of 100,000 cSt (97.8 Pa.s). It shows a similar behavior, but the critical shear rate at the transition between the two behaviors is only 193 s$^{-1}$.

Of course, the power-law description of the shear-thinning behavior has some limitation because the transition between the Newtonian behavior and the shear-thinning behavior, satisfying the power-law, is not always sharp, as shown in Figure 6. But this is an improvement over the analyses which consider that the liquid is only Newtonian.

## 4. RESULTS

The data concerning the spreading of silicone oils of viscosity 10,000 and 100,000 cSt on clean glass and FDS treated glass are presented in Figure 7. The equality between the energy dissipation, $E_d$, and the excess capillary energy, $E_c$, producing the liquid spreading is satisfactory.

The dotted line in Figure 7 represents the theoretical equality between log $E_d$ and log $E_c$. The solid line represents the linear fit. The slope of this line (0.9858) is very close to the expected value of 1.

Therefore, if the liquid becomes shear-thinning above a critical shear rate, which means that the liquid becomes more fluid under high shear rates, there is no divergence of the braking force and no need to introduce a "cut-off" or slip distance. Equality between the excess capillary energy and the viscous energy dissipation is perfectly verified as predicted by the hydrodynamic approach for liquid spreading. The main result is that the spreading phenomenon can be described only from the liquid rheological properties and wetting parameters.

The shear-thinning transition is observed experimentally with polymeric fluids, such as silicone oils, for which some type of structuring appears at reduced shear rates (Figure 2). However, this behavior is also predicted with shorter molecules such as alkanes, but at much higher shear rates. It has been shown from molecular simulations that the viscosity of liquid alkanes shows a shear-thinning behavior at extremely high shear rates [8, 9]. For example, squalane (hexamethyl tetracosane) becomes shear-thinning above 2 x 10$^9$ s$^{-1}$. Such extreme values correspond to conditions of lubrication in automotive engines and are well above the accessible experimental measurements. However, in liquid spreading, this transition may be relevant only in the case of high speeds of displacement of the triple line (typically greater than 1 m/s) for which other phenomena have to be considered (viscosity of the ambient air and related flow [21], i.e., liquid inertia).

The transition from the Newtonian to the shear-thinning behavior indicates some type of ordering in the fluid under shear stress. This transition can be predicted from the inverse of a characteristic relaxation time [8, 10]. The relaxation time depends on molecular size and temperature, T.

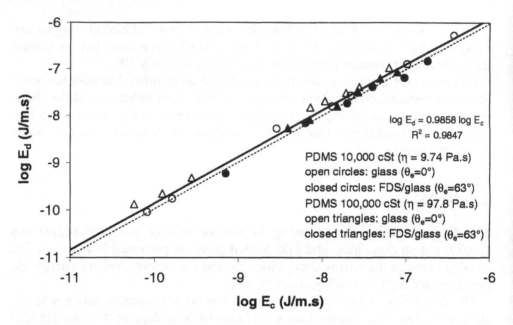

**Figure 7.** Results of spreading experiments of silicone oils with viscosities10,000 cSt and 100,000 cSt on clean glass and FDS treated glass.

## 5. CONCLUSIONS

When a liquid drop spreads on a rigid solid, the spreading speed is controlled by the viscous dissipation due to the flow inside the droplet and by the excess capillary energy. Several hypotheses exist that alleviate the divergence problem of the energy dissipation at a moving triple line. We propose a new approach for a fully explicit calculation of the energy dissipation under classical hydrodynamic approximation in liquid spreading. It is based on the fact that a liquid, such as silicone oil often considered as a model of Newtonian fluid, may become shear-thinning under high shear rates, which alleviates the divergence problem near the triple line. The Newtonian/shear-thinning transition in the moving liquid wedge is located near the triple line and the solid surface. The hypothesis is supported by experimental measurements of the viscosity of silicone oils.

The model was tested with spreading experiments on two silicone oils on bare and silanized glass. For both substrates and each liquid, the equality between the excess capillary energy and the energy dissipated by viscous shearing was found to be very satisfactory. Therefore, the proposed analysis of the viscous dissipation phenomenon in a liquid wedge moving on a solid substrate during drop spreading correlates with the experimental data, without requiring any adjustable parameters. The spreading kinetics is described from only the rheological properties of the liquid and wetting parameters, without the need to introduce other parameters such as a "cut-off" or slip length.

## REFERENCES

1. T.D. Blake and J.M. Haynes, *J. Colloid Interface Sci.* **30**, 421 (1969).
2. S. Glasstone, K.J. Laidler and H. Eyring, *The Theory of Rate Processes*, McGraw-Hill, New York (1941).
3. P.G. de Gennes, *Rev. Mod. Phys.* **57**, 827 (1985).
4. C. Huh and L.E. Scriven, *J. Colloid Interface Sci.* **35**, 85 (1971).
5. R.G. Cox, *J. Fluid Mech.* **168**, 169 (1986).
6. F. Brochard-Wyart and P.G. de Gennes, *Adv. Colloid Interface Sci.* **39**, 1 (1992).
7. A. Carré and P. Woehl, *Langmuir* **18**, 3603 (2002).
8. J.D. Moore, S.T. Cui, P.T. Cummings and H.D. Cochran, *AIChE J.* **43**, 3260 (1997).
9. S.T. Cui, P.T. Cummings, H.D. Cochran, J.D. Moore and S.A. Gupta, *Int. J. Thermophysics* **19**, 449 (1998).
10. M. Modello and G.S. Grest, *J. Chem. Phys.* **103**, 7156 (1995).
11. T. Young, *Philos. Trans. R. Soc.* **A95**, 65 (1805).
12. E.B. Dussan V and S.H. Davis, *J. Fluid Mech.* **65**, 71 (1974).
13. L.M. Hocking, *J. Fluid Mech.* **77**, 209 (1977).
14. W.B. Hardy, *Philos. Mag.* **38**, 49 (1919).
15. D. Bangham and S. Saweris, *Trans. Faraday Soc.* **34**, 561 (1938).
16. M.E.R. Shanahan, *Langmuir* **17**, 3997 (2001).
17. M.E.R. Shanahan, *Langmuir* **17**, 8229 (2001).
18. A. Carré and F. Eustache, *Langmuir* **16**, 2936 (2000).
19. A. Carré, in *Contact Angle, Wettability and Adhesion*, Vol. 2, K.L. Mittal (Ed.), pp. 417-434, VSP, Utrecht (2002).
20. B. Arkles and G. Vogelaar, in *Metal Organics Including Silicones. A Survey of Properties and Chemistry*, B. Arkles (Ed.), p. 397, Gelest Inc., Tullytown, PA (1995).
21. P. Bourgin and S. Saintlos, *Intl. J. Non-Linear Mech.* **36**, 585 (2001).

Contact Angle, Wettability and Adhesion, Vol. 3, pp. 441–462
Ed. K.L. Mittal
© VSP 2003

# Adsorption isotherms of cationic surfactants on bitumen films studied using axisymmetric drop shape analysis

M.A. RODRÍGUEZ-VALVERDE,[1] A. PÁEZ-DUEÑAS,[2]
M.A. CABRERIZO-VÍLCHEZ[1] and R. HIDALGO-ÁLVAREZ[*,1]

[1] *Biocolloid and Fluid Physics Group, Department of Applied Physics, University of Granada, Campus de Fuentenueva, E-18071 Granada, Spain*
[2] *Headquarters of Technology, Repsol-YPF S.A., Ctra. de Extremadura, km 18, E-28931 Móstoles, Madrid, Spain*

**Abstract**—The most known method to measure adsorption of a surfactant in oil-in-water emulsions is based on the depletion of surfactant in the bulk after the adsorption. This approach requires an optimal separation of phases from the emulsion. A promising alternative is the interfacial method based on wetting the adsorbent with different concentrations of surfactant solutions. Not only does it avoid the utilization of emulsions, but if the adsorbent is a thermoplastic material like bitumen, then a smooth film can be obtained just by warming. This type of surface guarantees the symmetry of revolution in any sessile drop placed on it. Hence, Axisymmetric Drop Shape Analysis-Profile (ADSA-P) is the best-suited technique allowing to access simultaneously the interfacial tension and contact angle from a side view of the drop. In the present work, this methodology was applied with success on a soft bitumen and four commercial emulsifiers, and respective adsorption isotherms were obtained on the bitumen/water interface at room temperature. The effective areas per molecule at the bitumen/water interface were compared to the areas at the water/air interface obtained using complementary pendant drop experiments. This *in situ* method allows to study dynamic as well as structural aspects of adsorption on bitumen and other hydrophobic substrates of technological interest.

*Keywords*: Adsorption isotherms; bitumen film; sessile drop; ADSA-P; contact angle.

## 1. INTRODUCTION

During the last three decades, emulsions and dispersions have been the subject of increasing interest. The excess energy associated with the large interfacial area created in droplets formation makes dispersions metastable. Consequently, emulsions are prepared by shearing two immiscible liquids in the presence of a surfactant.

Oil-in-water emulsions are widely used in a variety of applications (e.g., painting, paper coating, road surfacing, lubrication, etc.) because of their ability to

---
[*]To whom all correspondence should be addressed. Phone: (34) 958-24-32-13,
Fax: (34) 958-24-32-14, E-mail: rhidalgo@ugr.es

transport or disperse hydrophobic substances in a water continuous phase. Emulsion technology enables the handling of many hydrophobic materials, which are almost like solids at room temperature due to their high viscosity. For example, the _bitumen_ (also known as _asphalt_ in the U.S.A.) used in road construction [1] is very difficult to manipulate at room temperature. Consequently, it has to be heated to lower the viscosity in order to pour it easily during application. However, a bitumen-in-water emulsion is quite easy to handle due to its low viscosity at room temperature. This is the basis of "cold" technology in road paving. Besides, the utilization of emulsions provides a significant energy saving and offers an environment-friendly strategy at the same time. Also emulsions improve simultaneously the adhesion of bitumen to acidic rocks such as granite, so that the road becomes more resistant to rain. Bitumen emulsions can even be applied to wet surfaces or under raining conditions.

Bitumen-in-water emulsions are polydisperse bitumen droplets (from several tenths of a micrometer to several micrometers in diameter) dispersed in an aqueous phase that contains an ionic (usually cationic) surfactant. Bituminous emulsions are designed to "break" and release the bitumen when they are in contact with wetted gravel like flints, silicates and limestones.

As mentioned above, a third component (ionic surfactant) is required during the emulsification of bitumen in water. Initially, the surfactant acts as an emulsifier enabling an optimal dispersion and at the same time works as a stabilizing agent prolonging the life of the dispersed state of the oil phase. Ionic surfactants (surface-active agents) are amphipathic substances consisting of a long-chain hydrocarbon tail and a polar head. One unique property of surfactants is that at sufficiently high concentrations in an aqueous medium, surfactant molecules begin to form molecular assemblies known as micelles [2]. The concentration at which this occurs is known as the critical micelle concentration (_cmc_). The driving force for this phenomenon is the "hydrophobic interaction". This interaction explains the tendency of the hydrocarbon tails to join in the presence of water. The same interaction is responsible for the tendency of surfactant molecules to be concentrated at interfaces, _i.e._, to adsorb on interfaces like a buoy floats on the sea. If one immerses a buoy (surfactant molecule) inside the sea (bulk), it rapidly rises to the surface (interface) by ("hydrophobic") buoyancy.

Adsorption consists in the exchange of water molecules by surfactant molecules at the interface decreasing the interfacial tension and forming an adsorbed film on the oil droplet surface. The first effect promotes emulsification while the second insures the stability of the emulsion by preventing coalescence due to electrostatic repulsion between the ionic groups of surfactant molecules [3]. A third role of surfactant is to promote the adhesion between bitumen and rocks after the breaking process. The mineral surfaces of rocks can become hydrophobic substrates if an ionic surfactant is electrostically adsorbed on them.

The adsorption of surfactant molecules at the bitumen-water interface is key during the emulsification process because the interfacial properties of a bitumen emulsion are mainly determined by the emulsifier used [4]. Adsorption phenomenon on solid interfaces is usually analysed by applying the material balance between the overall surfactant concentration $c_0$ and the concentration of non-adsorbed surfactant [5, 6], *i.e.*, the equilibrium concentration $c_{eq}$. The former is known whilst the latter is unknown. This approach is called the *depletion method* and is expressed in mathematical form as:

$$\Gamma_{SL} = \frac{V}{A}\left(c_0 - c_{eq}\right) \tag{1}$$

where $\Gamma_{SL}$ stands for the surface excess concentration, $A$ the area of finely divided adsorbent and $V$ the volume of dispersion. The equilibrium concentration is determined by potentiometric titration of the supernatant resulting from phase separation. To achieve a complete separation between the aqueous and the organic phases of the bituminous emulsion is the most tedious task in the estimation of the amount of remaining emulsifier (non-adsorbed emulsifier). This is usually achieved by centrifugation, but as the densities of the two phases are very close, the temperature and the shear speed must be very controlled in the extreme [7]. Furthermore, with this method it is very difficult to obtain information on the interface at low concentrations of emulsifier, because it is not possible to produce a bitumen emulsion at a highly reduced emulsifier concentration. In this case, a simple dilution is not viable because the emulsifier concentration at the interface is probably not in equilibrium with the aqueous phase due to the slow desorption from the interface which is important at room temperature [5].

To optimize the bitumen adhesion to rocks, the surfactant concentration used in the emulsion should be slightly lower than the critical micelle concentration because this way, the presence of free surfactant in the bulk is ensured after adsorption onto bitumen. Nevertheless, a surfactant concentration higher than its *cmc* is necessary to increase the stability of bitumen emulsions [8, 9]. For this reason the emulsifier concentration used in bitumen emulsions is far beyond the *cmc*, which implies maximum coverage of the bitumen-water interface. To enable to determine the role of the bitumen, it is preferred to use a method where the bitumen-water interface can be characterized at much reduced surfactant concentrations, or even in the absence of emulsifier.

## 2. THEORY

Viscosity, storage stability and breaking behaviour are the main properties that define the quality of bitumen emulsions for road applications. These properties depend on the emulsion formulation, the emulsification process as well as on the bitumen used. To characterize the quality of some types of bitumen for emulsion applications, one must bear in mind that the emulsion properties are particularly determined by the interfacial properties of the bitumen-water system [4]. It is known that very low quantities of surface-active components are sufficient to modify these properties. Consequently, analysis of the bulk bitumen itself is of limited use in predicting the behaviour of the bitumen in its emulsified state.

The usual quantity related to an interface is the interfacial tension. There are many methods available for the measurement of a liquid surface tension [10]. Traditionally, instruments utilizing the du Noüy ring and Wilhelmy plate methods have been used throughout the industry and research laboratories although these can be quite time consuming. The interfacial tension of bitumen is often obtained using the du Noüy method. However, it is necessary to decrease the bitumen viscosity by dilution but this can change the colloidal structure of the bitumen, and the interfacial properties thus observed are not representative of the emulsion properties. A modified Wilhelmy plate [11, 12], which allows temperature scanning measurements, is used in studies of oil sands processing where the bitumen is recovered from rocks extracted of natural deposits. This is precisely the opposite of the purpose of the breaking of bitumen emulsion on rocks: to cover them with bitumen.

Many technological processes, which employ solids, involve interfacial phenomena such as wetting, coating and/or adhesion. Owing to the experimental inaccessibility of interfacial energies associated with solid phases, contact angle appears as a simple, useful and sensitive tool for quantifying the surface energy of different materials in contact with pure water or aqueous surfactant solutions.

Contact angles can be measured directly on macroscopic, smooth, nonporous, planar substrates by merely placing a drop of the liquid or solution on the substrate and determining the contact angle by means of different techniques [10, 13]. This is the principle of the sessile drop method. In this approach, contact angle is strictly defined as the angle formed by the orthogonal directions associated with the solid-vapour and liquid-vapour interfaces along the three-phase line between the solid, the liquid, and the vapour (Fig. 1a).

The Young equation relates the interfacial tensions $\gamma$, defined by the three phases, to the contact angle $\theta$ [13]:

$$\gamma_{LV} \cos\theta = \gamma_{SV} - \gamma_{SL} \tag{2}$$

where the subscripts, L, V and S refer to the liquid in the drop, the vapour in the surrounding medium and the solid surface, respectively. It should be noted that the solid is assumed to be perfectly smooth, isotropic and rigid, and the liquid is chosen so that it does not swell the surface of the solid nor does it react with the surface.

The left hand term of Eq. (2) is known as *adhesion tension* [2]. Note that $\gamma_{SV}$ is the solid-vapour interfacial energy and not the true surface energy of the solid $\gamma_S$. The difference $\gamma_S - \gamma_{SV}$ is the equilibrium spreading pressure, which is a measure of the energy released through adsorption of the vapour onto the surface of the solid.

As mentioned in the Introduction, a direct determination of the amount of surfactant adsorbed per unit area of solid-liquid interface is not carried out because of the difficulty in isolating the bulk phase from the interfacial region for pur-

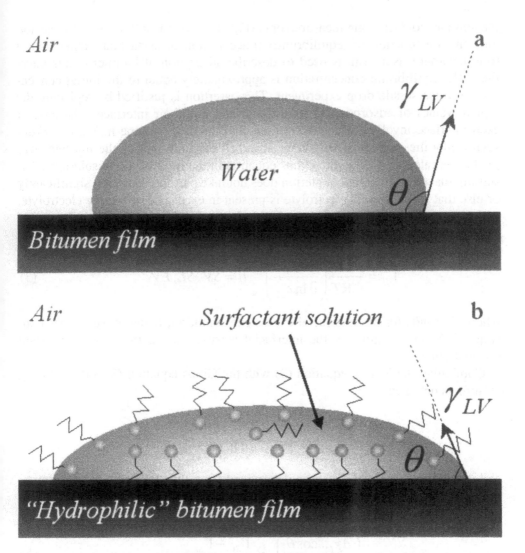

**Figure 1.** Effect of adsorption on wetting. a) Sessile drop of water on a bitumen film. The contact angle, $\theta$, is greater than 90 degrees due to the hydrophobic nature of bitumen. b) Sessile drop of a surfactant solution on a bitumen film. In this case the contact angle decreases by two concurrent effects. Bitumen surface becomes hydrophilic by the orientation of surfactant molecules with their tails towards the bitumen phase. In addition, the liquid-vapour adsorption brings about the reduction of surface tension, $\gamma_{LV}$, and the resulting drop flattening.

poses of analysis. When a surface-active agent is present in a liquid droplet, it can adsorb onto the surface, modify the surface energy, and cause the liquid contact angle to change. This phenomenon, known as *autophobicity* for hydrophilic surfaces, is quite striking in wetting films on clean surfaces since they first advance across the surface and then retract via a stick-jump motion. So, the amount of material adsorbed per unit area of interface can be calculated indirectly from surface tension and contact angle measurements (Fig. 1b). As a result, a plot of adhesion tension as a function of (equilibrium) concentration of surfactant, rather than a true adsorption isotherm, is used to describe adsorption at bitumen-water interface. The equilibrium concentration is approximately equal to the initial concentration in the sessile drop experiment. This assertion is justified by replacing the typical values of adsorption at solid-liquid and liquid-air interfaces, the area of each interface involved in the experiment and the drop volume in Eq. (1). From such a plot the amount of surfactant adsorbed per unit area of the interface can readily be calculated using the Gibbs equation in the limit of dilute solutions. Assuming that the surfactant counterion does not take part in adsorption significantly or else that an indifferent electrolyte is present in excess (a *swamping* electrolyte, which severely compresses electric double layers and minimizes the influence of electric charges of large molecules) [2], then:

$$\Gamma_{ij} = -\frac{1}{RT}\left(\frac{\partial \gamma_{ij}}{\partial \ln c_{eq}}\right)_T \quad ij = SV, SL, LV \tag{3}$$

where $\Gamma_{ij}$ stands for the interfacial excess concentration, R the universal gas constant, $T$ the temperature, $\gamma_{ij}$ the interfacial tension and $c_{eq}$ the bulk equilibrium concentration.

Combining the Young equation (2) with the Gibbs equation (3), the following relation is obtained:

$$\Gamma_{SL} - \Gamma_{SV} = \frac{1}{RT}\left(\frac{\partial \gamma_{LV} \cos\theta}{\partial \ln c_{eq}}\right)_T \tag{4}$$

and, the difference $\Gamma_{SL}-\Gamma_{SV}$ can be determined from the slope of $\gamma_{LV} \cos \theta$-log $c_{eq}$ plot at constant temperature [14]. A second relation can be deduced using Eq. (4) and Eq. (3) for liquid-vapour interface:

$$\left(\frac{\partial \gamma_{LV} \cos\theta}{\partial \gamma_{LV}}\right)_T = \frac{\Gamma_{SV} - \Gamma_{SL}}{\Gamma_{LV}} \tag{5}$$

The slope of the plot of $\gamma_{LV} \cos \theta$ versus $\gamma_{LV}$ thus provides information on the interfacial (excess) concentrations of the surfactant at each interface. Thus, the set of Eqs. (4)-(5) constitutes a convenient way for analysing the relation between adsorption and equilibrium wetting [15-18].

Other methods to study the influence of adsorption on wetting are based on graphical approaches [19] and on a kinetic interpretation of the Young equation (2) where a slow enough adsorption is assumed [20, 21].

The process of autophobicity mentioned above can occur by adsorption from the vapour ahead of the contact line. Thus, a thin film, known as a *precursor film*, exists on the surface around the liquid drop. This pre-existing water layer present on high and intermediate surface energy solids is considerably less likely on highly hydrophobic substrates because the equilibrium spreading pressure is zero, prohibiting the creation of the surface tension gradient necessary for Marangoni flow and accounting for very low spreading on these surfaces. Spreading on these surfaces is also thermodynamically less favourable owing to a very low spreading coefficient [2]. On a non-polar, low-energy surface, the adsorption of surfactant molecules on the solid-vapour interface can only be a local effect, close to the three-phase contact line. Surfactant migration is not possible [18] although small vibrations in the shape of the drop can provide a means of depositing surfactant molecules outside the liquid-covered area. Therefore, if the solid substrate is hydrophobic like bitumen, the adhesion tension can be used to determine numerically the surfactant surface (excess) concentration at the solid-liquid interface, as adsorption at the solid-vapour interface is negligible. Then, Eq. (4) can be simplified to:

$$\Gamma_{SL} \simeq \frac{1}{RT}\left(\frac{\partial \gamma_{LV} \cos\theta}{\partial \ln c_{eq}}\right)_T \tag{6}$$

## 3. EXPERIMENTAL

### 3.1. Axisymmetric Drop Shape Analysis-Profile (ADSA-P)

Once the proper theoretical background is established and without losing sight of the technological application, we need to choose a suitable technique with sufficient accuracy and reproducibility. This technique would allow to carry out routine experiments like a multipurpose instrument used permanently by industry laboratories. Of all the methods which have been developed [10], the approach based on the shape of liquid-fluid interfaces is the most general experimental technique. In addition to the simplicity of using pendant and sessile drops for determining surface tensions and contact angles, these methods require only small quantities of liquid and small sample sizes. This approach is non-intrusive and has an excellent control of the environmental conditions, thus avoiding any risk of contamination.

On the basis of the Young-Laplace equation [13] describing the drop profile of sessile and pendant drops as well as ascending drops and captive bubbles, it is possible to calculate the surface tension and the contact angle from digitalized images of the interfacial meridian:

$$\Delta P = \gamma_{LV}\left(\frac{1}{R_1}+\frac{1}{R_2}\right) \tag{7}$$

where $R_1$ and $R_2$ are the principal radii of curvature, $\gamma_{LV}$ the surface tension and $\Delta P$ the pressure difference along the interface.

Drop shape analysis has usually been performed by photographing a drop or bubble so that the characteristic lengths of the liquid-fluid interface can be measured on the photographic prints. As video imaging facilities and mathematical co-processors for personal computers have become readily available, there is a great potential for improvement of this method. Axisymmetric Drop Shape Analysis (ADSA) techniques [22] are based on the minimization of an objective function which depends on experimental data and on axisymmetric numerical solution given by the Young-Laplace equation (7). The strategy of ADSA-Profile (ADSA-P) consists in fitting the meridian of an experimental drop to the theoretical drop profile using the surface (interfacial) tension $\gamma_{LV}$ as one of the fitting parameters. The best fit gives the correct surface tension and, in the case of a sessile drop, the contact angle.

### 3.2. Bitumen films

Bitumen is the heavy crude oil obtained by distillation under vacuum of crude petroleum and consists of a complex mixture of hydrocarbons of various chemical structures and molecular weights. The bitumen is considered as a colloidal system [23] whose dispersed phase is polar, composed of asphaltanes and resins (acting as stabilizing agents for the former) while the dispersing phase is made of non-polar saturated and aromatic compounds (maltenes).

The acquisition of thermodynamically meaningful and reproducible contact angle data depends strongly on the quality of the solid surface. Therefore, roughness and heterogeneity can disguise the true interfacial information. In this sense, the preparation of high quality surfaces is essential to ensure that the measurements reflect the true interaction between the solid and the liquid.

When the solid material can be dissolved in a volatile solvent, like bitumen, then *solvent casting* technique becomes an option for the preparation of smooth surfaces in order to carry out contact angle measurements [12, 24]. The main drawbacks of this technique are the presence of traces of solvent remaining in the bitumen and changes in the composition of bitumen arising from interaction with the solvent. So the surface properties of bitumen are modified.

Alternatively, a smooth and isotropic surface is easily obtained by warming a thermoplastic material like bitumen [11, 12, 24]. Moreover, bitumen is insoluble in polar liquids and is stiff enough at room temperature. Therefore, a film made of molten bitumen can be considered as a chemically inert and rigid surface although not strictly. The rigidity and the chemical inertness will be discussed below. In spite of the heterogeneous nature of bitumen, any sessile drop of water placed on a bitumen film is axisymmetric (Fig. 2a) and shows a meaningful and reproducible contact angle. The simultaneous effects of the hydrophobic and thermoplastic characters seem to disguise the microscopic chemical heterogeneity of bitumen.

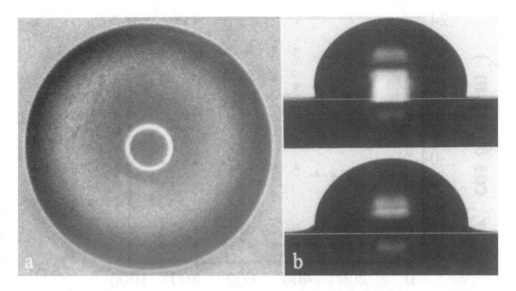

**Figure 2.** a) Top view of a sessile water drop on a bitumen film. Notice the clear axisymmetry of the drop. The bright ring inside the drop is produced by the annular illumination. b) Surface deformation of a bitumen film after the deposition of the sessile drop. The upper image was acquired at 21 min and the lower at 6 h 42 min. The distortion noticed will depend on the initial contact angle, the spreading rate and the bitumen aging although it becomes important approximately half hour after the drop deposition.

Owing to the thermoplastic nature of bitumen, bitumen films are pseudo-rigid surfaces because their surface is deformed over time by the unbalanced vertical strains which appear along the contact line of sessile drop (Fig. 2b). On soft surfaces a circular ridge is raised at the periphery of the drop; on harder solids there is no visible effect, but the stress is there. Actually, surface stiffness makes sense as an ideal concept because any solid surface with a sessile drop on it will be deformed until the true force balance is reached (it has been suggested that the contact angle is determined by the balance of surface stresses rather than that of surface free energies). Therefore, rigidity is a property related to the time-scale. In this sense, bitumen films are considered in this work as stiff surfaces because their characteristic times of deformation are longer than measurement times shown in Fig. 3. Nevertheless, a bitumen film could be considered as a link between a completely rigid surface and a liquid surface.

Likewise, the presence of natural surfactants [25, 26] dissolved in the drop from bitumen becomes important long after the measurement time. In spite of these natural surfactants existing in bitumen and the fact that non-polar liquids such as hydrocarbons cannot be used in the characterization of bituminous surfaces due to the mutual solubility of the phases, bitumen can be still used as a chemically inert phase. Consequently, a bitumen film seems to be a good model solid surface.

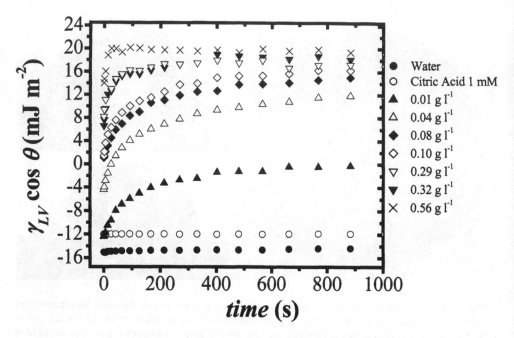

**Figure 3.** Instantaneous adhesion tension of a surfactant (E2) at different concentrations on bitumen films. Citric acid (1 mM) is used as buffer for the different aqueous solutions of surfactant.

Bitumen studied in this work was supplied by Repsol-YPF S.A. (Spanish-Argentinean Corporation) with the following physical properties: penetration grade, 150/200 (this is an indication of the viscosity of bitumen using the sinking of a needle in the bitumen at 25°C); viscosity, 7000 Pa·s at 25°C; density, 980 kg·m$^{-3}$ at 150°C and 1020 kg·m$^{-3}$ at 25°C.

The bitumen was heated in an oven at 90°C in order to soften it and allowed to stand at this temperature for one hour. Next, bitumen was laid on previously cleaned glass slides and then these slides covered with bitumen were placed in an oven for one additional hour. Thus, the smoothness and the surface isotropy were ensured after the total spreading of bitumen. The warming of the bitumen causes the migration of the polar components of the bitumen bulk towards the bitumen-glass interface as well as a partial oxidation of the bitumen. All these effects are representative of the bitumen interfacial state after the emulsification process where bitumen is heated and mixed with warm acid water. In order to minimize any additional oxidation of the bitumen films before use, a special precaution was taken to avoid overheating the bitumen as well as an immediate cooling during film preparation. Finally, the bitumen samples were stored in a dark place in closed containers.

Sessile drop and captive bubble methods involve three phases and three interfaces. Therefore, three different kinetics of adsorption will occur if the liquids are surfactant solutions. Even though solid-vapour adsorption can be neglected, the other two adsorptions always take place. This fact increases *a priori* the complexity of the method as compared with those methods involving only one interface like the spinning and pendant drop techniques [4, 7, 27-29]. In these cases, the contact angle is not defined because only two phases are involved in the system; therefore, the adsorption is known from measurements of interfacial tension at a liquid-liquid interface in the same way as at a liquid-vapour interface. However, the main drawback of bitumen in using the former methods at ambient temperature is its low fluidity and this can be solved if measurements are done at a high temperature. This way, the adsorption isotherms can be measured under conditions similar to emulsion production, *i.e.*, the experiments are carried out at temperatures up to 90°C since this value is reasonably close to emulsification conditions in the colloidal mill.

The spinning drop method [27] depends on a mathematically rigorous analysis interrelating interfacial tension and shape parameters of the drop spinning at a given rate within a second liquid. The setup cannot be modified when one attempts to implement the technique at an elevated temperature.

On the other hand, an "inverted" pendant drop of bitumen can be formed at the end of a motor driven syringe in a thermostatted cuvette filled with water or surfactant solution at different concentrations [4, 7, 28, 29]. However, in spite of the extensive use of this setup, some problems related either to the method in general or to bitumen are noticed. Working at high temperatures for longer times may lead to a partial evaporation of the water-phase, and hence to uncontrolled variations in the actual surfactant concentration. The small density difference between the two liquid phases is the major factor affecting the accuracy of the measurement [30, 31]. Moreover, after the pendant drop has been formed into pure water, some time is required before the interfacial tension reaches a stable value. This interfacial stabilisation is associated with the migration of natural surfactants from bitumen which competes with the emulsifier molecules used. Finally, the drop shape can deviate from purely Laplacian profile at high temperatures owing to the semi-liquid state of bitumen.

## 3.3. Cationic surfactants

Fatty amines, amines salts, and quaternary ammonium compounds are used in the preparation of cationic bitumen emulsions. Salts of polyamines such as N-alkyl propylenediamine or ethylenediamine derivatives are the most frequently used for this purpose. Sources of hydrophobic groups for these cationic surfactants are natural fatty acids such as tallow.

The four emulsifiers (denoted as E#) used in this work were commercially available cationic surfactants. Except for octadecyltrimethylammonium chloride (E1, Fluka), the emulsifiers were mixtures of several components derived from fats and oils. These surfactants were used as received, without further purification. However, due to the significant differences in purity and composition between commercial and research-grade materials, extreme care must be taken not to overlook the effects of such differences on the action of a given surfactant in a specific application.

The aqueous solutions of surfactants were prepared according to the manufacturer's recommendations. Owing to their pasty aspect at room temperature, to dissolve the fatty amines in (double distilled) water, the solutions were carefully heated to 50°C for 10-15 min under stirring. Without a previous treatment, our surfactants exhibit a non-ionic behaviour in aqueous solution. Thus, the addition of hydrochloric acid (35%, Panreac) enables a complete ionization of surfactant molecules. This way, molecules perform the function of true cationic surfactants. To maintain the final acid *pH*, a buffer of citric acid (Panreac) was used. The existing anions in solution such as chloride and citrate ions guarantee the formation of swamping electrolytes in excess. This condition validates Eq. (3).

The main parameters for each surfactant are summarized in Table 1. Currently, these surfactants are used as emulsifiers in the production of cationic rapid and medium setting bituminous emulsions. The cationic rapid setting emulsions are characterized by a relatively low emulsifier level and are used for spray application where viscosity is particularly important.

**Table 1.**
Characteristic parameters of each surfactant. The third column contains the values of the effective charge per molecule in units of electron charge (valence), measured by potentiometric titration. The average molecular weights supplied by the manufacturer are shown in the fourth column. The values of the effective area per molecule at the liquid-vapour interface (fifth column) and the values of the *cmc* (last column) were obtained with the pendant drop method using aqueous solutions of the surfactants

| Surfactant | Compound | $z$ | $M_m$ (g mol$^{-1}$) | $A_{LV}$ (nm$^2$) | *cmc* (mM) |
|---|---|---|---|---|---|
| E1 | Quaternary ammonium salt | 0.975 | 348.05 | 0.34 | 0.092 |
| E2 | Alkyl propylenediamine | 1.680 | 340 | 0.37 | 0.556 |
| E3 | Polyamine | 2.172 | 440 | 0.50 | 1.73 |
| E4 | Diamine | 4.12 | 470 | 0.29 | 2.70 |

$z$: effective charge per molecule in units of electron charge (valence)
$M_m$: average molecular weight
$A_{LV}$: effective area per molecule at liquid-vapour interface

## 4. RESULTS

### 4.1. Adsorption isotherms

The adsorption isotherms on the bitumen/water interface were obtained from the wettability data on bitumen films with different aqueous surfactant solutions.

The classical sessile drop method is based on measuring the contact angle of a liquid drop placed on a solid surface. According to the direction of the liquid front motion, different contact angles can be found. The contact angle is labelled as the *advancing contact angle* during the initial drop spreading over the surface. After this advancing stage the drop is in equilibrium with the surface and if the liquid is withdrawn from the drop then the contact angle is known as the *receding contact angle*. An equally suitable method would be to use a captive bubble (an air bubble is placed against the solid surface) instead of a sessile drop. However, we have chosen the advancing contact angles measured from sessile drops, because of the expected zero surfactant adsorption at the solid-vapour interface, *i.e.*, "outside" the sessile drop.

Owing to the non-polar nature of hydrophobic surfaces, adsorption will occur principally by dispersion interactions between the surface and the hydrophobic tail of the surfactant. For this reason the surfactant molecules will be oriented with their tails on the surface and the hydrophilic portions directed towards the solution. Therefore, adsorption will always result in an increase in the hydrophilic character of the surface (Fig. 1b). Such action is responsible for the increased dispersion of materials such as bitumen in aqueous systems. Hence, surfactants act properly as wetting agents for water on bitumen.

Although bitumen emulsions usually contain 50-70% bitumen and 0.1-2% cationic emulsifier by weight, the amount of emulsifier added to water in order to prepare the surfactant solutions used in this work was very small. Therefore, it was assumed that these solutions had the same density as pure water [32].

Dynamic surface tension and contact angles are defined as the non-equilibrium values of surface tension and contact angle during the spreading of the sessile drop on the solid surface. As the interfaces involved move towards equilibrium, either by adsorption or desorption of solute, the interfacial quantities change towards their static values. During an interfacial experiment, the time required for reaching the equilibrium can be from few hours to several days depending on the adsorption rate. Therefore, the equilibrium values of adhesion tension were computed from the respective values of surface tension and contact angle upon *pseudo-equilibrium* (Appendix).

The adhesion tension isotherms, plotted with the adhesion tension as ordinate, versus the concentration on a log scale (Fig. 4-6), show an increasing trend caused by the reduction in contact angle despite the fact that surface tension also decreases with increasing equilibrium concentration. A steep change of slope appears and from this point on adhesion tension stays uniform, which is common with typical isotherms at liquid-air interfaces. A possible maximum ("hump") in the isotherms is recognized as the consequence of contamination or the presence of hydrolysis products. Hydrolysis products can be even more surface active than some surfactants but these products are removed from the surface at higher concentrations by solubilization in micelles.

Assuming that the experimental data obey a Langmuir behaviour, Eq. (6) can be integrated. The solution is the following relation between the adhesion tension and the concentration [15]:

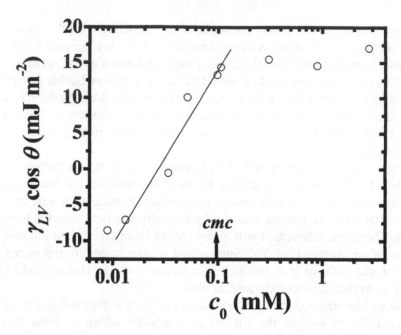

**Figure 4.** Adhesion tension, $\gamma_{LV} \cos \theta$, versus initial concentration, $c_0$, of surfactant E1 on a logarithmic scale. The equilibrium concentration is approximately equal to the initial concentration in a sessile drop experiment. The line is the fit of the experimental data to Eq. (9). The vertical arrow indicates the value of the *cmc* measured by the pendant drop method.

$$\gamma_{LV} \cos \theta = \gamma_0 \cos \theta_0 + \Gamma_{SL}^{\infty} RT \ln\left(1 + \frac{c_0}{a_{SL}}\right) \tag{8}$$

The first term on the right hand side is the adhesion tension without surfactant (buffer). $\Gamma_{SL}^{\infty}$ and $a_{SL}$ are the characteristic parameters of a Langmuir isotherm: the maximum amount adsorbed onto the interface (monolayer) and the concentration required to form half monolayer, respectively. The most practical parameter is $\Gamma_{SL}^{\infty}$ because it provides information on the situation of maximum adsorption, *i.e.*, when the saturation begins. For this reason, this parameter is closely related to the final size distribution in the emulsion. The reciprocal of $\Gamma_{SL}^{\infty}$ is proportional to the effective area per molecule of adsorbed surfactant at the interface. This area per molecule at the interface gives information about the degree of packing and the orientation of the adsorbed surfactant molecules. The limiting value of effective area is the geometrical area of the head group calculated from its molecular model. For example, this value is 0.18 nm² for the trimethylammonium group.

The linear relation between adhesion tension and concentration is observed only with highly dilute solutions ($c_0 << a_{SL}$), when the interfacial behaviour can be considered as ideal. For concentrations close to the point of maximum change in slope ($c_0 >> a_{SL}$), relation (8) can be reduced to:

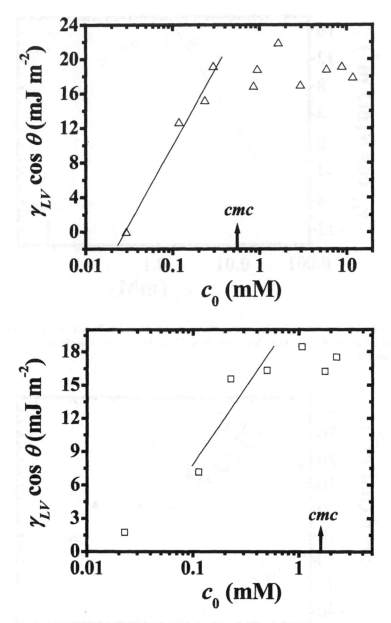

**Figure 5.** Adhesion tension-$c_0$ (on log scale) plots for surfactants E2 (top) and E3 (bottom).

$$\gamma_{LV} \cos\theta \simeq \gamma_0 \cos\theta_0 + \Gamma_{SL}^{\infty} RT \ln {}^{c_0}\!\big/\!_{a_{SL}} \qquad (9)$$

From this expression, $\Gamma_{SL}^{\infty}$ can be computed from the slope of a linear $\gamma_{LV} \cos\theta$-log $c_{eq}$ curve. These values and the respective effective areas per molecule are shown in Table 2 for each surfactant. Surfactant E1 is adsorbed on the bitumen-

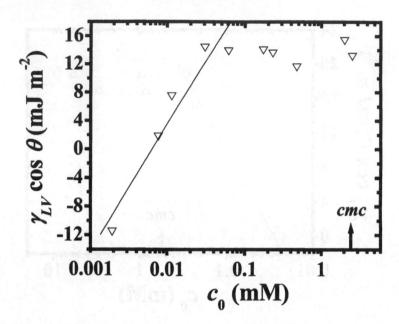

**Figure 6.** Adhesion tension-$c_0$ (on log scale) plots for surfactant E4. Notice the separation between the *cmc* and the point of maximum change in slope. This observation is likely due to the formation of surface micelles.

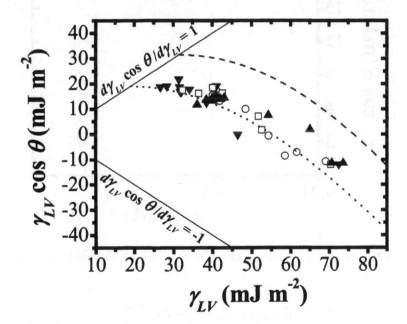

**Figure 7.** Adhesion tension plotted versus surface tension. Open circle -○-, emulsifier E1; inverted triangle -▼-, E2; open square -□-, E3; triangle -▲-, E4. Using Neumann's equation-of-state [43], two curves are plotted: dashed line ---, pure water in the receding mode (captive bubble) and dotted line ···, pure water in the advancing mode (sessile drop).

**Table 2.**
Surface excess concentration upon saturation, $\Gamma^\infty{}_{SL}$, and the effective area per molecule, $A_{SL}$, for adsorption of each surfactant on bitumen. The surface excess concentration values were computed from the respective slopes shown in Figs. 4-6 normalized by the factor $RT/\log e$

| Surfactant | $\Gamma^\infty{}_{SL}$ (µmol m$^{-2}$) | $A_{SL}$ (nm$^2$) |
|---|---|---|
| E1 | 4.69 | 0.35 |
| E2 | 3.26 | 0.51 |
| E3 | 2.09 | 0.80 |
| E4 | 3.91 | 0.44 |

water interface as on the water-air interface because its molecule has a simple tail and its head group is less charged as compared to other surfactants. These surfactants occupy greater area on bitumen than on a water-air interface due mainly to their branched tails although the electrostatic repulsion between head groups can also increase the distance between adsorbed molecules.

### 4.2. Verification of initial hypothesis

The validity of Eq. (6) depends on the condition $\Gamma_{SL} \gg \Gamma_{SV}$. This restriction was theoretically justified in Section 3.2 although it can be qualitatively verified using Eq. (5) and the experimental slope in Fig. 7. From Fig. 7 the following inequalities are deduced:

$$-1 \leq \left( \frac{\partial \gamma_{LV} \cos\theta}{\partial \gamma_{LV}} \right)_T < 0$$

This decreasing tendency observed in Fig. 7 confirms that the adsorption at the solid-vapour interface is, at least, less than the adsorption at the solid-liquid interface. A numerical verification is not viable even though $\Gamma_{LV}$ is experimentally accessible from Gibbs equation (3) for fluid interfaces using the same sessile drop experiment or a pendant drop experiment. We cannot know $\Gamma_{SL}$ without the hypothesis $\Gamma_{SL} \gg \Gamma_{SV}$. A complete verification of solid-vapour adsorption would be feasible from additional experiments like infrared measurements.

At high surface tensions, the slope is roughly uniform and equal to $-1$. This result shows that adsorption is approximately equal at non-polar solid-aqueous and aqueous-air interfaces [2] for dilute concentrations.

## 5. DISCUSSION

Most investigations of adsorption on bitumen have so far been concerned with equilibrium aspects of surfactant adsorption. In addition, these studies are based on *ex situ* and indirect methods. Correspondingly, there is a lack of information about the dynamic interfacial behaviour which definitely governs the overall adsorption process.

In the present paper, a direct and *in situ* method, based on the sessile drop method, is developed to directly determine the actual thermodynamic quantities governing the adsorption process. As main advantage, this method allows to study kinetics (Fig. 3) as well as structural aspects of adsorption on bitumen (Table 2). Besides, complete adsorption isotherms can be achieved.

The HLB (Hydrophilic-Lipophilic Balance) for selecting suitable emulsifiers to stabilize emulsions has been used since about five decades [33]. However, the emulsifiers are examined in isolation from the particular environment so that the HLB data are of dubious value. In this sense, adsorption isotherms are very useful for the purpose of comparing the appropriate performance of surfactants in bitumen-water adsorption. The concepts of efficiency and effectiveness can be applied to adsorption isotherms just like in liquid-air adsorption [2]. In this regard, efficiency of a surfactant is related to the bulk concentration of surfactant required to increase significantly the adhesion tension. And the effectiveness is defined in terms of the maximum increase in adhesion tension that can be obtained regardless of bulk concentration of surfactant.

Bitumen films have been used as model surfaces, thus providing the smoothness and structural homogeneity necessary for both reproducibility and reliable interpretation of interfacial thermodynamic quantities. Obviously, the heat treatment, required to prepare these films, introduces some changes in the bitumen surface properties due to a partial oxidation and the migration of polar components to the bitumen-glass interface. But the same interfacial alterations in bitumen take place during emulsification process in industrial stations. Bitumen is initially warmed at 90°C to inject it in the colloidal mill. Next, bitumen is mixed with hot water and the mixture is subjected to shear conditions for one hour. In this stage, bitumen can be rapidly dispersed in water even without emulsifier because some natural surfactants migrate to the bitumen-water interface. Finally, the bitumen emulsion exits from the mill moderately warm so that a sudden cooling is prevented. This way, a bitumen film undergoes a thermal history similar to that of the bitumen of an emulsion. The use of bitumen films as solid surfaces restricts the sessile drop method to temperatures close to the room temperature. However, this requisite makes sense for the purpose of "cold" technology. The limitation in temperature involves that surfactants are adsorbed onto bitumen at room temperature with their tails placed over bitumen surface instead of immersed in bituminous phase as it happens at higher temperatures.

The method discussed in this paper is clearly more time consuming during the material preparation than the typical plate or ring techniques. Preparation of bitumen films lasts for approximately 2 hours and many weeks are allowed to elapse at room temperature in order to ensure rigid enough surfaces. In spite of the fact that our method requires to record surface tension and contact angle as a function of time for 1 hour, a pseudo-equilibrium is reached after 20 minutes regardless of surfactant concentration.

The main advantage of this technique is that bitumen heating in the course of measurements or previous dilution of bitumen are not needed. Moreover, the sessile drop method provides a versatile approach for those research laboratories related to road construction because it allows to measure the wetting of bitumen emulsions on mineral surfaces with minor modifications in the setup [34].

## 6. SUMMARY

In the literature, adsorption isotherms on bitumen have been mainly carried out with bitumen fluidized by high temperature or by using solvents. The adsorption takes place in the interface between liquid phases. Only in ref. [35], wetting measurements were additionally included using bitumen emulsion drops on standard surfaces and films obtained by drying of emulsions. For this reason, there is no agreement between the measurements of adsorption on bitumen presented in this work and the published values measured by other interfacial techniques [4, 7, 27-29]. But this lack of agreement is also explained by differences in bitumen properties (like viscosity), type of surfactants (cationic, anionic and non ionic) and origin of bitumen (natural or by distillation). The validity of our methodology is confirmed by the agreement between the measured values of effective adsorption area at bitumen-water interface shown in Table 2 and those obtained at water-vapour interface (by pendant drop) for the same surfactants (Table 1). This fact points out formation of surfactant monolayer on bitumen films just like on bitumen droplets of an emulsion. In addition, the fact that the points of maximum change in the slope in adhesion tension plots (Fig. 4-5) are found at the *cmc* values obtained by pendant drop, ratifies the method except for surfactant E4 (Fig. 6). This anomalous shift of the *cmc* and the small effective area per molecule at the bitumen-water interface indicate the formation of micelles on the bitumen surface [36]. This arrangement of the hydrophobic tails on bitumen is more likely than the creation of a bilayer. The emulsifier E4, whose chemical structure is the most unknown, has the greatest effective charge in spite of its small area at the water-air interface. A molecule with different charges distributed along the hydrocarbon tail would explain these observations as well as the surface micellization mentioned.

To sum up, sessile drop method (ADSA-P) and bitumen films can be used to assess the affinity of emulsifiers to a given bitumen and conversely, the differences in the reactivity of different bitumens in the presence of an emulsifier, without requiring an emulsion. In addition, using this technique, the bitumen-water interface can be characterized at extremely low surfactant concentrations.

This method is equally suitable to other low surface energy, thermoplastic materials in the real-time investigations of adsorption dynamics (polymers, surfactants) and related events, specially in systems involving three-phase contacts.

## APPENDIX

The interfacial equilibrium state in a pure liquid is rapidly attained, within a few milliseconds. In solutions of surface-active solutes the rate at which equilibrium is reached is much lower as it depends on the rate of diffusion of solute molecules to the surface. The rate of diffusion is determined by the size of the molecule and its concentration. For conventional surface-active solutes, diffusion brings molecules to the surface within seconds; yet experience shows that some solutions take hours or even days to reach an equilibrium surface tension. Such long periods are due to highly surface-active impurities. When liquid-vapour and solid-liquid interfaces are involved, the adsorption rates at each interface will be controlled by diffusion as well as by surface interactions. In this case, the presence of a solid can even delay the final equilibrium state. Definitely we need to reduce the time of interfacial experiments by means of a practical criterion which defines a pseudo-equilibrium interfacial state. But this criterion must be consistent with the system studied [37]. For liquid-fluid interfaces, the criterion of pseudo-equilibrium is applied to dynamic interfacial tension [38-40]. Conversely, when solid-liquid interfaces are examined the phenomenon will be entirely described by contact angle and interfacial tension. Variations of these (dynamic) quantities with time are independently caused by adsorption on solid-liquid and liquid-air interfaces as well as by the inherent thermal fluctuations of any thermostat [41]:

$$\frac{dx}{dt} = \left(\frac{\partial x}{\partial t}\right)_T + \left(\frac{\partial x}{\partial T}\right)_t \frac{dT}{dt}$$

$x$ being a dynamic interfacial quantity such as contact angle or interfacial tension, i.e., the experimentally accessible quantities. From the triangular inequality:

$$\left|\frac{dx}{dt}\right| \leq \left|\left(\frac{\partial x}{\partial t}\right)_T\right| + \left|\left(\frac{\partial x}{\partial T}\right)_t\right|\left|\frac{dT}{dt}\right|$$

and applying the equilibrium condition, the following relation is derived:

$$\left|\left(\frac{dx}{dt}\right)_{eq}\right| \leq \left|\left(\frac{\partial x}{\partial T}\right)_{t_{eq}}\right|\left|\frac{dT}{dt}\right|$$

where the first partial derivative on the right hand side can be identified as the true state of thermodynamic equilibrium, i.e.,

$$\left(\frac{\partial x}{\partial T}\right)_{t_{eq}} = \frac{dx_{eq}}{dT}$$

The average thermal variations of contact angle ($x_{eq}=\theta$) [32] and surface tension ($x_{eq}=\gamma_{LV}$) of water [42] as well as the temporal fluctuations of temperature, $|dT/dt|$, in a commonly used thermostat ($\approx 10^{-3}$ K s$^{-1}$) will fix the respective pseudo-equilibrium values of contact angle and surface tension by the following inequalities:

$$\left|\left(\frac{d\theta}{dt}\right)_{eq}\right| \leq 10^{-4}\,\frac{\pi}{180}\,\text{rad}\,\text{s}^{-1}$$

$$\left|\left(\frac{d\gamma_{LV}}{dt}\right)_{eq}\right| \leq 10^{-4}\,\text{mJ}\,\text{m}^{-2}\,\text{s}^{-1}$$

(10)

## Acknowledgements

Financial support from Repsol-YPF S.A. and "Ministerio de Ciencia y Tecnología, Plan Nacional de Investigación Científica, Desarrollo e Innovación Tecnológica (I+D+I)", MAT2001-2843-C02-01 is gratefully acknowledged.

## REFERENCES

1. A.N.J Scott, in: *Theory and Practice of Emulsion Technology*, A.L. Smith (Ed.), pp. 179-200, Academic Press, New York (1976).
2. M.J. Rosen, *Surfactants and Interfacial Phenomena*. John Wiley & Sons, New York (1989).
3. J. Lyklema, *Colloids Surfaces A*, **91**, 25-38 (1994).
4. Y. Lendresse, M.F. Morizur, A. Cagna and G. Espósito, in: *Proc. 1st Eurasphalt & Eurobitume Congress*, **113**, European Asphalt Pavement Association (EAPA) and European Bitumen Association (Eurobitume), Brussels (1996).
5. J.E. Poirier, M. Bourrel, P. Castillo, C. Chambu and M. Kbala, *Prog. Colloid Polym. Sci.*, **79**, 106-111 (1989).
6. E. Unzueta, A. Páez, S. Torres and J. Sánchez, in: *Proc. 2nd International Symposium on Asphalt Emulsion Technology*, pp. 247-253, Asphalt Emulsion Manufactures Association, Annapolis, MD (1999).
7. B. Eckmann, K. Van Nieuwenhuyze, T. Tanghe and P. Verlhac, in: *Proc. 2nd Eurasphalt & Eurobitume Congress*, **114**, pp. 135-142, Foundation Eurasphalt, Breukelen, The Netherlands (2000).
8. N. Romero, A. Cárdenas, M. Henríquez and H. Rivas, *Colloids Surfaces A*, **204**, 271-284 (2002).
9. F. Leal-Calderón, J. Biais and J. Bibette, *Colloids Surfaces A*, **74**, 303-309 (1993).
10. S. Ross and I.D. Morrison, *Colloidal Systems and Interfaces*, pp. 113-122, John Wiley & Sons, New York (1988).
11. Z.M. Potoczny, E.I. Vargha-Butler, T.K. Zubovits and A.W. Neumann, *Alberta Oil Sands Technology and Research Authority (AOSTRA) J. Res.*, **1**(2), 107-115 (1984).
12. J. Drelich, K. Bukka, J.D. Miller and F.V. Hanson, *Energy & Fuels*, **8**, 700-704 (1994).
13. A.W. Adamson and A.P. Gast, *Physical Chemistry of Surfaces*, 6th Edition, John Wiley & Sons, New York (1998).
14. E.H. Lucassen-Reynders, *J. Phys. Chem.*, **67**, 969-972 (1963).
15. W.G. Rixey and D.W. Fuerstenau, *Colloids Surfaces A*, **88**, 75-89 (1994).
16. L.K. Koopal, T. Goloub, A. de Keizer and M.P. Sidorova, *Colloids Surfaces A*, **151**, 15-25 (1999).
17. T.B. Lloyd, *Colloids Surfaces A*, **93**, 25-37 (1994).

18. H. Haidara, L. Vonna and J. Schultz, *Langmuir*, **12**, 3351-3355 (1996).
19. E.A. Vogler, D.A. Martin, D.B. Montgomery, J. Graper and H.W. Sugg, *Langmuir*, **9**, 497-507 (1993).
20. H.J. Busscher, W. van der Vegt, J. Noordmans, J.M. Schakenraad and H.C. van der Mei, *Colloids Surfaces*, **58**, 229-237 (1991).
21. R. Miller, S. Treppo, A. Voigt, W. Zingg and A.W. Neumann, *Colloids Surfaces*, **69**, 203-208 (1993).
22. O.I. del Río and A.W. Neumann, *J. Colloid Interface Sci.*, **196**, 136-147 (1997).
23. J.Ph. Pfeiffer and R.N.J. Saal, *J. Phys. Chem.* **44**, 139-149 (1939).
24. E.I. Vargha-Butler, T.K. Zubovits, C.J. Budziak and A.W. Neumann, *Energy & Fuels*, **2**, 569-572 (1988).
25. S. Acevedo, X. Gutiérrez and H. Rivas, *J. Colloid Interface Sci.*, **242**, 230-238 (2001).
26. M. Salou, B. Siffert and A. Jada, *Fuel*, **77**, 343-346 (1998).
27. M. Di Lorenzo, H.T.M. Vinagre and D.D. Joseph, *Colloids Surfaces A*, **180**, 121-130 (2000).
28. R.A. Mohammed, M. Di Lorenzo, J. Mariño and J. Cohen, *J. Colloid Interface Sci.*, **191**, 517-520 (1997).
29. A. Seive, M.F. Morizur, B.G. Koenders, G. Durand and J.E. Poirier, in: *Proc. 2nd International Symposium on Asphalt Emulsion Technology*, pp. 256-263, Asphalt Emulsion Manufactures Association, Annapolis, MD (1999).
30. E.N. Stasiuk and L.L. Schramm, *Colloid Polym. Sci.*, **278**, 1172-1179 (2000).
31. A. Pandit, C.A. Miller and L. Quintero, *Colloids Surfaces A*, **98**, 35-41 (1995).
32. R.C. Weast (Ed.), *CRC Handbook of Chemistry and Physics*. 66th ed., CRC Press, Boca Raton, FL (1986).
33. A.M. Al-Sabagh, *Colloids Surfaces A*, **204**, 73-83 (2002).
34. M.A. Rodríguez-Valverde, M.A. Cabrerizo-Vílchez, P. Rosales-López, A. Páez-Dueñas and R. Hidalgo-Álvarez. *Colloids Surfaces A*, **206**, 485-495 (2002).
35. L. Loeber, G. Mueller, B. Héritier, Y. Jolivet and M. Malot, in: *Proc. 2nd Eurasphalt & Eurobitume Congress*, **254**, pp. 335-341, Foundation Eurasphalt, Breukelen, The Netherlands (2000).
36. T. Gu, B.Y. Zhu and H. Rupprecht, *Prog. Colloid Polym. Sci.*, **88**, 74-85 (1992).
37. A. Yeung, T. Dabros and J. Masliyah, *J. Colloid Interface Sci.*, **208**, 241-247 (1998).
38. C.H. Chang and E.I. Franses, *Colloids Surfaces A*, **100**, 1-45 (1995).
39. X.Y. Hua and M.J. Rosen, *J. Colloid Interface Sci.*, **124**, 652-659 (1988).
40. T. Svitova, Y. Smirnova and G. Yakubov, *Colloids Surfaces A*, **101**, 251-260 (1995).
41. M.A. Cabrerizo Vílchez, Z. Policova, D.Y. Kwok, P. Chen and A.W. Neumann, *Colloids Surfaces B*, **5**, 1-9 (1995).
42. M. de Ruijter, P. Kölsch, M. Voué, J. de Coninck and J.P. Rabe, *Colloids Surfaces A*, **144**, 235-243 (1998).
43. D.Y. Kwok and A.W. Neumann, *Colloids Surfaces A*, **161**, 31-48 (2000).

*Contact Angle, Wettability and Adhesion*, Vol. 3, pp. 463–478
Ed. K.L. Mittal

# Application of droplet dynamics analysis for assessment of water penetration resistance of coatings

LECH MUSZYŃSKI,* MAGNUS E.P. WÅLINDER, CIPRIAN PÎRVU,
DOUGLAS J. GARDNER and STEPHEN M. SHALER

*Advanced Engineered Wood Composites Center, University of Maine, Orono, ME 04469-5793*

**Abstract**—Moisture uptake in wood or wood-based products reduces their performance and durability. Development of improved, more effective coating systems that provide durable protection and thus extend the service life of wood products depends largely on a good understanding and accurate methods to assess their effectiveness in resisting water penetration into wood. The method presented in the paper is aimed at enhancing the development of new coated wood products through extending the current state of knowledge on water resistance of coatings applied onto wood products, and by providing a new robust evaluation tool. The primary objective was to investigate the applicability of droplet analysis methodology for comparative assessment of water resistance of coatings. The experimental method used for droplet dynamics analysis was extended so that water penetration and surface properties of coatings could be determined in the course of the same test. In this method, droplets of water were deposited on specimens of coated surfaces, and changes in droplet volume due to evaporation and penetration into the coatings were registered on a series of digital images. Digital image analysis techniques were used to determine changes in dimensions and volume of the droplet. Reference tests on impermeable surfaces enable an accurate determination of the amount of moisture penetrating the test surface by subtracting the evaporated volume. Advantages and shortcomings of the method are discussed.

*Keywords*: Coatings; wood protection; water; penetration; permeability; diffusion; wood composites; droplet analysis; digital image analysis.

## 1. INTRODUCTION

Wood is a strongly hygroscopic material, which means that it can easily exchange moisture with the environment [1, 2]. Moisture affects both dimensional stability and mechanical properties of wood products. Persistent high levels of moisture content in wood products can also contribute to its biological degradation.

One of the most common methods for wood protection against gaining moisture from the environment and extending its service life is to apply water repellent, waterproof or semi-waterproof coatings, seals or surface treatments [3-5].

---

*To whom all correspondence should be addressed. Phone: 207 581-2102, Fax: 207 581-2074,
E-mail: Lech.Muszynski@umit.maine.edu

Development of improved and more effective coating systems depends largely on a good understanding and accurate methods to assess their surface properties and effectiveness in resisting water penetration into the substrate [6].

The effectiveness of coatings is usually assessed based on the amount of moisture penetrating the coating exposed to moisture in specific conditions [7-14, 18]. In general, moisture may penetrate into a surface as liquid or vapor, and the process may be driven by diffusion and/or capillary transport and most often the two operate together. Chemical and physical changes in the coatings inflicted by exposure to service environment may affect the diffusion properties as well as the micro-capillary profile of the surface through appearance of micro-cracks, and can result in a significant increase in permeability. Moisture penetrating through the coating makes the wood swell and causes a mechanical stress at the coating/wood interface. In addition, intrusion of water breaks the chemical bonding between the wood and the coating film. On a macro-scale this phenomenon may be observed as peeling. In 1994, Williams estimated that about 1 billion square feet of plywood siding and another 1 billion square feet of hardboard siding were used annually in new construction in the U.S., and that the annual replacement costs of the siding could be close to 2 billion dollars [6].

The durability of coatings is often examined by comparing moisture penetration resistance of fresh-cured coatings with untreated specimens subjected to natural or artificial weathering/aging. One of the main problems in the standard test methods appears to be slow moisture transmission rate through coatings, which are essentially designed to resist penetration. Consequently, many test methods require substantial time to pass before measurable quantities of water or water vapor penetrates the coating [15-19].

Rapid measurement methods (measurements within hours rather than days) based on detection of very small volumes of water molecules require expensive high-resolution instruments (infra-red or water vapor capacitance detectors [20-22]). The methods are designed for testing free-film samples and are not suitable for bare surfaces and coatings applied onto substrates. A method using internal reflection Fourier transform IR spectroscopy for measurement of water diffusion in polymer films on substrates was described by Nguyen *et al.* [23] and Linossier *et al.* [24]. Rice and Phillips [25] investigated the moisture penetration resistance of coatings on wood using an acoustic emission method.

The in-service behavior and durability of a coating-wood system is strongly related to intermolecular forces (bonding) at the coating-wood interface. This adhesion mechanism is referred to as the wetting or adsorption theory [26, 27]. Subtle changes in the chemical and physical composition of surfaces might be detected by means of droplet dynamics analysis. Most commonly the techniques involve analysis of shape and contact angle of a single droplet of a reference liquid deposited on a test surface by means of a precision syringe. The analysis can be performed on a single image of a droplet when a pseudo-static equilibrium is reached where an arbitrary time since deposition is chosen (usually within first 2 seconds). From the contact angle analysis, wetting characteristics of surfaces are derived.

Usually, the shape of the droplet is recorded from a side view by single video or digital camera. Hence a strict axial symmetry of the droplet must be assumed to allow accurate measurements. On anisotropic surfaces such as wood, a water droplet will spread spontaneously along the grain, creating a non-axisymmetric droplet. Such wicking effects result in a distinctly different wood-water wetting behavior along than across the grain [28, 29]. Although some researchers have applied this method to evaluate the wettability of wood and other anisotropic surfaces (e.g. [30-31]) it should be stressed that the assumption of axial symmetry practically limits its use to isotropic surfaces with no directional preference for droplet advance.

The influence of evaporation on contact angle of water droplets on polymer surfaces was studied by Shanahan and Bourges [32], and Bourges-Monnier and Shanahan [33]. There are also studies on thermodynamics of droplet evaporation on hot metal surfaces where the reverse relation (i.e. effect of droplet shape on evaporation rate) is analyzed (e.g. [34, 35]). However, in most studies where surface properties of materials at ambient temperature are investigated by means of droplet analysis the influence of diffusion and evaporation on the droplet shape is simply neglected.

Although it is reasonable to assume that resistance to moisture penetration is correlated with surface properties of coatings, particularly when the micro-capillary transport is involved (see e.g. [36, 37]), it should be noted that such a correlation might not always be found (for instance, there are significant differences between characteristic surface energy of glass, Plexiglas and Teflon, where all of the materials are considered impermeable). If, however, a correlation between the resistance to moisture penetration and wetting properties of coatings is proven it could enable evaluation of coating durability from well documented analysis methodology. There is no evidence for much research in this area in the literature.

Correlation between the resistance to moisture penetration and surface properties of coatings was investigated by Hora [38] who compared changes in contact angle of water droplets measured on a range of coatings during short-term (up to 4 minutes) droplet dynamics analysis tests, with moisture uptake of coated wood specimens determined according to German standard DIN 52617 [39]. Both unweathered and weathered coatings were included in the investigation. A weak qualitative correlation was reported. However, in the experiment the effect of change in droplet volume due to evaporation and penetration into the substrate on the change in contact angle was totally neglected, which most likely distorted the results.

Tshabalala *et al.* [40] reported a good correlation between water vapor adsorption behavior of wood and its surface thermodynamic properties measured by means of inverse gas chromatography.

The primary objective of the project was to investigate applicability of droplet analysis methodology as a quick comparative assessment of water resistance of coatings, as well as investigate the potential correlation between water penetration resistance parameters of coatings (permeability, diffusivity, porosity) and parameters used to describe surface properties (e.g. contact angle, and surface energy). To achieve the goal, the experimental method used for droplet dynamics analysis was extended so that water penetration and surface properties of coatings could be determined in the course of the same test.

## 2. EXPERIMENTAL METHODS

In the present study, techniques used in surface wetting analysis were employed for accurate evaluation of changes in the volume of the droplet and determination of fraction of the volume that penetrated the substrate. A single digital camera with microscopic lenses was used to acquire a series of images of water droplets deposited on a range of isotropic surfaces with a precision syringe. Initial droplet volume was approximately 5 µl (equivalent to 5 mg). The substrates, surface treatments and coatings used in the study are presented in Table 1. Medium density fiberboard (MDF) was used as an untreated reference permeable surface that could be considered isotropic in the sense that the droplet deposited on the surface retained its axial symmetry throughout the test.

Generally both evaporation from the free droplet surface and penetration into the test surface are expected (Figure 1), therefore two practically impermeable reference surfaces: plain uncoated Plexiglas, and Parafilm coated glass slide, were used to evaluate the effect of evaporation.

The tests were conducted in a small environmental chamber. To avoid interference due to concurrent moisture exchange at the test surfaces around the droplet, all specimens were preconditioned in the chamber at constant environment (24°C, 56% RH) for more than 24 hours before testing so that equilibrium with the ambient environment could be assumed. The same conditions were maintained during the tests.

**Table 1.**
Substrate materials, coatings and surface treatments used in the study

| Specimen ID | Coating | Substrate | Comments |
|---|---|---|---|
| Plexi | Uncoated | Plexiglas | reference |
| Parafilm | Parafilm | Glass | reference |
| R | Uncoated | MDF | |
| WA | Wood Finishing Wax | WPC | Wood/PVC (50/50) |
| W | Wood Finishing Wax | MDF | |
| WS | Water sealant | WPC | Wood/PVC (50/50) |
| PE | Polyester | MDF | |

MDF = Medium Density Fiberboard
WPC = Wood/Plastic Composite

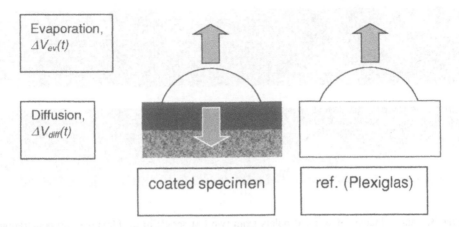

**Figure 1.** Balance of a test droplet volume on a hygroscopic/permeable substrate and on impermeable reference surface (e.g. Plexiglas, glass, Teflon) in constant environment (23°C and 55% RH).

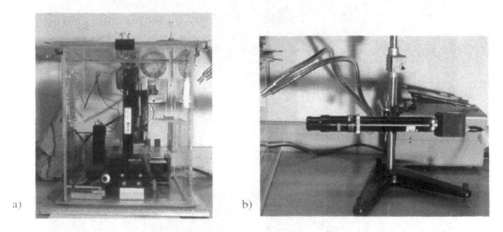

**Figure 2.** Test set-up: a) 3D adjustable stage enclosed in an environmental chamber; b) digital camera with microscopic lenses and lighting unit.

The test setup (Figure 2) consisted of a small translucent chamber, where saturated salt solution was used to maintain constant level of relative humidity. The specimens were supported on a manually operated stage with 3D linear micro-movement ability. A fiber-optic 'gooseneck' lighting unit was used for lighting the sample to minimize heating up the chamber atmosphere. The precision syringe was mounted in the top of the chamber. The syringe was operated manually and hence the shaft movement rate could not be precisely controlled, which resulted in variation in the initial volumes of the droplets. A single high-resolution (1296 x 1054 pixels) digital camera equipped with microscopic lenses was used to

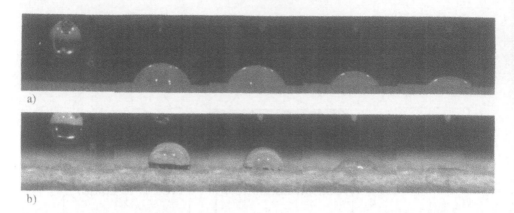

**Figure 3.** Sample images of water droplets from two test series: a) on Plexiglas (no penetration); b) on wood/PVC composite surface treated with wax.

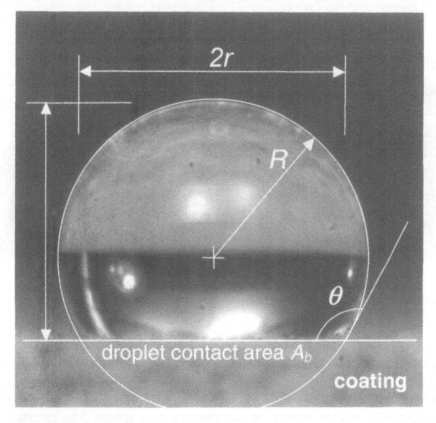

**Figure 4.** Quantities characterizing droplet shape: droplet radius ($R$); base or contact area radius ($r$); droplet height ($h$); contact angle ($\theta$). From this droplet free surface area ($A_s$), droplet contact area ($A_b$), and droplet volume ($V$) are calculated.

**Table 2.**
Formulae used for quantitative droplet geometry description. Droplet radius $(R)$ and height $(h)$ were measured on the digital images

| Quantity | Formula |
|---|---|
| Contact angle | $\theta = a\cos(1 - h/R)$ |
| Base radius | $r = R\sin\theta$ |
| Free droplet area (cap area) | $A_e = 2\pi Rh$ |
| Droplet base area | $A_b = 2\pi r$ |
| Droplet volume | $V = \frac{\pi}{3}(3R - h)h^2$ |

acquire images of the droplets during the tests. The camera was supported on a vertically adjustable stand. The optical system was calibrated with a micrometric calibrating glass placed on the stage once the position of the camera had been fixed. Digital images of the droplets were recorded on a PC with a camera controller and custom frame grabbing software at scheduled instants.

Digital image analysis (DIA) techniques were employed to quantify changes in shape of the droplet recorded on the digital images over time (Figure 3). The basic quantities characterizing the shape of the droplet are defined in Figure 4. These include: droplet radius $(R)$ and droplet height $(h)$, which were automatically measured from the digital images; as well as base (or contact) area radius $(r)$, contact angle $(\theta)$, droplet free surface area $(A_s)$, droplet contact area $(A_d)$, and volume $(V)$, which were calculated using simple geometrical formulas (Table 2). Axial symmetry of the droplet was assumed.

## 3. RESULTS AND DISCUSSION

Figure 5 and Figure 6 show, respectively, changes in base areas and contact angles of droplets deposited on the test surfaces until they disappeared due to evaporation and diffusion into the coating/substrate system. As axial symmetry of the droplets was assumed, the volume of the droplet at specific instants of time could be derived from the dimensions using simple geometric relationships. Droplet volume changes over time are shown in Figure 7. Points on the graphs represent averages of measured values. Tests with oval or very small droplets were excluded. Also outlying data points resulting from poor images were excluded. In addition, polynomial curves were fitted to the mean volumetric data to enhance evaluation of second-order quantities (e.g. volume change rates).

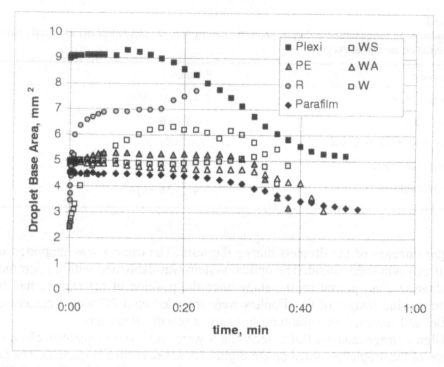

**Figure 5.** Evolution of droplet base area during experiments on different test surfaces.

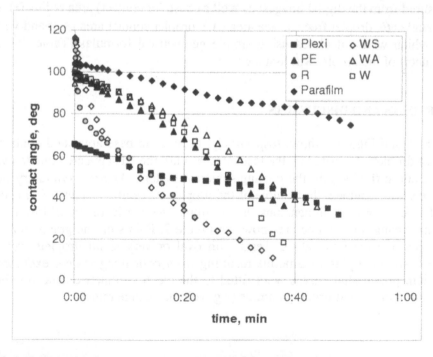

**Figure 6.** Evolution of contact angles during experiments on different test surfaces.

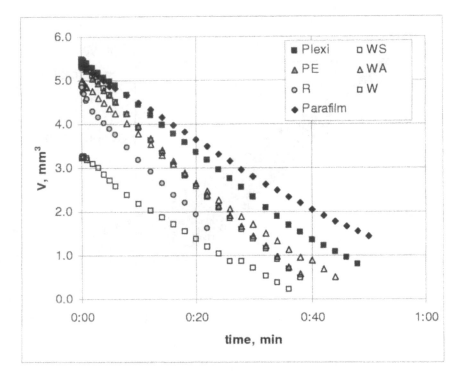

**Figure 7.** Change in volume of droplets deposited on different substrates.

## 3.1. Evaluation of surface penetration

The two top curves in Figure 7, characterized by the slowest volume change rates, represent evaporation from reference surfaces (Parafilm coated glass and Plexiglas), where no diffusion into the substrate is assumed. Steeper curves below represent faster volume change rates for semi-permeable surfaces. The differences in volume change rates between droplets of the same initial volumes subjected to the same environmental conditions but on different test surfaces indicate penetration of water into the coating/substrate system (Figure 8).

The general balance of the volume of the test droplet at any given instant of time (Figure 1) can be expressed as:

$$V(t) = V_0 - \left[ \Delta V_{ev}(t) + \Delta V_{diff}(t) \right] \tag{1}$$

where: $V(t)$ is the volume of the droplet at time $t$; $V_o$ is the initial volume or $V(t{=}0)$; $\Delta V_{ev}(t)$ and $\Delta V_{diff}(t)$ are, respectively, changes in volume due to evaporation and diffusion into the substrate at the time instant $t$.

In general, the volume penetrating the surface may be found by subtracting the volume changes of droplet evaporating from the reference impermeable surface. The evaporated volume was adjusted to address the differences in the initial droplet volumes as well as free surface areas of the droplets on test ($A_s^{sp}$) and refer-

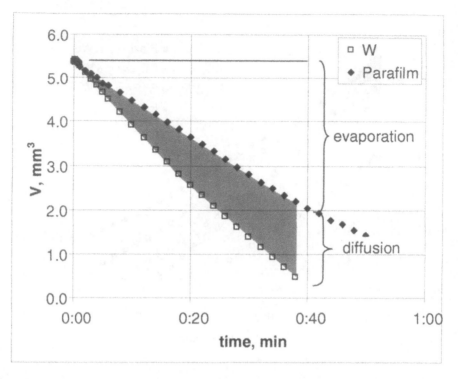

**Figure 8.** Separation of the evaporation and diffusion effects from total volume change measured on droplet deposited on wax treated MDF surface.

ence specimens ($A_s^{ref}$) from test to test. Theoretically, the volume evaporating from droplet on the test specimen ($\Delta V_{ev}$) at any given time instant $t$ can be calculated from the flux of evaporation or the evaporation rates from a unit surface area of the droplets deposited on the reference specimen. Assuming that the evaporation rate is uniform over the droplet area and does not depend on the shape of the droplet, the evaporation flux may be defined as:

$$f_{ev} = \frac{\dot{V}_{ev}}{A_s^{ref}}, \left[\frac{mm^3}{mm^2 h}\right], \tag{2}$$

where the dotted symbol denotes time derivative. Then the volume evaporating from the surface of the droplet on the test surface could be calculated by integrating the flux function over the surface area and elapsed time. Since all data were recorded using the same time schedules a simplified formula for the fraction of volume penetrating the surface ($\Delta V_{diff}$) by time instant $t$ is given as:

$$\Delta V_{diff}(t) = \Delta V(t) - \frac{A_s^{sp}}{A_s^{ref}} \Delta V_{ref}(t), \tag{3}$$

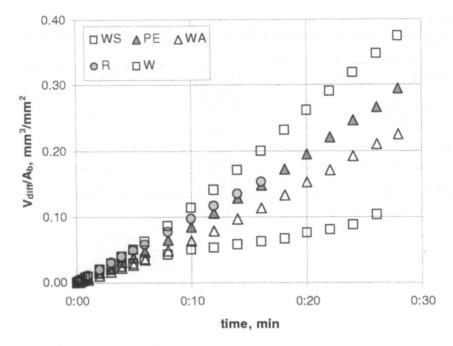

**Figure 9.** Volume of water penetrating the substrate through unit base area.

where $\Delta V$ is the total volume change of the droplet deposited on the test surface, $A_s^{sp}$ and $A_s^{ref}$ designate surface areas of the droplets on the test and reference surfaces, respectively and $\Delta V_{ref}$ is the change in volume measured at the same time on the reference surface and attributed exclusively to evaporation (Figure 8). This simplification is valid only when the evaporation rate is assumed uniform over the droplet surfaces regardless of their shape.

Once the volume penetrating a particular substrate is known, the relative water resistance of the substrates can be evaluated either by comparing volumes penetrating the surfaces through unit base areas of the droplets ($A_b^{sp}$), or characteristic volume fluxes of the penetrating water ($f_{diff}$), defined as:

$$f_{diff} = \frac{\dot{V}_{diff}}{A_b^{sp}}, \left[\frac{mm^3}{mm^2\,h}\right],$$

(4)

where the dotted symbol designates time derivative.

Volume flows of water penetrating the sample surfaces through unit base areas are compared in Figure 9. The curves appear quite linear. The average characteristic fluxes of water penetrating the surfaces (slopes of curves from Figure 9) are summarized in Table 3.

**Table 3.**
Average diffusion fluxes $f_{diff}$ (as defined in Eq. 4) and linear fit correlation coefficients of curves in Figure 9

| Specimen | $f_{diff}$ mm$^3$/mm$^2$*h | $R^2$ – |
|---|---|---|
| Uncoated MDF | 0.577 | 0.9998 |
| Waxed WPC | 0.457 | 0.9899 |
| Waxed MDF | 0.775 | 0.9947 |
| Water Sealant on WPC | 0.237 | 0.9688 |
| Polyester on MDF | 0.592 | 0.9925 |

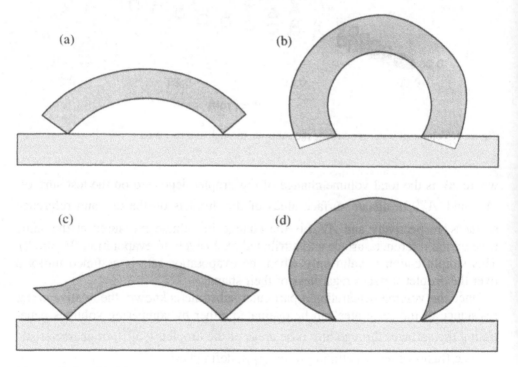

**Figure 10.** Idealized illustration of (a)-(b) the assumed (uniform) and (c)-(d) hypothetical (non-uniform) distribution of the evaporation flux over droplet free surface. The assumed (uniform) evaporation rates do not account for the effect of droplet shape.

Although the general trends, i.e., low water penetration rates into water-sealed and waxed wood plastic composites (WPC) and high water penetration rate for untreated MDF surface, seem to be in agreement with general intuition, the high penetration rates obtained for waxed and polyethylene (PE) coated MDF are not, which requires a deeper analysis of the methodological assumptions. One of the potential problems may be the fact that the evaporation rate from the droplet unit free surface is not necessarily uniform, which may make evaporation rates from droplets of different shapes different even when their free areas and all other conditions are identical. Chandra *et al.* [35] showed, for instance, that at elevated temperatures most of evaporation from "flat droplets" took place at the outer edge. It is also reasonable to assume that the evaporation rates near the edges of droplets of higher contact angles are lower than for the rest of the surface because of higher local partial pressure of vapor in the air filling the space near the droplet edge (Figure 10). The reference surface was chosen so that the initial contact angles for the reference and tested samples were similar. However, Figure 5 and Figure 6 show that even for similar initial contact angles the dynamics of the droplet on different surfaces may differ significantly. Significance of these phenomena for the practical assessment of the evaporation rates from reference specimens needs further investigation.

## 3.2. Features of droplet dynamics

It is interesting to note that the total volume change for all observed droplets is characterized by almost steady rates (Figure 7). At the same time, droplet radii and contact angles change in distinct phases indicating contact angle hysteresis (Figure 5 and Figure 6). The hysteresis may be explained as follows. Shortly after deposition of a droplet on the test surface it is supposed to reach an equilibrium (or pseudo-equilibrium) state characterized by an "equilibrium contact angle", which is a characteristic of a given three-phase system (surface, liquid, gas/vapor). If, however, the volume of the droplet changes (e.g. decreases as a result of evaporation or diffusion or being aspirated out by an experimental device, or increases as a result of condensation or mechanical "feeding" the droplet), the contact angle may change until it reaches a limiting value and the droplet begins either to recede or advance without further change in the contact angle (Figure 11). The limiting values of contact angle are called receding and advancing contact angles. The contact angle hysteresis is the difference between the advancing and receding contact angles in otherwise stable conditions.

It may be observed from the graphs in Figure 5 and Figure 6 that only the droplets deposited on the impermeable reference surfaces (Plexiglas and Parafilm) reach a distinct receding value of the contact angle (45° and 83°, respectively), even though some additional decrease in the contact angle below the limiting value was observed in the later stage of the experiment. This kind of droplet behavior was theoretically described by Shanahan as "Stick-Slip" wetting hysteresis [41]. Droplets on waxed WPC and PE coated MDF seem to reach the receding contact angle at a later stage of the experiment (40° and 42°, respectively), however the contact angles did not level off completely. The receding contact angles were not observed on waxed MDF, WPC treated with water sealant and on the untreated MDF surface. In fact, droplets deposited on WPC treated with water sealant and on the untreated MDF surface apparently did not attain equilibrium in the initial stage of the experiment and continued advancing. The initial (equilib-

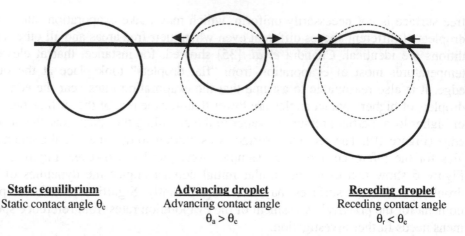

| **Static equilibrium** | **Advancing droplet** | **Receding droplet** |
|:---:|:---:|:---:|
| Static contact angle $\theta_c$ | Advancing contact angle | Receding contact angle |
| | $\theta_a > \theta_c$ | $\theta_r < \theta_c$ |

**Figure 11.** Illustration of contact angle hysteresis.

**Table 4.**
Initial volumes, initial, receding and "final" contact angles measured on surfaces tested. "Final" contact angles refer to droplets reduced to 20% of their initial volume

| Specimen surface | Initial volume, mm³ | Contact angle, deg | | |
|---|---|---|---|---|
| | | Initial | Receding | @ 20% of $V_o$ |
| Plexiglas | 5.51 | 65 | 45 | 37 |
| Parafilm | 5.41 | 104 | 83 | 70* |
| Uncoated MDF | 4.72 | 113 | N/D | 20* |
| Waxed WPC | 4.93 | 97 | 40 | 42 |
| Waxed MDF | 5.37 | 99 | N/D | 37 |
| Water Sealant on WPC | 3.22 | 116 | N/D | 19 |
| Polyester on MDF | 5.29 | 99 | 42 | 40 |

*Extrapolated values

rium), receding and "final" (observed when only about 20% of the initial droplet volume was left) contact angles are summarized in Table 4. The "final" values of contact angle for Parafilm and untreated MDF surface shown in the table were extrapolated assuming that the trend of the curves continued beyond the range of recorded data.

## 4. CONCLUSIONS

This study demonstrated that techniques used in dynamic contact angle analysis might also be applied for accurate evaluation of small changes in the volume of a water droplet due to both evaporation and penetration into the substrate.

The primary advantages of the proposed test method are short test duration (about 60 minutes, which may be considered "rapid" when compared to other available test methods) combined with relatively high resolution. The method is capable of accurate detection of very small changes in droplet volume (below $0.1 \text{ mm}^3$ or 0.01 mg). The use of an appropriate optical system makes resolution of the method practically independent of droplet size. The technique provides sufficient accuracy to determine the volume fraction that penetrates the surface.

Significant contact angle hysteresis was observed not only on hygroscopic materials, but also on Plexiglas. This is attributed to: 1) water sorption at the surface; and 2) polymer surface dynamics (reorientation of functional groups at the liquid-substrate interface).

Lower evaporation rates at higher contact angles were observed, which leads to the conclusion that a non-uniform evaporation flux from the free surface seems to be an important factor in accurate assessment of the evaporation effect. This observation makes reconsideration of some assumptions necessary.

The significance of the non-uniform evaporation flux distribution from surfaces of droplets of different shapes for the practical assessment of the evaporation rates from reference specimens needs further investigation.

## REFERENCES

1. J.F. Siau, *Transport Processes in Wood*. Springer-Verlag, Berlin (1984).
2. C. Skaar, *Wood-Water Relations*. Springer-Verlag, Berlin (1988).
3. W. Feist and D.N.-S. Hon, in: *The Chemistry of Solid Wood*. R.M. Rowell (Ed.), Adv. Chem. Ser. No. 207, chapter 11, American Chemical Society, Washington, DC (1984).
4. R.S. Williams and W.C. Feist, Gen. Tech. Rep. FPL-GTR-109. USDA, Forest Service, Forest Products Laboratory, Madison, WI, 12 p. (1999).
5. P.R. Blankenhorn, S. Bukowski, J.A. Kainz and M. Ritter, in: *Proceedings of Pacific Timber Engineering Conference*, Rotorua, New Zeland, 240-248 (March 1999).
6. R.S. Williams, Amer. Paint Coatings J. **79(11)**, 29-30 (1994).
7. M. Hulden and C.M. Hansen, Prog. Organic Coatings, **13(3-4)**, 171-194 (1985).
8. N.L. Thomas, Progr. Organic Coatings, **19**, 101-121 (1991).
9. H. Derbyshire and E.R. Miller, J. Inst. Wood Sci. **14(1)**, 40-47 (1996).
10. H. Derbyshire and E.R. Miller, J. Inst. Wood Sci. **14(4)**, 162-168 (1997).
11. H. Derbyshire and E.R. Miller, J. Inst. Wood Sci. **14(4)**, 169-174 (1997).
12. H. Derbyshire and D.J. Robson, Holz als Roh- und Werkstoff, **57**, 105-113 (1999).
13. M. de Meijer and H. Militz, Holz als Roh- und Werkstoff, **58**, 354-362 (2000).
14. M. de Meijer and H. Militz, Holz als Roh- und Werkstoff, **58**, 467-475 (2001).
15. E. Nilson, C.M. Hansen, J. Coatings Technol., **53 (680)**, 61-64 (1981).
16. ASTM E 96-00 "Standard Test Methods for Water Vapor Transmission of Materials."
17. ASTM D 5795-95 (2000), "Standard Test Method for Determination of Liquid Water Absorption of Coated Hardboard and Other Composite Wood Products Via "Cobb Ring" Apparatus."
18. ASTM D 2065-96 "Standard Test Method for Determination of Edge Performance of Composite Wood Products Under Surfactant Accelerated Moisture Stress."
19. ASTM D 1037-99 "Standard Test Methods for Evaluating Properties of Wood-Based Fiber and Particle Panel Materials." See related paragraphs: Water Absorption and Thickness Swelling (chapters 100-107); Linear Variation with Change in Moisture Content (chapters 108-111); Accelerated Aging (chapters 112-118); Cupping and Twisting (chapter 119).

20. ASTM F 372-99 "Standard Test Method for Water Vapor Transmission Rate of Flexible Barrier Materials Using an Infrared Detection Technique."
21. ASTM F 1249-90(1995) "Standard Test Method for Water Vapor Transmission Rate Through Plastic Film and Sheeting Using a Modulated Infrared Sensor."
22. ASTM F 1770-97e1 "Standard Test Method for Evaluation of Solubility, Diffusivity, and Permeability of Flexible Barrier Materials to Water Vapor."
23. T. Nguyen, D. Bentz and E. Byrd, J. Coatings Technol., **67(844)**, 37-46 (1995).
24. I. Linossier, F. Gaillard and M. Romand, J. Appl. Polym. Sci., **66**, 2465-2473 (1997).
25. R.W. Rice and D.P. Phillips, Wood Sci. Technol. **34 (6)**, 533-542 (2001).
26. A.J. Kinloch, *Adhesion and Adhesives: Science and Technology*. Chapman and Hall, London (1987).
27. J. Schultz and M. Nardin, in: *Adhesion Promotion Techniques: Technological Applications*, K.L. Mittal and A. Pizzi (Eds), pp. 1–26, Marcel Dekker, New York (1999).
28. S.Q. Shi and D.J. Gardner, Wood Fiber Sci. **33(1)**, 58-68 (2001).
29. D.J. Gardner, N.C. Generella, D.W. Gunnells and M.P. Wolcott, Langmuir **7,** 2498-2502 (1991).
30. M.A. Kalnins, C. Katzenberger, S.A. Schmieding and J.K. Brooks, J. Colloid Interface Sci. **125**, 344-346 (1988).
31. M.A. Kalnins and W.C. Feist, Forest Products J. **43(2)**, 55-57 (1993).
32. M.E.R Shanahan and C. Bourges, Int. J. Adhesion Adhesives. **14 (3)**, 201-205 (1994).
33. C. Bourges-Monnier and M.E.R. Shanahan, Langmuir, **11**, 2820-2829 (1995).
34. M. di Marzo, P. Tartarini, Y. Liao, D. Evans and H. Baum, Int. J. Heat Mass Transfer, **36**, 4133-4139 (1993).
35. S. Chandra, M. di Marzo, Y.M. Qiao and P. Taitorini, Fire Safety J., **27**, 141-158 (1996).
36. S. Gerdes, "Dynamic Wetting of Solid Surfaces. Influence of Surface Structures and Surface Active Block Copolymers." Doctoral Dissertation. Division of Physical Chemistry 1, Lund University, Lund, Sweden (1998).
37. M.E.P. Wålinder and D.J. Gardner, J. Adhesion Sci. Technol. **13**, 1363-1374 (1999).
38. G. Hora, J. Coatings Technology **66 (832)**, 55-59 (1994).
39. DIN 52617, 1987-05: "Determination of the Water Absorption Coefficient of Construction Materials."
40. M.A. Tshabalala, A.R. Denes and R.S. Williams, J. Appl. Polym. Sci., **73**, 399-407 (1999).
41. M.E.R. Shanahan, Langmuir, **11**, 1041-1043 (1995).

*Contact Angle, Wettability and Adhesion*, Vol. 3, pp. 479–500
Ed. K.L. Mittal
© VSP 2003

# Indirect measurement of the shrinkage forces acting during the drying of a paper coating layer

G.M. LAUDONE,[1] G.P. MATTHEWS[*, 1] and P.A.C. GANE[2]

[1]*Environmental and Fluid Modelling Group, University of Plymouth, Plymouth PL4 8AA, UK*
[2]*Omya AG, CH-4665 Oftringen, Switzerland*

**Abstract**—A coated paper consists of a mineral coating (e.g., clay, calcium carbonate, titanium dioxide or silica) containing a binder (e.g., starch, proteins or latex) to fix it to the fibrous paper substrate. The coating is applied to the paper surface as a slurry. The shrinkage occurring while the coating layer dries has been measured by observing the deflection of strips of a substrate coated with calcium carbonate and different binders. The force acting on the surface of the strips to cause a given deflection has been calculated using the Elementary Beam Theory. The porous structure generated by the dry coating layer has been studied using mercury porosimetry and Pore-Cor, a software package able to generate a model network structure of porous materials using data derived from mercury porosimetry.

Within this simulated structure, both the water distribution during drying and the dynamic wetting can be studied spatially in a virtual reality environment. The experiments and simulation provide a better understanding of these processes which, together with the polymeric binder film forming behaviour, are considered responsible for the shrinkage of the coating layer.

*Keywords*: Paper coating; binder film formation; shrinkage; drying of porous structures; capillary force modelling; dynamic wetting.

## 1. INTRODUCTION

The market demand for and consequent production of coated paper, used in products such as "glossy" magazines, advertising, art papers and packaging, is rapidly increasing. The natural consequence of such a growth in this market is the increasing interest for new developments and a better understanding of the basic processes of this production, in particular with respect to the effect of surface uniformity on print performance. The application of a pigmented coating layer to a base sheet of paper or board improves its optical and printing properties such as uniformity in appearance, gloss or matt finish, opacity, ink absorption with con-

---

*To whom all correspondence should be addressed. Phone and fax: +44 1752 233021,
E-mail: pmatthews@plymouth.ac.uk

trolled ink spread etc., and gives enough capillarity to allow ink setting within the time-scale of a modern printing press.

A coating colour formulation typically consists of:

- water: the coating is applied as an aqueous particulate suspension;
- pigments: amongst others calcium carbonate (ground or precipitated), clays, polymeric pigments, titanium dioxide, silica or talc can be used;
- binders: needed to provide good cohesion of the porous structure formed by pigments and adhesion of the coating to the substrate.

The usual composition of a coating colour formulation on a dry basis is 80-90% (w/w) mineral pigment and 10-20% (w/w) binder. The solid content in the water dispersion is usually between 50 and 70% (w/w). Other compounds, such as dispersants, are used in the formulation of the coating colour in lower percentages to act as stabilising agents and to make the components compatible in water suspensions. The particle diameters for the mineral pigments usually range from 0.01 to 10 µm. A thin layer (5-10 µm) of coating colour formulation is metered on the surface of the base-paper and is dried thermally.

The binders are usually divided into two types: natural (like starch or protein) and synthetic (like styrene-butadiene, styrene-acrylic latex, or poly(vinyl acetate)). The natural binders often give poor results in terms of gloss and light scattering, partly due to the coating shrinkage during film-forming upon drying. Synthetic binders suffer less from this problem, and their use, as a result, is becoming more popular in the coated paper industry.

Attempts to follow the process of consolidation of the coating layer have been made by various authors using several different techniques [1-5]. They all follow the initial approach proposed by Watanabe and Lepoutre [1], who divided the drying process into three stages. The application of the coating is followed by a first phase of water evaporation at the liquid–air interface. This phase is unaffected by the solid content in the liquid phase. At the First Critical Concentration (FCC), a three-dimensional network is formed and particle motion is greatly restricted. The water–air interfaces recede into the surface capillaries, creating a capillary pressure that causes a shrinkage of the network. This continues until the Second Critical Concentration (SCC) is reached, at which the network is fixed and air enters the rigid structure.

Practically little is known about the distribution of the liquid–air menisci in a porous network undergoing drying, and especially the forces related to the capillarity acting at the free liquid front. The role of these forces in combination with the film-forming and compressibility characteristics of the binders has not previously been addressed. The evaporation, however, of a liquid from a porous medium has been studied with an interesting approach by Laurindo and Prat [6-8]. They studied evaporation as a displacement between two immiscible fluids. As evaporation results in the invasion of the porous media by a non-wetting phase (vapour), the similarity with percolation is evident. They created a bi-dimensional network model describing isothermal evaporation, controlled mainly by mass transfer, as they considered the effect of the heat transfer to be negligible in their experimental conditions. They also considered the viscous forces to be negligible upon drying. The latter approximation is acceptable if the evaporation is slow, and/or if the length of pore features is short, i.e., the aspect ratio of the pores is low, as is the case with isotropic calcite particles.

Our aim was to reach a better understanding of the shrinkage phenomena which occurred during drying. This was achieved by considering the relative importance of the capillary forces acting on the coating colour formulation while the water receded into the porous structure formed by the pigment particles, and the shrinkage force of the film-forming binders/polymers.

The dry samples were analysed with mercury intrusion porosimetry and their porous structures were modelled with a network simulation software. The future aim of the study is to determine the distribution and areas of the free menisci in the porous structure. This will allow the calculation of the meniscus forces as the derivative of the total free energy of the liquid-mediated interface [9, 10]. These calculated forces will then be compared with those observed experimentally.

## 2. EXPERIMENTAL

### 2.1. Method and materials

The characterisation of the forces acting during the drying of the thin coating layer applied onto paper cannot be carried out directly. This is due to the complexity and small dimensions of the system. Thus it is necessary to find an indirect approach. Therefore, we decided to measure such forces by observing the deflection of an elastic material upon drying. Many different materials, mainly polymeric films, were tested and Synteape (Arjo Wiggins, Issy Le Moulineaux, France), a synthetic laminate substrate made from a stretched calcium carbonate-filled polypropylene (Fig. 1), was chosen. Its elastic behaviour and its slightly rough surface, able to "accept" the coating colour formulation, made it the ideal substrate for our experiment. The precise formulation and preparation method of Synteape is a trade secret. It is possible to apply the coating layer, allow it to dry progressively and measure the deformation of the Synteape as a function of time, and hence as a function of solids concentration.

By approximating a strip of Synteape to an elastic bending beam, it is possible to use the standard theory (as detailed in the Appendix) which relates the forces acting on the beam with the actual deflection of the beam itself. The forces acting on the surface, causing the deformation, can thus be calculated through the Young's modulus of the Synteape.

It is not possible to derive these forces from coatings on a fibrous paper substrate because paper has a plastic response behaviour: its fibres, under the forces caused by the drying coating layer, rearrange, especially under high-moisture conditions. This rearrangement causes a deformation in the plane of the paper

**Figure 1.** SEM micrograph of the cross section of Synteape showing the laminated structure and $CaCO_3$ particles. Scale bar 50 μm.

displaying no measurable bending other than the action of wetting on the built-in stress relaxation properties of the fibre mat formation. This planar deformation, however, cannot act to alleviate the non-uniformities of the coating caused by the shrinkage phenomenon.

In this kind of approach we made three main assumptions:

- the weight of the beam was negligible when compared with the forces acting on its surface, the Synteape strip being 4.5 cm long, 0.5 cm wide and 0.01 cm thick;
- the weight of the strip was assumed to be negligible when compared with the shrinkage forces;
- the mechanical properties of the system did not change during the drying of the coating layer, which means that we considered the coating layer just like a stress deliverer.

The mechanical properties of the Synteape are realistically assumed to be elastic and acting within the Hookean elastic limit. However, the coating laminate layer does change properties from plastic, through a viscoelastic structure to a brittle structure as drying proceeds. These changes, however, relate to the effective deformation and indeed contribute to the mechanical behaviour of the system. Therefore, the exact mechanical transitions between these states of the coating are

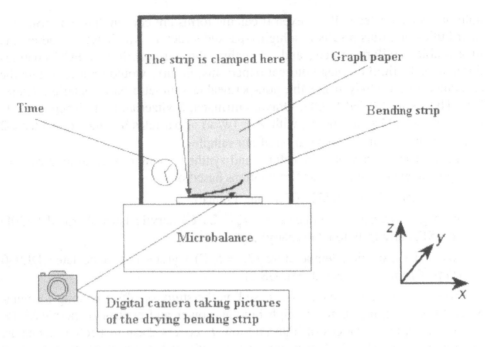

**Figure 2.** Diagram of the experimental apparatus for the indirect measurement of the shrinkage forces.

not the target of investigation, rather the roles of the various force components as a function of time and concentration which lead to the transitions.

Taking several pictures of the sample during the drying process and measuring its weight loss during the process makes it possible to find the forces acting on the surface of the substrate as a function of weight (water) loss. A diagram of the experimental apparatus is presented in Fig. 2. In order to understand which part of the shrinkage forces is due to the capillary forces only and which is due to other effects, such as the shrinkage of the drying binder, it was necessary to create different structures, which would mimic and separate individual components of the combined force process. These consisted of:

- pigment and stabilisers only (capillary forces acting only during the drying);
- complete coating formulations with pigment and binders.

## 2.2. Formulation components

The method described above makes it possible to measure the forces that drying pigmented structures exert on the surface of the elastic substrate. The different coatings used in these experiments are based on Ground Calcium Carbonate (GCC), to avoid the problems presented by the anisotropy of clay platelets and the acicularity of aragonitic Precipitated Calcium Carbonate (PCC). Adsorption of starch [11, 12] onto clay together with particle anisotropy leads to poorly definable orientation effects. PCC, even if calcitic forms are chosen, has a narrow size distribution and thus results in a highly-packed structure which acts to reduce the measurable capillary effects, and, in reality, the permeability of PCC coatings dominates the fluid drainage on real paper: this, in turn, would tend to negate the relevance of the study using the non-permeable synthetic base material. Therefore, Hydrocarb 90 (HC90OG; Omya, Oftringen, Switzerland), a dispersed GCC limestone from Orgon, France, with 90% (w/w) of particles having a diameter < 2 μm, was chosen for the preparation of our samples.

The different behaviours of natural and synthetic binders was studied by preparing different coating colour formulations based on:

- maize starch C-film 07321 (Cerestar, France)
- high glass transition temperature ($T_g$ = 23°C) acrylic latex Acronal S320D (BASF, Ludwigshafen, Germany) and
- low glass transition temperature ($T_g$ = 5°C) styrene butadiene latex DL930 (Dow Chemical, Midland, MI, USA)

The choice of latex binders was based on the study of offset coating formulations. Offset printing demands high binding power and control of permeability. The use of different latices of high and low $T_g$ covers these respective demands. The impact of the chemical structures of the synthetic binders on coating structure is assumed to be negligible as both latex types are designed for stability in the presence of calcium ions.

On a dry basis, the composition of our coating colour formulations was 25% (w/w) binder and 75% (w/w) pigment. This is a higher percentage of binder than in a typical industrial paper coating, but the level was chosen in order to maximise the effect of the binder shrinkage in the following observations. A preliminary series of experiments showed that the effect of binder was proportional to its content within this dosage range, and so the increased binder quantity is a useful experimental parameter to employ without influencing the basic mechanisms themselves.

In order to investigate the effect of the amount of coating colour formulation applied, the strips were coated using two different draw-down coating rods. These rods have a wire winding applied to the surface to define a given application volume in contact with the substrate. The diameter of the wire winding controls this volume. The rods were labelled "rod 2" (applying about 10 g/m² of dry coating, corresponding to a dry coating layer thickness of about 5 μm) and "rod 3" (applying about 20 g/m² of dry coating, corresponding to a dry coating layer thickness of about 10 μm).

## 2.3. Mercury porosimetry

Mercury porosimetry has been used for the characterisation of the porous structure formed by the dried coating [13, 14]. An evacuated sample is immersed in mercury and the external pressure is gradually increased. The amount of mercury intruding the void space of the sample is measured as a function of pressure: the amount increases as the pressure increases and the mercury is forced into smaller voids. The result is a mercury intrusion curve, which, knowing the interfacial tension and contact angle of mercury with the solid, can be converted into a void size distribution by the Laplace equation, assuming that the structure consists of equivalent capillary elements of diameter $D$:

$$D = \frac{-4\gamma \cos \theta}{P} \tag{1}$$

where $\gamma$ is the interfacial tension between mercury and air (0.485 N/m), $P$ is the applied pressure and $\theta$ is the contact angle. A similar procedure with pressure reduction allows the drainage curve to be obtained as the mercury emerges from the sample. The intrusion and extrusion curves differ and show hysteresis caused by the pressure needed to force the mercury into a larger void space (pore) through a narrower void space (throat) being greater than that at which it will come out when the pressure is decreasing, i.e. small throats can shield large pores.

Mercury porosimetry can be used to determine the pore size distribution of porous materials, even when these are supported by a laminate or fibrous substrate. The technique described by Ridgway and Gane [15] makes it possible to correct the results of the mercury porosimetry in order to subtract the contribution of the substrate.

The mercury porosimetry tests made on our dry samples were carried out using a Micromeritics Autopore III mercury porosimeter, able to reach a pressure of 414 MPa.

### 2.4. Modelling

#### 2.4.1. Pore-Comp

Pore-Comp is a precursor software for Pore-Cor (both from the Environmental and Fluids Modelling Group, University of Plymouth, Plymouth, UK). It corrects the mercury intrusion curves for:

- mercury compression and penetrometer expansion;
- compressibility of skeletal solid phase of the sample.

The correction for the compression of mercury, expansion of the penetrometer and compressibility of the solid phase of the sample is calculated using the following equation from Gane *et al.* [16]:

$$V_{int} = V_{obs} - \delta V_{blank} + \left[ 0.175(V_{bulk}^1)\log_{10}\left(1 + \frac{P}{1820}\right) \right] - V_{bulk}^1(1 - \Phi^1)\left(1 - \exp\left[\frac{(P^1 - P)}{M_{ss}}\right]\right) \tag{2}$$

$V_{int}$ is the corrected volume of mercury intruded into the sample, $V_{obs}$ the experimentally observed volume of intruded mercury, $\delta V_{blank}$ the change in the blank run volume reading, $V_{bulk}^1$ the sample bulk volume at atmospheric pressure, $P$ the applied pressure, $\Phi^l$ the porosity at atmospheric pressure, $P^l$ the atmospheric pressure and $M_{ss}$ the bulk modulus of the solid sample.

### 2.4.2. Pore-Cor

The Pore-Cor model [17-19] simulates the pore-level properties of a porous medium. It generates a three-dimensional structure, consisting of a repeated unit cell containing a regular $10\times10\times10$ array of cubic pores interconnected by cylindrical throats, which matches the experimentally determined porosity and void size distribution, such that the percolation characteristics measured by mercury porosimetry are reproduced. The simulation is performed by optimising sets of adjustable parameters, such as connectivity, pore and throat skew — each affecting, in turn, the relevant size distribution. More than one structure can be generated by the model to fit an experimental intrusion curve, since the pores in a modelled unit cell can be arranged in different ways. This creates simulated pore structures with different best-fit values of the adjustable parameters. The model creates different families of structures called "stochastic generations". Each stochastic generation is created by using a different set of pseudo-random starting numbers. If the unit cell is the same size as the Representative Elementary Volume (REV) of the sample, and its complexity is equivalent to the complexity of the experimental void structure, then different stochastic generations will have the same properties, as the random effects will average out across the unit cell. Frequently, however, the unit cell is smaller than the REV of the sample. In this case the problem can be overcome by studying the properties of different stochastic generations, but this is outside the scope of this work.

Pore-Cor was used to generate the following simulated measurements: throat and pore-size distribution, fluid and gas absolute permeabilities, connectivity and dynamic liquid absorption measurements.

Using the Virtual Reality Mark-up Language, Pore-Cor was used to generate simulated structures in a virtual reality environment. These can be explored at the web page www.pore-cor.com/virtual_reality.com

## 3. RESULTS AND DISCUSSION

### 3.1. Bending strips

A detailed image analysis of the bending beam viewed in the $xz$ plane (Fig. 2) and the application of the software package Table Curve 2D (SPSS, Chicago, IL, USA) made it possible to confirm the initial hypothesis of circular deformation of the strips and the negligibility of the weight of the samples.

The stiffness of the beam $EI$ (as defined in the Appendix) was calculated for the Synteape substrate, and data were collected from the microbalance as the curling of the samples was measured. The value of $\tau$ (written tau in the figures), which is the force per unit of area within the cross section of the coating, was plotted as a function of the weight loss. $\tau$ is equal to $T$ (as defined in the Appendix) divided by the surface area of the coated strips.

The results are shown in Figs 3-6, where the continuous lines represent the samples coated with a lower weight of coating per unit of surface and the dotted ones represent the samples coated with a higher weight of coating per unit of surface.

There is no information available for the samples coated with $CaCO_3$ slurry alone using rod 2 because the bending of the strip and the following relaxation was too rapid to be observed with the technique used. Fig. 3, therefore, does not show any continuous line.

From Fig. 3 it is possible to observe the effect of the capillary forces only acting upon drying, i.e., in the absence of binder. The samples bend as soon as the drying begins and the water recedes into the porous structure created by the particles of calcium carbonate, reaching a peak of about 50-60 Pa. The more the water leaves the structure, the weaker the capillary forces are. This leads to a complete relaxation of the sample with no stress retention at all. This case of the purely meniscus-driven deformation, which disappears on complete drying, provides the control experiment. The effect of the presence of binder in the coating colour formulation can be observed by comparing the results from the control experiment with the results obtained when complete coating colour formulations were used.

It is interesting to notice (Fig. 4) how hard, high-$T_g$, latex-based samples show almost no deflection. The maximum stress measured on these samples is 10 Pa, lower than that measured on the samples coated with $CaCO_3$ only. It appears that this kind of latex acts to fill the structure but the spheres are non-deformable and so shrinkage cannot occur even by the capillary forces.

Soft, low-$T_g$ latex (Fig. 5) allows such deformation and the capillary forces act causing deflection of the strip. The fixation of the capillary-induced structure takes place where the binder is present in the structure. The binder allows the deformation to take place and serves to hold the deformation during the adhesional bonding between itself and the pigment particles.

The force acting while a starch-based coating colour formulation is drying is far larger (Fig. 6) than in the case of latex-containing formulations, although the results for this formulation are more scattered. Inclusion of starch acts to flocculate the pigment due to partial adsorption of starch onto the $CaCO_3$ particles and osmotic effect. This reduces the effective number of voids in the dried coating but simultaneously tends to make those that are present larger. The unit size for flocculated pigment and starch is close to that of the layer thickness for rod 2 and slightly smaller for rod 3, i.e., the effective particle size in relation to film thickness is an important parameter when considering the rearrangement ability of the particles in the layer undergoing shrinkage. A slight upward gradient of the force curve as the coating dries indicates a real shrinkage force originating from starch film formation.

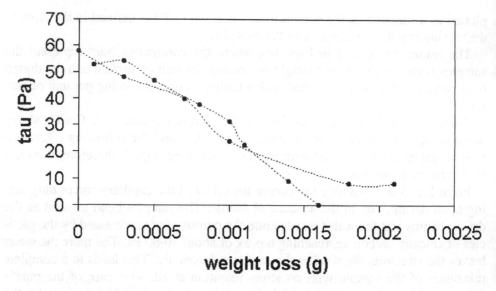

**Figure 3.** Stress acting on two replicate samples coated with calcium carbonate only. The maximum stress measured is 58 Pa.

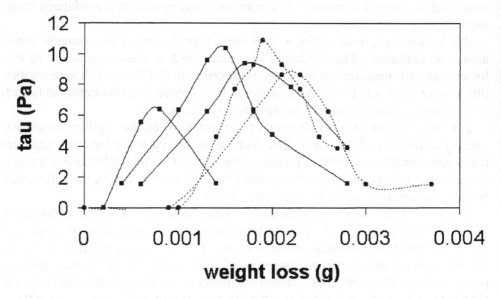

**Figure 4.** Stress acting on samples coated with a high-$T_g$ latex-based coating colour formulation. The continuous lines represent the samples coated with rod 2, while the dotted ones represent rod 3. The maximum stress measured is lower than in the case of $CaCO_3$ slurry with no binder.

To confirm such interpretation it is necessary to characterise the porous structure created by the drying coating colour formulation using mercury porosimetry and model the porous structure using Pore-Cor.

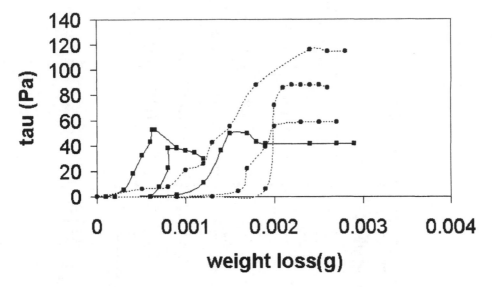

**Figure 5.** Stress acting on samples coated with a low-$T_g$ latex-based coating colour formulation. The continuous lines represent the samples coated with rod 2, while the dotted ones represent rod 3. The stress measured is larger than in the case of $CaCO_3$ slurry only and part of the stress is retained by the binding action of the latex.

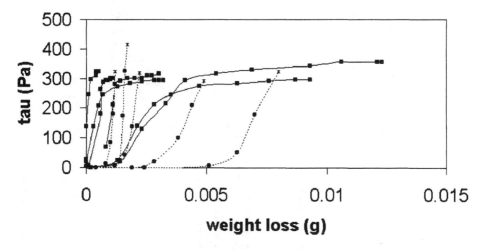

**Figure 6.** Stress acting on samples coated with a starch-based coating colour formulation. The continuous lines represent the samples coated with rod 2, while the dotted ones represent rod 3. The stress measured is far larger than in the case of latex-based coating colour formulations.

## 3.2. Mercury porosimetry

A series of dry samples coated with the same coating colour formulations used for the preparation of the bending strips was analysed with the mercury porosimetry technique.

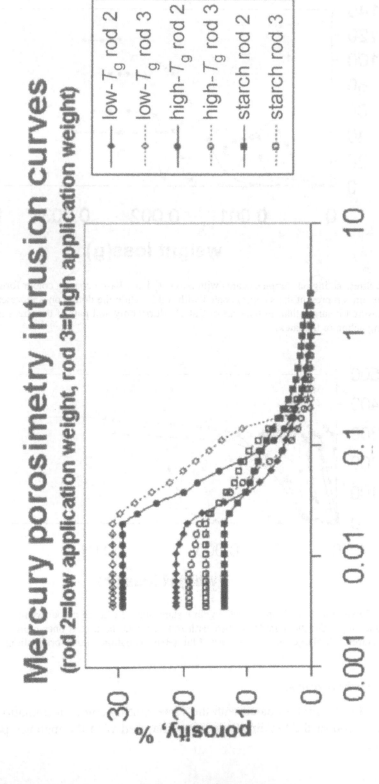

**Figure 7.** Mercury porosimetry intrusion curves.

**Table 1.**
Fully corrected porosities (%)

| low-$T_g$ latex | | high-$T_g$ latex | | starch | |
| --- | --- | --- | --- | --- | --- |
| rod 2 | rod 3 | rod 2 | rod 3 | rod2 | rod 3 |
| 21.0 | 30.8 | 29.2 | 18.9 | 13.5 | 16.3 |

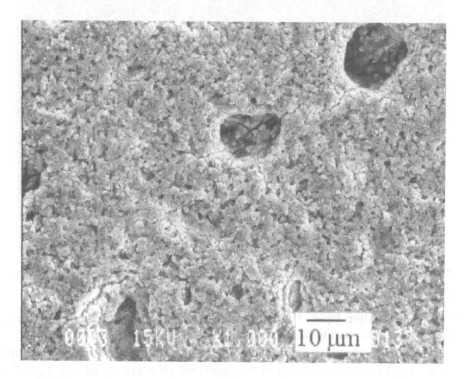

**Figure 8.** SEM micrograph of a starch-based sample.

The results of the mercury porosimetry on the dried samples were first corrected with the use of Pore-Comp. The correction of the intrusion curves for the compressibility of the mercury and the expansion of the glass chamber at high pressure and for the compressibility of the sample itself led to the following values (Table 1) of fully corrected porosity.

In Fig. 7 the results of the mercury porosimetry are presented. The intrusion curves for the starch-based samples confirm the interpretation given in the previous section: the starch-based samples show a lower porosity but bigger pores (especially the coating applied with rod 2), suggesting that the flocculating action of the starch causes the aggregation of the pigment itself, making the structure more permeable with bigger pores but having lower overall porosity. This can be visually confirmed by the electron microscopy image for a starch-based sample, shown in Fig. 8. It is clear from such a picture that big pores are present at the

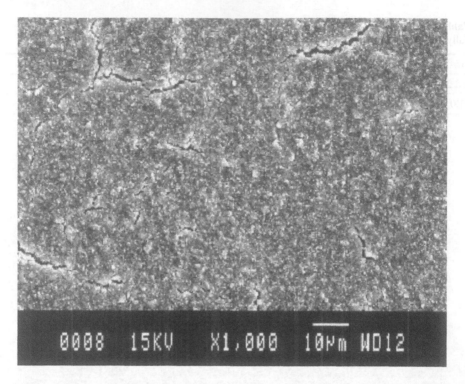

**Figure 9.** SEM micrograph of "mud cracks" on the surface of a low-$T_g$ latex-based coating layer.

surface of the coating layer. This phenomenon also explains the bimodality of the intrusion curves for starch-based coating colour formulations.

It is interesting to notice the trend in the values of porosity: the samples most prone to shrinkage can be placed in the order: starch > low-$T_g$ latex > high-$T_g$ latex. The most shrinkage is expected (and observed in the case of starch and low-$T_g$ latex) from a higher coating weight, but this does not always lead to a lower porosity, as would be intuitively expected. Both low-$T_g$ latex- and starch-based coating colour formulations show a higher porosity for increasing amount of coating colour formulation applied onto the surface. This can be explained in terms of the presence of a "mud cracking" effect on the surface of shrinking binder-based coating formulation at higher weight of coating per unit of surface applied. This is represented by the slight bimodality shown in the intrusion curve for the low-$T_g$ latex applied with rod 3. The presence of "mud cracks" on the surface of the coating layer can be observed in Fig. 9.

The high-$T_g$ latex-based coatings, however, do follow the expected trend with respect to decreasing porosity as a function of coating weight. High-$T_g$ latex samples do not manifest shrinkage, and the "mud cracking" effect is not present on such non-shrinking samples.

The coating colour formulations prepared using high-$T_g$ latex show significant intrusion for pore diameters of ~ 0.2 μm and smaller. The latex particles have a diameter of ~ 0.15 μm. Thus, we can suggest why the samples coated with these coating colour formulations do not bend while the coating dries: the hard and non-deformable particles of high-$T_g$ latex fill the porous structure formed by the calcium carbonate, making it impossible for the capillary forces to cause deformation and shrinkage.

## 3.3. Pore-Cor

The values generated by Pore-Cor for our samples are shown in Table 2. The visual representations of two of the simulated structures are shown in Figs 10 and 11.

It is interesting to observe how the thinner latex-based coating layer has a lower connectivity and a more random structure; the high-$T_g$ rod 2 has a completely random structure, as shown in Fig. 10 and by its value of correlation level in Table 2. This can once again be interpreted taking into account the diameters of the biggest pigment particles, which are ~ 5 μm, in relation to the thickness of the coating layer. In the case of the thin coating layer, the dimension of the biggest particle is larger or equal to the thickness of the layer; this limits the freedom of movement and rearrangement of the particles while the coating layer is drying, leading to a random structure. If the coating layer is thicker than the biggest particle of calcium carbonate, the structure is free to rearrange and this leads to more order and, therefore, to a higher correlation level.

Pore-Cor can simulate the behaviour of fluids as these structures dry. The distribution of air as water evaporates is assumed to be the same as if air was a non-wetting fluid displacing the water by percolation. In Fig. 12 the results of this algorithm applied to the starch-based sample coated with rod 2 are shown, in which the percolation intrusion occurs from above.

The subsequent wetting of the dried structure by a wetting fluid can also be simulated using a "wetting algorithm", based on the Bosanquet equation [20]. In Fig. 13 it is possible to observe the flow of water into the structure from above after 1 ms. It is interesting to observe how the water flows deeper into the structure following preferential paths.

**Table 2.**
Pore-Cor results

| Sample | Experimental and simulated porosity (%) | Simulated connectivity | Simulated liquid permeability (milliDarcy) | Simulated correlation level |
|---|---|---|---|---|
| Low-$T_g$ latex rod 2 | 21.0 | 3.62 | 3.37E-7 | 0.16 |
| Low-$T_g$ latex rod 3 | 30.8 | 5.13 | 9.03E-6 | 0.20 |
| High-$T_g$ latex rod 2 | 29.2 | 4.03 | 1.52E-6 | 0.03 |
| High-$T_g$ latex rod 3 | 18.9 | 5.56 | 4.31E-7 | 0.31 |
| Starch rod 2 | 13.5 | 5.30 | 6.87E-7 | 0.37 |
| Starch rod 3 | 16.3 | 5.03 | 6.91E-7 | 0.28 |

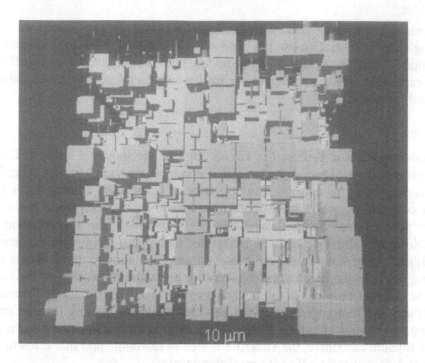

**Figure 10.** Modelled structure of high-$T_g$ latex-based coating applied with rod 2: the structure results show it to be completely random.

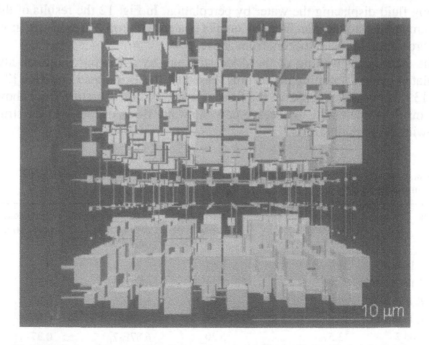

**Figure 11.** Modelled structure of high-$T_g$ latex-based coating applied with rod 3 showing central zone of small throats and pores.

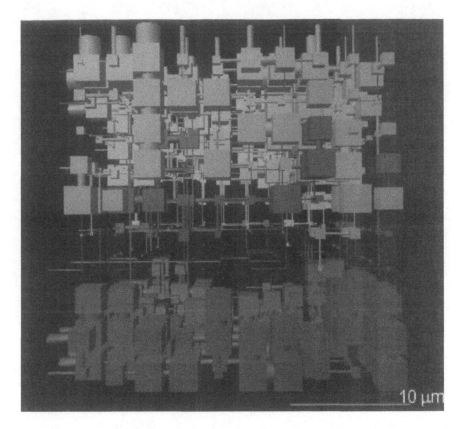

**Figure 12.** Simulation of evaporation from the starch-based rod 2 modelled structure using the percolation algorithm (acting from above) of an effective non-wetting fluid to describe the air/water distribution during the evaporation process.

Balancing these wetting and percolation properties can be used to describe the time dependency and rearrangement of fluid in a given porous structure, which will be key to the understanding of evaporative drying and associated shrinkage phenomena.

## 4. CONCLUSIONS AND FUTURE RESEARCH WORK

The shrinkage-related behaviour of different binders in a coating colour formulation is a problem well known to the paper-making industry. In this paper, we indirectly quantified the force acting upon drying, and the relative importance of capillary forces and film-forming polymer shrinkage forces. Since the substrate used in our experiments does not absorb water and the dewatering can only take place by evaporation, the values of the surface forces measured during these experiments may not compare well with the forces acting on coated paper. However,

*G.M. Laudone* et al.

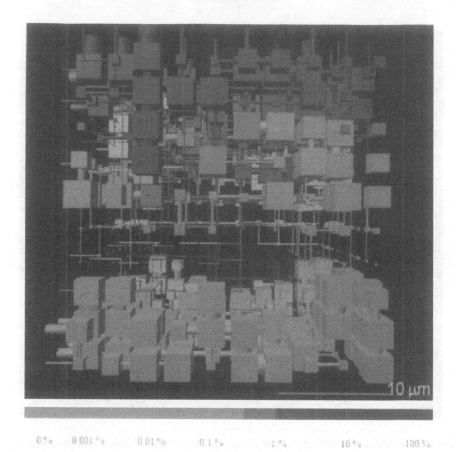

0 %    0 001 %    0 01 %    0.1 %    1 %    10 %    100 %

**Figure 13.** Simulation of wetting into the starch-based rod 2 modelled structure.

they can give important information about the mechanism of drying for different binders.

The mercury porosimetry analysis on the dried porous structure, formed by the pigment and the binder, showed interesting trends, enabling us to explain the difference in behaviour for the different binders. The Pore-Cor network modelling software package provided model 3D structures representing the porous structures and their pore-level properties. The models indicated that distinctive banding of thin layer structures occurred during drying, in which the middle layers exhibited reduced porosity due to particle rearrangement during shrinkage.

Using these 3D modelled structures with the percolation algorithm used to simulate the mercury intrusion, we aim to simulate in more detail the water evaporation from the porous structure, as shown in Fig. 12, and its redistribution in the structure itself as a function of time using the wetting algorithm as shown in Fig. 13. This will finally allow the calculation of the meniscus forces in the simulated network structures and their comparison with the experimental data presented in this paper.

# REFERENCES

1. J. Watanabe and P. Lepoutre, *J. Appl. Polym. Sci.* **27**, 4207-4219 (1982).
2. S.X. Pan, H.T. Davis and L.E. Scriven, *Tappi J.* **78**, 127-143 (1995).
3. P. Bernada and D. Bruneau, *Tappi J.* **79**, 130-143 (1996).
4. P. Bernada and D. Bruneau, *Drying Technol.* **15**, 2061-2087 (1997).
5. R. Groves, G.P. Matthews, J. Heap, M.D. McInnes, J.E. Penson and C.J. Ridgway, *Proceedings of the 12th Fundamental Research Symposium : Science of Papermaking*, The Pulp and Paper Fundamental Research Society, PITA, pp. 1149-1182 (2001).
6. J.B. Laurindo and M. Prat, *Chem. Eng. Sci.* **51**, 5171-5185 (1996).
7. J.B. Laurindo and M. Prat, *Drying Technol.* **16**, 1769-1787 (1998).
8. J.B. Laurindo and M. Prat, *Chem. Eng. Sci.* **53**, 2257 (1998).
9. C. Gao, *Appl. Phys. Lett.* **71**, 1801-1803 (1997).
10. C. Gao, P. Dai, A. Homola and J. Weiss, *ASME Trans. J. Tribol.* **120**, 358-368 (1998).
11. J.C. Husband, *A Comparison of the Interactions of Various Starch Derivatives and Sodium Carboxymethyl Cellulose with Kaolin and Paper Coating Latex Suspensions*, Ph.D. Thesis, University of Manchester Institute of Science and Technology (1997).
12. J.C. Husband, *Colloid. Surf. A* **131**, 145-159 (1998).
13. R.W. Johnson, L. Abrams, R. Maynard and T.J. Amick, *Tappi J.* **82**, 239-251 (1999).
14. L. Abrams, W. Favorite, J. Capano and R.W. Johnson, *Proceedings of the 1996 Tappi Coating Conference*, Tappi Press, Atlanta, GA, pp. 185-192 (1996).
15. C.J. Ridgway and P.A.C. Gane, *Nordic Pulp Paper Res. J.* (2002) in press.
16. P.A.C. Gane, J.P. Kettle, G.P. Matthews and C.J. Ridgway, *Ind. Eng. Chem. Res.* **35**, 1753-1764 (1995).
17. J.P. Kettle and G.P. Matthews, *Proceeding of the 1993 Tappi Advanced Coating Fundamentals Symposium*, Tappi Press, Atlanta, GA, pp. 121-126 (1993).
18. C.J, Ridgway, P.A.C. Gane and J. Schoelkopf, *J. Colloid Interf. Sci.* **252**, 373-382 (2002).
19. C.J. Ridgway, J. Schoelkopf, G.P. Matthews, P.A.C. Gane and P.W. James, *J. Colloid Interf. Sci.* **239**, 417-431 (2001).
20. J. Schoelkopf, C.J. Ridgway, P.A.C. Gane, G.P. Matthews and D.C. Spielmann, *J. Colloid Interf. Sci.* **227**, 119-131 (2000).
21. R.T. Frenner, *Mechanics of Solids*, pp. 314-388, Blackwell Scientific Publications, London (1989).

# APPENDIX

*Beam theory*

The elementary beam theory [21] gives:

$$M = \frac{EI}{R} \tag{A-1}$$

where $M$ is the bending moment, $E$ is the elastic modulus, $I$ is the second moment of inertia related to the $z$ axis, $R$ is the radius of curvature of the deformed beam. Equation A-1 is considered valid when the deformed beam is circular.

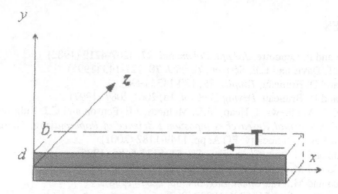

**Figure 14.** Beam and traction acting on its surface.

## Effect of the coating layer

It is possible to assume that the coating gives rise to a total surface traction $T$ (Fig. 14) and, hence, to a bending moment $M=T{\cdot}d$, where $d$ is half the thickness of the beam. From equation A-1:

$$\frac{EI}{R}=T\cdot d \tag{A-2}$$

where $I = b(2d)^3/12$, $b$ is the width of the beam and the product $EI$ is called the "beam stiffness".

## Radius of curvature of the bending beam

For a generic function $y = f(x)$ the radius of curvature is

$$\frac{1}{R}=\frac{y''}{\sqrt{1+y'^2}} \tag{A-3}$$

where $y' = dy/dx$ and $y'' = d^2y/dx^2$. For small $y'$, $1/R \cong y''$. If we have a circular deformation through $(x_0, y_0)$, then:

$$\frac{1}{R}=\frac{2y_0}{x_0^2+y_0^2} \tag{A-4}$$

Equations A-3 and A-4 are consistent if $y_0^2>>x_0^2$, i.e. for small deformation. If a deflection $y_0$ is measured, substituting from equation A-2:

$$T\cdot d=\frac{2EIy_0}{x_0^2+y_0^2} \tag{A-5}$$

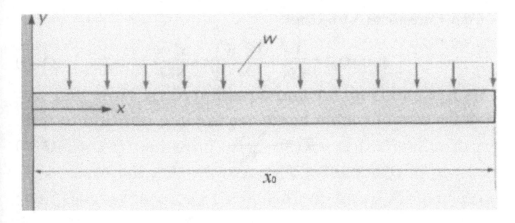

Figure 15. Beam with uniformly distributed weight. The deflection of the beam due to its own weight can be used to calculate *EI*.

If $x_0$, $y_0$ and *EI* are known, *T* can be calculated from equation A-5. The stress acting on the surface of the beam can be calculated as $\tau = T / b \cdot x_0$.

Since $T \propto I \propto b$, the stress $\tau$ is independent of the strip width *b*.

*Evaluation of the stiffness of the beam (EI)*

The standard beam theory is also helpful if the stiffness of the beam is unknown. Performing a simple experiment (Fig. 15) and measuring the deflection of the strip due to its own weight (which is negligible when compared to the stress caused by the drying coating layer, but enough to cause a small deflection), it is possible to determine the value of the product *EI*.

Let us call *w* the weight per unit of length and setting $1/R \cong y''$ equation A-1 gives:

$$EI \frac{d^4 y}{dx^4} = -w \tag{A-6}$$

$$\therefore \quad EI \cdot y(x) = -\frac{w \cdot x^4}{24} + \frac{w \cdot x_0 \cdot x^3}{6} - \frac{w \cdot x_0^2 \cdot x^2}{4} + A \cdot x + B \tag{A-7}$$

The boundary conditions are:

$$y(0) = y'(0) = 0 \tag{A-8}$$

This leads to:

$$y(x) = \frac{1}{EI} \left\{ -\frac{w}{24} \left( x^4 + 6x_0^2 \cdot x^2 - 4x_0 \cdot x^3 \right) \right\} \tag{A-9}$$

For $x = x_0$, equation A-9 becomes:

$$y_1 = y(x_0) = \frac{1}{EI}\left(-\frac{3x_0^4 \cdot w}{24}\right) = -\frac{x_0^4 \cdot w}{8EI} \tag{A-10}$$

And finally:

$$EI = -\frac{w \cdot x_0^{\,4}}{8y_1} \tag{A-11}$$

*Contact Angle, Wettability and Adhesion*, Vol. 3, pp. 501–519
Ed. K.L. Mittal
© VSP 2003

# Effect of the physical chemistry of polymeric coating surfaces on fouling and cleanability with particular reference to the food industry

L. BOULANGÉ-PETERMANN,*, 1 C. DEBACQ,1 P. POIRET1 and B. CROMIÈRES2

1 *Arcelor Innovation, Research Center of Isbergues, BP 15, 62 330 Isbergues, France*
2 *Arcelor Innovation, Centre d'Etudes et de Développement, 60 160 Montataire, France*

**Abstract**—The cleanability of carbon steels covered by organic coatings, commonly used in hygienic applications and, more particularly, for the wall panel in the cold storage rooms in food industry, hospitals and in public buildings was investigated here using various industrial organic coatings. They were also assessed after aging in conditions close to the end application. Two types of coating materials can be defined: (i) materials with a polar surface free energy component such as polyester and poly(ethylene terephthalate) and (ii) coatings displaying low surface free energy (less than 35 mJ/m$^2$) such as poly(vinyl chloride) and poly(vinylidene fluoride). After aging in water, the surface energy of the PVC coating is increased and could be explained by an organic component being released from the surface. Surface fouling was performed by splashing a nutritious oil on the organic coatings. Then, the cleaning kinetics was investigated in a laminar flow cell using an alkaline detergent. The cleaning kinetics decreases exponentially with time. The kinetic constant was investigated with regard to the surface properties of the organic coatings. A surface with an improved cleanability after an oil spray displays a high polar component of surface free energy. This could be explained by favorable interactions between the coating material surface and the surfactant, thus enhancing the cleaning efficiency.

*Keywords*: Steel; polymeric coating; oil; fouling; cleaning; detergent.

## 1. INTRODUCTION

Carbon steels covered by organic coatings are commonly used in hygienic applications and, more particularly, for the wall panel in the cold storage rooms in food industry, hospitals and in public buildings [1, 2]. Foodstuffs are only stored for a short time in such cold rooms, and during their handling, some splashing can occur on the wall-panel surfaces. This surface fouling can then constitute a favor-

---

*To whom all correspondence should be addressed. Phone: (33-3) 2163-5604,
Fax: (33-3) 2163-2056, E-mail : laurence.boulange@ugine-alz.arcelor.com

able environment for further bacterial development. As a consequence, proper cleaning is essential [3, 4].

Various organic coatings for wall-panel manufacturing are proposed. Numerous studies on corrosion resistance and chemical inertness of these coatings have been published [5–7], but only a few investigations dealing with the hygienic and/or cleanability properties have been performed. It is thus important to propose pertinent tests to assess the cleanability of different coating materials.

The removal of oils from a solid surface is specific as some foodstuff components are water-soluble, whereas others, such as oil or grease, are emulsifiable [8]. This solubilization/emulsification behavior will dictate the choice of the detergent to use. A particular attention should be paid to the nature of the detergent which can be neutral, acidic, alkaline, oxidant and/or enhanced by using builders [9]. The interactions between the detergent and the solid material should also be taken into account as the solid wettability by the detergent is necessary.

As far as the solid material cleanability is concerned, currently there are empirical tests which compare the cleanability of various solid materials with reference to food stain [10]. Overall, the surface cleanability refers to the removal rate of oil from the material surface under experimental cleaning conditions and/or to the variation of contact angles on as-received and soiled solid surfaces [11]. The quantification of the organic materials needs very precise methods, such as X-ray photoelectron spectroscopy (XPS) analysis. Nevertheless, on a pilot scale, it is impossible to assess the material cleanability by using such sophisticated technique. One solution proposed is to add micro-organisms in the fouling solution. After a Cleaning-In-Place cycle, the residual microorganisms are quantified and directly related to material cleanability [12].

Only a very few studies have been performed on fouling by foodstuffs and the cleaning kinetics on steel coatings, assessed in as-received condition and after aging. Actually, the coating properties can be modified after exposure to UV, cleaning products, or high relative humidity. In practice, a coating degradation due to blister formation is very often observed [5]. This is the reason why the materials must be assessed both in an as-received condition and after an accelerated aging test.

The aim of this work was to relate the surface energetic characteristics of coating materials with fouling and cleanability mechanisms. Secondly, the cleaning kinetic constant was investigated with regard to the surface properties of the organic coatings in order to propose the most promising industrial coatings with an improved cleanability after a fatty acid spray. The organic coatings selected were poly(ethylene terephthalate) (PET), polyester (PE), poly(vinyl chloride) (PVC) and poly(vinylidene fluoride) (PVDF), which are commonly used in industrial wall panels. Surface fouling was performed by splashing fatty acids on the organic coatings.

## 2. MATERIALS AND METHODS

### 2.1. Solid surfaces

#### 2.2.1. Selection of polymeric coatings

Galvanized carbon steels, coated by an organic product were selected as solids in this study. The polymeric coatings are applied on steel surfaces according to the following process [13]. At the first stage of activation and passivation, the steel surfaces are cleaned and prepared. A first coat named primer with a thickness 5–30 μm is then applied with a roll coat, to provide anti-corrosive and adhesion properties. The coated steel is heated in a curing oven at 250°C where the solvent removal occurs and is finally exposed to a cooling section. A second coat named topcoat is applied with a second coating machine (roll-coat). The second coat generally has a thickness from 20 to 200 μm. As previously described, the coated steel is heated in a curing oven and then cooled.

The following coatings (topcoats) were investigated: poly(ethylene terephthalate) (PET), polyester (PE), poly(vinyl chloride) (PVC) and poly(vinylidene fluoride) (PVDF). Two types of PVC were studied: one containing linear alkyl groups as plasticizer and named PVC (1), whereas the second, PVC(2), contained phthalates as plasticizer.

The thickness of the polymeric coatings was typically from 20 to 200 μm. The chemical composition of the polymeric coatings (determined by FT-IR spectroscopy), as well as their average thickness are presented in Table 1. The surface roughness of these coatings is also reported in this table. The topographic parameters selected were the arithmetic average roughness ($S_a$), the maximum peak-to-valley height ($S_t$) and the skewness $S_{sk}$ expressed in μm [14]; the number of peaks per mm$^2$ is also presented. These parameters were deduced from an optical profiler (scans of area 100×100 μm$^2$) using the Surfvision software.

#### 2.2.2. Cleaning of coated steels

The as-received coated steels were first degreased by ethanol to eliminate traces due to sticky tape peeling. Then, the samples were cleaned in a soft alkaline detergent. The choice of the cleaning agent was based on the following recommendations [15]: the pH of the cleaning solution must be between 4 and 9, the temperature lower than 30°C and the contact time less than 30 min. We chose a commercial soft detergent (Procter & Gamble) containing anionic surfactants (less than 5%) and non-ionic surfactants between 5 and 15% (wt). The final concentration was 1.2 wt% which corresponds to a pH of 8.2. The cleaning was performed at 20°C for 10 min. Finally, the samples were rinsed five times in distilled water at 20°C and dried with a tissue paper.

#### 2.2.3. Surface aging

In order to assess the coating durability, it was necessary to simulate surface aging. In practice, the coating degradation is very often observed due to blister for-

**Table 1.**
Main characteristics of industrial polymeric coatings investigated

| Polymer | Chemical composition IR band assignments | Thickness (μm) | $S_a$ (μm) | $S_t$ (μm) | $S_{sk}$ | Peaks/mm² |
|---|---|---|---|---|---|---|
| Poly (ethylene terephthalate) (PET) | 3055 cm⁻¹($\nu_a$=C–H); 2969 cm⁻¹($\nu_a$–CH₂); 2907 cm⁻¹($\nu_s$–CH₂); 1713 cm⁻¹ (Free C=O stretch); 1243 ((O=)C–O–C stretch); 1097 ($\nu_a$ C–O–C); 1018 cm⁻¹ (C–C–O out of plane); 722 cm⁻¹(δ C–H) | 60 | 0.04 | 1.68 | 3.17 | 2500 |
| Polyester (PE) | 3054 cm⁻¹($\nu_a$=C–H); 2968 cm⁻¹($\nu_a$–CH₂); 2907 cm⁻¹($\nu_s$–CH₂); 1712 cm⁻¹ (Free C=O stretch); 1250 ((O=)C–O–C stretch); 1096 ($\nu_a$ C–O–C); 1017 cm⁻¹ (C–C–O out of plane); 722 cm⁻¹(δ C–H) | 25–35 | 0.18 | 1.48 | 0.853 | 7000 |
| Poly (vinylidene fluoride) (PVDF) | 3030 cm⁻¹($\nu_a$=C–H); 2933 cm⁻¹($\nu_a$CH₂); 2870 cm⁻¹($\nu_s$ CH₃); 1718 cm⁻¹ ((free C=O stretch); 1580 cm⁻¹($\nu_a$ C=C); 1547 cm⁻¹($\nu$CN,δNH); 1474 cm⁻¹ (δ CH₂); 1373 cm⁻¹(δ CH₃); 1164 ($\nu_a$ C–N–C), 1136 (C–C stretch) and 1074 cm⁻¹ (C–C skeleton vibration); 915 cm⁻¹(C–C–O in plane); 813 cm⁻¹ (δ triazine); 721 cm⁻¹ (δ C–H) | 35 | 0.26 | 4.15 | 1.03 | 9000 |
| Poly(vinyl chloride) (PVC(1)) | 2955 cm⁻¹($\nu_a$CH₃); 2928 cm⁻¹($\nu_s$CH₂); 2855 cm⁻¹($\nu_s$CH₂); 1726 cm⁻¹((Free C=O stretch); 1401 cm⁻¹ and 1343 cm⁻¹ (δ CH₃); 1066 cm⁻¹(C–C skeleton vibration); 1150 cm⁻¹(δ CF) | 140 | 0.40 | 3.51 | 0.175 | 12000 |
| Poly(vinyl chloride) (PVC(2)) | 3020 cm⁻¹($\nu_a$=C–H); 2956 cm⁻¹($\nu_a$CH₃); 2921 cm⁻¹($\nu_a$CH₂); 2870 cm⁻¹ ($\nu_s$ CH₃); 1712 cm⁻¹((Free C=O stretch); 1586 cm⁻¹ (δ =C–H); 1424 cm⁻¹ (δ CH₃); 1254 and 1122 cm⁻¹(Cl–CH strech) | 100 | 0.22 | 2.89 | −0.475 | – |

$S_a$ is the arithmetic average roughness, $S_t$ the maximum peak-to-valley height and $S_{sk}$ the skewness.

mation [5]. We chose to vertically immerse samples for 113 h in sterile water at 40°C and then samples were dried at room temperature for 24 h.

The sample edges were protected by silicone (Rhodorsil™, Rhône-Poulenc, France) before immersion in order to avoid water infiltration. Prior to the fouling test, the surfaces were cleaned by a soft alkaline detergent as previously described.

### 2.2.4. Chemical inertness

The chemical inertness of coated steels was based on the European standards [16, 17]. The materials were immersed for 240 h at 5°C in 15% or 95% ethanol or for 24 h in isooctane at 5°C. The global migration of components from the polymeric coatings was measured in these solutions by weight loss and expressed in mg/dm².

### 2.3. Solid surface free energy ($\gamma_s$)

Contact angles were measured on coating surfaces (S) by the sessile drop technique using diiodomethane, formamide and water (L). The contact angle was directly measured with a Krüss goniometer G-10. The solid surface free energy was estimated from contact angle measurements by the least-squares fitting of the data to [18, 19]:

$$\gamma_{LV}(\cos\theta+1)/2(\gamma_L^d)^{1/2} = (\gamma_S^p)^{1/2}(\gamma_L^p/\gamma_L^d)^{1/2}+(\gamma_S^d)^{1/2} \tag{1}$$

in which $\gamma^d$ denotes the apolar Lifshitz–van der Waals component and $\gamma^p$ the polar component including ionic, hydrogen, acid–base and covalent interactions. The solid surface free energy is expressed in mJ/m².

### 2.4. Liquid surface free energy ($\gamma_L$)

The surface energy characteristics of the pure liquids (diiodomethane, formamide and water) were taken from the literature [20] and are reported in Table 2. The surface free energy of the sunflower oil, as well as the commercial detergent were determined by the du Noüy ring method using a Krüss tensiometer (K6). The $\gamma_L^d$ and $\gamma_L^p$ components were calculated using contact angle measurements of sunflower oil and detergent on Parafilm™ (3M) and PTFE, which are considered to be completely apolar ($\gamma_S^p=0$) and have a surface free energy ($\gamma_S=\gamma_S^d$) of 26 mJ/m² and 19 mJ/m², respectively at 20°C [21].

The critical micelle concentration (cmc) of the commercial detergent was also determined by measuring the surface tension by the du Noüy ring method of solutions of various concentrations expressed in wt%. As illustrated in Fig. 1, the cmc of the commercial detergent is 0.1 wt%. Consequently, the usual concentration in our experiments was above the cmc. Nevertheless, as we used a commercial detergent with a proprietary chemical composition, we can only give a relative cmc expressed in wt% and not in molarity.

**Table 2.**
Liquid surface tensions and their components (in mJ/m²)

| Liquid | $\gamma_L$ | $\gamma_L^d$ | $\gamma_L^p$ |
|---|---|---|---|
| Water | 72.8 | 21.8 | 51 |
| Formamide | 58 | 39 | 19 |
| Diiodomethane | 50.8 | 50.8 | 0 |
| Sunflower oil | 34.6 | 34.6 | 0 |
| Detergent | 31.4 | 31.2 | 0.2 |

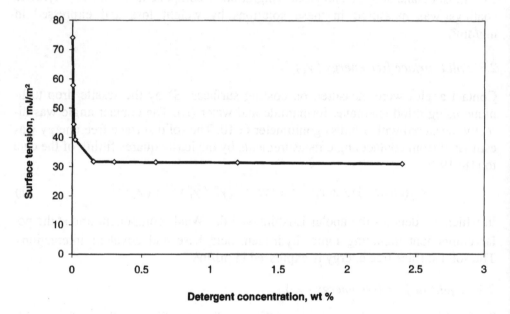

**Figure 1.** Determination of the cmc of the commercial detergent expressed in wt%.

## 2.5. Free energies of solid–oil ( $\gamma_{SO}$ ), detergent–oil ( $\gamma_{LO}$ ) and solid–detergent ( $\gamma_{SL}$ ) interfaces

The solid–oil interfacial free energy ( $\gamma_{SO}$ ), the detergent (L)–oil interfacial free energy ( $\gamma_{LO}$ ) and the solid–detergent interfacial free energy ( $\gamma_{SL}$ ) were computed according to the geometric mean equation. The interfacial free energy ( $\gamma_{12}$ ) between two materials (1 and 2) can be calculated according to the geometric-mean expression [22] as:

$$\gamma_{12} = \gamma_1 + \gamma_2 - 2\left(\gamma_1^d \gamma_2^d\right)^{1/2} - 2\left(\gamma_1^p \gamma_2^p\right)^{1/2} \tag{2}$$

where $\gamma^d$ denotes the apolar Lifshitz–van der Waals component and $\gamma^p$ the polar component including ionic, hydrogen, acid–base and covalent interactions.

## 2.6. Fouling tests

### 2.6.1. Oil characteristics

A nutritious sunflower oil (Lesieur, France) was used in this test. It is mainly composed of fatty acids (13 wt%), monounsaturated fatty acid (oleic acid) (22 wt%) and polyunsaturated acids (65 wt%). A fluorescent colorant (Yellow 131SC™, Morton, USA) was added into the oil at a final concentration of 0.05 wt% to enhance the contrast between the oil and the coated steel. The optimal excitation wavelength for this fluorescent marker is in the UV range (365 nm).

### 2.6.2. Surface splashing

The oil was sprayed on as-received or aged surfaces using a brush that was first soaked in the colored oil. The coated surfaces were completely covered by small droplets with a diameter not exceeding 1 μm. The sample size was 13 cm². The oil quantity left by the brush on the surface was 8.5 mg (± 1.5 mg) and was controlled by weighing the sample on a precision balance (Mettler AE 240). This gave a fouling density of 0.70 mg/cm² (± 0.15 mg/cm²). The oil fouling was observed with a color CCD camera (TK-C1380, JVC) equipped with a zoom lens (magnification ×25) and a filter (filter 400FG03-50S, Andover Corporation, USA) and an ultraviolet lamp (lamp B-100EP, UVP, USA) with a 100 W bulb (H44GS-100, Sylvania).

The camera was coupled to an image analyzer. An image covered a surface area of 3 cm². After each experiment, the number of droplets were counted and expressed as the number of droplets /cm². Every experiment was independently repeated four times.

## 2.7. Cleanability kinetics

The surface cleaning was carried out in a laminar flow cell (hydraulic diameter 3.9 mm) [23]. This laminar flow cell can be considered as two flat plates between which the commercial detergent at a final concentration of 1.2 wt% was circulating, the bottom plate was the sample to clean (Fig. 2). During the cleaning operation, the flow was equal to $1.7 \times 10^{-2}$ dm³ s⁻¹ corresponding to a shear stress of 0.64 Pa. The cleaning was realized in 10-s sequences at 20°C. There was no intermediate water rinse. Between every step of cleaning in the laminar flow cell, the sample was dried and weighed. Based on weight difference, it was possible to evaluate the oil quantity removed by the cleaning procedure, where $w_o$ is the initial

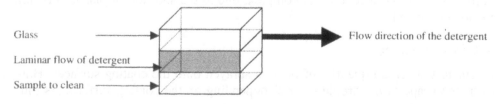

Glass

Laminar flow of detergent

Sample to clean

Flow direction of the detergent

**Figure 2.** Principle of the flow cell.

weight of the fouled surface by the fatty acids and $w_t$ the weight measured after cleaning sequences at time $t$.

The cleaning kinetics decreases exponentially with time. The kinetic constant was graphically determined by plotting $\ln(w_t/w_o)$ as a function of time. The kinetic constant ($k$) corresponds to the slope and is expressed in $s^{-1}$. Each experiment was repeated four times. The variation in the slope determination was less than 20%.

## 3. RESULTS

The surface properties of polymeric coatings were determined on as-received and aged samples. Then, the oil fouling as well as the cleaning kinetics were investigated in a laminar flow cell.

### 3.1. Surface properties of polymeric coatings

As indicated by the roughness parameters $S_a$ and $S_t$ (Table 1), the coatings used display rough surfaces, except for PET. A positive value of the skewness ($S_{sk}$) indicates the presence of peaks on PET, PE, PVDF and PVC(1), whereas a negative one corresponds to the presence of holes on the PVC(2) surface.

Tables 3 and 4 summarize, respectively, the contact angles and the solid surface free energies derived from these contact angles on as-received and aged coatings. Two types of surfaces can be defined: (i) high-energy surfaces, such as PE and PET, with a surface free energy ($\gamma_s$) of 40 mJ/m² and (ii) low-energy surfaces, such PVC and PVDF, with a $\gamma_s$ value less than 35 mJ/m².

It should be noted that after aging, the $\gamma_s$ value considerably increased for the PVC(2) coating. Polar coatings, such as PE and PET, display similar surface energetic characteristics after aging, whereas PVDF and PVC(1) coating surfaces are modified during the water aging. The polar component $\gamma_s^p$ of PVC(1) is twice as high after aging.

After immersion in 95% ethanol and isooctane, liquids simulating foodstuffs, a migration of components was measured only from the PVC(2) coating (Table 5).

The higher values of the surface free energy ($\gamma_s$) and of the polar component ($\gamma_s^p$) of PET and PE-based coatings could be explained by the presence of C=O and O–C=O groups in the coatings. On the other hand, the PVDF and PVC coatings possess $CH_2$–$CF_2$ and $CH_2$–$CHCl$ groups, respectively. Nevertheless, the composition of these coatings is complex, due to the addition of plasticizers, pigments and/or additives.

### 3.2. Surface fouling

In our test, a certain quantity of oil was sprayed onto the coating surfaces. However, the shape of the droplets varies depending on the coating surface. As illus-

**Table 3.**
Contact angles on as-received and aged coatings (in degrees)

| Liquid | PET | | PE | | PVDF | | PVC (1) | | PVC (2) | |
|---|---|---|---|---|---|---|---|---|---|---|
| | As-received | Aged | As-received | Aged | As-received | Aged | As-received | Aged | As-received | Aged |
| Water | 74 (2) | 76 (2) | 79 (2) | 76 (2) | 86 (1) | 84 (1) | 89 (2) | 82 (2) | 94 (2) | 86 (2) |
| Formamide | 55 (2) | 61 (3) | 63 (1) | 58 (2) | 70 (2) | 72 (2) | 70 (2) | 66 (2) | 74 (2) | 65 (2) |
| Diiodomethane | 36 (2) | 40 (1) | 39 (2) | 41 (3) | 50 (2) | 52 (4) | 46 (2) | 48 (2) | 49 (2) | 22 (2) |

The measurement errors are shown in parentheses.

**Table 4.**
Surface free energies and their components obtained on as-received and aged coatings (in mJ/m$^2$)

| Material | $\gamma_s$ | | $\gamma_s^d$ | | $\gamma_s^p$ | |
|---|---|---|---|---|---|---|
| | As-received | Aged | As-received | Aged | As-received | Aged |
| PET | 43.6 (5.7) | 40.5 (5.4) | 38.2 (4.2) | 35.4 (3.9) | 5.4 (1.5) | 5.1 (1.5) |
| PE | 39.9 (5.0) | 41.0 (5.4) | 36.0 (4.0) | 35.9 (4.0) | 3.9 (1.0) | 5.1 (1.4) |
| PVDF | 33.8 (4.6) | 32.8 (4.8) | 31.1 (3.7) | 29.2 (3.6) | 2.7 (0.9) | 3.6 (1.2) |
| PVC(1) | 35.2 (4.4) | 36.0 (5.0) | 33.6 (3.9) | 32.2 (3.8) | 1.6 (0.5) | 3.8 (1.2) |
| PVC(2) | 33.1 (4.1) | 43.6 (4.7) | 32.3 (3.8) | 42.6 (4.4) | 0.8 (0.3) | 1.0 (0.3) |

The error limits are shown in parentheses.

**Table 5.**
Lixiviation (expressed in mg/dm$^2$) in various liquids simulating foodstuffs in contact with polymeric coatings

| Material | 15% Ethanol, 240 h | 95% Ethanol, 240 h | Isooctane, 24 h |
|---|---|---|---|
| PET | 0.3 | 1.3 | 0.7 |
| PE | 0.3 | 7.6 | 1.4 |
| PVDF | 0.3 | 1.5 | 0.4 |
| PVC(1) | 0.3 | 0.8 | 0.5 |
| PVC(2) | 3.5 | 302 | 268 |

trated in Fig. 3a, 3b and 3c and in Table 6, various shapes of oil droplets can be seen on polymeric coatings. On as-received PVC(2) and PE coatings, a large number of small individual droplets are observed.

On as-received PET and aged PVC(2)-based coatings, the oil droplets were larger and more diffuse and coalesced. An intermediate case was observed on as-received and aged PVC(1) and PVDF coatings. A noticeable difference in the

**Table 6.**
Number of oil droplets/cm² and oil contact angle values (in degrees) on polymeric coatings

| | Coating | | | | | | | | | |
|---|---|---|---|---|---|---|---|---|---|---|
| | PET | | PE | | PVDF | | PVC (1) | | PVC (2) | |
| | As-received | Aged | As-received | Aged | As-received | Aged | As-received | Aged | As-received | Aged |
| Number of droplets/cm² | 420 | 260 | 1100 | 1010 | 720 | 650 | 850 | 700 | 1500 | 280 |
| Oil contact angle (degrees) | 15 (2) | 15 (2) | 18 (2) | 14 (2) | 19 (2) | 18 (2) | 26 (2) | 24 (2) | 38 (2) | 18 (2) |

The errors in contact angle measurements are shown in parentheses.

a. case of small droplets     b. case of coalescing droplets   c. intermediate case

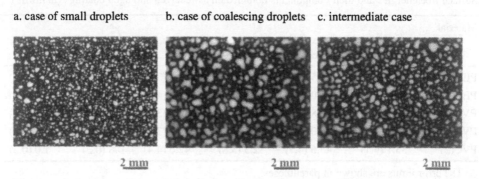

**Figures 3.** Illustration of oil droplets on coatings. (a) Case of individual oil droplets. (b) Case of coalescing droplets. (c) Intermediate case.

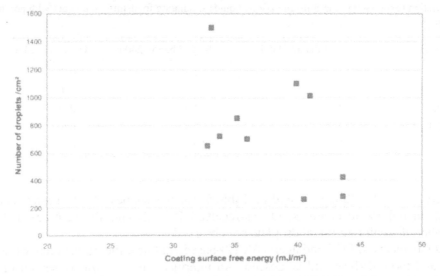

**Figure 4.** Evolution of the number of visible oil droplets/cm² as a function of the coating surface free energy, $\gamma_s$ (in mJ/m²).

density of individual droplets was observed on as-received and aged PVC(2) coatings.

There is a relation between the number of individual oil droplets and solid surface free energy (Fig. 4). The number of individual droplets increases with a decrease in the solid surface free energy. Moreover, the wettability of the solid by the oil is very high as the oil contact angle is between 15 to 38°.

## 3.3. Cleanability kinetics

As shown in Fig. 5a, as-received PE and PET-based coatings initially fouled by oil droplets are easier to clean by an alkaline detergent than PVDF and PVC coatings. On the latter two coatings, only 20% of oil is removed after 90 s of cleaning. On aged surfaces (Fig. 5b), the cleaning kinetics is quite similar except for PVC(1) and PVDF. The cleaning kinetics on as-received and aged materials decreases exponentially with time (Fig. 6a and 6b). Consequently, the kinetics obeys first order. The kinetic constant ($k$) values graphically determined are summarized in Table 7.

The kinetic constant varies from 2 to $37 \times 10^{-3}$ s$^{-1}$. Two types of kinetics can be distinguished: (i) slow kinetics, with a constant $k$ varying from 2 to $4 \times 10^{-3}$ s$^{-1}$, and (ii) faster kinetics, with a constant $k$ 10-times higher than the previous ones. In the latter case, $k$ varies from 20 to $40 \times 10^{-3}$ s$^{-1}$.

Overall, on as-received samples, two types of coatings can be distinguished: (i) coatings with an easy cleanability, such as PET and PE, and (ii) coatings with a more difficult cleaning such as PVC and PVDF.

The surface cleanability was also studied on coatings after 100-h aging in a severe environment simulating high relative humidity. On aged surfaces, the constants ($k$) are quite similar. It should be noted that the constant $k$ increases by a factor from 3 to 4 on PVC(1) and PVDF-based coatings. For other materials, the constant remained stable after aging.

**Table 7.**
Cleaning kinetic constants $k$ (expressed in s$^{-1}$)

|  | As-received | | After 100 h aging | |
|---|---|---|---|---|
|  | $k$ | $R^2$ | $k$ | $R^2$ |
| PET | $37 \times 10^{-3}$ | 0.9862 | $29 \times 10^{-3}$ | 0.9774 |
| PE | $23 \times 10^{-3}$ | 0.9895 | $28 \times 10^{-3}$ | 0.9972 |
| PVDF | $2.4 \times 10^{-3}$ | 0.9780 | $10 \times 10^{-3}$ | 0.9847 |
| PVC(1) | $4.3 \times 10^{-3}$ | 0.9608 | $12 \times 10^{-3}$ | 0.9347 |
| PVC(2) | $2.1 \times 10^{-3}$ | 0.9733 | $3.5 \times 10^{-3}$ | 0.9168 |

$R^2$ represents the regression coefficient obtained from the graphical determination.

**(a)**

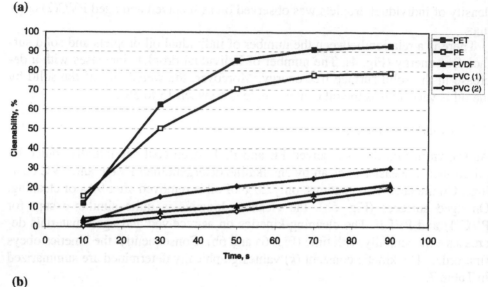

**(b)**

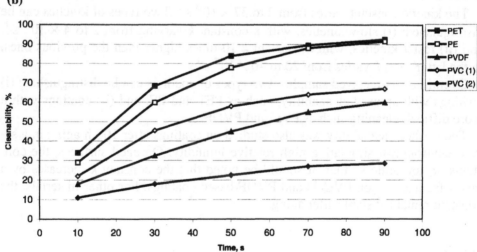

**Figure 5.** Cleaning kinetics (% cleanability *vs.* time) after an oil spray on as-received polymeric coatings (a) and after aging (b).

To summarize, the cleanability of PET and PE-based coatings is high and remains constant after 100-h aging. The aged PVC(1) and PVDF coatings display an improved surface cleanability, whereas the PVC(2) shows a low cleanability even after aging.

There is a relation between the cleaning kinetic constant ($k$) and the polar component of the polymeric coating (Fig. 7). The higher the $\gamma_s^p$ value, the better the surface cleanability.

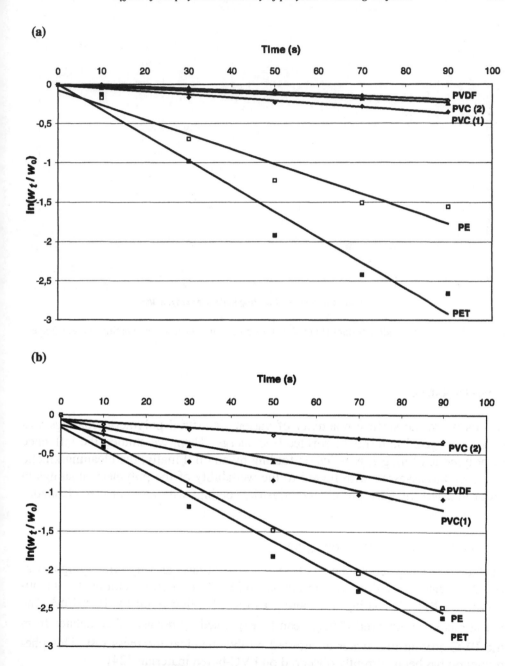

**Figure 6.** Graphical determination of the cleaning kinetic constant ($k$ in s$^{-1}$) on as-received coating materials (a) and after aging (b). The constant $k$ corresponds to the slope, $w_o$ is the initial weight of the fouled surface and $w_t$ is the weight after cleaning time $t$.

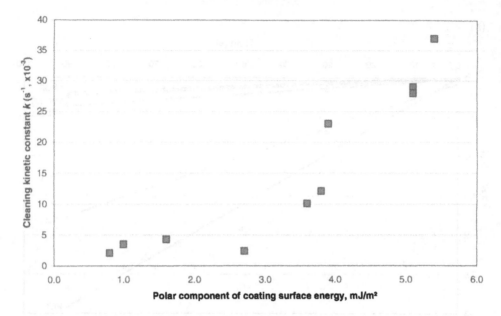

**Figure 7.** Cleaning kinetic constant ($k$) in s$^{-1}$ *versus* polar component of the coating surface free energy.

## 4. DISCUSSION

In this work, the surface reactivity of various industrial polymeric coatings was first investigated in terms of surface free energy. Then, we studied the influence of polymeric coating energetics on fouling by an oil. Finally, the cleaning kinetic constant was assessed with regard to the wettability of the polymeric coatings in order to select the most promising industrial coatings with an improved cleanability.

### 4.1. Solid surface reactivity on aging

Surface free energy components of polymeric coatings can vary after aging in water. Noticeable variations in the apolar surface free energy component are observed on PVC(2) coating after aging as the $\gamma_s^d$ value is up by 10 mJ/m². The variations in $\gamma_s^d$ on aged PVC(2) can be explained by release of phthalates from the PVC coating into water as attested by the chemical inertness test. This phenomenon has been currently observed on PVC-based materials [24].

Conversely, PVDF and PVC(1) display a polar component twice as high after aging. These modifications in the surface properties after aging of polymeric coatings can be attributed to water adsorption at the external surface of the polymeric coating. Water could accumulate at the coating/steel interface [5] as we observed a weight increase of samples after aging (data not shown). Nevertheless, no blister effect was observed at this stage of aging.

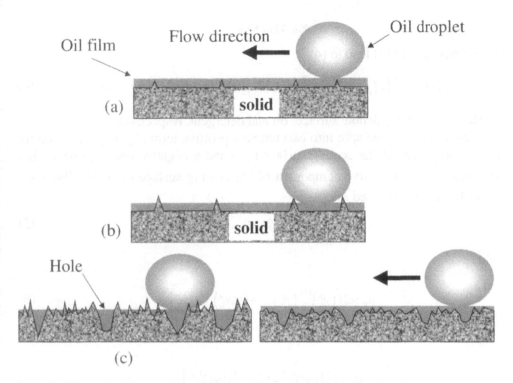

**Figure 8.** Effect of coating surface topography on the droplet shape. (a) Case of a smooth coating, (b) case of coating displaying large peaks and (c) case of coating after aging.

### 4.2. Influence of surface wettability on fouling and surface cleanability

As previously mentioned, the number of individual droplets increases with a decrease of the solid wettability by oil. It should be noted that only oil droplets visible to the naked eye were monitored. However, a microscopic oil film could be formed between the visible droplets [25].

This difference in the droplet shape could be explained by the surface topography. As shown in Fig. 8, the PET surface is the smoothest, displaying the lowest number of peaks. In this case, the coalescence of droplets is possible. On other materials, where the surface topography is more uneven due to the presence of numerous peaks, droplet sliding is not possible. Finally, on the PVC(2) surfaces that initially display holes, the droplet cannot slide easily.

As far as the coating cleanability is concerned, the most efficient cleaning was observed on coatings displaying the highest polar component. To remove the oil from the surface, the detergent must minimize the oil–water and solid–water interfacial energies, thereby causing removal to become spontaneous.

It is well documented that the cleaning process proceeds by one of three primary mechanisms, namely solubilization, emulsification and roll-up. In order to achieve roll-up, one requires the surface free energies for oil detachment to obey [25, 26]:

$$\gamma_{SO} \geq \gamma_{LO} + \gamma_{SL} \tag{3}$$

Equations (2) and (3) lead to (4)

$$2\left[\left(\gamma_L^p\right)^{1/2}\left(\gamma_S^p\right)^{1/2}+\left(\left(\gamma_L^d\right)^{1/2}-\left(\gamma_O^d\right)^{1/2}\right)\left(\gamma_S^d\right)^{1/2}+\left(\gamma_O^d\gamma_L^d\right)^{1/2}-\gamma_L\right]\geq 0 \tag{4}$$

Here, S, O and L denote surface, oil and detergent, respectively.

Equation (4) can be split into two terms: a positive term $(E_R)$ depending on the polar component of the coating surface $(\gamma_S^p)$ and a negative one $(E_A)$ which is a function of the dispersive component of the coating surface $(\gamma_S^d)$. Finally, equation (4) can be expressed as:

$$E_A + E_R \geq 0 \tag{5}$$

where

$$E_R = 2\left[\left(\gamma_L^p\right)^{1/2}\left(\gamma_S^p\right)^{1/2}+\left(\gamma_O^d\gamma_L^d\right)^{1/2}-\gamma_L\right] \tag{6}$$

and

$$E_A = 2\left[\left(\left(\gamma_L^d\right)^{1/2}-\left(\gamma_O^d\right)^{1/2}\right)\left(\gamma_S^d\right)^{1/2}\right] \tag{7}$$

In our particular case, the surface free energies of the oil, the solid and the detergent, as well as their components, were experimentally determined leading to the numerical expressions:

$$E_R = 0.9\cdot\left(\gamma_S^p\right)^{1/2}+2.8$$

and

$$E_A = -0.6\cdot\left(\gamma_S^d\right)^{1/2}.$$

The variations of the cleaning constant ($k$ in s$^{-1}$) versus $E_R$ and $E_A$ are shown in Fig. 9a and b, respectively. In our case, a coated surface with a high polar component would be easier to clean after fouling by oil splashing. Irrespective of the coating considered, the non-polar or dispersive interactions between the solid and the greasy oil remain constant, whereas the cleaning kinetics is directly related to the polar interactions. As illustrated previously, the cleaning is more efficient on coating materials displaying a polar component of the surface free energy. This could be explained by favorable interactions between the surfactants and the solid surface enhancing the solubilization and/or the emulsification mechanism.

The anionic and/or non-ionic surfactants from the alkaline detergent possess the general polar/non-polar character. In view of the concentration used in our experimental conditions (higher than the cmc), it can be assumed that there are mi-

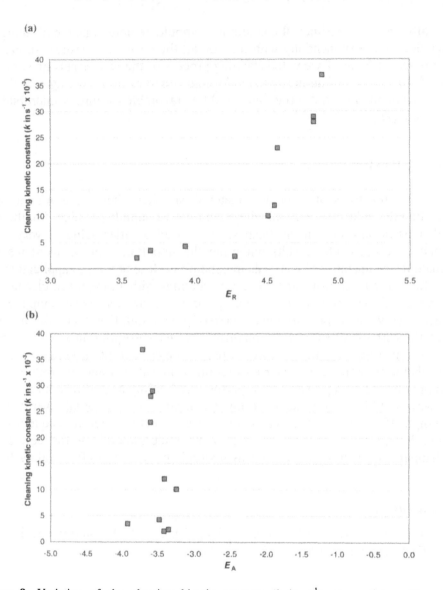

**Figure 9.** Variation of the cleaning kinetic constant ($k$ in $s^{-1}$) *versus* the positive term $E_R = 2\left[\left(\gamma_L^p\right)^{1/2}\left(\gamma_S^p\right)^{1/2}+\left(\gamma_O^d\gamma_L^d\right)^{1/2}-\gamma_L\right]$ (in mJ/m²). (b) Variation of the cleaning kinetic constant ($k$ in $s^{-1}$) *versus* the negative term $E_A = 2\left[\left(\left(\gamma_L^d\right)^{1/2}-\left(\gamma_O^d\right)^{1/2}\right)\left(\gamma_S^d\right)^{1/2}\right]$ (in mJ/m²).

celles present. It is suggested that oily oil can be more easily incorporated in detergent micelles when in contact with polar organic coatings. To explain the difference obtained on polar and non-polar polymeric coatings, it is proposed that the detergent micelles orient their polar heads towards the solid surface. This enhances their adsorption onto polar materials [27].

Finally, in our experimental conditions, it should be noted that the cleaning kinetics decreases exponentially with time on all the coatings investigated. From a more practical point of view, the cleaning process of the wall panels must be optimized. In the case of an incomplete cleaning, some residual fouling can always take place on the material surface and could be favorable for further microbial development [28].

## 5. CONCLUSIONS

The surface reactivity of an organic material can alter when aging in conditions simulating the cold storage room where the relative humidity is high. So, the material must be assessed in the fresh state, as well as after aging in very well-defined conditions. Our results underline the importance of pertinent tests to evaluate the cleanability of materials in conditions close to the end application.

As far as the most promising polymeric coatings with an improved cleanability are concerned, only coating with a high polar and a low dispersive component of surface energy can be proposed in the case of greasy soil. However, after aging in water, the corrosion resistance, explained by water absorption in some industrial coatings under assessment, decreases, whereas the cleanability is enhanced by the polar character brought by water absorption. Controlled water absorption in the coating will help to improve both properties: cleanability and corrosion resistance. One way could be to propose a two-layer coating with an external layer displaying a high affinity for water and an internal one protecting steel from aqueous corrosion. Future work will be devoted to the understanding of the detergency mechanisms on hydrophilic as well as hydrophobic coatings after fouling by an oil.

*Acknowledgements*

We thank Dr. F. Bruckert (CEA, Grenoble, France), Pr. B. Baroux and Pr. J.C. Joud (INP-Grenoble) for helpful discussions.

## REFERENCES

1. Z.W. Wicks, Jr., F.N. Jones and S.P. Pappas (Eds.), *Organic Coatings Science and Technology, Vol 2: Applications, Properties and Performance*, Wiley Interscience, New York, NY (1992).
2. M. Schmitthenner, *Eur. Coat. J.* No. 9, 618 (1998).
3. E.A. Zottola, *Food Technol.* **48**, 107 (1994).
4. L. Boulangé-Petermann, *Biofouling* **10**, 275 (1996).
5. E. Deflorian, L. Fedrizzi and P.L. Bonora, in: *Organic Coatings for Corrosion Control*, G.P. Bierwagen (Ed.), pp. 92-105, American Chemical Society, Washington, DC (1998).
6. R. Lambourne, in: *Paint and Surface Coatings: Theory and Practice, second edition.* R. Lambourne and T.A. Strivens (Eds.), pp. 658-693, Woodhead Publishing, Cambridge (1993).
7. G.R. Hayward, in: *Paint and Surface Coatings, Theory and Practice, second edition.* R. Lambourne and T.A. Strivens (Eds.), pp. 725-766, Woodhead Publishing, Cambridge (1993).

8. B.P. Binks and J.H. Clint, *Langmuir* **18**, 1270 (2002).
9. J. Vincent, in: *Nettoyage, désinfection et hygiène dans les bio-industries,* J.Y. Leveau and M. Bouix (Eds.), pp. 167-204, Tech&Doc, Paris (1999) (in French).
10. R.A. Stevens and J.T. Holah, *J. Appl. Bacteriol.* **75**, 91 (1993).
11. J. Yang, J. McGuire and E. Kolbe, *J. Food Protection* **54**, 879 (1991).
12. C. Faille, L. Dennin, M.N. Bellon-Fontaine and T. Benezech, *Biofouling* **14**, 143 (1999).
13. G.R. Pilcher, *Eur. Coat. J.* No. 3, 62 (1999).
14. E.S. Adelmawla, M.M. Koura, T.M.A Maksoud, I.M. Elewa and H.H. Soliman, *J. Mater. Proc. Technol.* **123**, 133 (2002).
15. French Standards, NFP 75-401 (2000).
16. European Standards, XP ENV 1186-3 (1995).
17. European Standards, XP ENV 1186-14 (1999).
18. H.J. Busscher, M.N. Bellon-Fontaine, N. Mozes, H.C. van der Mei, J. Sjollema, A.J. Leonard, P.G. Rouxhet and O. Cerf, *J. Microbiol. Methods* **12**, 101 (1990).
19. M.N. Bellon-Fontaine and O. Cerf, *J. Adhesion Sci. Technol.* **4**, 475 (1990).
20. R.J. Good, in: *Contact Angle, Wettability and Adhesion*, K.L. Mittal (Ed.), pp. 3-36. VSP, Utrecht, The Netherlands (1993).
21. A. Vernhet and M.N. Bellon-Fontaine, *Colloid. Surf. B: Biointerf.* **3**, 255 (1995).
22. D.K. Owens and R.C. Wendt, *J. Appl. Polym. Sci.* **13**, 1741 (1969).
23. I.H. Pratt-Terpstra, A.H. Weerkamp and H.J. Busscher, *J. Gen. Microbiol.* **133**, 3199 (1987).
24. K. Bouma and D.J. Schakel, *Food Addit. Contam.* **19**, 602 (2002).
25. A.W. Adamson and A.P. Cast, *Physical Chemistry of Surfaces, Sixth Edition*, John Wiley, New York, NY (1997).
26. P.G. de Gennes, F. Brochart-Wyart and D. Quéré (Eds.), *Gouttes, Bulles, Perles et Ondes*, Belin, Paris (2002) (in French).
27. H. Haidara, L. Vonna and J. Schultz, in: *Apparent and Microscopic Contact Angles*, J. Drelich, J.S. Laskowski and K.L. Mittal (Eds.), pp. 475-485. VSP, Utrecht, The Netherlands (2000).
28. E. Robine, L. Boulangé-Petermann and D. Dérangère, *J. Microbiol. Methods* **49**, 225 (2002).